Cladocera: the Biology of Model Organisms

Developments in Hydrobiology 126

Series editor
H. J. Dumont

Cladocera: the Biology of Model Organisms

Proceedings of the Fourth International Symposium on
Cladocera, held in Postojna, Slovenia, 8–15 August 1996

Edited by

A. Brancelj, L. De Meester & P. Spaak

Reprinted from Hydrobiologia, volume 360 (1997)

Springer-Science+Business Media, B.V.

Library of Congress Cataloging-in-Publication Data

A C.I.P. Catalogue record for this book is available from the Library of Congress.

ISBN 978-94-010-6084-4   ISBN 978-94-011-4964-8 (eBook)
DOI 10.1007/978-94-011-4964-8

Printed on acid-free paper

Hydrobiologia **360**: v–vii, 1997.
A. Brancelj, L. De Meester & P. Spaak (eds), Cladocera: the Biology of Model Organisms.

Contents

Behaviour

Trophic Interactions and Community Structure

Invited Contributions: Personal Views on the Future of Cladoceran Research

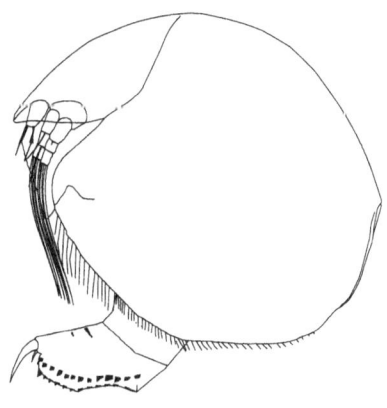

Illustration on cover: *Alona Stochi*, the third blind cave-dwelling cladoceran. This new species was found in a small cave in central Slovenia. See the paper by Brancelj (pp. 47–54) for further details.

Hydrobiologia **360:** ix–x, 1997.
A. Brancelj, L. De Meester & P. Spaak (eds), Cladocera: the Biology of Model Organisms.

Preface

Cladocera: the Biology of Model Organisms

This volume contains 29 regular papers presented at the Fourth International Symposium on Cladocera held in Postojna (Slovenia), 8–15 August 1996. This symposium was the fourth in a series of symposia started in Budapest (Hungary), 1985 and followed by Tatranska Lomnica (former Czechoslovakia), 1989 and Bergen (Norway), 1993.

The symposium was organized by Anton Brancelj of the National Institute of Biology in Ljubljana, in co-operation with the Ecological Association of Slovenia. The organizer wishes to thank the many colleagues and students who helped with the practical organization of the symposium. The main sponsor of the symposium was the Slovenian Ministry of Science and Technology.

Compared to the previous symposium in Bergen, the number of participants in Postojna was relatively low. There were 55 participants from 19 different countries, who presented 32 lectures and 17 posters. The lectures were structured into seven sessions reflecting the diversity of research done on cladocerans. The topics were: community analyses, morphology and systematics, grazing-predation-competition, life history, population dynamics, behaviour and paleolimnology and genetics. There was an organized visit to the spectacular cave Postojnska Jama, a whole-day excursion to Southwestern Slovenia visiting karst phenomena and former saltworks protected as a bird sanctuary, and as many as three common dinner parties.

No doubt, cladocerans are increasingly used as model organisms for many research areas in modern biology, ranging from molecular to population and community studies. The elegancy with which the characteristics of a given model organism can be exploited in such research is, however, dependent upon the quality and detail of the knowledge of the specific biology of the organism, including its morphology, systematics and ecology. We feel that the present volume is especially important in this respect, because it provides novel information on various levels and topics of the biology of the cladoceran. Although, so far, it has mainly been members of the genus *Daphnia* that have served as model organisms for research in ecology, evolutionary biology, genetics and physiology, we are pleased that the symposium in Postojna hosted several contributions on other cladoceran taxa, including moinids, bosminids and chydorids.

The present volume contains a special section with invited contributions of well-known cladoceran researchers presenting their personal opinion on the future of research on cladocera. We hope that these are inspiring, and contribute to the future developments by stimulating the application of new techniques and approaches to tackle old and new questions.

ANTON BRANCELJ
LUC DE MEESTER
PIET SPAAK

Hydrobiologia **360:** xi–xii, 1997.

Participants in the Fourth International Symposium on Cladocera

For legend see page xii.

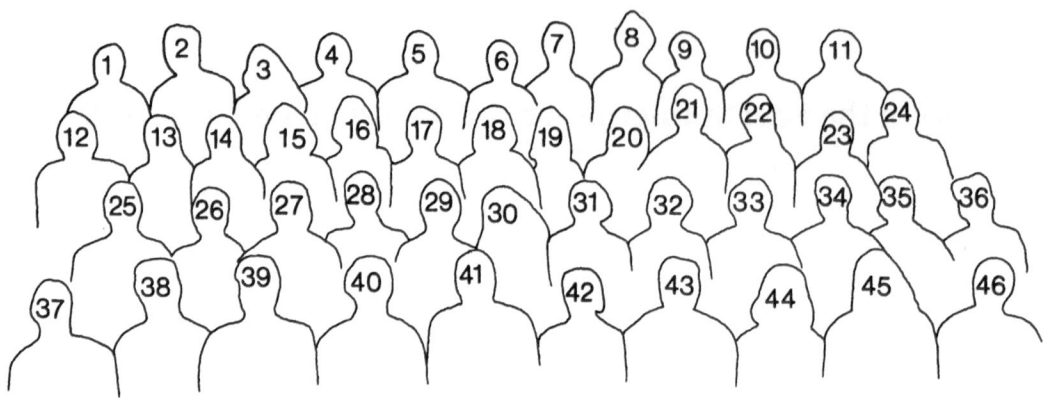

1. Jan KOKSVIK	17. Jacobus VIJVERBERG	32. Maria Rosa CARAMUJO
2. Henri J. DUMONT	18. Tatjana SIMČIČ	33. Laszlo FORRO
3. Dolores BORONAT	19. Tjaša POGAČNIK	34. Janos KORPONAI
4. Ulrik Ib ROEN	20. Heike MUMM	35. Catherine DUIGAN
5. Milijan ŠIŠKO	21. Piet SPAAK	36. Krystyna SZEROCZYNSKA
6. Iris ZELLMER	22. Catherine F. FULLWOOD	37. Anton BRANCELJ
7. Erik Van GOOL	23. Ilona POLCYN	38. Igor HUDEC
8. Steven DECLERCK	24. Wojciech JURASZ	39. Nikolai KOROVCHINSKY
9. Anke WEBER	25. Petter LARSSON	40. Joost VANOVERBEKE
10. Luc De MEESTER	26. Eugenij BURAK	41. Euzen STUCHLIK
11. Maarten BOERSMA	27. Fernando MARTÍNEZ-JERÓNIMO	42. Ann Kristin Lien SCHARTAU
12. Alexey KOTOV	28. Radka PICHLOVA	43. Ben FLIK
13. Maria Rosa MIRACLE	29. Cristina CRISPIM	44. Tanja ČELHAR
14. Elías Manuel GUTIERREZ	30. Celia BAIÂO	45. Irina TRUBETSKOVA
15. Veronika SACHEROVÁ	31. Tineke REEDE	46. Ramesh D. GULATI
16. Alexandra DEČEVSKA		

Participants in the Fourth International Symposium on Cladocera.

Hydrobiologia **360:** 1–11, 1997.
A. Brancelj, L. De Meester & P. Spaak (eds), Cladocera: The Biology of Model Organisms.
©1997 *Kluwer Academic Publishers.*

On the history of studies on cladoceran taxonomy and morphology, with emphasis on early work and causes of insufficient knowledge of the diversity of the group

N. M. Korovchinsky
A. N. Severtsov Institute of Ecology and Evolution of Russian Academy of Sciences, Leninsky Prospect 33, 117071 Moscow, Russia

Key words: cladocerans, history of studies, taxonomic diversity

Abstract

The long history of cladoceran morphological and taxonomic studies beginning in the second half of the 17th century to nowadays is divided into seven subsequent periods. Each of them is characterized by domination of specific trends and a noticeable increase of the number of taxa described. In spite of the multitude of studies the taxonomic diversity of cladocerans, especially at species level, remain to be insufficiently known. The history of hydrobiology is briefly reviewed in order to reveal the causes of this insufficient knowledge among which historical factors and a long domination of an integral approach to studies of continental water bodies are thought to be most important.

Introduction

The general history of cladoceran studies was briefly reviewed in many publications, but only in two recent papers by Frey (1982a, 1995) the development of taxonomic insight in the group has been analysed in more detail. This author divided the history into four periods, first of which began in the second half of the 17th century, and the last one continues nowadays. This scheme is generally acceptable but insufficient because its author was mostly interested in the development of the idea of cosmopolitanism of the distribution of cladocerans, especially Chydoridae. A brief review of the history of cladoceran taxonomy may thus be useful for specialists that often work with only a few model species and are little informed on cladoceran morphological and taxonomic diversity considered in the long-term aspect. Futhermore, the present review may help to reveal causes of the insufficient knowledge of taxonomic diversity of the group that has been revealed by some recent works (Frey, 1982a, 1987; Korovchinsky, 1996). One of the causes is undoubtedly the fact that it was long accepted that many cladocerans have a cosmopolitic distribution, as has already been pointed out

by Frey (1982a). Other causes are to be found in the history of hydrobiology.

Each of historic periods of cladoceran studies that will be defined in the present paper was characterized by sufficient qualitative changes, domination of specific trends which were evident from principal publications. In particular, the changes were also indicated by a noticeable increase of the number of taxa at the genus or species level. This increase was not only due to an expansion of the geographic scale, but was largely the result of new approaches and methods of investigation.

It is not possible here to describe the history of cladoceran studies in detail. My aim is to show just the scheme of the process as well as to reveal some regularities.

The historic periods of morphological and taxonomic studies

An overview of the seven periods of taxonomic research on cladocerans is given in Table 2.

Initially, the European scientists, the first of which were Goedardo (1662) and Swammerdam (1669), observed isolated species of the genera *Daphnia*

O. F. Müller, *Polyphemus* (L.) and possibly *Simo-cephalus* Schoedler, *Scapholeberis* Schoedler and *Moina* Baird (Table 1). These authors used differ-ent names (often *Monoculus*) and did not separate the cladocerans taxonomically from other microcrus-taceans ('Conchostraca', Ostracoda, Copepoda). Their figures were usually schematic or even allegoric (e.g. Joblot, 1754), but Schaeffer (1763) presented a rela-tively detailed description of *Daphnia magna* Straus, including thoracic limbs and inner structures. He fig-ured juveniles, parthenogenetic eggs and ephippium as well as epibionts on the body surface. Other good early descriptions were presented by Geoffroy (1764) and DeGeer (1778).

The Danish naturalist O. F. Müller (1776) was the first who divided the Cladocera into the genera *Daph-nia* (*Daphne*), *Lynceus* and *Polyphemus* and gave them binominal species names. According to modern classi-fication, Müller (1785) described 18 species of belong-ing to 15 genera (Table 1). As such, his work provided the first evidence of the diversity of the group. The sig-nificance of O. F. Müller's works was so high that for the 30 following years carcinologists only mentioned species of Cladocera described by him and often copied his figures (e.g. Latreille, 1802). Original data were presented by Jurine (1820) and Straus (1820). Jurine described *Daphnia* (possibly *D. pulex* (L.)) in detail including maxillule, mandibles, all thoracic limbs, and structures of the males and ephippial females. In addi-tion, he described embryogeny and copulation. Straus described new genera *Sida* and *Latona*. By the end of the 1820s the Cladocera were recognized as an inde-pendent group (Latreille, 1829, cited in Fryer, 1987a). Milne-Edwards (1840) summarized the results of this period and listed 21 species of 6 genera (*Daphnia, Sida, Latona, Lynceus, Polyphemus, Evadne*). From the modern point of view, the differentiation of the material mostly did not exceed the genus level. The representatives of different families were not separat-ed properly (the genera *Lynceus* and *Daphnia* included chydorids, bosminids and daphniids, macrothricids, moinids, sidids, respectively).

The next noticeable step in the study of cladoceran diversity was initiated by Baird (1843) who proposed 8 new subgenera (then genera) of the genus *Lynceus* (Chydoridae and Macrothricidae of the modern sys-tem). This differentiation could be obtained due to much more detailed morphological investigations. Lat-er on, he established the new genera *Bosmina* (Baird, 1845), *Moina* and *Daphnella* (= *Diaphanosoma* Fisch-er, 1850) (Baird, 1850). In the latter extensive mono-graph, Baird described 29 species of Cladocera includ-ing 13 new ones. His family Lynceidae was properly delimited, whereas his Daphniadae comprised repre-sentatives of five families (sidids were separated as the subfamily Sidina).

Somewhat later, the descriptions of other new genera such as *Ceriodaphnia* (Dana, 1853), *Simo-cephalus, Scapholeberis* (Schoedler, 1858) and *Holo-pedium* (Zaddach, 1855) were published. The data of the latter author were especially detailed. Monographs by Fischer (1850), Lilljeborg (1853) and Leydig (1860) represented descriptions of many European species. In total, 19 new genera were described in 20 years, and the number of described species had reached approxi-mately 60.

All these studies paved the way for the subsequent period which is marked by the outstanding carcinol-ogist G. Sars. In 1861, Sars presented an extensive manuscript on Norwegian Cladocera which was pub-lished only 130 years later (Sars, 1993). In the last cen-tury, only part of the manuscript with descriptions of Ctenopoda (families Sididae and Holopediidae) (Sars, 1865) as well as two earlier papers with brief Latin diagnoses of species without figures (Sars, 1861, 1862) were published. This incompletion of important basic research was the reason for the subsequent confusion in cladoceran taxonomy (see Frey, 1982b, 1987).

Sars' data may be considered as a summit of the results on the morphology and taxonomy of cladocer-ans for about 200 years up to recent time. In total, 72 species of 33 genera were described and illus-trated in the mentioned manuscript, with 30 species and 11 genera (*Limnosida, Ophryoxus, Drepanothrix, Streblocerus, Ilyocryptus, Anchistropus, Monospilus, Alonella, Graptoleberis, Alonopsis, Harporhynchus* (=*Rhynchotalona*)) being new to science. The repre-sentatives of 16 genera were described in detail, includ-ing all thoracic limbs. However, as mentioned above, most of the data have not been published, and published parts were written in inaccessible Danish which also negatively affected the perception of the results. The latter was also characteristic for some other important works (e.g. Müller, 1867; Lund, 1870). Probably due to these factors, subsequent cladoceran studies were characterized by less high standards than might have been.

Among works published during this period, those by Kurz (1875), Claus (1876), Schoedler (1877) and Hellich (1877) should be cited. Kurz (1875) paid much attention to the description of males, whereas Claus (1876) and Hellich (1877) made drawings of head

Table 1. Chronology of the study of cladoceran diversity at the initial stages of their exploration.

Author	Original names of taxa of genera level (number of species)	Modern interpretation of taxa
Swammerdam, 1669	*Pulex arborescens* (1)	*Daphnia*
Linnaeus, 1744	*Monoculus pulex arborescens*	
	M. pediculus (2)	*Daphnia, Polyphemus*
Joblot, 1754	Pou aquatique (2?)	*Moina?, Scapholeberis?*
Schaeffer, 1763	Geschwanzten zackiger Wasserflöh, Ungeschwanzten zackigen Wasserflöh (2–3)	*Daphnia, Simocephalus?*
Geoffroy, 1764	Perroquet d'eau (1)	*Daphnia*
Müller, 1776	*Lynceus, Daphne*	*Alona, Pleuroxus, Camptocercus*
	Polyphemus	*Chydorus, Eurycercus, Daphnia, Simocephalus, Moina, Scapholeberis, Sida, Latona, Bosmina, Macrothrix, Polyphemus*
Müller, 1785	'——'	'——'
Leach, 1816	*Chydorus*	*Chydorus*
Straus, 1820	*Sida, Latona*	*Sida, Latona*
Loven, 1836	*Evadne*	*Evadne*
Baird, 1843	*Macrothrix, Eurycercus, Camptocercus, Acroperus, Alona, Pleuroxus, Peracantha*	*Macrothrix, Eurycercus, Camptocercus, Acroperus Alona, Pleuroxus*
Baird, 1845	*Bosmina*	*Bosmina*
Baird, 1850	*Moina, Daphnella*	*Moina, Diaphanosoma*

pores of the Chydoridae which later were used as a very important diagnostic feature in this family (Frey, 1959). The publications by Weismann (1877, 1880) presented a lot of significant information on inner structures, especially on reproductive system. At this time, a number of structures that were subsequently often overlooked were described, such as the proximal segment of the lower antennal branch of *Sida* (Fischer, 1850) and the functional mechanism with which this peculiar crustaceans attach to a substrate (Sars, 1865). Most of known European species were described by the end of the 1880s (Frey, 1982a). The development of knowledge of diversity of the group led to taxonomic difficulties mainly connected with strong morphological variability of *Daphnia* and *Bosmina* (Frey, 1982b). This resulted in considerable complication of the classification system (e.g. Sars, 1890) and which signified the end of 'Linnean' era in the cladoceran taxonomy characterized by the requirement of distinct species differences (Mayr, 1963).

In the second half of the nineteenth century, some authors started active cladoceran studies outside Europe as well, first of all in North America (Birge,

1879, 1893; Herrick, 1879, 1884), and subsequently in Australia (King, 1853; Sars, 1885, 1888) and tropical Asia (Richard, 1894; Daday, 1898) but they made a small share in the bulk. For example, there are only 16 papers (5%) concerning non-European forms on a total of approx. 300 publications from 1861 till 1894 (Klocke, 1894). 22 new genera were described during this period, 6 of them from the tropics and subtropics (*Latonopsis* Sars, *Pseudosida* Herrick, *Moinodaphnia* Herrick, *Grimaldina* Richard, *Guernella* Richard, *Bosminopsis* Richard). The number of known European species surpassed 100, and the world fauna counted of approx. 150 species. At this time, the descriptive canon was formed that has been used for many years, until present. Description focused on general lateral body form and a few other especially visible structures (postabdomen, antennules, etc.). This descriptive scheme did not achieve the level of the best of earlier description, but the average quality of the work during this period was high and sometimes exceeded that of present time.

The publication of Lilljeborg's (1901) monograph on Swedish cladoceran can be considered the apotheo-

4

Table 2. The main historic periods of studies on cladoceran taxonomy and the number of described taxa

Years	Number of taxa	
	Described genera	Known species
1662–1776	–	5–8
1776–1843	6	20–25
1843–1861	19	c. 60
1861–1901	22	c. 150
1901–1952	11	c. 400
1952–1973	8	c. 450
1973–present	14	c. 600

sis of the preceding period of their study, and at the same time was a powerful stimulus to further studies on cladoceran taxonomy. Nothing similar was published on the Cladocera before. The first extensive taxonomic reviewes by Richard (1895, 1896) were not notable for high quality. Lilljeborg described and figured 102 species in detail and paid much attention to phenotypic variability of *Daphnia* and *Bosmina*. It is not surprising that it became a model for imitation up to recent time, to the extent that it was republished in 1982. The volume also become the permanent source of drawings and a strong stimulus of the idea of cladoceran cosmopolitanism (Frey, 1982a). The latter became apparent from a number of manuals published on the cladocerans of Germany (Keilhack, 1909; Storch, 1926; Wagler, 1927, 1937), United States (Birge, 1918), Italy (Parenzan, 1932), Japan (Ueno, 1937) and Caucasus (Behning, 1941). They clearly developed the trend of 'copying', showing the aspiration 'to find' species and forms described by Lilljeborg.

By the beginning of our century, the interest in general cladoceran morphology had reduced. On the other hand, researchers paid more attention to the problem of variability of bodyform, cyclomorphosis, and the taxonomy of the pelagic representatives of the group (e.g. Wesenberg-Lund, 1926, 1939; Woltereck, 1920, 1932; Wagler, 1927, 1937; Rühe, 1912, 1913). The suggested theories have been reviewed by Hutchinson (1967) and Kiselev (1969). The variability of other cladocerans, with a few exceptions (Burckhardt, 1899; Werner, 1924; Rammner, 1927), was much neglected.

During the 1910–30s there was a lot of interest in the structure and functioning of the thoracic limbs (Behning, 1912; Storch, 1929; Cannon, 1933; Eriksson, 1934) though they mostly concerned Ctenopods

and Daphniids, and a number of their results needed subsequent correction (Fryer, 1987b, 1991).

The study of 'exotic' faunas was stimulated by abundant material from different regions of the world. Many carcinologists treated such materials, including von Daday (1898, 1905, 1910), Ekman (1900, 1908), Gurney (1927), Stingelin (1900, 1905) and Sars (1885, 1888, 1901, 1916). Later on Gauthier (1928, 1951), Rammner (1937) and Ueno (1938, 1939, 1966) joined them. Brehm (1933, 1952, 1959) published numerous papers with descriptions of new taxa, later often synonymized or fallen into incertae sedis.

In general, the level of taxonomic descriptions of this period was lower, and only a few traditional diagnostic traits were used. The works by Ocioszynska-Bankierova (1933a, b) may be cited as a rare example demonstrating innovations.

On the whole, 11 new genera were described up to the 1950s though most of them in the beginning of the century (1901–1912). The number of known species increased up to about 400.

The next, relatively short historical period may be regarded as a transitional period because it was mostly dedicated to summing up the results of the previous periods resulting in the publication of numerous monographs of regional (Johnson, 1952; Brooks, 1957, 1966; Sramek-Husek et al., 1962; Herbst, 1962; Røen, 1962; Olivier, 1962; Manuilova, 1964; Flössner, 1972) or world-wide significance on separate families (Goulden, 1968; Smirnov, 1971a). In most cases, traditional diagnostic traits were used for species descriptions, though a number of new ones, including male structures, have been proposed in the revisions of the Moinidae of the world (Goulden, 1968) and *Daphnia* of England (Scourfield, 1942; Johnson, 1952) and North America (Brooks, 1957). On the other hand, new traits were considered in the classification at the superspecies level, as exemplified by the use of head pores in the classification of Chydoridae and *Bosmina* (Frey, 1959; Goulden & Frey, 1963; Korinek, 1971). Two new families, the Podonidae and the Cercopagidae were described in the Onychopoda (Mordukhai-Boltovskoi, 1968).

A number of interesting works were published on functional morphology of the poorly studied Chydoridae and Macrothricidae (Fryer, 1963, 1968, 1974; Smirnov, 1971a and previous papers). These studies also had taxonomic consequences, including the description of *Pseudochydorus*, *Phryxura* (=*Disparalona*) by Fryer (1968). The representatives of other new genera were discovered in studies on previously poorly known areas (Vasiljeva & Smirnov, 1969; Fryer & Paggi, 1972). In addition, evolutionary trends were discussed based on new observations (Geodakyan & Smirnov, 1968; Smirnov, 1969, 1971b; Fryer, 1968).

The considerable progress in morphological studies allowed to reveal adequate diagnostic traits and to describe 8 new genera during a comparatively short period of time (20 years), though not all of them, like *Biapertura* N. N. Smirnov were accepted by all specialists (Frey, 1987; Alonso, 1996).

The reviewing and revisions of collected data also continued later on (Smirnov, 1976; Pennak, 1978; Chiang & Du, 1979; Negrea, 1983; Margaritora, 1985; Mordukhai-Boltovskoi & Rivier, 1987). Since the beginning of 1970s, however, another direction focused on thorough reevaluation of cladoceran species. The paper by Frey (1973) on species of *Eurycercus* and the many others that followed it clearly indicated that the diversity of the cladocerans is much greater than accounted for by the number of species currently recognized. This information could only be obtained through the application of new approaches and methods of taxonomic investigation (Frey, 1982a, 1987, 1995; Korovchinsky, 1996). It was also found that detailed morphological data are in some cases quite congruent with genetic ones (Wolf & Mort, 1986; Hebert, 1987; Benzie, 1988a, b but see also Hebert & Taylor, 1997).

The species number has by now increased to about 600, among which the portion of sufficiently described species has become appreciable. Meanwhile, however, it has been shown that many of the currently known cladoceran species have vague taxonomic status from the modern point of view (Korovchinsky, 1996). The view on the geographic distribution of cladoceran has also much changed. The idea of cosmopolitanism of cladoceran species is refuted by more and more observations. In addition, some peculiar species occurring in ground waters (Dumont, 1983), caves (Brancelj, 1990, 1992) and even terrestrial habitats (Frey, 1980) have been discovered.

The number of new genera has also increased, partly because of comprehensive morphological studies (*Ephemeroporus* Frey, *Notoalona* Rajapaksa & Fernando, *Estatheroporus* Alonso, *Celsinotus* Frey, *Sarsilatona* Korovchinsky), and partly because of a detailed investigation of the endemic Australian fauna (Smirnov & Timms, 1983; Frey, 1991; Smirnov, 1995). The families of Macrothricidae (Smirnov, 1992), Sididae and Holopediidae (Korovchinsky, 1992) were revised thoroughly.

Recently, the number of studies on the influence of biotic factors (mainly predators) on cladoceran morphology has much increased (Krueger & Dodson, 1981; Havel, 1987; Black & Slobodkin, 1987; Boikova & Korovchinsky, 1995 and many others). Other morphological studies focused on fine morphology of thoracic limbs (e.g. Crittenden, 1981; Geller & Müller, 1981; Hessen, 1985). The representatives of the Daphniidae were reevaluated from the functional-morphological point of view, partly correcting for former errors of interpretation (Fryer, 1987a, 1991). On the whole, however, the proportion of pure morphological studies was small relative to rapidly increasing mass of studies on ecology, eco-morphology and genetics.

Finally, the cladocerans have been considered as a phylogenetically and taxonomically artificial group, and their representatives were attributed to different orders within Branchiopoda (Fryer, 1987a, b) or even to different subclasses within the Crustacea (Starobogatov, 1986).

Brief notes on the history of hydrobiology

It was noted more than once that hydrobiology (especially when considered as a pure ecological discipline) is young compared with a number of other branches of biology (Zernov, 1949; Winberg, 1975). This is particularly true for the study of the microfauna of continental water bodies. As it was mentioned above, the isolated descriptions of the first European cladocerans appeared in the second half of the 17th century but intensive studies started only in two centuries later and the level of species differentiation was rather low. The first American aquatic invertebrates were described in 1830 (Ruth, 1977), and the first cladocerans only in the end of 1870s (Birge, 1879; Herrick, 1879). In Russia, the first freshwater Crustacea (7 species, of which 5 cladocerans) were listed in the beginning of the 19th century (Dwigubsky, 1892), and their descriptions appeared only 45 years later (Fischer, 1848). Only by the end of the 1880s the freshwater fauna of Central Russia began to be studied rather intensively (Zernov, 1921). The example of Lake Baikal is quite significant: the start of its studies seems to be connected with the politic exile of its future researchers B. Dybowski and V. Godlewski during the late 1860s. The first mass microcrustacean species *Epishura baikalensis* Sars (Copepoda) was described from the lake only in

1900, and detailed investigation of other species awaited the 1930s (Kozhow, 1962).

Comparing with the chronology of studies of the ichthyofauna and especially the terrestrial fauna (e.g. insects, birds and mammals), all dates mentioned above seem to be much retarded. The faunistic studies of many groups of organisms, covering large areas, were almost completed at the beginning of the 20th century (Sabrosky, 1950; Mayr et al., 1953; Ivanov, 1967), and the taxonomists of these groups could start their investigations of intraspecies units and their search for taxonomic traits that are important for differentiation of closely related species earlier than freshwater invertebrate zoologists. For comparison: at the beginning of the 19th century, there were only 7 known species of Crustacea compared to 116 of birds and more than 700 of insects in the Moscow district (Dwigubsky, 1892).

During the second half of the 19th century, after the publication of Darwin's theory, there was an inward interest in comparative morphology, comparative embryology and phylogeny. However, most attention was paid to the marine fauna (Plavilshikov, 1941), and marine biological stations were established 20 years earlier and attracted most scientists than the fresh-water stations (Lampert, 1900).

By the time the studies of continental water bodies and their fauna began to accelerate, the interests of biologists shifted to ecology, biocenology and fish management. As a result, the taxonomic and faunistic investigations were pushed aside again and never regained their due position. Owing to these early trends, hydrobiology became one of the most developed branches of ecology, having a central role in the development of the theory of biological productivity (Winberg, 1967, 1988) as well as some other important directions (Edmondson, 1990).

The continental water bodies began to be considered as integral objects ('microcosms') at early stages of their investigation (Forbes, 1887; Forel, 1892–1904). According to this point of view, lakes are considered natural entities, 'geographic individuals' that require an investigation in all their complexity. They are interpreted as organisms of highest order (biocenoses and lakes are organisms of second and third order respectively), and hydrobiology was considered to be the science studying their physiology (see Thienemann, 1926; Winberg, 1975). This perception of hydrobiology unfortunately narrowed the perspectives of the discipline. In the wake of the dominant ideas (Welch, 1935; Winberg, 1975, 1988), hydrobiol-

ogy was actually reduced to applied science, the main aims of which were the problems of water quality and biological productivity.

Due to the influence of the factors described above, the taxonomy and faunistics of the cladocerans, as well as of many other microcrustaceans of continental waters, had no time to build up the 'critical mass' of knowledge needed for a correct following elaboration of the species concept that was developed during the first half of our century and led to the 'new systematics' (Huxley, 1940; Mayr, 1963). Hydrobiology deviated much from basic zoology and botany and lost the susceptibility to new ideas in these fields. A negative role was probably also played by crises of the species concept at the start of the 20th centuries (Zavadsky, 1968) and a lack of the equipment needed for fine morphological investigations. It is thus not surprising that new taxonomic practice was just recently established with respect to aquatic microinvertebrates, including cladocerans, and is yet little known to most of the hydrobiologists who frequently have an obsolete notion about the real species composition of their samples (Korovchinsky, 1991, 1996).

Conclusion

In the light of the above considerations, it is possible to conclude that the insufficient knowledge of cladoceran taxonomic diversity is caused both by their often fine morphological differentiation, phenotypic variability and by historical factors. Intensive taxonomic and faunistic studies of cladocerans and other microinvertebrates of continental waters started comparatively late, beginning in the 1840–70s. These studies soon met difficulties connected with the variability of *Daphnia* and *Bosmina* unresolved up to now. The fast accumulation of abundant materials and the high diversity of this material, analysed by a low number of specialists led to superficial descriptions and identifications. The latter was also the result by the popular idea of cosmopolitan distribution of many aquatic invertebrates. The late publication of the outstanding work of Sars and the publication of some other basic information in obscure languages also retarded the progress of cladoceran studies. As already mentioned the crisis in systematics and the reduced interest for this basic biological branch at the beginning of the present century, also had a negative influence on the development of this field.

Cladoceran taxonomy began to approach its current high level only by 1970s, mainly due to an increase of interest of researchers for local faunas, more detailed morphological descriptions, and consideration of genetic data.

The beginning of a new epoch in hydrobiology was realized with the transition from an integral to a population-centred approach (see Ghilarov, 1981, 1988) characterized by studies on particular species interactions. Within this framework, it may be hoped that the interest for taxonomic and faunistic research of aquatic invertebrates will increase. The recognition of the latter as a basic part of hydrobiology, as it was considered to be originally (Zaharias, 1891; Gajewskaya, 1948) should favour the further development of all this complex field of knowledge.

Acknowledgments

I am much indebted to Professor N. N. Smirnov (Institute of ecology and evolution) for his help with rare literature sources and correction of the English text. Valuable comments by three anonymous reviewers and Dr U. Røen (Zoological Museum, University of Copenhagen, Denmark) greatly improved the manuscript as well. This study was supported by the Russian Foundation for Basic Research (grants 96-04-48063 and 96-04-58609). Dr A. Brancelj (National Institute of Biology, Ljubljana, Slovenia) promoted the participation of the author in the 4th International Symposium on Cladocera.

References

Alonso, M., 1996. Crustacea Branchiopoda. Fauna Iberica 7, Museo Nacional de Ciencias Naturales, Consejo Superior de Investigaciones Cientificas, Madrid: 486 pp.

Baird, W., 1843. The natural history of the British Entomostraca. No. VI. Ann. Mag. Nat. Hist. 11: 81–95.

Baird, W., 1845. Arrangement of the British Entomostraca, with a list of species, particularly noticing those which have yet been discovered within the bounds of the Club. Hist. Berwickshire Naturalist' Club 2: 145–158.

Baird, W., 1850. The natural history of the British Entomostraca. The Ray Soc., London, 364 pp.

Behning, A., 1912. Studien über die vergleichende Morphologie sowie über temporale und Lokalvariation der Phyllopodenextermitation. Int. Revue ges. Hydrobiol. und Hydrogr., Biol. Ser 4, 1: 1–70.

Behning, A., 1941. Kladotsera Kavkaza. Tbilisi, 384 pp. (in Russian) (Cladocera of Caucasus).

Benzie, J. A. H., 1988a. The systematics of Australian Daphnia (Cladocera: Daphniidae). Species descriptions and keys. Hydrobiologia 166: 95–161.

Benzie, J. A. H., 1988b. The systematics of Australian Daphnia (Cladocera: Daphniidae). Electrophoretic analyses of the Daphnia carinata complex. Hydrobiologia 166: 183–197.

Birge, E. A., 1879. Notes on Cladocera. Trans. Wisc. Acad. Sci., Arts and Lett. 4: 77–109.

Birge, E. A., 1893. Notes on Cladocera. III. Trans. Wisc. Acad. Sci., Arts and Lett. 9: 275–317.

Birge, E. A., 1918. The water fleas (Cladocera). In Ward, H. B. & G. C. Whipple (eds), Freshwater Biology, Wiley, New York: 676–740.

Black, R. W. & B. Slobodkin, 1987. What is cyclomorphosis? Freshwat. Biol. 18: 373–378.

Boikova, O. S. & N. M. Korovchinsky, 1995. On the intrapopulation polymorphism in Daphnia cristata Sars, 1862 (Crustacea, Daphniiformes): A new approach to the cyclomorphosis of the species. Arthropoda Selecta 4: 25–32.

Brancelj, A., 1990. Alona hercegovinae n.sp. (Cladocera: Chydoridae), a blind cave-inhabiting cladoceran from Hercegovina (Yugoslavia). Hydrobiologia 199: 7–18.

Brancelj, A., 1992. Alona sketi sp.n. (Cladocera, Chydoridae), the second cave-inhabiting cladoceran from former Yugoslavia. Hydrobiologia 248: 105–114.

Brehm, V., 1933. Die Cladoceren der Deutschen limnologischen Sunda-Expedition. Arch. Hydrobiol. 11: 631–771.

Brehm, V., 1952. Diaphanosoma hydrocephalus nov.spec. eine eigenartige Sididae aus Vorder-Indien. Zool. Anz. 149: 138–140.

Brehm, V., 1959. Cladoceren und Calanoide Copepoden aus New Guinea. Nova Guinea, n.s. 10: 1–10.

Brooks, J. L., 1957. The systematics of North American Daphnia. Mem. Conn. Acad. Arts Sci. 13: 1–180.

Brooks, J. L., 1966. Cladocera. In Edmondson W. T. (ed.), Freshwater Biology, 2nd edn., Wiley, New York: 587–656.

Burckhardt, G., 1899. Faunistische und systematische Studien über das Zooplankton der grosseren Seen der Schweiz und ihrer Grenzgebiete. Rev. suisse Zool. 7: 353–713.

Cannon, H. G., 1933. On the feeding mechanism of the Branchiopoda. Phil. Trans. r. Soc. London, B 222: 267–352.

Chiang, S. C. & N. S. Du, 1979. Freshwater Cladocera. Fauna Sinica, Crustacea. Peking, 297 pp. (in Chinese).

Claus, C., 1876. Zur Kenntnis der Organisation und des feineren Baues der Daphniden und verwandter Cladoceren. Z. wiss. Zool., 27: 362–402.

Crittenden, R. N., 1981. Morphological characteristics and dimension of the filter structures from three species of Daphnia (Cladocera). Crustaceana 41: 233–248.

Daday, E. von, 1898. Mikroskopische Susswasserthiere aus Ceylon. Termes. Fuzetek. 21: 1–123.

Daday, E. von, 1905. Untersuchungen über die Susswaser-Mikrofauna Paraguays. Zoologica 44: 1–374.

Daday, E. von, 1910. Untersuchungen über die Susswasser-Mikrofauna Deutsch-Ost-Afrikas. Zoologica 59: 1–316.

Dana, J., 1853. Crustacea. II. Daphnioides. United States Exploring expedition during the years 1838–1842 under the command of Charles Wilkes 14: 1–11.

DeGeer, C., 1778. Memoires pour servir à l'histoire des Insectes 7, Stockholm, 470 pp.

Dumont, H. J., 1983. Discovery of groundwater-inhabiting Chydoridae (Crustacea: Cladocera), with the description of two new species. Hydrobiologia 106: 97–106.

8

Dwigubsky, J., 1892. Primitiae Faunae Mosquensis, seu Enumeration animalium, quae fronte circa Mosquam vivunt, 2nd edn. Moscow, 135 pp.

Edmondson, W. T., 1990. Perspectives in plankton studies. Mem. Ist. ital. Idrobiol. 47: 331–361.

Ekman, S., 1900. Cladoceren aus Patagonien, gessamelt von der schwedischen Expedition nach Patagonien 1899. Zool. Jb., Syst., Geogr. Biol. Tiere 14: 62–84.

Ekman, S., 1908. Cladoceren und Copepoden aus antarktischen und subantarktischen Binnengewassern, gesammelt von der schwedischen antarktischen Expedition 1901–1903. Wiss. Ergeb. Schwedischen Sudpolar-Exped. 5: 1–40.

Eriksson, S., 1934. Studien über die Fangapparate der Branchiopoden nebst einigen phylogenetische Bemerkungen. Zool. Bidr. Uppsala 15: 23–287.

Fischer, S., 1848. Über die in der Umgebung von St. Petersburg Vorkommenden Crustacea aus der Ordnung der Branchiopoden und Entomostraceen. Mem. Acad. Sci. St. Petersb. 6: 159–198.

Fischer, S., 1850. Erganzungen, Berichtungen und Fortsetzung zu der Abhandlung über die in der Umgebung von St. Petersburg Vorcommenden Crustaceen etc. Mem. Acad. Sci. St. Petersb. 7: 1–14.

Flössner, D., 1972. Kiemen- und Blattfusser. Branchiopoda. Fishlause. Branchiura. Die Tiervelt Deutschlands 60, Jena, 501 pp.

Forbes, S. A., 1887. The lake as microcosm. Bull. Sci. Ass. Peoria.

Forel, F. A., 1892–1904. Le Leman. Monographie Limnologique I–III: 540 pp., 650 pp., 306 pp.

Frey, D. G., 1959. The taxonomic and phylogenetic significance of the head pores of the Chydoridae (Cladocera). Int. Revue ges. Hydrobiol. 44: 27–50.

Frey, D. G., 1973. Comparative morphology and biology of three species of Eurycercus (Chydoridae, Cladocera), with a description of Eurycercus macrocanthus sp.nov. Int. Revue ges. Hydrobiol. 58: 221–267.

Frey, D. G., 1980. The non-swimming chydorid Cladocera of wet forests, with descriptions of a new genus and two new species. Int. Revue ges. Hydrobiol. 65: 613–641.

Frey, D. G., 1982a. Questions concerning cosmopolitanism in Cladocera. Arch. Hydrobiol. 93: 484–502.

Frey, D. G., 1982b. G. O. Sars and the Norwegian Cladocera: continuing frustration. Hydrobiologia 96: 267–293.

Frey, D. G., 1987. The taxonomy and biogeography of the Cladocera. Hydrobiologia 145: 5–17.

Frey, D. G., 1991. The species of Pleuroxus and of three related genera (Anomopoda, Chydoridae) in Southern Australia and New Zealand. Rec. Aust. Mus. 43: 291–372.

Frey, D. G., 1995. Changing attitudes toward chydorid anomopods since 1769. Hydrobiologia 307: 43–55.

Fryer, G., 1963. The functional morphology and feeding mechanism of the Chydorid Cladoceran Eurycercus lamellatus (O. F. Müller). Trans. r. Soc. Edinb. 65: 335–381.

Fryer, G., 1968. Evolution and adaptive radiation in the Chydoridae (Crustacea: Cladocera): a study in comparative functional morphology and ecology. Phil. Trans. r. Soc. London, B, 254: 221–385.

Fryer, G., 1974. Evolution and adaptive radiation in the Macrothricidae (Crustacea: Cladocera): a study in comparative functional morphology and ecology. Phil. Trans. r. Soc. London, B, 269: 137–274.

Fryer, G., 1987a. Morphology and the classification of the so-called Cladocera. Hydrobiologia 145: 19–28.

Fryer, G., 1987b. A new classification of the branchiopod Crustacea. Zool. J. Linn. Soc. 91: 357–383.

Fryer, G., 1991. Functional morphology and adaptive radiation of Daphniidae (Branchiopoda: Anomopoda). Phil. Trans. r. Soc. London, B, 331: 1–99.

Fryer, G. & J. C. Paggi, 1972. A new Cladoceran genus of the family Macrothricidae from Argentina. Crustaceana 23: 255–262.

Gajewskaya, N. S., 1948. Trofologitcheskoye napravlenie v gidrobiologii, ego objekt, nekotorye osnovnye problemy i zadachi. Pamjati akademika A. S. Zernova, Moscow –Leningrad, 27–47. (in Russian) (Trophologic direction in hydrobiology, its main problems and tasks).

Gauthier, H., 1928. Recherches sur la faune des eaux continentales de l'Algérie et de la Tunicie. Alger, 419 pp.

Gauthier, H., 1951. Contribution à l'étude de la faune des eaux douces au Sénégal (Entomostracés). Alger, 169 pp.

Geller, W. & R. Müller, 1981. The filtration apparatus of Cladocera: filter mesh sizes and their implication on food selectivity. Oecologia 49: 316–321.

Geodakyan, V. A. & N. N. Smirnov, 1968. Polovoy dimorphism i evolutsia nizshih rakoobraznyh. Problemy evolutsii 1: 30–36. (in Russian) (Sexual dimorphism and problems of Entomostraca evolution).

Geoffroy, E. L., 1764. Histoire abregée des Insectes, dans lequelle ces Animaux sont rangés suivant un ordere méthodique 2, Paris, 651 pp.

Ghilarov, A. M., 1981. Metodologitcheskye problemy sovremennoi ekologii. Smena vedushih koncepcyi. Priroda 9: 96–103. (in Russian) (Methodological problems of modern ecology. Change of leading conceptions).

Ghilarov, A. M., 1988. Sootnoshenie organicizma i redukcionizma kak osnovnyh metodologicheskyh podhodov v ekologii. J. obschey biol. 49: 202–217. (in Russian) (Correlation of organicism and reductionism as the main methodological approaches in ecology).

Goedardo, J., 1662. Metamorphosis Naturalis. Metamorphosis et Historia Naturalis Insectorum. Cum commentariis D. Joannus de Mey ecclesiasts Medioburgensis ac Doct. Med. & duplici ejusdem appendice, una de hemerobiis, altera de natura comentarum, & vanis ex iis divinationibus. Medioburgi.

Goulden, C. E., 1968. The systematics and evolution of the Moinidae. Trans. Am. Phil. Soc., Philadelphia, N.S. 58: 1–101.

Goulden, C. E. & D. G. Frey, 1963. The occurrence and significance of lateral head pores in the genus Bosmina (Cladocera). Int. Revue ges. Hydrobiol. 48: 513–522.

Gurney, R., 1927. Some Australian freshwater Entomostraca reared from dried mud. Proc. zool. Soc. Lond.: 59–71.

Havel, J. E., 1987. Predator-induced defences: a review. In Kerfoot, W. C. & A. Sih (eds), Predation: Direct and Indirect Impacts on Aquatic Communities, Hanover, N.H.: 263–278.

Hebert, P. D. N., 1987. Genotypic characteristics of the Cladocera. Hydrobiologia 145: 183–193.

Hebert, P. D. N. & D. Taylor, 1997. The future of cladoceran genetics: methodologies and targets. Hydrobiologia 360: 295–299.

Hellich, B., 1877. Die Cladoceren Bohmens. Die Arbeiten der Zoologischen Abtheilung der Landesdurchforschung von Böhmens. Arch. naturw. Land. Bohmen. Prag. 3, Sect. 4: 1–131.

Herbst, H.-V., 1962. Blattfusskrebse (Phyllopoden: Echte Blattfüsser und Wasserflöhe). Stuttgart, 130 pp.

Herrick, C. L., 1879. Microscopic Entomostraca of Minnesota. 7th Annu. Rep. min. geol. nat. Hist. Surv.: 81–123.

Herrick, C. L., 1884. A final report on the Crustacea of Minnesota included in the orders Cladocera and Copepoda. 12th Annu. Rep. min. geol. nat. Hist. Surv.: 1–191.

Hessen, D. O., 1985. Filtering structures and particle size selection in coexisting Cladocera. Oecologia 66: 368–372.

Hutchinson, G. E., 1967. Treatise on Limnology 2, N.Y., 1115 pp.

Huxley, J. S. (ed.), 1940. The New Systematics. Clarendon Press, Oxford, 583 pp.

Ivanov, A. I., 1967. Faunistika i zoogeographia sushy. Razvytie biologii v SSSR, Nauka, Moscow: 165–174. (in Russian) (Terrestrial faunistics and zoogeography).

Joblot, C., 1754. Observations d'Histoire naturelle, faites avec le Microscope 1, Paris.

Johnson, D. S., 1952. The British species of the genus *Daphnia* (Crustacea, Cladocera). Proc. Zool. Soc. London 122: 435–462.

Jurine, L., 1820. Histoire des *Monocles* qui se trouvent aus environs de Genève. Genève et Paris, 258 pp.

Keilhack, L., 1909. Phyllopoda. In Brauers, A. (ed.), Die Susswasserfauna Deutschlands 10, Jena, 112 pp.

King, R. L., 1853. On Australian Entomostraca. Pap. Proc. r. Soc. Tasmania 2: 253–263.

Kiselev, I. A., 1969. Plankton morey i kontinentalnyh vodoyemov 1, Nauka, Leningrad: 657 pp. (in Russian) (The plankton of marine and continental waters).

Klocke, E., 1894. Zur Cladocerenfauna Westfalens (Verzeichnis der Litteratur über Cladoceren von 1669–1894). Westfal. Prov.– Ver. Wiss. und Kunst 22: 1–21.

Korinek, V., 1971. Comparative studies of head pores in the genus *Bosmina* Baird (Crustacea, Cladocera). Vest. cs. spolec. zool. 35: 275–296.

Korovchinsky, N. M., 1991. Naskolko nam izvesten vidovoy sostav zooplanktona 'horosho izutchennogo' ozera? Bull. MOIP, ser.biol., 96: 17–29. (in Russian) (How well do we know the species composition of zooplankton of 'well studied' lake?).

Korovchinsky, N. M., 1992. Sididae and Holopediidae. Guides to the identification of the microinvertebrates of the continental waters of the World 3, SPB Acad. Publ., The Hague, 82 pp.

Korovchinsky, N. M., 1996. How many species of Cladocera are there? Hydrobiologia 321: 191–204.

Kozhow, M. M., 1962. Biologia ozera Baikal. Nauka, Moscow, 315 pp. (in Russian) (Biology of Lake Baikal).

Krueger, D. A. & S. I. Dodson, 1981. Embriological induction and predation ecology in *Daphnia pulex*. Limnol. Oceanogr. 26: 219–223.

Kurz, W., 1875. Dodekas neuer Cladoceren nebst einer kurzen Übersicht der Cladocerenfauna Böhmens. Sitzunsber. math.-naturw. Akad. Wiss. 70: 7–88.

Lampert, K., 1900. Zhizn presnyh vod. St. Peterburg, 880 pp. (in Russian) (The life of freswaters).

Latreille, P. A., 1802. Histoire naturelle generale et particuliare de Crustacés et Insectes 4, Paris, 348 pp.

Leach, W. E., 1816. Annulosa. Supplement to the 4th edition of Encyclopedia Britannica, 406.

Leydig, F., 1860. Naturgeschichte der Daphniden. Tubingen, 252 pp.

Lilljeborg, W., 1853. Om de ihom Skane forekomande Crustaceer of ordningarne Cladocera, Ostracoda ech Copepoda. Lund, 222 pp.

Lilljeborg, W., 1901. Cladocera sueciae oder Beitrage sur Kenntnis der in Schweden lebenden Krebstiere von der Ordnung der Branchiopoden und der Unterordnung der Cladoceren. Nova acta Req. soc. sci. Upsal., ser. 3, 19: 1–701.

Linnaeus, K., 1744. Systema Naturae, sive Regna tria Naturae systematicae proposita per Classes, Ordines, Genera et Species, etc. Lugduni.

Loven, S., 1836. *Evadne Nordmanni* et hittils okandt Entomostracon. Kong. Vet.– Acad. Handl. for ar 1835: 1–73.

Lund, L., 1870. Bidrag til cladocernes morphologie og systematic. Naturv. Tids. 3: 129–174.

Manuilova, E. F., 1964. Vetvystousye rachki fauny SSSR. Opredeliteli po faune SSSR 88, Moscow – Leningrad, 327 pp. (in Russian) (Cladocera of the USSR).

Margaritora, F. G., 1985. Cladocera. Fauna d'Italia 23, Bologna: 399 pp.

Mayr, E., 1963. Animal Species and Evolution. The Belknap Press, Harvard Univ. Press, Cambridge, 797 pp.

Mayr, E., E. G. Linsley & R. L. Usinger, 1953. Methods and Principles of Systematic Zoology. N. Y., Toronto, London, 352 pp.

Milne-Edwards, A., 1840. Histoire naturelle des Crustacés, comprenant l'anatomie, la physiologie et la classification de ces animaux 3, Paris, 638 pp.

Mordukhai-Boltovskoi, F. D., 1968. On the taxonomy of the Polyphemidae. Crustaceana 14: 197–209.

Mordukhai-Boltovskoi, F. D. & I. K. Rivier, 1987. Hishnye vetvystousye Podonidae, Polyphemidae, Cercopagidae i Leptodoridae fauny mira. Nauka, Leningrad, 180 pp. (in Russian) (The carnivorous Cladocera of the families Podonidae, Polyphemidae, Cercopagidae and Leptodoridae of the world fauna).

Müller, O. F., 1776. Zoologiae Danicae prodromus seu Animalium Daniae et Norvegiae indigenarum characteres, nomina, et synonyma imprimis popularium. Havniae, 282 pp.

Müller, O. F., 1785. Entomostraca seu insecta testacea quae in aquis Daniae et Norvegiae reperit. Lipsiae et Havniae, 135 pp.

Müller, P. E., 1867. Danmarks Cladocera. Natur. Tids. 3: 53–240.

Negrea, S., 1983. Cladocera. Fauna Republici Socialiste Romania 4, 12. Bucuresti, 399 pp.

Ocioszynska-Bankierova, J., 1933a. Über den Bau der Mandibeln bei *Daphnia magna* Straus. Ann. Mus. Zool. Pol. 7: 33–40.

Ocioszynska-Bankierova, J., 1933b. Über den Bau der Endkrallen bei der Cladoceren-Gattung *Daphnia* und die damit in Verbindung stehenden systematischen Probleme. Ann. Mus. Zool. Pol. 9: 382–410.

Olivier, S. R., 1962. Los Cladoceros Argentinos. Revista Mus. La Plata, Zool. 7: 173–269.

Parenzan, P., 1932. Cladocera, sistamatica e corologia dei Cladoceri Limnicoli italiani ed appendice sui Cladoceri in generale. Minist. agricol. forest. Lab. idrobiol. pesca. Mem. sci. 8, ser. B: 1–340.

Pennak, R. W., 1978. Fresh-water Invertebrates of the United States. 2nd edn., N.Y., 803 pp.

Plavilshikov, N. N., 1941. Ocherki po istorii zoologii. Moscow, 296 pp. (in Russian) (Essays on the history of zoology).

Rammner, W., 1927. Zur Lokalvariation von *Scapholeberis mucronata* und deren Abhangigkeit von der Gewassergrosse. Zool. Anz. 72: 218–224.

Rammner, W., 1937. Beitrag zur Cladocerenfauna von Java. Int. Revue ges. Hydrobiol. Hydrogr. 35: 35–50.

Richard, J., 1894. Entomostracés recueillis dans lac Toba (Sumatra). Ann. Mus. civ. Stor. Natur. 14: 565–578.

Richard, J., 895. Révision des Cladocerés. I. Ann. Sci. Nat. Zool., ser. 7, 18: 279–389.

Richard, J., 1896. Révision des Cladocerés. II. Ann. Sci. Nat. Zool., ser. 8, 2: 187–363.

Røen, U., 1962. Studies on freshwater Entomostraca in Greenland. Medd. Gronland 170: 1–249.

Rühc, F., 1912. Monographie des Genus *Bosmina*. A. *Bosmina coregoni* im Baltischen See Gebiet. Zoologica 63: 1–141.

Rühe, F., 1913. Biologie und Verbreitung der Bosminen und deren Beziehungen zur Eiszeit. Int. Revue ges. Hydrobiol. Hydrogr. 6: 77–95.

Ruth, P., 1977. The changing scene in aquatic ecology. Acad. Natur. Sci. Phil. Spec. Publ. 12: 205–222.

Sabrosky, C. W., 1950. Taxonomy and ecology. Ecology 31: 151–152.

Sars, G. O., 1861. Om de omegnen af Christiania forekommende cladocerer. Forh. Vidensk Selsk. Krist. 1861: 1–25.

Sars, G. O., 1862. Hr. studios. medic. G. O. Sars fortsatte sit foredrag over de afham i omegnen af Christiania iagttagne Crustacea cladocera. Forh. Vidensk Selsk. Krist. 1861: 250–302.

Sars, G. O., 1865. Norges Ferskvandskrebsdyr. Forste afnir Branchiopoda. 1. Cladocera Ctenopoda (Sididae & Holopediidae). Efter det Acad. Collegium, Christiania, 79 pp.

Sars, G. O., 1885. On some Australian Cladocera raised from dried mud. Forh. Vidensk Selsk. Christiania: 1–46.

Sars, G. O., 1888. Additional notes on Australian Cladocera raised from dried mud. Forh. Vidensk Selsk. Christiania: 1–74.

Sars, G. O., 1890. Oversight af Norges Crustaceer med forelobige Bemaerkninger over de nye eller mindre bekjendte Arter. II. Branchiopoda-Ostracoda-Cirripedia. Forh. Vidensk Selsk. Christiania: 1–80.

Sars, G. O., 1901. Contributions to the knowledge of the freshwater Entomostraca of South America, as shown by artificial hatching from the dried material. Arch. math. og naturv. 23: 1–102.

Sars, G. O., 1916. The freshwater Entomostraca of Cape Province (Union of South Africa). Part I. Cladocera. Ann. South Africa Mus. 15: 303–351.

Sars, G. O., 1993. On the freshwater Crustaceans occurring in the vicinity of Christiania. Bergen, 197 pp.

Schaeffer, J. C., 1763. Die grunen Armpolypen, Die geschwangen und ungeschwanzten, zackigen Wasserflöhe und eine besondere Art kleiner Wasseraale. Zweyte Auflage, Regensburg.

Schoedler, J. E., 1858. Der Branchiopoden in der Umgegend von Berlin. Jahresber. Louisentist. Realschule. Berlin: 1–28.

Schoedler, J. E., 1877. Über die Cladoceren Australiens. Sitz. Gesell. naturf. Berlin, Jahr. 1877: 11–14.

Scourfield, D. J., 1942. The 'pulex' forms of Daphnia and their separation into two distinct series represented by D. pulex (DeGeer) and D. obtusa Kurz. Ann. Mag. Nat. Hist. 9: 202–219.

Smirnov, N. N., 1969. Morpho-functional grounds of mode of life of Cladocera. III. Oligomerization in Cladocera. Hydrobiologia 34: 235–242.

Smirnov, N. N., 1971a. Chydoridae fauny mira. Fauna SSSR, Racoobraznye 1, 2, Nauka, Leningrad, 531 pp. (in Russian) (Chydoridae of the world fauna).

Smirnov, N. N., 1971b. Morfo-funkcionalnye osnovy obraza zhizni vetvistousyh rakoobraznyh. IV. Zakon gomologitcheskyh rjadov u Cladocera calyptomera. J. obschey biol. 32: 82–86. (in Russian) (Morpho-functional grounds of Cladoceran life. IV. The law of homological rows in Cladocera caliptomera).

Smirnov, N. N., 1976. Macrothricidae i Moinidae fauny mira. Fauna SSSR, Rakoobraznye 1, 3, Nauka, Leningrad, 327 pp. (in Russian) (Macrothrycidae and Moinidae of the world fauna).

Smirnov, N. N., 1992. Macrothricidae of the world. Guides to the identification of the microinvertebrates of the continental waters of the World 1. SPB Acad. Publ., The Hague, 143 pp.

Smirnov, N. N., 1995. Check-list of the Australian Cladocera (Crustacea). Arthropoda Selecta 4: 3–6.

Smirnov, N. N. & B. V. Timms, 1983. Revision of the Australian Cladocera (Crustacea). Rec. Austral. Mus. 1: 1–132.

Sramek-Husek, R., M. Straskraba & J. Brtek, 1962. Lupenonozci-Branchiopoda. Fauna CSSR 16, Praha, 470 pp.

Starobogatov, Ya. I., 1986. Systema rakoobraznyh. Zoologitchesky J. 65: 1769–1781.

Stingelin, T., 1900. Beitrag zur Kenntnis der Süsswasserfauna von Celebes. Entomostraca. Rev. Suisse Zool. 8: 193–207.

Stingelin, T., 1905. Untersuchungen über die Cladocerenfauna von Hinterindien, Sumatra und Java, nebst einem Beitrage zur Clado-

ceren Kenntnis der Hawaii-Inseln. Zool. Jahrb., Syst., Geogr. und Biol. der Tiere 21: 327–370.

Storch, O., 1926. Cladocera. Wasserflöhe. Biol der Tiere Deutschlands 15: 23–102.

Storch, O., 1929. Analyse der Fangapparate niederer Krebse auf Grund von Mikro-Zeitlupenaufnahmen. I. Mitteilung der Fangapparat von Sida crystallina O.F.M. Biol. General. Wien 5: 1–62.

Straus, E., 1820. Mémoire sur les Daphnia de la classe des Crustacés. Mem. Mus. Hist. Nat. Paris 6: 149–162.

Swammerdam, J., 1669 (1685). Histoire générale des Insectes. Ou l'on expose clairement la manière lente & presq'insensible de l'accroissement de leurs membres, & ou l'on découvre évidemment l'erreur ou l'on tombe d'ordinaire au sujet de leur prétendue transformation. Autrecht. Chez Jean Ribbius (The original version of this in Latin was published in 1669).

Thienemann, A., 1926. Limnologie. Eine Einfuhrung in die biologischen Probleme der Süsswasserforschung. Jedermanns-Bucherei, Abt. Biologie, Breslau: 1–108.

Ueno, M., 1937. Order Branchiopoda (Class Crustacea). Fauna Nipponica 9, Tokyo, 135 pp.

Ueno, M., 1938. Notes on the Cladocera of Dalai-nor and its neighbouring waters. Annot. zool. jap. 17: 1–6.

Ueno, M., 1939. Manchurian fresh-water Cladocera. Annot. zool. jap. 18: 219–231.

Ueno, M., 1966. Notes on some Cladocera from Iran. Japan. J. Limnol. 26: 146–151.

Vasiljeva, G. L. & N. N. Smirnov, 1969. Chydoridy Baikala. Zool. J. 48: 184–196 (in Russian) (Chydorida of Lake Baikal).

Wagler, E., 1927. Crustacea: Branchiopoda, Phyllopoda-Kiemenfussler. In Kukenthal, W., Handbuch der Zoologie 3: 305–398.

Wagler, E., 1937. Crustacea. Die Tierwelt Mitteleuropas II, 2a: 224 pp.

Weismann, A., 1877. Beitrage zur naturgeschichte der Daphnoiden. Teil II, III und IV. Z. wiss. Zool. 28: 93–254.

Weismann, A., 1880. Beitrage zur naturgeschichte der Daphnoiden. Abhandl. VI und VII. Z. wiss. Zool. 33: 55–270.

Welch, P. S., 1935. Limnology. New York, London, 471 pp.

Werner, F., 1924. Variation analytische Untersuchungen an Chydoriden. Versuch einer quantitativen Morphologie der Cladoceren-Schale. Z. Morph. Okol. Tiere 2: 58–188.

Wesenberg-Lund, C., 1926. Contributions to the biology and morphology of the genus Daphnia. Kgl. Danske vid. selskab. Skr., ser. 8, 11: 91–250.

Wesenberg-Lund, C., 1939. Biologie der Süsswassertiere (Wirbellose Tiere), ubersetzt von O. Storch. Wien: 1–817.

Winberg, G. G., 1967. Osobennosty vodnyh ekologitcheskyh system. J. obschey biol. 28: 538–545. (in Russian) (The peculiarities of aquatic ecological systems).

Winberg, G. G., 1975. Gidrobiologia. Istoria biologii s nachala XX veka do nashyh dney. Nauka, Moscow: 231–248. (in Russian) (Hydrobiology – The history of biology from the beginning of the 20th century to nowadays).

Winberg, G. G., 1988. Konceptualnye osnovy, perspektivnye zadachi i voprosy kadrovogo obespechenyja gidrobiologitcheskyh issledovanyi. Gidrobiol. J. 24: 3–30. (In Russian) (Conceptual grounds, perspective tasks and questions of selection of personel for hydrobiological studies).

Wolf, H. G. & M. A. Mort, 1986. Inter-specific hybrydization underlies phenotypic variability in Daphnia populations. Oecologia 68: 507–511.

Woltereck, R., 1920. Variation und Artbildung. Analytische und experimentelle Untersuchungen an pelagischen Daphniden und anderen Cladoceren. Erster Teil. Morphologische, entwicklungs geschichtliche und physiologische Variations-Analyse. Int. Revue ges. Hydrobiol. Hydrogr. 9: 1–156.

Woltereck, R., 1932. Races, Associations and Stratification of pelagic Daphnids in some lakes of Wisconsin and other regions of the United States and Canada. Trans. Wisconsin Acad. Sci., Arts and Letters 27: 487–522.

Zaddach, E. G., 1855. *Holopedium gibberum*, ein neues Crustaceum aus der Familie der Branchiopoden. Arch. Naturgesch. 21: 159–188.

Zaharias, O., 1891. Die Tier- und Pflanzenwelt des Süsswassers. Leipzig.

Zavadsky, K. M., 1968. Vid i vidoobrazovanie. Nauka, Leningrad: 400 pp. (in Russian) (Species and speciation).

Zernov, S. A., 1921. Opyt synhronycheskoy tablitsy po razvytiju gidrobiologii, ihtiologii i drugih blizhaishyh nauk. Russ. hydrobiol. J. 1: 1–7. (in Russian) (Synchronous table on development of hydrobiology, ichthyology and related sciences).

Zernov, S. A., 1949. Obschaja gidrobiologia. Moscow-Leningrad: 587 pp. (in Russian) (General hydrobiology).

Hydrobiologia **360**: 13–23, 1997.
A. Brancelj, L. De Meester & P. Spaak (eds), Cladocera: The Biology of Model Organisms.
©1997 *Kluwer Academic Publishers.*

Subgeneric differences in head shield and ephippia ultrastructure within the genus *Bosmina* Baird (Crustacea, Cladocera)

Vladimír Kořínek[1], Veronika Sacherová[1] & Ladislav Havel[2]

[1] *Department of Parasitology and Hydrobiology, Charles University, Viničná 7, Praha 2, CZ-128 44, Czech Republic*
[2] *Water Research Institute T.G.M., Podbabská 30, Praha 6, CZ-160 00, Czech Republic*

Key words: Bosmina, head pores, ephippium

Abstract

Using S.E.M. techniques, we compared head pore and ephippia ultrastructures in 115 different populations of the four subgenera: *Bosmina, Eubosmina, Neobosmina* and *Sinobosmina*. We differentiated three types of head pores: median, lateral and frontal pores. The pattern of lateral pores conforms to previously published descriptions. The frontal pore within the subgenus *Eubosmina* is horseshoe-shaped, except in *B. longispina* and *B. oriens*, in which they are oval. In the subgenus *Bosmina*, the frontal pore is oval with the longer axis perpendicular to the animal plane of symmetry. Within the subgenera *Neobosmina* and *Sinobosmina*, the frontal pores are either circular or slightly oval. Ephippial surface sculpture does not differ noticeably among species within the genus. In species with pronounced reticulation of the carapace, the same pattern appears on the ephippial surface. The function of the median and frontal pores are unknown. The pores seem to be connected with the supraoesophageal ganglion. The lateral pores are connected to a tissue of glandiform cells, and a narrow band or rope-like protrusion from the pore is sometimes visible in preserved specimens.

Introduction

Taxonomic decisions within the genus *Bosmina* are difficult because of the paucity of morphological characters that are not influenced by environmental conditions. Phenotypic plasticity (cyclomorphosis; Kerfoot, 1975; Lieder, 1983a) is common and introgressive hybridisation is suspected in some populations (Lieder, 1953, 1983c). Since Goulden & Frey (1963) and Uéno (1968) proposed the use of the lateral pore pattern as a diagnostic character, only a few new morphological traits have been examined for use as taxonomic indicators: postabdomen, first limb of the male, shell spine, antennules, shape of rostrum, shape of frontal head pore (Kořínek, 1971; Deevey & Deevey, 1971; Paggi, 1979; Lieder, 1983a, b).

Goulden & Frey (1963) observed the presence of a small pore on each side of the head shield in the region of the mandibular articulation and a small pore in the median line above the compound eye. Another pore located above the rostrum, between the two frontal setae, was identified by Kořínek (1971), who compared the lateral head pore (LHP) pattern within fourteen presumable species of the genus. Uéno (1968) described a species specific LHP position in the East Asian species *Bosmina fatalis* and Paggi (1979) used head pores as one of the differential characters among Argentinean species of the genus. Havel (1978) undertook one of the most detailed studies of LHP using specimens of numerous bosminid populations from different continents, but his dissertation was not published in any scientific journal.

Previous studies of the ultrastructure of the carapace in the region of the ephippium in cladocerans have demonstrated that these structures can be useful for taxonomic evaluations in some of the anomopod genera: *Ceriodaphnia* (Greenwood et al., 1991), *Daphnia* (Alonso, 1985) and *Moina* (Goulden, 1966). An earlier study of the ephippium in the genus *Bosmina* (Scourfield, 1901) revealed a thick dorsal carina and a lateral narrow band of chitin running obliquely across each valve, from the anterior dorsal angle to the posterior margin. Kerfoot & Peterson (1980) described differences in honey-comb patterning of the ephippial surface of *Bosmina longirostris* in Union Bay, Washington, U.S.A.. According to their results, the ultrastructure of the ephippium can indicate the differences in life strategy of species exposed to predatory pressure. Electrophoretic analysis of allozymes was first applied for studies of bosminid taxonomy by DeMelo & Hebert (1994a, 1994b, 1994c). Genetic analysis detected eight assemblages within the North American Bosminidae, and the morphological analysis distinguished three additional taxa. This demonstrates that further morphological studies are necessary to identify characters useful to bosminid taxonomy. The present study was aimed at comparing three types of head pores within the genus *Bosmina* using scanning electron microscopy, and detecting possible differences in the ultrastrastructure of the ephippium.

Material and methods

The specimens used come from the collection of the Department of Hydrobiology, Charles University, Prague. We examined 115 populations belonging to 20 different species or subspecies. Only adult females were used for the study of head pores, with the exception of the study of the so-called juvenile pore. In collections labelled *, all head pores were studied, those marked + contained ephippial females. Remaining specimens were used for the study of LHP only.

Bosmina berolinensis

Poland: Lake Zywy, 9.11.1961*+, Stradunskie lake; 13.8.1963; Plociczno lake, 8.1963; Kamiene lake, 19.8.1965; Zdrezno lake, 12.8.1965, K. Patalas.

Bosmina brehmi

Nigeria: Kainji Reservoir, 28.1.1981*, C. Tudorancea.

Bosmina chilensis

Chile: Lago di Villarica, 31.3.1899, Silvestri, Daday's coll. Hungarian Nat. Museum Budapest; Lago Calafquén, 6.12.1953, K. Thomasson.

Bosmina coregoni

Canada: Lake Ontario, 9.1967; Lake Erie, 18.10.1969, D. Schindler. Czech Rep.: Slapy Reservoir, 9.8.1960, M. Straškraba; Želivka Res., 3.6.1973, M. Jarešová; Želivka Res., 6.11.1974*+, V. Houk; Lipno Res. 6.1965, Z. Brandl; Knínicky Res., 28.9.1972; Mostiště Res., 22.9.1973, I. Přikryl. Estonia: Peipsi lake, 26.7.1962, A. Mäemets. Germany: Molfsee, 11.5.1921*+, A. Thienemann; Dieksee, 15.8.1944, M.P. Inst. Limnol., Plön; Nehnitz See, 9.1972; Stechnlin See, 4.9.1972, J. Fott; Grebiner See, 6.1962, W. Ohle. Norway: Oremark sjö, 9.1973; Bornevann, 9.1973; Goksjö, 9.1973, J. P. Nilsen. Poland: Mlynskie lake, 4.8.1962; Klosowskie lake, 22.8.1961; Biale lake, 23.8.1966; Kielskie lake, 27.8.1965; Dziadkovo lake, 11.8.1965; Zamiec lake, 18.8.1968; Plociczno lake, 8.1963, K. Patalas. Russia: Otradnoe lake, 15.6.1960, Zool. Inst. A. Sc., St. Peteresburg. Sweden: Lake Mälaren, 8.1970; Lake Väringen, 22.8.1969, B. Grönberg; Orsjön, 19.7.1939; Torserydsjön, 3.9.1937, Limnol. Inst., Uppsala.

Bosmina crassicornis

Estonia: Vagula lake, 8.9.1969, A. Mäemets. Germany: Pulsee, Zool. Mus. Berlin, cat. nr. # 10704, 9398. Poland: Szurpily lake, 12.10.1966, K. Patalas; Kortowo lake, 12.10.1960, L. Szlauer; lake Mikolajskie, 29.9.1962*+, J. Hrbáček.

Bosmina fatalis

Japan: Kawaguchi lake, 20.8.1919, I. Amenomiya, V. Brehm's coll. Russia: Lake Chanka, 15.6.1937*; Lake Bolon, 8.9.1933*, Zool. Inst. A. Sc., St. Petersburg.

Bosmina freyi

Canada: Lake # 302 nr. Ericsson, Man., 27.9.1971 and 14. 9. 1971*; White Mud lake, Duck. Mnt. Prov. Park., Man., 26.10.1971, V. Kořínek; Lake # 239, E.L.A., Kenora Distr., Ont., D. Schindler,

Bosmina gibbera

Poland: Saszyn lake, 31.8.1966; Mausz lake, 26.6.1966; Głębokie lake, 1.9.1964*; Kielskie lake, 27.8.1965, K. Patalas. Sweden: Gläppen lake, 21.8.1969, B. Grönberg.

Bosmina hagmanni

Brazil: Lago Castanho, Amazon Basin, 28.12.1974*, G. O. Brandorff. U.S.A.: Ring Res., 16.3.1961, G. Deevey.

Bosmina kessleri

Slovakia: Bukovec Reservoir, 4.10.1989*, I. Hudec.

Bosmina liederi

Canada: Lake # 81, Kenora Distr., Ont., 3.9.1971, V. Kořínek; N. Scotia, East River Sheetharbor Lake, 7246, 7.10.1984*+, D. G. Frey.

Bosmina longicornis

Poland: Kamien lake, 15.8.1963; Lubowidzkie lake, 17.8.1966*, K. Patalas. Sweden: Vedasjön, 12.7.1942; Bysjön, 8.1943, Limnol. Inst. Uppsala; lake Mälaren, 18.8.1970, B. Grönberg.

Bosmina longirostris

Belgium: Mirwart pond, 10.10.1969*+, J. Hrbáček. Czech Rep.: Želivka Res., 3.6.1973, M. Jarešová; Pond Slaný, Blatná, V. Kořínek; Lipno Res., 5.6.1965, Z. Brandl; V. Arazimova Pool, 3.1975, L. Havel. Germany: Ukleisee, 20.10.1936, M. P. Inst. Limnol., Plön. Japan: Karagon Dam, 28.8.1971, V. H. Jolly. Norway: Svalbard, Alpe Arresjön, 3.8.1993*. Poland: Hancza lake, 8.1966, K. Patalas. Russia: Pond at Jakot, Moscow, 27.6.1967, N. Smirnov. Slovakia: Vinianske lake, Vihorlat Mnts., 2.9.1960, V. Kořínek. Sweden: Krattret lake, 11.8.1913, Limnol. Inst. Uppsala.

Bosmina longispina

Austria: Wörthersee, 8.1959l; Ossiachersee, 12.1958*+, SIL Congress collection; Lunzer Untersee, 5.1931, O. Jírovec. Canada: Babine lake, 5.1968, K. Patalas. Mongolia: Dod-tsagan-nur, 20.7.1969, K. Pivnička. Poland: Niedziegel lake, 9.8.1965; Napruszewo lake, 23.8.1966, K. Patalas. Russia: Otradnoe lake, 1.7.1960, Zool. Inst. A. Sc., St. Petersburg. Sweden: Askoker lake, 18.7.1939, Limnol. Inst. Uppsala; Straken lake, 15.8.1955, K. Thomasson; Sattisjaure, 5.8.1972*+, A. Nauwerck.

Bosmina meridionalis

Australia: Lake Burragorang, 8.8.1965, V.H. Jolly; Mardi Dam, 12.5.1968, B. Timms; Lake Toomba, 20.6.1974*+, D. G. Frey. Ghana: Volta lake, 27.2.1964*, T. Petr. New Zealand: Ngapouri lake, 10.1970*, V. H. Jolly. Thailand: Khao-laem reservoir, 2.1993*, J. Machek.

Bosmina oriens

Canada: N. Scotia, Hwy 349, S. at Halifax, Sheenan Lake, # 7325, 30.10.1984*+, D. G. Frey.

Bosmina reflexa

Poland: Dadaj lake, 6.1962* and 8.1966; Hancza lake, 8.1966; Leleski lake, 8.1966, K. Patalas.

Bosmina thersites

Estonia: Peipsi lake, 26.7.1962, A. Mäemets. Germany: Wandlitz See, cat. nr. 10707; Schermützelsee, cat. nr. 10710; Gr. Müggelsee, cat. nr. 11191, Zool. Mus. Berlin; Gr. Krampe See, 30.10.1986*+; Gr. Müggelsee, 29.10.1986*+, V. Kořínek. Norway: Oremark lake, 9.1973; Iunevann lake, 9.1973, J. P. Nielssen; Björnstadvand, cat. nr. F 17236; Tunhovand, cat. nr. 17235, G.O. Sars, Zool. Mus. Oslo. Poland: Buszew lake, 23.8.1961; Upinek lake, 4.7.1961; Wizajny lake, 13.8.1965, K. Patalas; Gardynskie lake, 5.9.1961, J. Lellák.

Bosmina tubicen

Brazil: Lago Castanho, Amazon Basin, 27.5.1975, G. O. Brandorff. Cuba: Pond at El Dique, 13.4.1965*, N. Martinez; Laguna Sabanilla, 21.2.1966, M. Legner. U.S.A.: Florida, Lake Kingsley, 19.1.1974*, W. M. Lewis jr.

Bosmina sp.

India: Tamil Nadu, Nilgiris Mnts., Kundha Dam, 28.4.1977*, C. H. Fernando's coll., Univ of Waterloo, Canada.

Preparation for scanning electron microscopy (S.E.M.) followed the method described by Inoué & Osatake (1988). Preserved specimens were washed in distilled water to remove the preservative, post-fixed with 1% osmium tetroxide for 1 hour, washed in distilled water to remove the remains of osmium, then dehydrated in 100% ethanol for 45 minutes during which the alcohol was changed three times. The specimens were subsequently transferred to 100% t-butyl alcohol (TBA) (again three changes of alcohol). Because of the high melting point of the TBA (25.6 °C), it was necessary to heat the alcohol above 26 °C. With the third change of TBA, the specimens were placed in a refrigerator for 10 minutes. Containers with frozen specimens were transferred to an extraction glass flask that was evacuated with an oil rotary pump. The frozen TBA completely sublimated. Dry specimens were fixed on brass S.E.M. stubs with double-sided adhesive tape, and coated with gold.

For each species, all three types of head pores and the ultrastructure of the ephippium, in populations containing ephippial females, were documented in photographs. For measurements, we used the photograph with the best image of the pores. The listed measurements and distance therefore do not represent the mean value for the population.

Results

Morphology of the head pores

The median head pore (MHP) is situated on the dorsal surface of the head, in the plane of animal symmetry, just above the compound eye (see Figure 1A). In populations of all the subgenera studied, MHPs were round and small, with an average diameter of 1.4 μm (range 0.7–2.3) (see Figure 1A, C–F). The MHPs seem to be connected to the supraeosophageal ganglion with a cord-like structure (Kořínek, 1971).

The lateral head pores (LHP) are situated on both sides of the head shield in the region of the fornix, and close to the anterior margin covering the base of the antennae (see Figure 1B). The cuticle around the pore is thick, forming a ridge. The pore is connected with a funnel-shaped channel directed inwards. Just below the surface of the cuticle, the channel narrows and turns in a dorsal direction. The shape of the channel is noticeable in specimens treated with hot lactic acid or 10% potassium hydroxide, as it is well sclerotized due to the invagination of the cuticle. Male LHP patterns and shapes are similar to those of adult females.

The frontal head pore (FHP) is located in midway between the two frontal sensory setae (see Figure 1A, B). The thick bordering ridge is not fully developed.

The pore does not seem to be connected to inner tissue by a channel-like tube. Instead, a cord-like structure extends to the supraoesophageal ganglion (Kořínek, 1971).

In all species studied, an additional 'pore' was found in the first instar, above the suture with the carapace, in the region of the so-called dorsal organ. However, no perforation of the cuticle was found using S.E.M., and the structure is probably only a shallow pit in the cuticle which disappears in the second instar.

Differentiation of LHP and FHP among subgenera

Subgenus Bosmina

Eleven populations of *Bosmina longirostris* (O. F. Müller, 1785) were studied. Their LHP pattern conforms with the description provided by Goulden & Frey (1963) (see Figure 2A). The distance from the ventral head shield margin to the pore is 1 to 3 μm. The longer axis of the pore is about twice as long as the shorter one (4.5 × 2.1 μm).

The FHP is oval with the longer axis, which measures 3 μm, perpendicular to the plane of symmetry of the body. The pore is situated between the frontal sensory setae, or slightly shifted to the tip of the rostrum (see Figure 3A).

Subgenus Eubosmina

Seventy nine populations identified as species or subspecies were studied; they comprised the following species: *B. longispina* Leydig, 1860; *B. reflexa* Seligo, 1900; *B. berolinensis* Imhof, 1888; *B. longicornis* Schoedler, 1865; *B. kessleri* Uljanin, 1874; *B. coregoni coregoni* Baird, 1857; *B. coregoni gibbera* Schoedler, 1863; *B. coregoni thersites*, Poppe, 1887; *B. crassicornis* Lilljeborg, 1887 and *B. oriens* DeMelo & Hebert, 1994.

In all populations studied, the LHP pattern is similar to the description given by Goulden & Frey (1963) (see Figure 2B). The shape of the pore is circular or slightly oval with a diameter of about 2 μm, except in *Bosmina coregoni thersites* in which the diameter measures about 1.4 μm and in *B. reflexa*, about 3 μm. The pear-shaped thickening of the cuticle surrounding the pore, observed under the light microscope by Goulden & Frey (1963) and Kořínek (1971), is hardly visible on S.E.M. photographs. The distance from the ventral head shield margin to the pore is about 50 μm.

The FHP is quite narrow, laterally compressed, horseshoe-shaped pore. It is nearly circular or oval

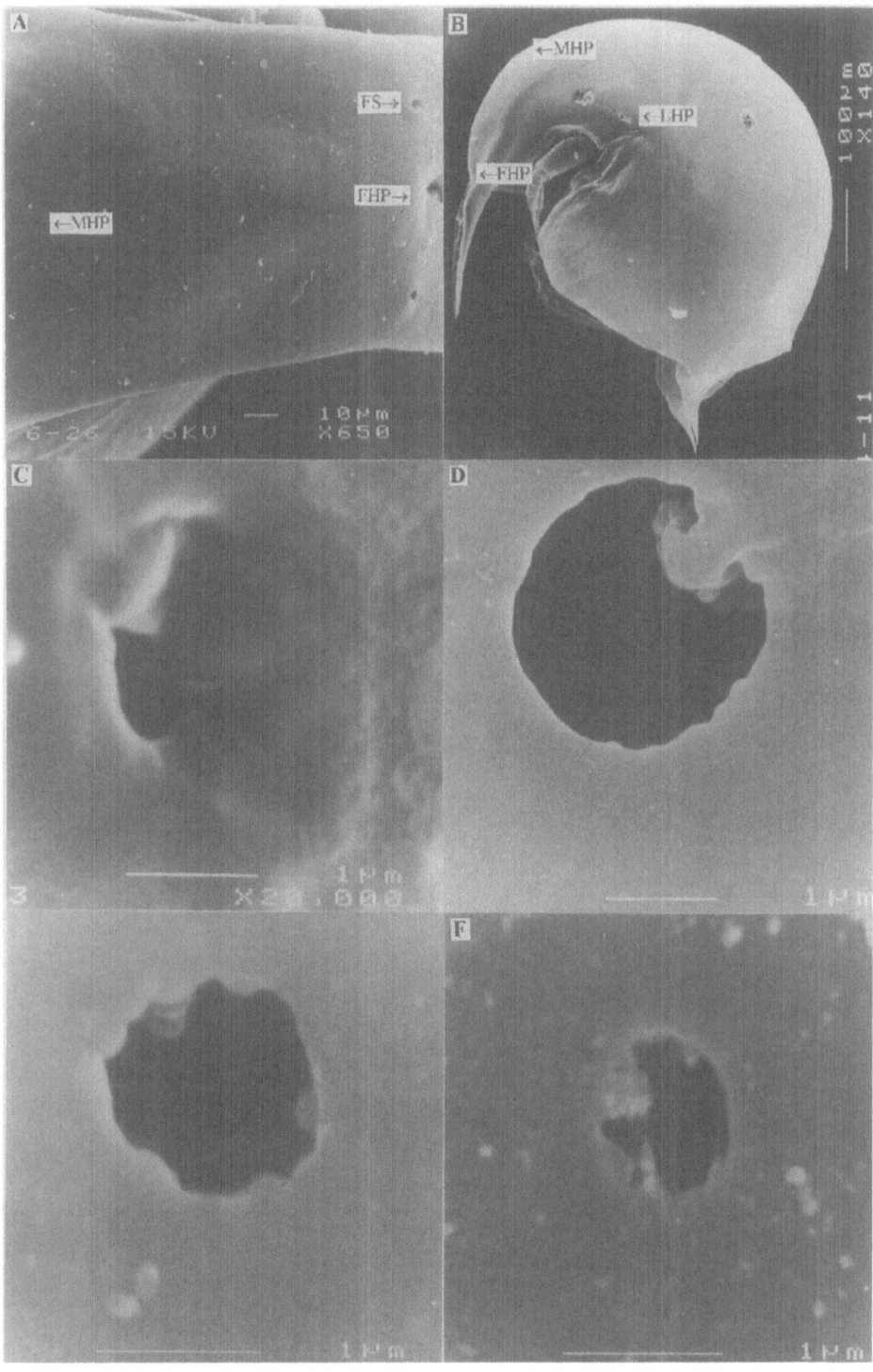

Figure 1. A. Median and frontal pores, dorsal aspect, *Bosmina (Eubosmina) coregoni thersites,* Germany, Gr. Krampe See, 30.10.1986, B. lateral head pore, lateral aspect, *B. (E.) reflexa,* Poland, Dadaj lake, 6.1962. Austria, Ossiachersee, 12.1958, C. median head pore – *Bosmina (Bosmina) longirostris,* Belgium, Mirwart pond, 10.10.1969, D. median head pore – *Bosmina (Eubosmina) reflexa,* Poland, Dadaj lake, 6.1962, E. median head pore – *Bosmina (Neobosmina) tubicen,* U.S.A., Lake Kingsley, 19.1.1974, F. median head pore – *Bosmina (Sinobosmina) fatalis,* Russia, Lake Chanka, 15.6.1937.

18

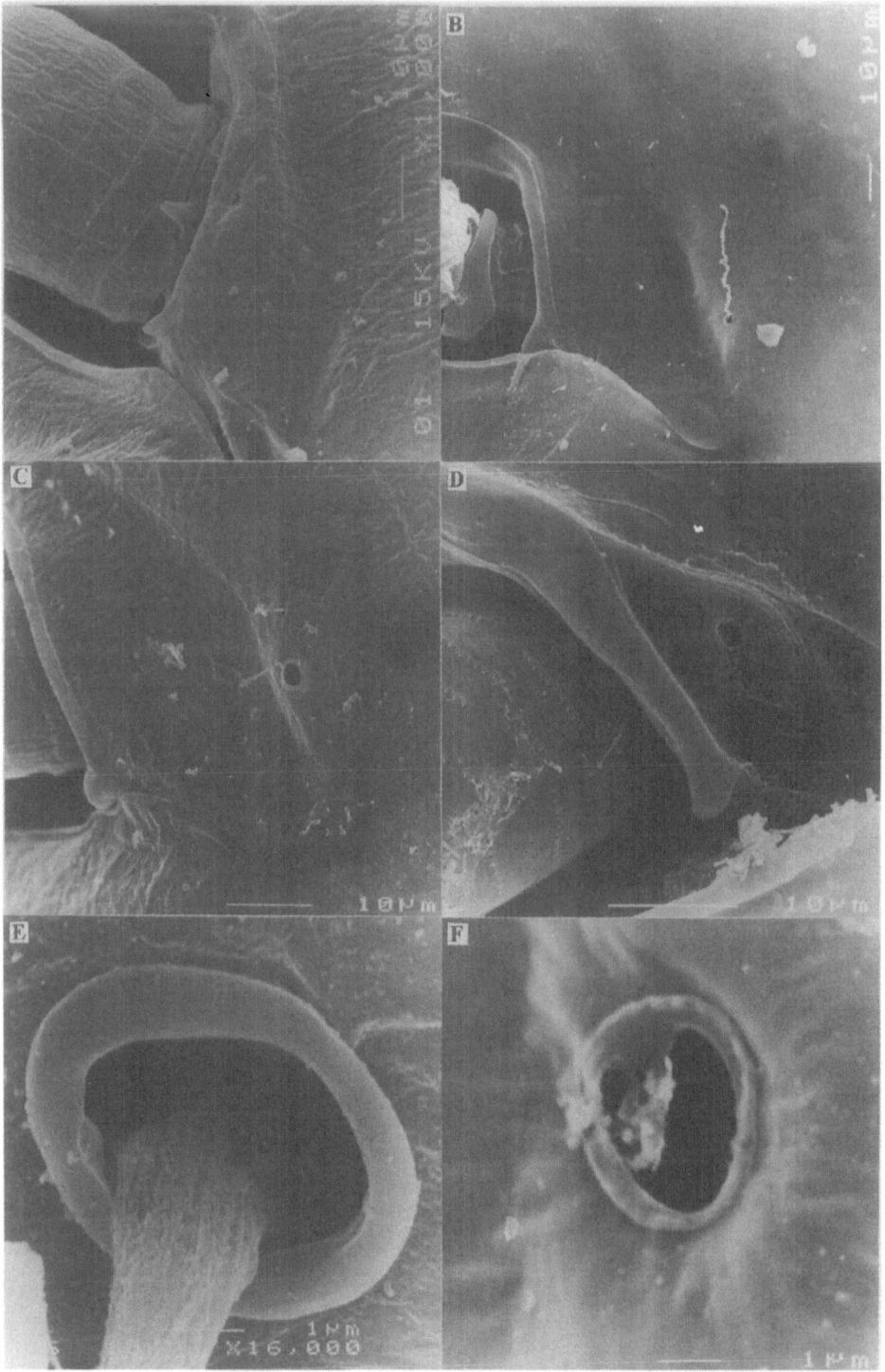

Figure 2. Lateral pores. A. *Bosmina (Bosmina) longirostris,* Belgium, Mirwart pond, 10.10.1969, B. *Bosmina (Eubosmina) kessleri,* Slovakia, Bukovec Reservoir, 4.10.1989, C. *Bosmina (Neobosmina) meridionalis,* New Zealand, Ngapouri lake, 10. 1970, D. *Bosmina (Sinobosmina) fatalis cyanopotamia,* Russia, Lake Bolon, 8.9.1933, E. material secreted from lateral pore, *Bosmina (Eubosmina) reflexa,* Poland, Dadaj lake, 6.1962, F. *Bosmina (Sinobosmina) fatalis cyanopotamia,* Russia, Lake Bolon, 8.9.1933.

19

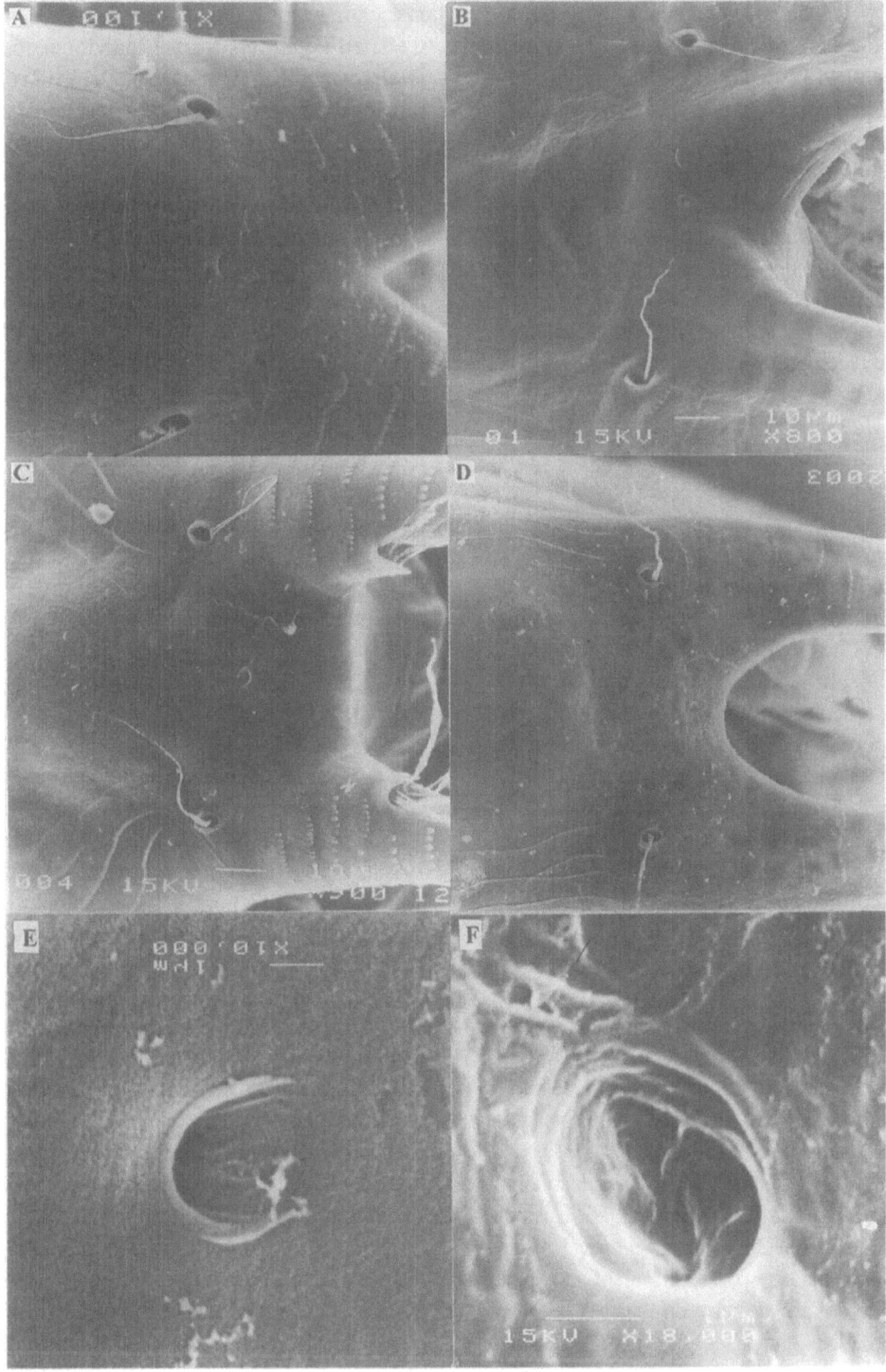

Figure 3. Frontal pores. A. *Bosmina (Bosmina) longirostris,* Belgium, Mirwart pond, 10.10.1969, B. *Bosmina (Eubosmina) reflexa,* Poland, Dadaj lake, 6.1962, C. *Bosmina (Neobosmina) hagmanni,* Brazil, Lago Castanho, 28. 12. 1974, D. *Bosmina (Sinobosmina) liederi,* Canada, N.Sc. East River Sheetharbor Lake, 7. 10. 1984, E., F. – details, the tip of the rostrum on the right, E. *Bosmina (Eubosmina) coregoni gibbera,* Poland, Głębokie lake, 1.9.1964, F. *Bosmina (Neobosmina) brehmi,* Nigeria, Kainji Reservoir, 28.1.1981.

in *B. longispina* and *B. oriens*. The horseshoe image is produced by the dorsal ridge surrounding the top and sides of the FHP (see Figure 3 B). The ridge is not well developed in oval pores. The width of the horseshoe-shaped pores varies from 2.2 to 2.7 μm. In *B. longispina* and *B. oriens,* the longer axis of the pore is 1.9 μm 3.1 μm, respectively. In *B. berolinensis, B. coregoni thersites, B. crassicornis, B. reflexa,* the pores are situated midway between the two frontal setae, whereas in *B. coregoni coregoni, B. coregoni gibbera, B. kessleri, B. longicornis, B. longispina,* and *B. oriens,* the pores are shifted a little towards the tip of the rostrum (see Figure 3E).

Subgenus Neobosmina

Fifteen populations were examined belonging to five species: *B. tubicen* Brehm, 1953; *B. brehmi* Lieder, 1962; *B. chilensis* Daday, 1902; *B. hagmanni* Stingelin, 1904 and *B. meridionalis* Sars, 1903.

The LHP pattern is similar to that found in the subgenus *Eubosmina* (Goulden & Frey, 1963) (see Figure 2C). Pores are broadly oval (longer axis varies from 2.8 to 3.4 μm) with the exception of the population of *B. hagmanni* from Lago Castanho in Brazil, in which the pore is noticeably elongated and enlarged (longer axis about 4,5 μm). The pore is surrounded by a distinctly reticulated cuticle. The distance from the pore to the ventral head shield margin is about 50 μm.

The FHP is circular, with a diameter of 1.8 to 2.6 μm, and shifted towards the tip of the rostrum (see Figure 3C, F).

Subgenus Sinobosmina

Ten populations were examined, belonging to five (sub)species: *B. fatalis* Burckhadt, 1924; *B. fatalis cyanopotamia* Burckhardt, 1924; *B. liederi* DeMelo & Hebert, 1994; *B. freyi* DeMelo & Hebert, 1994; and a new, not yet described species from India.

The LHP is situated between the posterior branches of the bifurcated fornix line (Uéno, 1968, Kořínek, 1971) (see Figure 2D). Pores in *B. fatalis, B. fatalis cyanopotamia* and *B. sp.* are circular, and located approximately halfway between the two posterior fornix branches (see Figure 2F). Pores in *B. freyi* and *B. liederi* are oval and shifted more to the ventral branch of the fornix bifurcation. The longer axis of the LHP varies in length: in *B. liederi* it measures1.5 μm, in *B. fatalis* and *B. fatalis cyanopotamia* 2.1 and 2.3 μm, respectively, in *B. sp.* 2.5 μm, and in *B. freyi* 4.2 μm.

The distance from the pore to the ventral head shield margin is 15 to 20 μm.

The FHP differs among individual species. In *B. fatalis cyanopotamia* and *B. sp.*, the FHPs are transversally oval (longer axis 2.0 and 2.1 μm, respectively); in *B. freyi* the FHP is circular with a diameter of 1.5 μm (see Figure 3D); *B. fatalis* and *B. liederi* have broadly horseshoe-shaped pores, with diameters of about 2.2 and 1.9 μm, respectively.

Ephippium

Ephippial surface relief does not differ among species. In species with pronounced reticulation of the carapace, the same pattern appears on the ephippial surface (see Figure 4A–F). Across each valve there is a band-shaped thickening of chitin from the anterior dorsal angle to the posterior margin of the carapace (see Figure 4A–D). The thickening differs in shape among species, but there are no significant differences in the ephippium among subgenera.

Discussion

The more or less intuitive division of the genus *Bosmina* suggested by Lieder (1957, 1962) has recently been supported by genetic studies (DeMelo & Hebert, 1994a, 1994b, 1994c). The morphology of the lateral head pores, viewed under the light microscope, also provides supporting evidence for such a classification. Application of S.E.M. techniques reveals differences in the shape of pores, not only at the subgeneric, but also at the species level. Intrapopulation variability of pores has not yet been satisfactorily studied. Havel (1978) found that the size of the LHP does not depend on the body size of adult females. Since a small number of individuals from one population are generally used in S.E.M. studies, most of the measurements have only an approximate value. Information about the variability of pore size for species with continental patterns of distribution is also lacking.

While pores on the head shield within the family Chydoridae are widely used for the identification of genera and individual species (Frey, 1959 and 1965; Megard, 1967), only a few studies adopted such a practice in the genus *Bosmina* (Paggi, 1979). We think that one of the problems is insufficient knowledge of the morphology of the pores. DeMelo & Hebert (1994c) suggested that the lateral head pores in *Bosmina longirostris* (O.F. Müller) are circular and their lateral

21

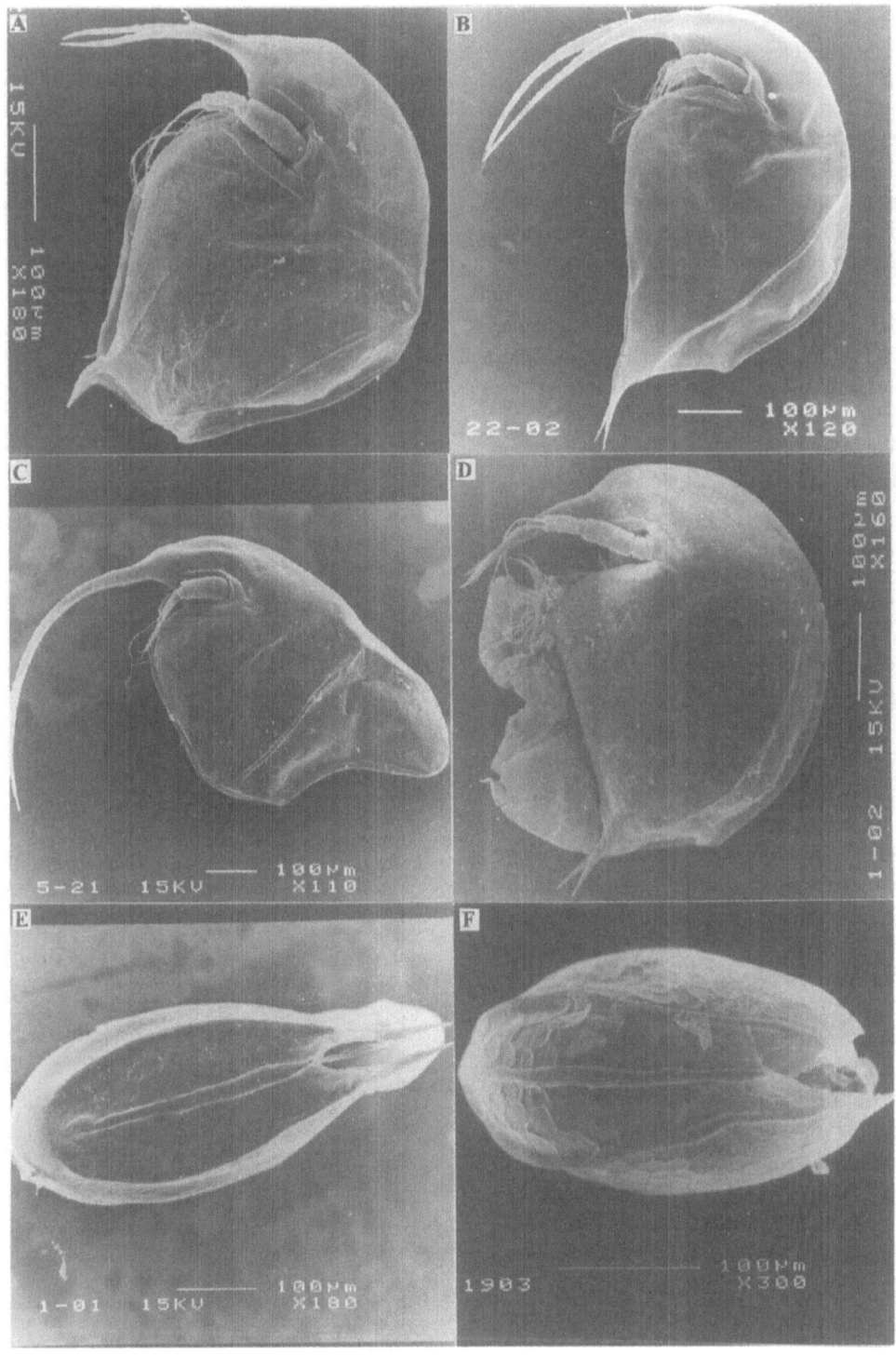

Figure 4. Ultrastucture of ephippia. A, B, C, D. lateral aspects, A. *Bosmina (Bosmina) longirostris,* Belgium, Mirwart pond, 10.10.1969, B. *Bosmina (Eubosmina) berolinensis,* Poland, Lake Zywy, 9.11.1961, C. *Bosmina (Eubosmina) coregoni thersites,* Germany, Gr. Krampe See, 30.10.1986, D. *Bosmina (Neobosmina) tubicen,* U.S.A., Lake Kingsley, 19.1.1974, E, F. Dorsal aspects, E. *Bosmina (Neobosmina) tubicen,* U.S.A., Lake Kingsley, 19.1.1974, F. *Bosmina (Neobosmina) meridionalis,* Australia, Lake Toomba, 20.6.1974.

projection in the microscope produces an elongated shape as an apparent image. These authors based their opinion on one population from a Californian lake. They called the population *Bosmina longirostris* without any comparison with populations from the region of the original description, Europe. Their reference to Deevey & Deevey (1971) is not valid either. The latter authors used the name *Bosmina longirostris* for a population from Nova Scotia belonging probably to *B. freyi* or *B. liederi* (see the shape of the male clasper, Deevey & Deevey's Fig. 2b, page 203). Our results from European populations show an undistorted image of an elongated pore, its ventral margin nearly straight, and dorsal margin vaulted. There is a high probability that the population from California belongs to another species. We refer to this case in detail because it illustrates the deceptive value of generalisation from locally limited data.

Thus far, only a descriptive morphology of the head pores has been published. However, we often observed in our specimens a minute string-like, solid excretion from lateral head pores. Preliminary results of studies conducted by Havel suggest their possible function. Havel (1978) used transmission electron microscopy to study the tissue connected to lateral pores, and found a complex of large cells forming a hyaline and lobose structure situated along the digestive tube, reaching anteriorly to the region of the LHPs and posteriorly to the ovary. The cytoplasm contains granules in various stages of disintegration, suggesting a glandiform tissue. The material secreted was not observed in living animals, and it is not clear if it solidifies after contact with water, or after preservation with formaldehyde (see Figure 2E). In preserved specimens, the secretion is not dissolved in alcohol or acetone, and after treatment with hot 20% sodium hydroxide, it is not fully washed out. This suggests that the secretion is composed of a polysaccharide.

There are remarkable differences in the morphology of the ephippium in some anomopod genera, e.g. in *Moina* (Goulden, 1968), *Daphnia* (Alonso, 1985) and *Ceriodaphnia* (Greenwood et al., 1991). Goulden (1966) and Kerfoot & Peterson (1980) suggested the possible importance of the ephippial surface structure for specific life strategies. In the populations studied, we did not find remarkable differences in surface ultrastructure. The shape of the sclerotinised ridge is species-specific in some cases, but no clear differences were found among subgenera.

Acknowledgements

We are indebted to all who provided some of the populations studied. We appreciated the introduction to Japanese preparative methods for S.E.M. provided by Dr Milan Richter, 1st Faculty of Medicine, Charles University, Prague. Two anonymous reviewers helpfully commented the earlier version of the manuscript. We thank Ms. Paula Johnson for the editing of our English. The study was supported by the Faculty of Science, Charles University, Prague.

References

Alonso, M., 1985. *Daphnia (Ctenodaphnia) mediterranea*: A new species of hyperhaline waters, long confused with *D. (C.) dolichocephala* Sars, 1895. Hydrobiologia 128: 217–228.

Deevey, E. S. Jr. & G. B. Deevey, 1971. The American species of *Eubosmina* Seligo (Crustacea, Cladocera). Limnol. Oceanogr. 16: 201–218.

DeMelo, R. & P. D. N. Hebert, 1994a. Allozymic variation and species diversity in North American Bosminidae. Can. J. Fish. aquat. Sci. 51: 873–880.

DeMelo, R. & P. D. N. Hebert, 1994b. Founder effects and geographical variation in the invading cladoceran *Bosmina (Eubosmina) coregoni* Baird 1857 in North America. Heredity 73: 490–499.

DeMelo, R. & P. D. N. Hebert, 1994c. A taxonomic reevaluation of North American Bosminidae. Can. J. Zool. 72: 1808–1825.

Frey, D. G., 1959. The taxonomic and phylogenetic significance of the head pores of the Chydoridae (Cladocera). Int. Rev. ges. Hydrobiol. 44: 27–50.

Frey, D. G., 1965. Differentiation of *Alona costata* Sars from two related species (Cladocera, Chydoridae). Crustaceana 8: 159–173.

Goulden, C. E., 1966. Co-occurrence of moinid Cladocera and possible isolating mechanisms. Verh. int. Ver. Limnol. 16: 1669–1672.

Goulden, C. E., 1968. The systematics and evolution of the Moinidae. Trans. Am. Phil. Soc. 58: 1–101.

Goulden, C. E. & D. G. Frey, 1963. The occurrence and significance of lateral head pores in the genus *Bosmina* (Cladocera). Int. Rev. ges. Hydrobiol. 48: 513–522.

Greenwood, T. L., J. D. Green & M. A. Chapman, 1991. New Zealand *Ceriodaphnia* species: identification of *Ceriodaphnia dubia* Richard, 1894 and *Ceriodaphnia* cf. *pulchella* Sars, 1862. New Zealand J. mar. freshwat. Res. 25: 283–288.

Havel L., 1978. Ultrastructure and taxonomical value of head pores in the genus *Bosmina* (Cladocera). Dissertation. Charles University, Prague, 55 pp. (In Czech).

Inoué, T. & H. Osatake, 1988. A new drying method of biological specimens for scanning electron microscopy: the t-butyl alcohol freeze-drying method. Arch. Histol. Cytol. 1: 53–59.

Kerfoot, W. Ch., 1975. Seasonal changes of *Bosmina* (Crustacea, Cladocera) in Frains Lake, Michigan: laboratory observations of phenotypic changes induced by inorganic factors. Freshwat. Biol. 5: 227–243.

Kerfoot, W. Ch. & C. Peterson, 1980. Predatory copepods and *Bosmina*: replacement cycles and further influences of predation upon prey reproduction. Ecology 61: 417–431.

Kořínek, V., 1971. Comparative study of head pores in the genus *Bosmina* Baird (Crustacea, Cladocera). Věst. Čs.společ. Zool. 35: 275–296.

Lieder, U., 1953. Baiträge zur Kenntnis des Genus *Bosmina*. II. Über Bastarde zwischen einigen Formentypen des *Coregoni*-Kreises. Arch. Hydrobiol. 47: 453–469.

Lieder, U., 1957. Beiträge zur Kenntnis des Genus *Bosmina* (Crustacea, Cladocera). IV. Versuch einer Monographie der Untergattung *Eubosmina* Seligo 1900. Dissertation, Humboldt Universität Berlin, 247 pp.

Lieder, U., 1962. Beschreibung einer neuen Bosminen-Art, *Neobosmina brehmi* n. sp., aus Äquatorialafrika und über die aus dem Orinoko beschriebene *Neobosmina tubicen* (Brehm), (Crustacea, Cladocera). Int. Rev. ges. Hydrobiol. 47: 313–320.

Lieder, U., 1983a. Die Arten der Untergattung *Eubosmina* Seligo, 1900 (Crustacea: Cladocera, Bosminidae). Mitt. zool. Mus. Berl. 59: 195–292.

Lieder, U., 1983b. Revision of the genus *Bosmina* Baird, 1845 (Crustacea, Cladocera). Int. Revue ges. Hydrobiol. 68: 121–139.

Lieder, U., 1983c. Introgression as a factor in the evolution of polytypical plankton Cladocera. Int. Rev. ges. Hydrobiol. 68: 269–284.

Megard, R. O., 1967. Three new species of *Alona* (Cladocera, Chydoridae) from the United States. Int. Rev. ges. Hydrobiol. 52: 37–50.

Paggi, J. C., 1979. Revision de las especies argentinas del genero *Bosmina* Baird agrupadas en el subgenero *Neobosmina* Lieder. (Crustacea: Cladocera). Acta Zool. Lilloana 35: 137–163.

Scourfield, D. J., 1901. The ephippium of *Bosmina*. Journ. Quekett microsc. Club 48: 51–56.

Uéno, M. 1968. Lateral head pores and other characteristics of *Bosmina fatalis* Burckhardt. Annot. Zool. Jap. 41: 159–162.

Hydrobiologia **360**: 25–32, 1997.
A. Brancelj, L. De Meester & P. Spaak (eds), Cladocera: The Biology of Model Organisms.
©1997 *Kluwer Academic Publishers.*

Structure of thoracic limbs in *Bosminopsis deitersi* Richard, 1895 (Anomopoda, Branchiopoda)

Alexey A. Kotov
A. N. Severtsov Institute of Ecology and Evolution of Russian Academy of Sciences, Leninsky Prospekt 33, 117071 Moscow, Russia

Key words: Bosminopsis deitersi, Bosmina, thoracic limbs, comparative morphology, Bosminid monophyly

Abstract

A study of the structure of the limbs of *Bosminopsis deitersi* Richard, 1895 was performed. The structure is similar in populations from different regions. According to Meissner (1903), limbs of *Bosminopsis* differ from those of *Bosmina* strongly enough to separate *Bosminopsis* in a special family. However, our observations indicate that the structure of the limbs of *Bosmina* and *Bosminopsis* is similar, and conform to the general plan of Bosminid limbs. The first limb is strongly simplified, and the first limb of males is unique among anomopods. Limbs III–V are less oligomerized than limbs I–II, and are organized according to the chydorid (and macrothricid) type. The degree of oligomerization is different in the two genera. Comparative information on the limb structure in one genus may contribute to the reconstruction of the evolution of a corresponding limb in the other genus. The uniformity of limb structure within each genus, and their similarity between both genera is thought to be a reflection of the monotonous life conditions in open water.

Abbreviations

dag – distal armature of gnathobase; **db** – distal brush of setae of limb I; **dse** – distal setae of exopodite; **end** – endopodite; **eh** – ejector hooks of limb I; **ep** – epipodite; **ex** – exopodite; **fp** – filter plate of gnathobase; **fpd** – flat projection of endopodite; **fpe** – flat projection of exopodite; **frs** – fringe of setules; **fs** – field of setules; **gn** – gnathobase; **hls** – hook-like spines; **hok** – hook of male clasper; **ies** – innermost spines of endopodite III; **ise** – innermost seta of endopodite I; **lee** – long element of exopodite; **lse** – lateral setae of exopodite; **mp** – maxillar process of limb I; **odl** – outer distal lobe of limb I; **ofp** – outer filter plate; **ose** – outermost soft setae of limb II; **pep** – preepipodite; **pws** – powerful seta; **rofp** – rudimentary outer filter plate of *Bosmina*; **sdl** – inner subdistal lobe of limb I; **seg** – sawform element of distal armature of gnathobase II; **sen** – sensilla; **sss** – solitary shaggy seta; **sis** – sawlike innermost seta of limb II; **ths** – thick setules

Introduction

Earlier, a comparative study on the limb structure of females of several *Bosmina* species was performed, and a uniformity in morphology in all investigated representatives was demonstrated (Kotov, 1995). So far, however, the limbs of *Bosminopsis* have not been in detail. There was only one attempt at description of the limbs in *Bosminopsis zernovi* Linko, 1901, by Meissner (1903) (at present, *B. zernovi* is considered to be a subspecies of *B. deitersi* Richard, 1895). According to Meissner (1903), the limbs of *Bosminopsis* differ from those of *Bosmina* very strongly, and he proposed to assign the genus *Bosminopsis* to a different family. Meisner's publication remained unknown to western investigators, but in Russian literature, this question was recognized as unsolved (Rylov, 1950; Manujlova, 1964; Korovchinsky, 1992). The 'naïve' (in the words of N. N. Smirnov) drawings and descriptions of Meissner do not comply with the recent standard of knowledge on the Anomopod limbs. The present paper is aimed at a detailed study of thoracic limbs of *Bosminopsis deitersi* and a comparison of the morphology of thoracic appendages in representatives of two genera of the Bosminidae. I studied the 'cosmopolitan' *B. deitersi* from different regions, because many Anomopoda once claimed to be cosmopolitan are now considered series of separate species (Frey, 1987).

Materials and methods

Preserved *B. deitersi* were selected from samples of the collection of the Zoological Museum of Moscow State University or supplied by N. N. Smirnov and N. M. Korovchinsky from: 1. 12 different sites in Amazonia (Kotov, 1997) (coll. B. Braun, G.-O. Brandorff, M. V. Mina); 2. Surinam (sample without detailed marking); 3. Thailand: Boena Lake, PSU reservoir (coll. P. Pholpunthin); 4. Cambodia: Tonlesap River; 5. Far East of Russia: the River Amur, Lake Bolon (samples from the collection of 'Amur expedition'). Limbs of *Bosminopsis* were dissected and studied using a light microscope. Material from Amazonia was the most abundant, with over 100 specimens examined. Males were found only in samples from Lago Timbó and Lago Castancho (Brazil).

Structure of thoracic limbs of *Bosminopsis deitersi*

There are five pairs of limbs in *Bosminopsis* (*versus* six pairs in *Bosmina*, the sixth thoracic appendage being only an epipodite). Each limb is organized uniformly in representatives from different regions and localities. The set of elements on each limb is equal, does not vary within populations and does not depend on instar or on external conditions. All drawings (Figures 1–3) represent different limbs of Bosminidae in posterior view. There are two seta series on the Anomopod limb: the marginal row (called stiff setae by Smirnov [1974], and scrapers or 'internal' setae by Fryer [1963]), and a more posterior (and medial) row (called soft setae by Smirnov [1974], sensory or 'external' setae by Fryer [1963]), The former are designated by +, the latter by * on the figures. The distal armature of the gnathobase is the continuation of the marginal row, and its filter plate is a continuation of the medial row (Cannon, 1933).

The first thoracic limb is massive, widened distally (Figure 1A), and with two small projections on its distal end: a large outer (odl) (with two setae of different size, feathered by sparse, long, robust hairs) and a small inner subdistal lobe (sdl) (with one seta, densely fringed by delicate setules). Anteriorly, there are two robust ejector hooks (Figure 1A: eh), strongly different in size and armed with short denticles. A dense bunch of long setules (fs) is present near the hooks. On the inner edge, three soft setae (db) are situated: two longer elements feathered by long, strong hairs, and one shorter one fringed by soft setules from the base to the tip. The maxillar process (mp), a derivative of gnathobase I (Kotov, 1996), with long shaggy seta, is located near the base of the limb. The epipodite (ep) of the first limb is specific: it bears two long finger-like projections, unlike any of the first (or other) limbs in other Anomopods (this part of limb I was correctly pictured by Meissner, 1903).

The first limb of the adult male bears a large inner distal lobe (clasper) (Figure 1B: idl) with a powerful denticulated hook (hok) and two setulated setae of different size.

The second limb is relatively small. Its endopodite (Figure 1C: end) bears six powerful scrapers (according to the terminology of Fryer, 1963). Both the element nearest to the gnathobase (sis), and the outermost one stand out by size and armature. On the back of the appendage, there is a seta near the gnathobase (ise) and another near the proximal end of the limb (ose). They are remainders of a row of soft setae, which still occurs in primitive anomopods (e.g. *Eurycercus* [Fryer, 1963; Smirnov, 1974]). Gnathobase II (gn) is distally crowned with three setae of different armature (dag). The filter plate (fp) consists of five long setulated setae, each of which terminates in a flexible 'whip'. There is no exopodite on this limb. The globular epipodite has a finger-like projection.

The third limb (Figure 1D) consists of three large parts of approximately the same size. The external distal portion corresponds to the exopodite of other Branchiopoda (ex) (terminology of Behning, 1912). It is rectangular and bears two lateral setae (lse) of different size and a distal group of five similar elements (dse) on its proximal boundary. All these setae are fringed with long, delicate hairs. Endopodite III (after Behning, 1912) is subdivided in two portions (basal and distal), with two rows of setae. The series of seven soft setae (outer filter plate, after Fryer, 1963) is present medially on the back side (ofp). Probably, a specific solitary shaggy seta (sss), sitting on a small projection of the external edge of the endopodite, is a derivative of this row too. Scrapers have marginal positions on this limb. One short and two long spines are situated

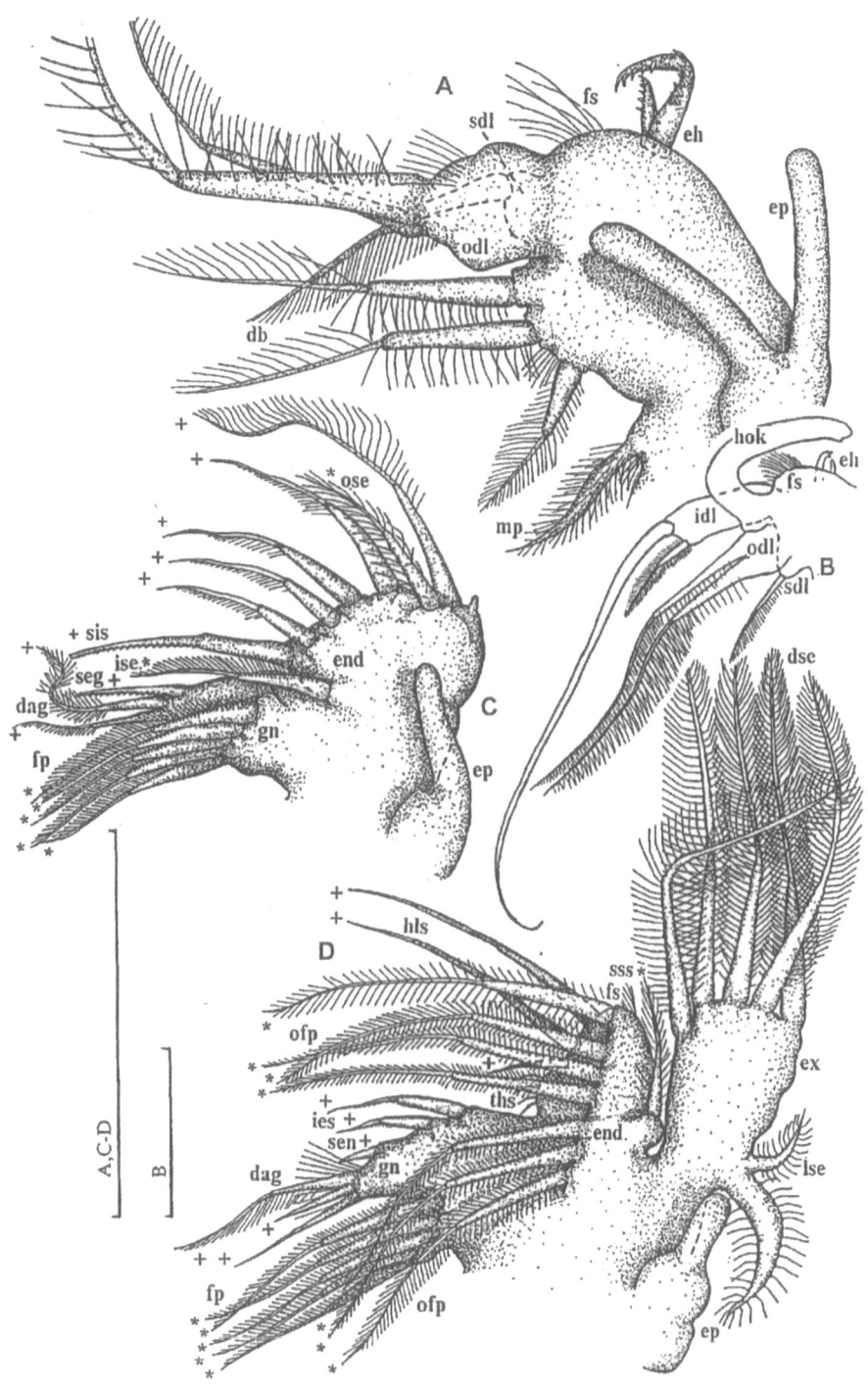

Figure 1. First thoracic limb of female (A), distal portion of limb I of male (B), second (C) and third (D) thoracic appendages of *Bosminopsis deitersi* in posterior view. All scales: 50 μm.

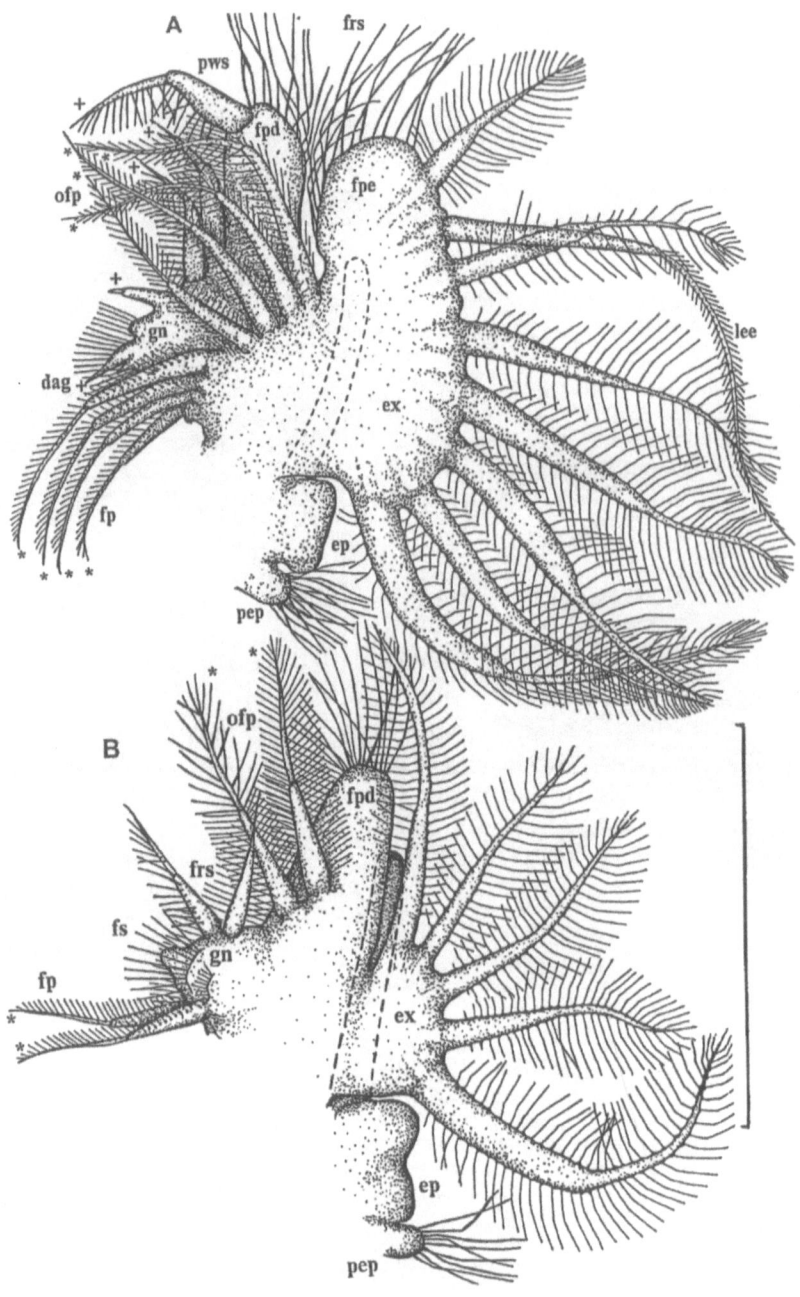

Figure 2. Fourth (A) and fifth (B) limbs of *Bosminopsis deitersi* in posterior view.

on the boundary of the distal portion of the endopodite (hls). Basally, there is a series of five thick setules (ths). There are two strong spines (ies) on the basal part of the endopodite, and a sensilla (sen) between the endopodite and the gnathobase. Gnathobase III is armed with specific distal series of elements: two spines and one longer setulated setae. The filter plate has five setae, similar to the soft setae of endopodite III. The epipodite is at the base of the limb, on its external part.

The fourth limb (Figure 2A) has a large circular exopodite, bearing long and powerful setae organized as those of the third limb, except for one long element (lee) with particularly short feathering. The distalmost part of exopodite carries a flat projection (fpe), with a setulated internal edge (frs). The distalmost portion of the endopodite has at its tip powerful seta (pws) with long hairs. Two analogous (but smaller) setae and one short spine are found at its base. The back side carries four long soft setae, similar to those of the third limb. The distal armature of the gnathobase is strongly reduced. The filter plate consists of four elements. The epipodite, with a particularly long finger-like projection, is situated at the base of the external side. More basally, there is a small preepipodite (pep) with scanty thick and long setules.

The fifth limb (Figure 2B) has an exopodite with five large setae (the lateral ones being longer and more powerful), and a kind of blade, of which the distal extremity and internal edge are armed with a dense row of long hairs. There are two soft setae on the posterior side of endopodite (called vertical setae by Fryer, 1963). The status of two other setae on the inner side of the endopodite is not clear. Gnathobase V is very poorly expressed and supplied with only two setae as the filter plate. There is a typical epipodite and preepipodite near the base on the external side of this limb.

Limbs of *Bosmina*

For comparison I provide drawings of the limbs of *Bosmina* (Figure 3). I have pictured the thoracic appendages of *B. (Sinobosmina) fatalis* Burckhardt, 1924, and my drawings are the first ones for a representative of the *Sinobosmina* subgenus (Figure 3). The only difference between its appendages and those of the previously described *B. (Eubosmina) longispina* (Figure 3K) and *B. (Eubosmina) coregoni kessleri* (Kotov, 1995) is a more reduced system of small elements of the fourth limb. The uniformity of limbs in all representatives of theis genus is also confirmed by the most recent drawings made of *B. (B.) longirostris*, given by Alonso (1996).

Discussion

Several authors have pointed out that the Chydoridae and Bosminidae form a particular evolutionarily branch apart from the Daphniidae and Moinidae

(Smirnov, 1974; Fryer, 1974, 1995). The third, fourth and fifth limbs of *Bosminopsis* are organized typically for their evolutionarily branch, and they are very slightly oligomerized. The degree of oligomerization of limb V is greater in *Bosmina* than in *Bosminopsis*: there are no setae on the endopodite, the filter plate of the gnathobase is absent, the setae of the exopodite are divided into a distal and a lateral group, and one of them (lee) has a special structure. In contrast, the endopodite and gnathobase III and IV are more oligomerized in the *Bosminopsis*. There are five setae of the filter plate on both these limbs compared six in *Bosmina*. On the third limb, one soft seta (sss) is situated far apart of the main row. In *Bosminopsis*, only two spines on the basal part of endopodite are located near the gnathobase (*versus* three in *Bosmina*), and one of proximal spines is rather small. On limb IV of *Bosminopsis*, there are four soft medial setae, in *Bosmina* (*Eubosmina*) these four elements are only represented by small rudiments (Figure 3K: rofp), of which in *B. (Sinobosmina)* only three remain (Figure 3E). Comparative information on the limb structure in one genus may contribute to the reconstruction of evolution of the homologous limb in the another genus. Evolutionarily transformations of Bosminid limbs III–V are analogous to those in advanced groups of chydorids. Smirnov (1974) and Fryer (1968) described that in chydorids too, there is a reduction of the number of elements in the filter plates of gnathobases of all limbs as species get more specialized, and a gradual disappearance of soft setae on the fourth limb and stiff setae of the third limb. Some chydorids have the same set of the setae of exopodite III–V as Bosminidae: 7-8-5. The complete reduction of soft setae on the second limb is very typical for Chydoridae and Macrothricidae (Fryer, 1968, 1974). The sixth limb is frequently absent in evolutionarily advanced chydorids too (Smirnov, 1974). However, in the structure of the second and especially the first limbs, there are some features which sharply distinguish the Bosminidae from the Chydoridae and Macrothricidae.

1. A main feature of the Bosminidae limb structure is the simplification of limb I. The inner distal lobe of this appendage in the female of *Bosmina* is very small and does not bear any setae. The lobe is even absent in females of *Bosminopsis*. In the Chydoridae and Macrothricidae, the inner distal lobe is well developed and usually bears 3–4 setae. In these animals, which are associated with bottom sediments or vegetation, the inner distal lobe I is important for scrambling and crawling (Fryer, 1968, 1995). Apparently, pelag-

30

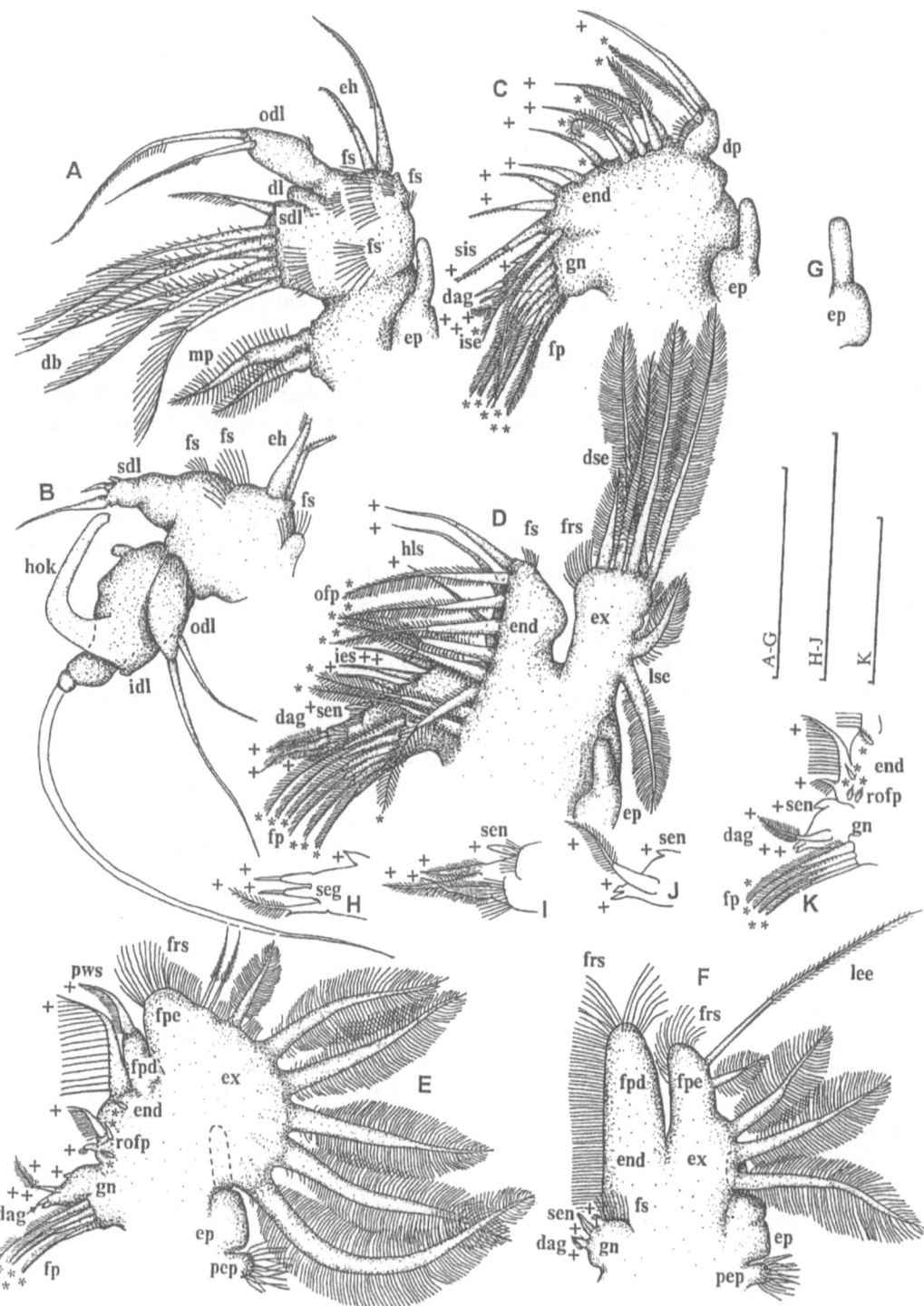

Figure 3. First limb of female (A), distal portion of one of male (B), second (C), third (D), fourth (E), fifth (F), sixth (G) limbs, distal armature of gnathobase II (H), III (I), IV (J) of *Bosmina (Sinobosmina) fatalis*; gnathobase IV of *B.(Eubosmina) coregoni kessleri*. Scales: 50 μm.

ic bosminids can do without it. The number of setae on the internal edge of limb I is reduced, especially in *Bosminopsis*. According to the descriptions of *B. negrensis* Brandorff, 1976 and *B. brandorffi* Rey & Vasques, 1989, their first limb is also strongly simplified, thougt we can not entirely trust the drawings by Brandorff (1976) and Rey & Vasques (1989) completely, since ejector hooks and epipodites are not shown. It is interesting that a similar simplification of the first thoracic appendage is observed in *Ilyocryptus* (Fryer, 1974), which are not similar to bosminids with respect to other features.

2. A uniform modification of the male limb I in comparison to the female limb is also a typical feature of the Bosminidae In the male the large inner distal lobe (clasper) is developed with a hook and one (*Bosmina*) or two (*Bosminopsis*) setae of very different length. The inner subdistal lobe on limb I of the *Bosmina* male is separated, and strongly increased in size in *Bosminopsis*. This thesis is confirmed by analysis of a drawing in Brandorff (1976) for *B. negrensis*. Such sharp distinction of the limb structure of males and females is not observed in other anomopod groups.

3. There are specific features in the structure of the Bosminid limb II too. The innermost setae of the medial row of the endopodite are directed inwards. The nearest element to the gnathobase is particularly powerful and armed with strong serration (sawform). Gnathobase II is separated from endopodite II by a deep incision. There is a special seta of the endopodite near the gnathobase edge, which is a remainder of the disrupted row of medial soft setae, and a functional continuation of the filter plate. Limb II of *Ophryoxus* (Sergeev, 1970) is the most similar to that of *Bosmina*, but the two genera are not closely related. The persistence in more (*Bosmina*) or less (*Bosminopsis*) reduced state of the remainders of the series of soft setae on limb II is a very primitive feature, which sharply distinguishes the Bosminidae from many evolutionarily advanced genera of the Macrothricidae and from all Chydorinae and Aloninae (contrary to Wolterek, 1920). Among the setae of the distal armature of gnathobase II of the Bosminidae, there is one element which is similar to the sawform seta on the endopodite.

The outer distal lobe of the anomopod limb I is traditionally are not homologized with the exopodite (Claus, 1876; Fryer, 1963). In my preliminary investigation on the *Bosmina* limbs (Kotov, 1995), I followed the tradition (Behning, 1912; Graf, 1930; Alonso, 1996) and named the distal projection of limbs I and II of *Bosmina* 'exopodite'. Now the application of

this term may be unjustified. The exopodite of a branchiopod limb is a flat lobe with a single row of marginal setae, but on the globular distal lobe of bosminid limbs I and II we may find elements, which may be continuations of two rows of setae that occurs on the endopodite. In my oppinion, the external ramus on these limbs of Bosminidae is completely reduced. So, I refered to this structure with the neutral name 'distal projection' without any attempts to homologize one.

In my oppinion, the features mentioned above, especially the modification of the male limb I of Bosminidae, which unique among anomopods, are evidence for the monophyly of the family. The extreme uniformity of appendages in different species of *Bosmina* and the similarity of those to limbs of *Bosminopsis deitersi* are a direct consequence of the monotony of conditions of their life. All these animals are typical inhabitants of the open water and consume a relatively uniform food: pelagic (or littoral) phytoplankton, suspended in the water. The great radiation of the Macrothricid and (particularly) the Chydorid limbs (Fryer, 1968, 1974) is a reflection of the great variety of ecological niches on the bottom of water bodies. The weak distinctions in the structure of the limbs in different species and even in different genera (which are strongly distinguished by overall body morphology and a series of other features) are observed too in the Daphniidae, another group of Anomopoda which was adapted to a 'truly planktonic lifestyle' (Fryer, 1995). The independent invasion of the open water by the bosminids led to a number of analogies with the daphniids.

Acknowledgments

I thank Prof. N. N. Smirnov and Prof. H. J. Dumont for discussion on the principal points of the manuscript and for help in the preparation of the English version, Dr N. M. Korovchinsky and Dr F. Margaritora for valuable criticism, and Dr K. A. Mikrjukov and Dr K. G. Michailov for help in my work at the Biological Faculty and Zoological Museum of Moscow State University. The study was partly supported by the Russian Foundation of Basic Research (96-04-48063).

References

Alonso, M., 1996. Crustacea, Branchiopoda. Fauna Iberica 7: 1–486.
Behning, A. L., 1912. Studien über die vergleichende Morphologie sowie über die temporale und Lokalvariation der Phyllopodenextremitäten. Int. Rev. ges. Hydrobiol., Biol. Suppl. 15: 1–70.

32

Brandorff, G. O., 1976. A new species of *Bosminopsis* (Crustacea, Cladocera) from the Rio Negro. Acta Amazonica 6: 109–114.

Cannon, H. G., 1933. On the feeding mechanism of the Branchiopoda. Phil. Trans. r. Soc., Lond. B. 222: 267–339.

Claus, C., 1876. Zur Kenntnis des Organisation und des feineren Baues der Daphniden und verwandter Cladoceren. Z. wiss. Zool. 27: 362–402.

Frey, D. G., 1987. The taxonomy and biogeography of the Cladocera. Hydrobiologia 145: 5–17.

Fryer, G., 1963. The functional morphology and the feeding mechanism of the chydorid cladoceran *Eurycercus lamellatus* (O. F. Müller). Trans. r. Soc. Edinb. 65: 335–381.

Fryer, G., 1968. Evolution and adaptive radiation in the Chydoridae (Crustacea: Cladocera): a study of comparative functional morphology and ecology. Phil. Trans. r. Soc., Lond. B. 254: 221–385.

Fryer, G., 1974. Evolution and adaptive radiation in the Macrothricidae (Crustacea: Cladocera): a study of comparative functional morphology and ecology. Phil. Trans. r. Soc., Lond. B. 269: 137–274.

Fryer, G., 1995. Phylogeny and adaptive radiation within the Anomopoda: a preliminary exploration. Hydrobiologia 307: 57–68.

Graf, H., 1930. Der Fangapparat in *Bosmina*. Z. Morph. Okol. Tiere 19: 381–369.

Korovchinsky, N. M., 1992. Sovremennoye sostoyaniye i problemi sistematiki vetbistousikh rakoobraznikh. In N. N. Smirnov (ed.), Sovremennie problemi izucheniya vetvistousikh rakoobraznikh. Gidrometeoizdat Publ., S-Pb: 4-45 [Modern state and problems of systematics of Cladocera (Crustacea)].

Kotov, A. A., 1995. Structure of the thoracic limbs in *Bosmina* Baird, 1845 (Crustacea Anomopoda). Arthropoda Selecta 4: 41–50.

Kotov, A. A., 1996. Fate of the second maxilla during embryogenesis in some Anomopoda Crustacea (Branchiopoda). Zool. J. Linn. Soc. 116: 393–405.

Kotov, A. A., 1997. Study on the morphology and variability of Amazonian *Bosminopsis deitersi* Richard, 1895 (Anomopoda: Bosminidae). Arthropoda Selecta 6 (in press).

Manujlova, E. F., 1964. Vetvistoysie rachki fauni SSSR. Opredeliteli po faune SSSR 88: 1–327. [The cladocerans of fauna of the USSR].

Meisner, V. I., 1903. Materiali k faune nizshich racoobraznich reki Volgi. Raboty Volzhskoi biol. stantsii 2: 159–201. [Materials on the entomostracan fauna of River Volga].

Rey, J. & E. Vasques, 1989. *Bosminopsis brandorffi* n.sp. (Crustacea, Cladocera) une nouvelle espèce de Bosminidae des systèmes Amazone et Orénoque. Ann. Limnol. 25: 215–218.

Rylov, V. M., 1950. Vetvistousie rakoobraznie (Cladocera). In Zhadin, V. I. (ed.), Zhizn presnich vod, 'Nauka' Publ., Moscow, Leningrad: 331–357. [Cladocerans (Cladocera)]

Sergeev, V. N., 1970. Feeding mechanism, feeding behaviour and functional morphology of *Ophryoxus gracilis* G. O. Sars (Macrothricidae, Cladocera). Int. Rev. ges. Hydrob. 55: 245–279.

Smirnov, N. N., 1974. Chydoridae. Fauna of the USSR. Crustacea 1. Keter Publishing House, Jerusalem. 531 pp.

Wolterek, R., 1920. Variation und Artbildung. Analytische und experimentelle Untersuchungen an pelagischen Daphniden und anderen Cladoceren. Francke Verl., Berlin, 151 ss.

Hydrobiologia **360**: 33–46, 1997.
A. Brancelj, L. De Meester & P. Spaak (eds), Cladocera: The Biology of Model Organisms.
©1997 *Kluwer Academic Publishers.*

Sensory and glandular equipment of the trunk limbs of the Chydoridae and Macrothricidae (Crustacea: Anomopoda)

Henri J. Dumont & Marcelo Silva-Briano
Laboratory of Animal Ecology, University of Ghent, Ledeganckstraat 35, 9000 Gent, Belgium

Key words: sense organs, glands, trunk limbs, Chydoridae, Macrothricidae, Anomopoda, Crustacea, functional morphology, mouth

Abstract

Each of the five pairs of trunk limbs of the Chydoridae and Macrothricidae contains one or several cuticular pores, some of which connect to glands which secrete to the filter chamber. They are situated in a fixed or a variable position, usually on the endite of the limb.

Limbs 2 to 5 are also equipped with small sensory organs which may be apically perforated or not. Relatively large, perforated sensilla, often with a species- or genus-specific morphology occur on the endites of P3, P4 and sometimes on P5; unperforated papilliform and batilliform sensilla occur in various positions on P2, P3, and P4. In the *Macrothrix laticornis*-group, three of the six setae of the posterior row of the endite of P3, which are filtratory in other species groups, are modified into rod-like structures resembling the esthetascs of the first antenna.

We conclude that the filter chamber in the families studied here combines a sensory and a glandular function with the filtratory one. These functions, typical of a mouth, are localised behind the masticatory apparatus in vertebrates. In these Cladocera, however, they are physically situated in front of the mandibles.

Introduction

Even the earliest workers were impressed with the complexity of the 'filtering basket' of the cladocerans, and stressed the likely functional and taxonomical importance of its fine structure, while recognising the technical problems obstructing their observation (e.g. Sars, 1861, published 1993: 'the examination of the legs is one of the most difficult tasks regarding the anatomy of the cladocerans'). Sars' remark is still pertinent, and most authors at present still grossly underuse this 'organ'. Even when the trunk limbs are correctly pictured, as in Alonso (1996), not much use is made of their structure in species or genus demarcation. A notable exception has been the work of Smirnov (1966a,b, 1968) on the trunk limbs in several chydorid genera. However, instrumental limitations caused even these studies to neglect most of the finer structures, and focus on the easier visible characters.

Much recent work has specifically concentrated on the food gathering function of the basket, especially in various pelagic *Daphnia* species. A critical evaluation of much of that literature is given by Fryer (1991).

Beyond any doubt, the most thorough study of trunk limb morphology is to be found in a series of monumental publications by G. Fryer, dealing with *Eurycercus* (Fryer, 1963), the Chydoridae (Fryer, 1968), the Macrothricidae (Fryer, 1974) and, finally, the Daphniidae (Fryer, 1991). This work unequivocally revealed the multiplicity of shape and function of the numerous, often extremely specialised, appendages of the various legs. Filtration, achieved by filter combs on some (not all) of the five (or six) pairs of legs was shown to be only one function of the limbs, but other modes of food collecting (scraping, raking, brushing...) were demonstrated to be equally widespread.

The most important discovery, however, was that some limbs are equipped to fulfill two other functions, viz. a secretory and a sensory one. Fryer (loc.cit.) discovered that glands producing 'entangling' secretions, opening via a duct to the oesophagus or the filtering basket, occur in the labrum (a mouth part), but also

34

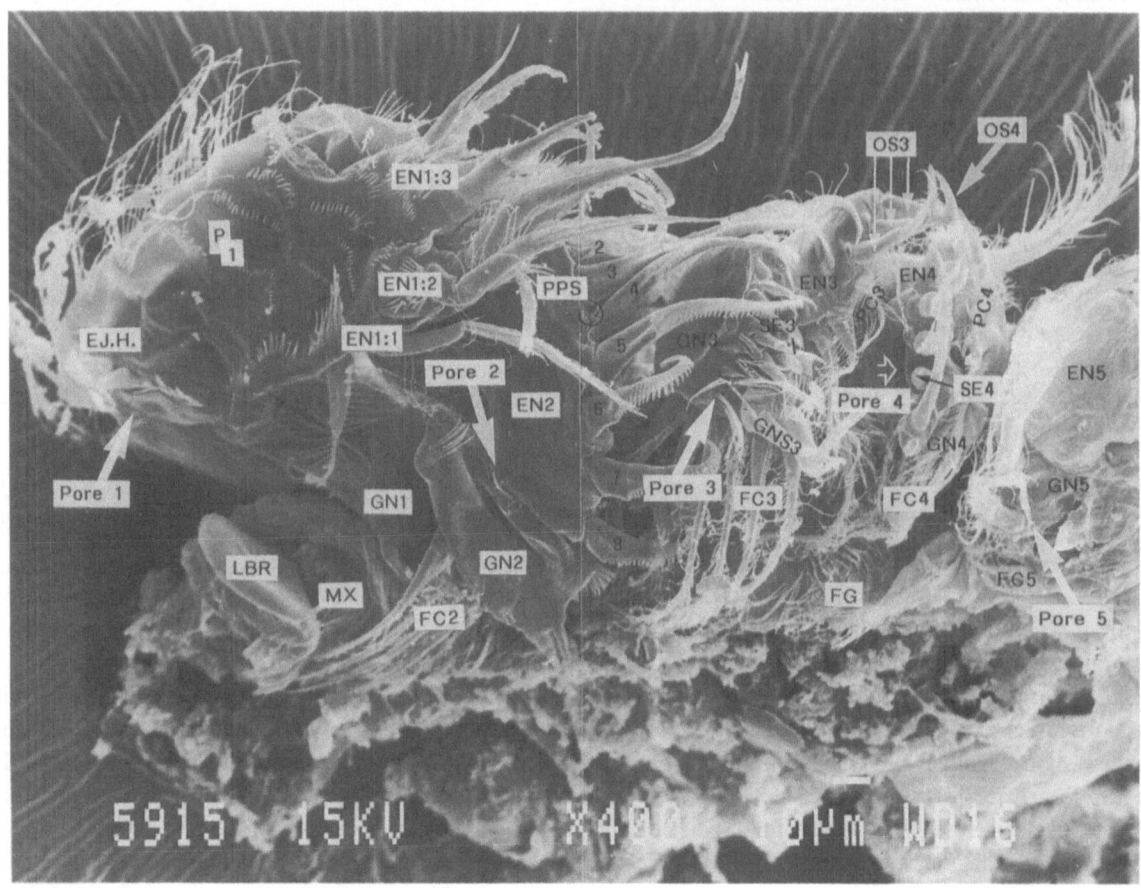

Plate 1. Alona affinis (Aguascalientes, Mexico), general view of the trunk limbs *in situ*. The food groove is in the foreground, such that the exopodites are not visible. The general position of the five pores, one on each trunk limb, is indicated, even though the pores are not revealed at this magnification (for details, see other plates).

in the first (P1) and fourth (P4) trunk limbs. He also identified sensillar structures on the P3 and P4 of Chydorids. Smirnov (1967), following Fryer's work on *Eurycercus*, identified sensillar structures on the P3 and P4 of 13 species of Chydorids, and on the P2 of *Macrothrix laticornis*. The latter were confirmed by Fryer (1974).

Recently, Kotov (1995) illustrated sensilla on P3, P4 and P5 of *Bosmina*, while Alonso (1996) refers to a sensillum on the P4 of *Ilyocryptus*, but pictures such structures on P3–P5. Sensillar structures have therefore now been identified on the P3, P4 and/or P5 of four families of the Anomopoda; they have not yet been recorded in the Daphniidae (Fryer, 1991), and in the Ctenopoda (Korovchinsky, 1992).

Much of the aforementioned work was based largely on light microscopy; here, we make extensive use of SEM techniques, and serial sections, to produce an atlas of the glandular and sensillar equipment of representatives of two families, the Chydoridae and the Macrothricidae.

Materials and methods

All animals examined were formaldehyde-preserved; they were dissected, with either all trunk limbs individually isolated, or one side of the animal left intact to reveal the relative position of the five pairs of limbs. Series of limbs were prepared for SEM by critical point drying and individually mounted on stubs for gold-coating and examination under a Jeol JSM840 SEM. Because of the small size and delicacy of some limbs, sets of individual limbs were prepared; as well, they were mounted either posterior or anterior side up, to examine the two surfaces of each limb individually.

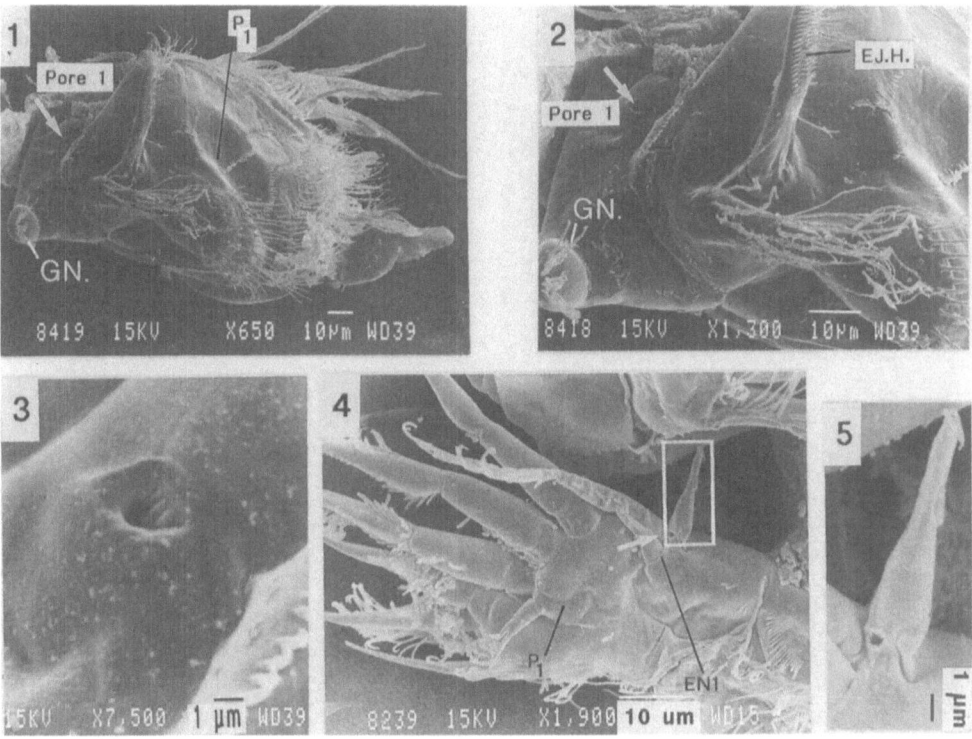

Plate 2. First trunk limbs: 1–3, *Camptocercus rectirostris* (Glubokoe lake, Russia), showing the position of the pore at progressively higher magnification; 4–5, *Macrothix* cf *laticornis* (Aguascalientes, Mexico), showing a pore at the base of seta 1 of endite 1, which is perhaps an artifact.

Of *Camptocercus rectirostris*, a whole animal was embedded and sectioned transversely at 2 μm intervals. Sections were observed and photographed under an Olympus BX 40 microscope using phase contrast optics.

Results

Plate 1 gives an overview of the arrangement of the five pairs of trunk limbs in *Alona affinis*, a representative of the Chydoridae, seen from the inside (the food groove is in the foreground, and the exopodites are not visible). As is typical in the family, trunk limbs P2, P3 and P4 each have an internal filter comb[1] (FC 2, 3, 4), which sweeps inside the food groove, while P3 and P4 also have a posterior filter comb (PC 3 and 4). For all other structures, see list of abbreviations. The plate further indicates the locations of five pores (not necessarily visible on the plate; for details see other plates), one or several on each trunk limb. At least some of these

represent the opening of a gland inside each of these legs. Also shown are two major sensilla, (SE3, SE4) one on P3 and one on P4, as well as a small papilliform sensillum between scrapers 3 and 4 of P2.

Plate 2 shows the pore on P1 in *Camptocercus*, at the base of the gnathobase, a little basad of the ejector hooks. The pore opening is slightly under 2 μm across, and the floor of its opening ribbed. In *Macrothrix*, a pore was found at the base of the external seta of endite 1 of P1 (Plate 2, figs 4–5); a similar pore is present at the base of the Fryer's fork in *Wlassicsia* (not shown). Their function is unknown.

Plate 3 deals with the P2, where a pore is present on the back side of the limb, in an identical position in a chydorid (1–2) and in a macrothricid (3–4), in what seems to be an articulating zone, such that the position of the limb might determine whether the underlying pore canal is open or closed. An additonal small pore (less than 0.5 μm across) was found in several *Chydorus* species at the base of scraper 5 (the scrapers are numbered from outside to inside) (plate 3, 5–6), while a papilliform sensillum (lacking a pore) was found between scrapers 3 and 4, as in *A. affinis*. The batil-

---

[1] Note that in some spp this comb is not a filter.

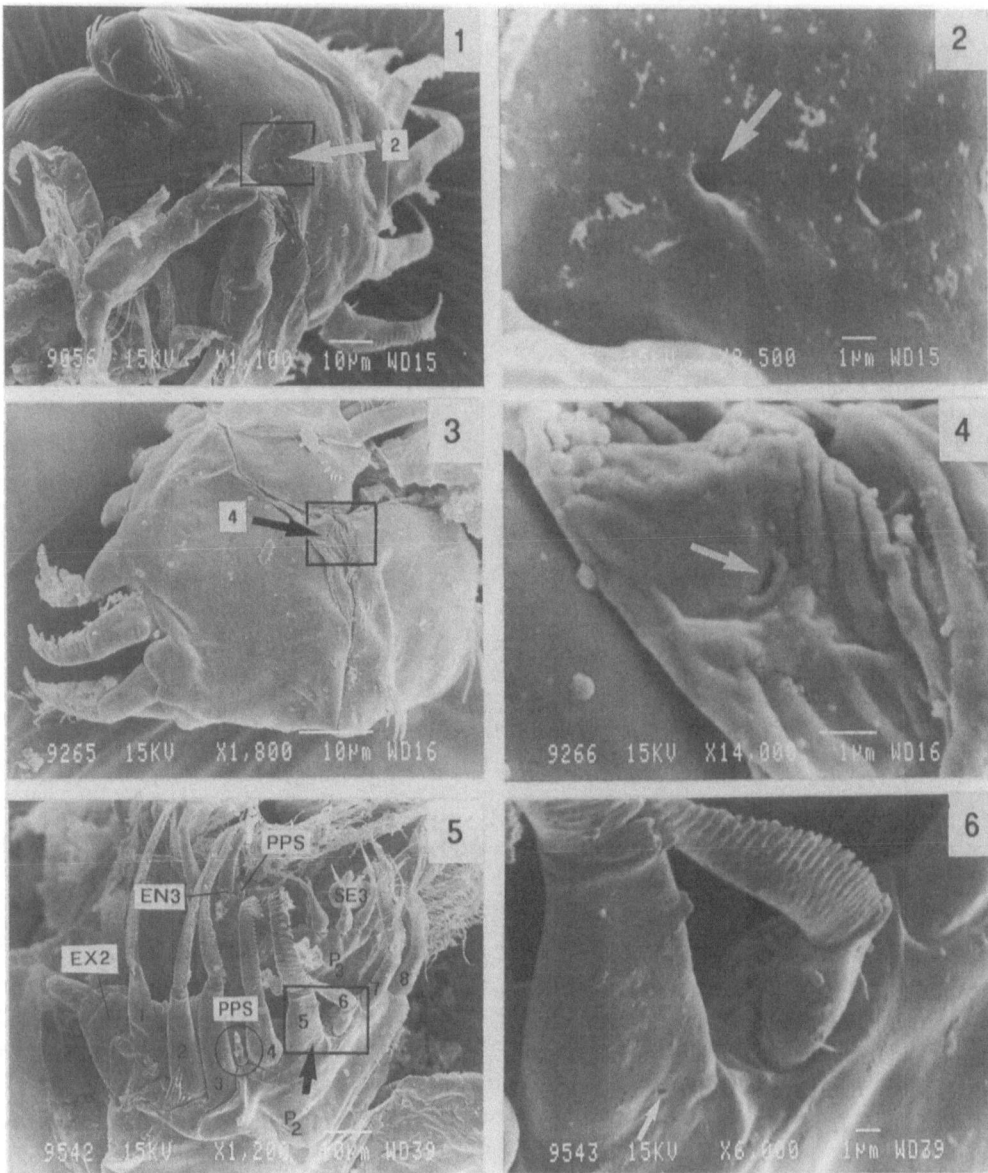

Plate 3. Second trunk limb in 1–2, *Alona affinis* (Mexico); 3–4, *Macrothrix laticornis* (Ild River, Russia); 5–6, *Chydorus* aff. *sphaericus* (Aguascalientes, Mexico): 1, P2 seen from behind (arrow: position of pore); 2, rounded pore, enlarged; 3, P2 seen from behind (arow: position of pore); 4. slit-shaped pore, enlarged; 5. P2, frontal view (with endite of P3 in background), arrow: basal zone of scrapers 5 and 6; 6, this zone enlarged, to show pore at foot of scraper 5 (arrow).

liform sensillum described by Smirnov and Fryer in *Macrothrix* was also found again (plate 4). It typifies the *laticornis*-group (Plate 4,1), and is absent in other groups, e.g. the *triserialis*-group (Plate 4,2). In other groups (e.g. the *hirsuticornis* group), it is replaced by a setulated seta (not shown).

Limb 3 consistently has a pore of almost 2 μm diameter on the anterior side of the endite, at the base of the gnathobase (Plates 5 and 6). In the *Macrothrix lati-*

cornis group, where no internal filtercomb is present, the other components of the endite armature are strongly modified. The four anterior setae are broad-based, with apical brushes. At the base of the outermost and smallest seta, a sensillum (SE3) is found, as well as two possibly papilliform sensilla (plate 5,2). The posterior row consists of three rod-shaped spines, split into two or three tubular parts apically, suggestive of a sense-organ function; the three inner ones are

37

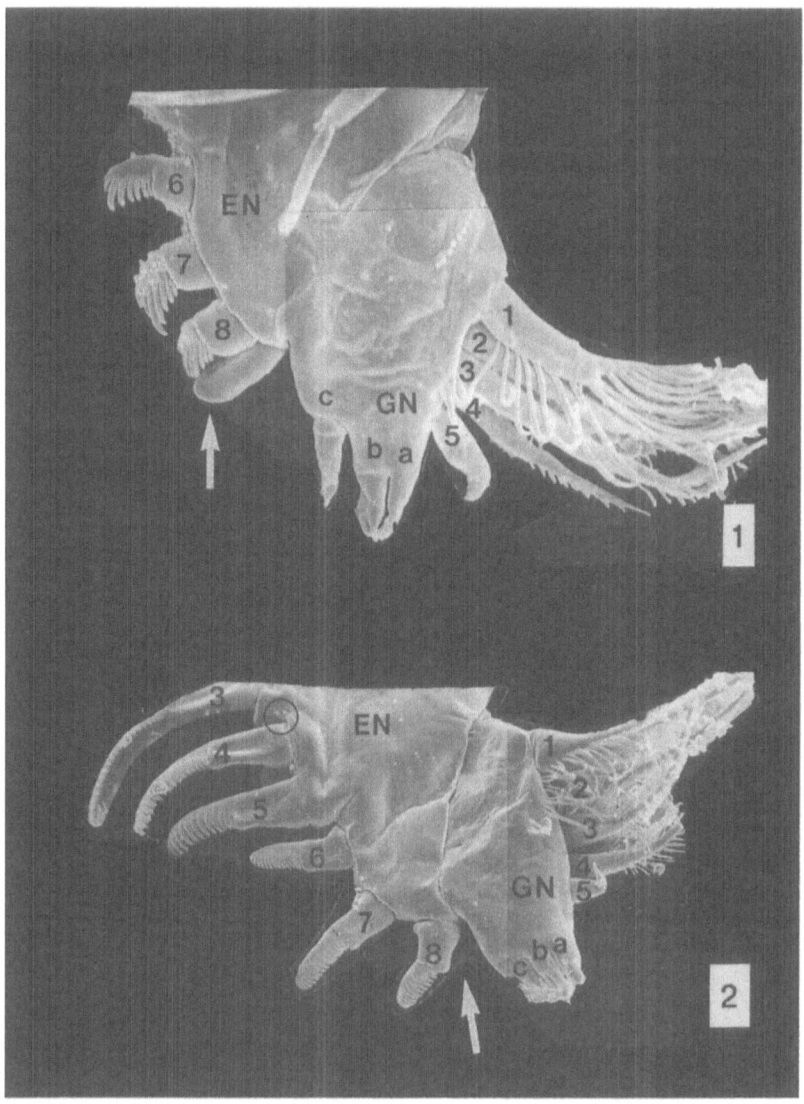

Plate 4. Second trunk limb, frontal view: 1, *Macrothrix laticornis* (Russia) and 2, *Macrothrix* cf *triserialis* (Aguascalientes, Mexico). Arrows point to presence or absence of a batilliform sensillum between scraper 8 and gnathobase; circle shows a papilliform sensillum between scrapers 3 and 4; a–b–c, gnathobasic appendages; 1–5 (right side of figure), setae of gnathobasic filtercomb.

mushroom-shaped, crowned by fine setules. In the *M. triserialis*-group (Plate 5, 4–5), where the pore is in the same position as in *M. laticornis*, the anterior setae are crowned with stiff setules, and the posterior row appears filtratory. In chydorids, the pore is in an identical position (Plate 6.1), again at the base of a screen of setae (Plate 6.2) and, in *Chydorus*, the anterior row of spines consists of scrapers. A large sensillum extends anteriad between scrapers 2 and 3. Its apex opens via a slit-shaped apical pore (Plate 6,3). In addition, two papilliform sensilla occur between the outer setae (OS 3) of the endite. The position and number of these

papilliform sensilla might be another generic character: in *Alona* aff. *karua*, there is only a single such sensillum (Plate 6, 5–6). In *Alona affinis*, where the setae of the anterior row are pencil-shaped (Plate 7, 1) and those of the posterior row filtratory, the sensillum is elongate, slender, and with an apical pore (Plate 7, 1–2; Plate 8,4), a situation similar to that in *A.* aff. *karua* (Plate 8, 1).

The endite of P4 has a denticulated outer seta in *Alona affinis*, and three barrel-shaped setae crowned with a group of undulating setules (Plate 7, 3–4). Between the innermost seta (S4:3) and the furry-like

38

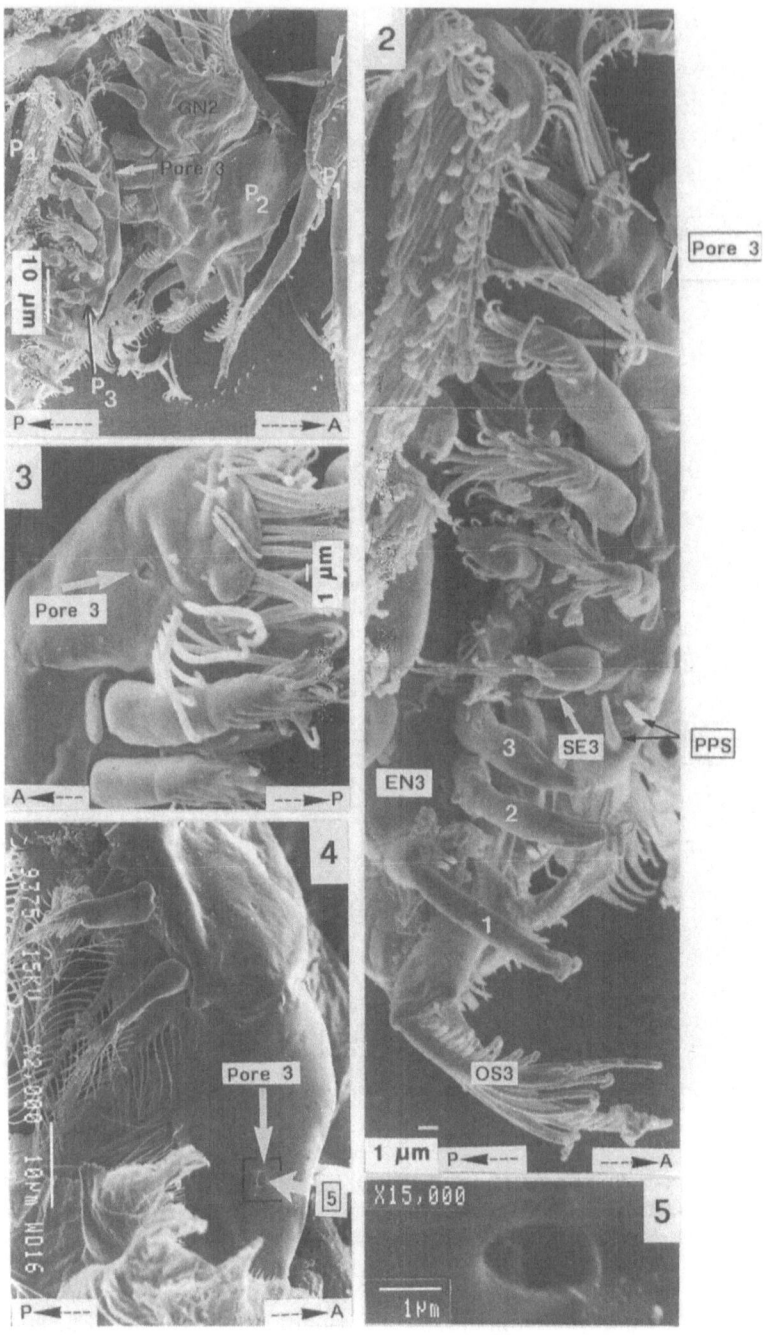

Plate 5. Third trunk limb in *Macrothrix*: 1–3, *M.* cf *laticornis* (Mexico); 4–5, *M.* cf. *triserialis* (Mexico). 1, overview of the zone of P1–P3; 2, endite of P3, enlarged. A sensillum (SE3) and possible papilliform sensillum identified by arrows, as well as the three (1–3) exterior setae of the posterior row, modified into esthetasc-like rods, apically bifid or trifid; 3, top of endite in another specimen, showing structure of pore aperture; 4, pore area (enlarged in 5) in *M.* cf. *triserialis*, where the setae of the endite are filtratory or brushing, not modified into sensilla. (A: anterior, P: posterior side of animal).

gnathobasic seta, a sensillum is found which is bottle-shaped, and caries a funnel-shaped pore at its apex (Plate 7, 5–6; Plate 8, 4). In *Camptocercus rectirostris*, the arrangement is similar (Plate 8: 7–8): a spinous outer seta is followed by three barrel-shaped setae, crowned with swollen setules. At the base of the

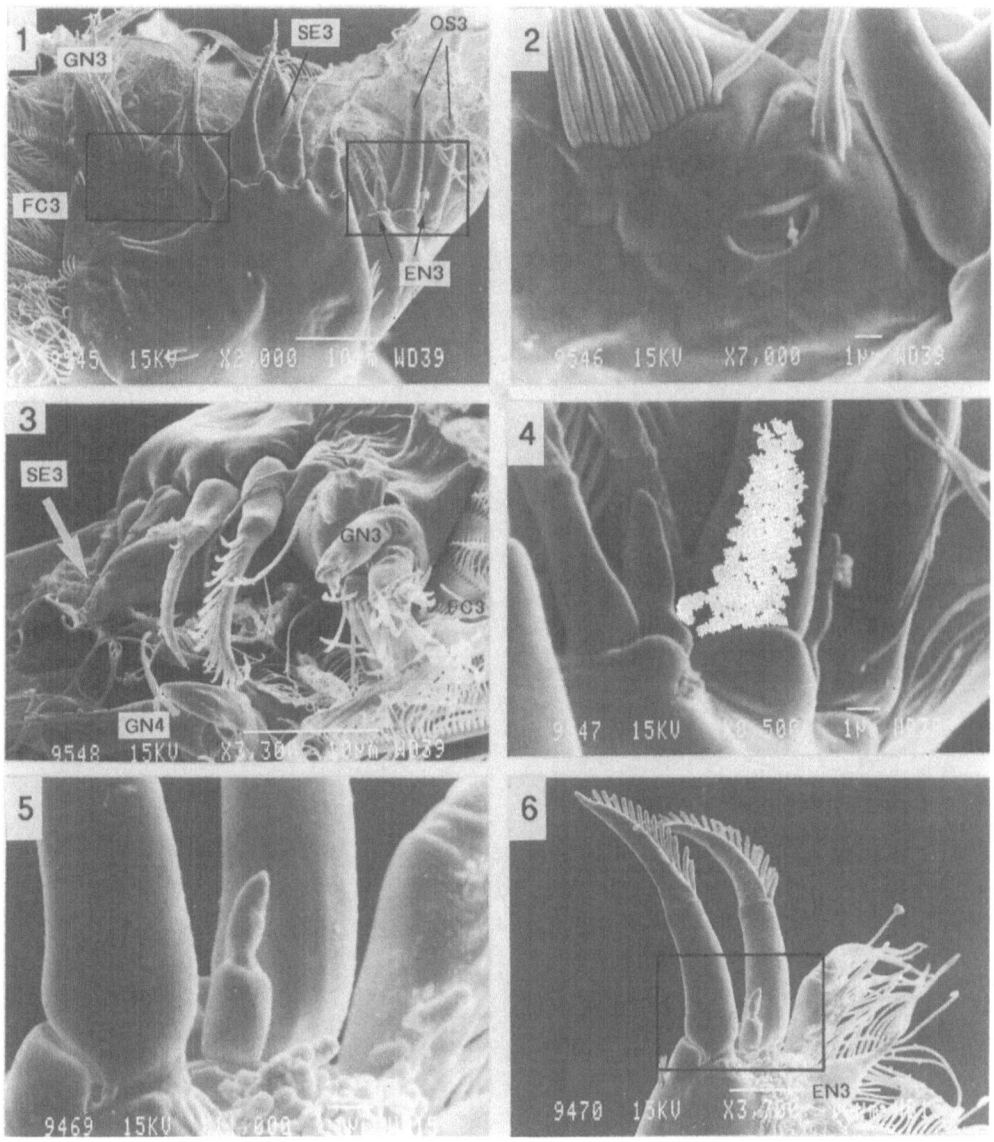

Plate 6. Trunk limb 3 in *Chydorus* aff. *sphaericus* (Mexico); 1, frontal view. Pore area enlarged; 3, endite viewed from above, showing slit-shaped apex of sensillum 3; 4, enlargement of outer setae of endite 3, showing two papilliform sensilla; 5, papilliform sensillum in *Alona* aff. *karua* (Socotra island); 6, position of the sensillum between the outer setae of the endite.

gnathobasic seta, there is a large sensillum (SE4) with a broad apical pore. Here, accessory sensilla without apical pores also occur (AS4). In *Alona* aff. *karua* (Plate 8, 1–3), the sensillum close to GNS4 is tubular, with a smallest apical aperture, and there are only two setae, with reduced apical ornamentation, between the sensillum and the outer seta OS4. In addition, there is a single pore, adjacent to SE4 and inserted on the rim of the endite in *Alona* aff. *karua* (Plate 8 2–3), or on the anterior surface of the body of the endite in *Alona affinis* (Plate 8, 5–6), or at the side of S4:3 in *Campto-*

cercus rectirostris (Plate 8:8), where the pore appears to be a pore-plate.

On P5, finally, a pore appears at the base of the endite (Plate 9).

All these pores are extremely hard to identify using classical microscopy, except, occasionally, the pore on P3. Conversely, the sensilla can usually be located without difficulty, although details of structure are often hard to identify.

Serial sections revealed that several pores on the trunk limbs are indeed connected to glands, the secre-

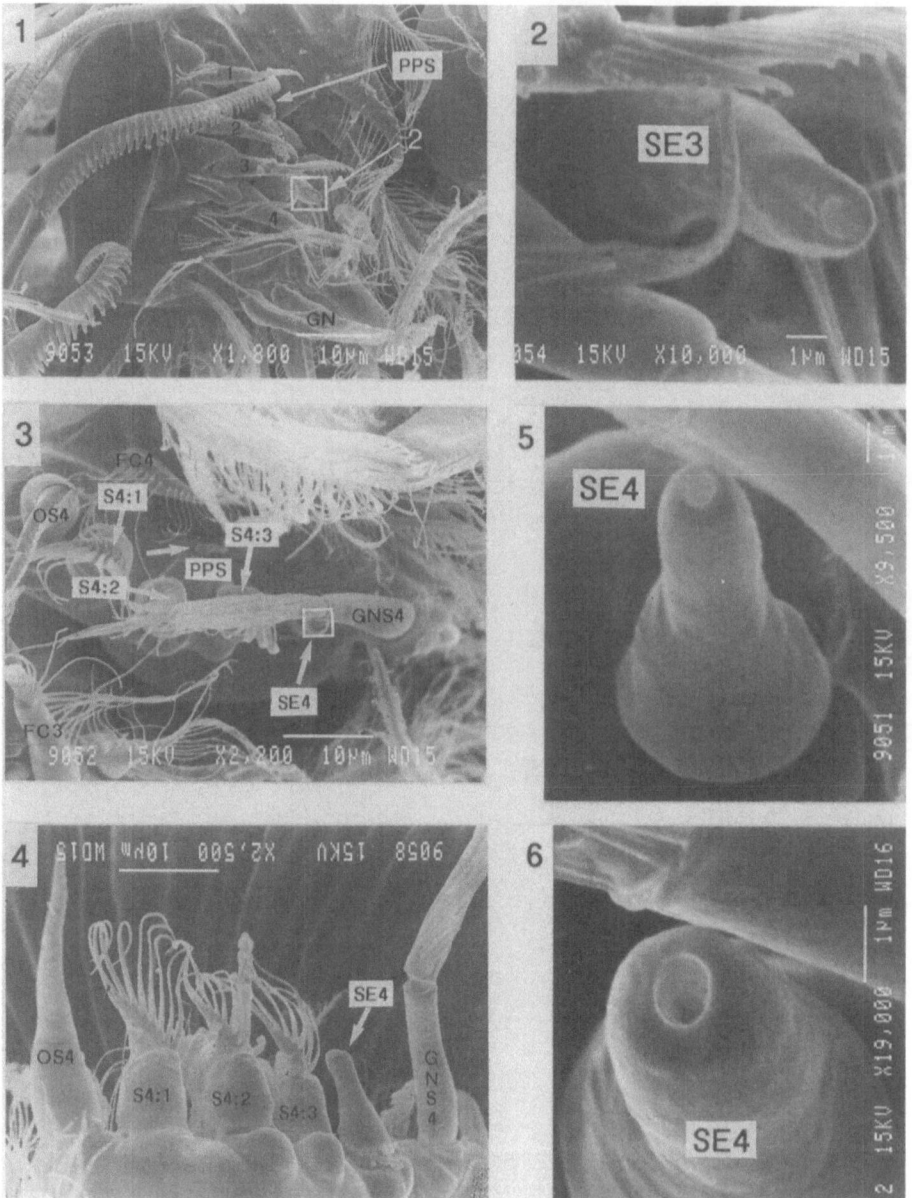

Plate 7. Third and fourth trunk limbs in *Alona affinis* (Mexico). 1, Overview of P3, showing the row of 4 anterior pencil-shaped setae with intervening sensillum. 2, sensillum enlarged; 3, endite of P4, showing (encased) the position of the bottle-shaped sensillum and (arrow), a small accessory sensillum; 4, lateral view of the anterior row of appendages of the endite; 5, 6, enlargements of the bottle-shaped sensillum with apical pore.

tion of which accumulates in reservoirs. Such reservoirs are here illustrated in the case of P1 (Plate 10), P3 (Plate 11, 3), P4 (Plate 12, 1–4), and P5 (Plate 13, 1–2). Plates 11, 3, and 12, 3 suggest that the internal structure of these reservoirs should be studied in more detail, as they might be internally partitioned. Plate 12, 2 also shows a cross-section of the sensillum on

P4, in which a central canal leads deep into the limb, unconnected to the reservoir beyond the secretory pore. Cross-sections reveal the small papilliform sensilla on the endites much better than dissected and pressurised limbs (e.g. Plate 11, 2, 4).

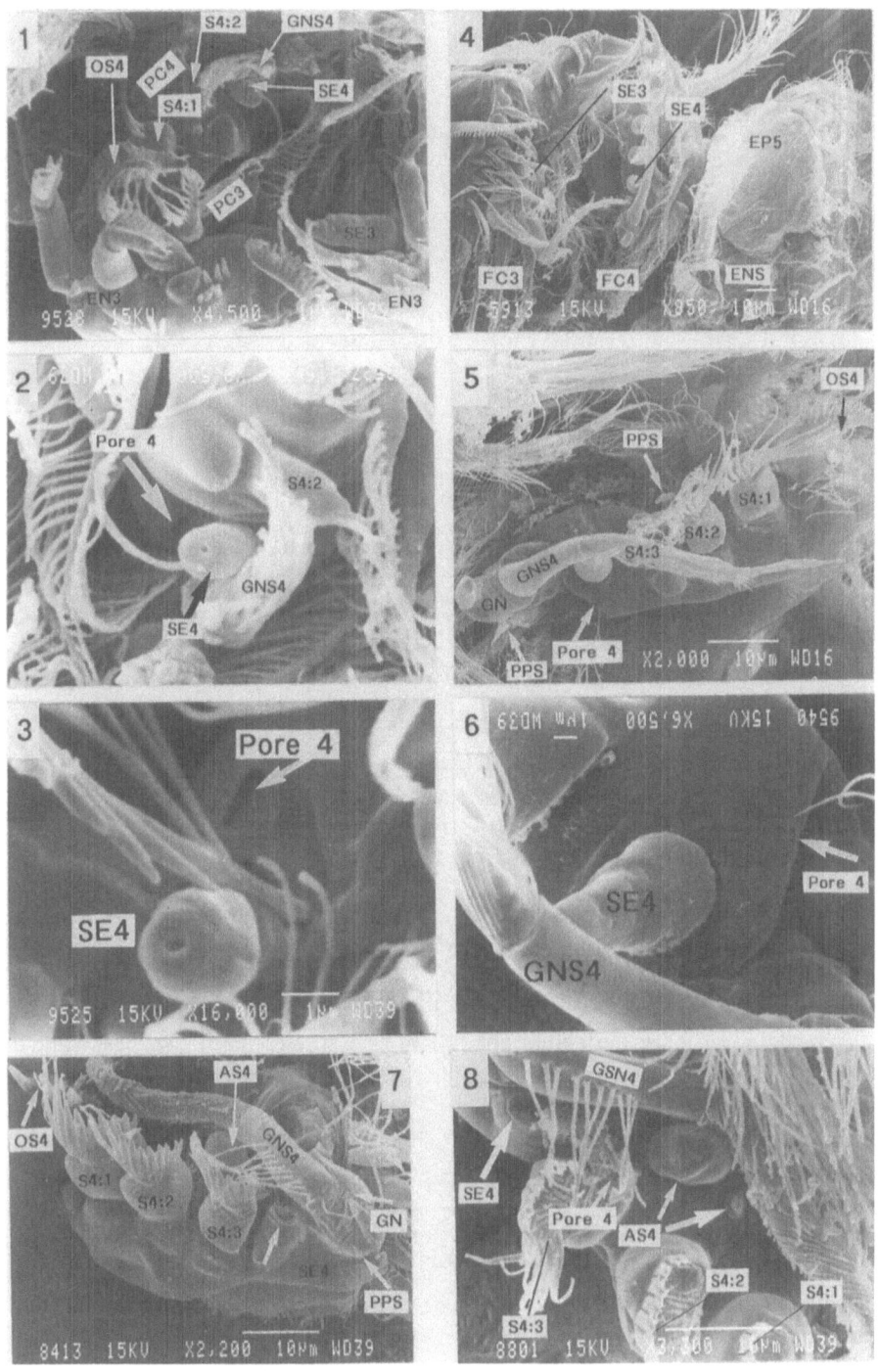

Plate 8. Fourth trunk limb in various chydorids. 1, endite of P4 in *Alona* aff. *karua* (Mexico), with only two, rudimentary setae in the frontal row, and a barrel-shaped sensillum with large apical pore; 2, 3, sensillum enlarged, and position of secretory pore indicated (2: Mexico, 3: Socotra Island); 4, region of P3–P5 in *Alona affinis* (Mexico); 5, endite of P4 obliquely from above, showing the three massive setae and the position of the secretory pore at the base of the sensillum. 6 sensillum and pores, enlarged; 7,8, endite of P4 in *Camptocercus rectirostris* (Glubokoe lake, Russia), showing the barrel-shaped sensillum, three massive setae, accessory sensilla, and position of the secretory pore, here on the flank of S4: 3.

42

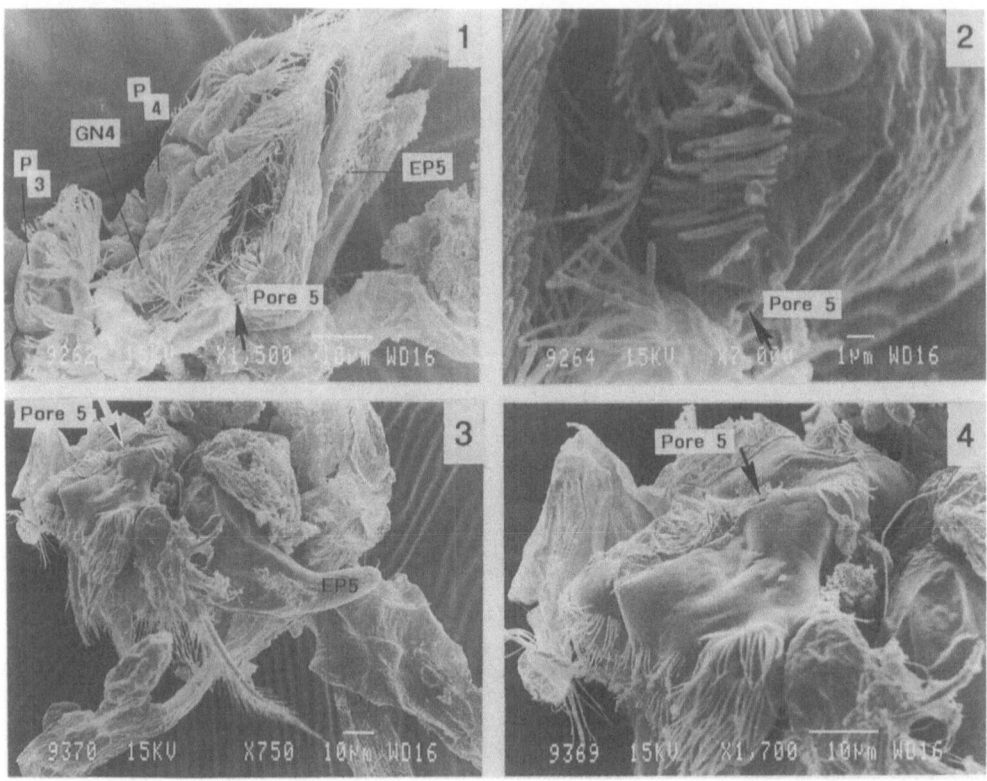

Plate 9. Fifth trunk limb in 1,2, *Macrothrix laticornis* (Ild River, Russia), and 3,4, *Macrothrix* cf. *triserialis* (Mexico), showing the position of the secretory pore at the base of the endite, at low (1,3) and high magnification (2,4).

Discussion

The sensillar and glandular structures identified earlier in selected trunk limbs by Fryer and Smirnov are not only confirmed, but it is shown that a gland may occur in each pair of legs (in addition to that in the labrum, not studied here). Fryer, in his various writings, calls their product entangling secretions, by which presumably a type of mucus is meant. This is almost certainly true, although we suggest that – as in saliva – some digestive enzymes might also be produced by these glands.

As to the sensilla, these too are shown to be much more numerous and to represent more types than hitherto suspected. Although we will refrain from even suggesting a specific function to each type, it is logical to expect the different types to play distinct roles, and at least some to be chemoreceptors, involved in judging the quality of incoming food particles. Others may be mechanoreceptors, collecting positional information on (parts of) the legs.

In spite of a remarkable constancy in the localisation of some of the pores, even across families (e.g. the pores on P2 and P3), suggesting a phylogenetic rela-

tionship, other pores (like the one on P4) may vary in position, and some (like the one at the base of scraper 5 in P2 of *Chydorus*) even seem species (or genus) specific. The same is true of the sensilla, where especially the large apically perforated ones on P3 and P4 seem to have species-specific shapes, while the papilliform sensilla on P3 may vary in both shape and in number between genera. The papilliform sensillum between scrapers 3 and 4 of P2 appears to represent another common feature between Chydoridae and Macrothricidae, and additional sensilla may occur on P2 (like the batilliform sensillum between scraper 8 and the gnathobase in the *M. laticornis* group).

Finally, it should be stressed that any seta, because presumably individually innervated, can become modified into a specialised sensillum. This is the case in the *Macrothrix laticornis*-group (as opposed to e.g. the *triserialis*-group), where half of the setae of the posterior row on the endite of P3 are modified into specialised, rod-shaped and multi-headed receptors, comparable in gross structure to the esthetascs of the first antenna.

We conclude that the trunk limbs of the two families of which representatives were studied here are much

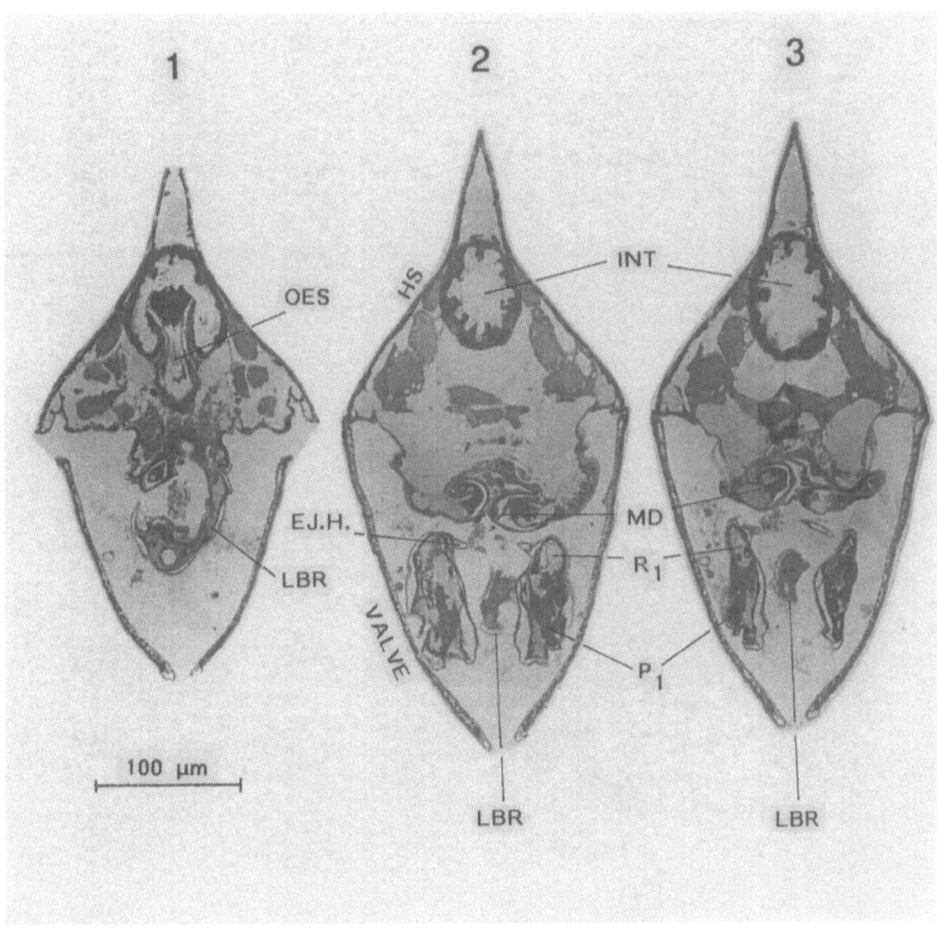

Plate 10. Transverse sections of the anterior region of *Camptocercus rectirostris* (low magnification). 1, through the labral region, immediately behind the mouth; 2–3, slightly further posteriad, showing the secretion reservoirs at the base of P1.

more than just a filtering apparatus: their glandular and sensillar equipment qualifies them as a functional mouth, but in which the secretory and 'tasting' funtions lie ahead of the masticatory apparatus (the mandibles). This contrasts with most vertebrates, and especially with mammals, where these two functions lie behind the masticatory apparatus.

Acknowledgements

We thank Mrs S. Wellekens for assistance with the SEM work, Dr G. Criel for sectioning the specimen of *C. rectirostris* for us, and Dr G. Fryer for constructive remarks.

References

Alonso, M., 1996. Crustacea Branchiopoda. Fauna Iberica 7, 486 pp. Museo Nacional de Ciencias Naturales, Madrid.

Fryer, G., 1962. Secretions of the labral and trunk limb glands in the Cladoceran Eucycercus lamellatus. Nature (London) 195: 97.

Fryer, G., 1963. The functional morphology and feeding mechanism of the chydorid cladoceran Eurycercus lamellatus (O.F. Muller). Trans. r. Soc. Edinb. 65: 335–381.

Fryer, G., 1968. Evolution and adaptive radiation in the Chydoridae (Crustacea: Cladocera): a study of comparative functional morphology and ecology. Phil. Trans. r. Soc. Lond. 254 B: 221–385.

Fryer, G., 1974. Evolution and adaptive radiation in the Macrothricidae (Crustacea: Cladocera): a study of comparative functional morphology and ecology. Phil. Trans. r. Soc. Lond. 269 B: 137–274.

Fryer, G., 1991. Functional morphology and the adaptive radiation of the Daphniidae (Branchiopoda: Anomopoda). Phil. Trans. r. Soc. Lond. 331 B: 1–99.

Korovchinsky, N. M., 1992. Sididae and Holopediidae (Crustacea: Daphniiformes). In H. J. Dumont (ed.), Guides to the Identifica-

44

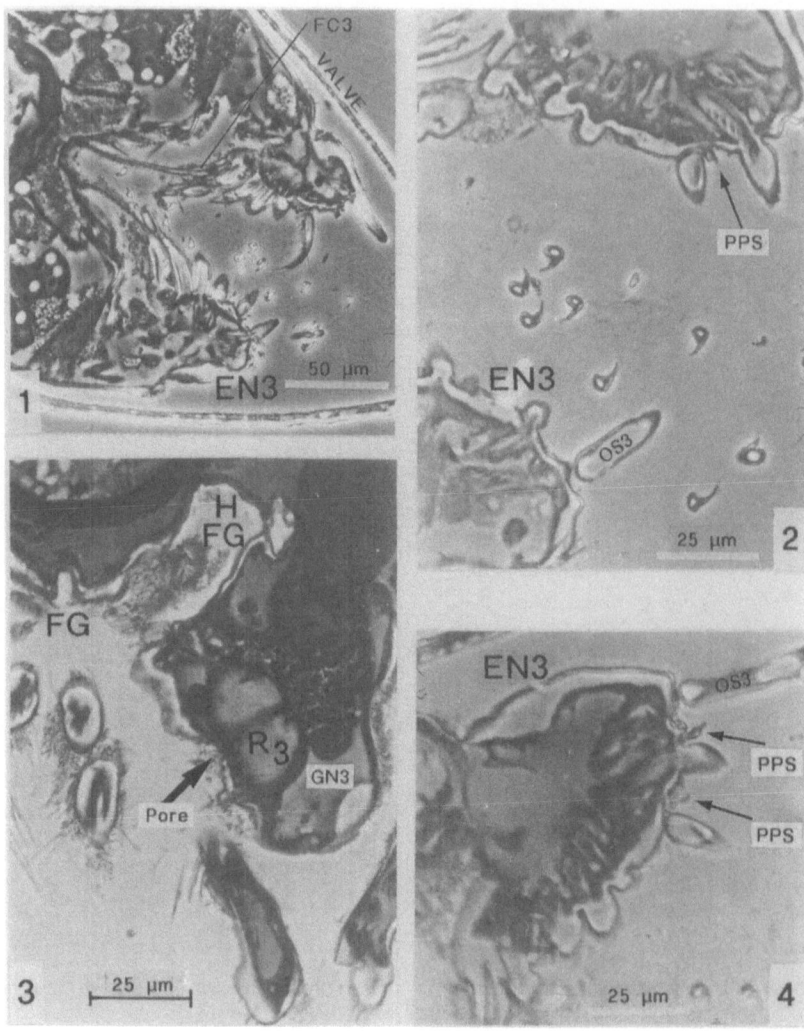

Plate 11. Camptocercus rectirostris, transverse sections through P3. 1, general view, to show the setae of the filter comb directed into the food groove, and the endite and its appendages; 2, endite enlarged, showing the implant of a papilliform sensillum (two such sensilla are shown on 4); 3, section through the endite near its junction with the gnathobase, at the site of the opening of the reservoir; 4, endite, showing two papilliform sensilla.

tion of the Microinvertebrates of the Continental Waters of the World, 3, 82 pp. SPB Publishers, The Hague.

Kotov, A. A., 1995. Structure of the thoracic limbs in Bosmina Baird, 1845 (Crustacea: Anomopoda). Arthropoda selecta 4: 41–50.

Sars, G. O., 1861 (published 1993). On the freshwater crustaceans occurring in the vicinity of Christiania. University of Bergen, 197 pp. + 113 plates.

Smirnov, N. N., 1966a. The taxonomic significance of the trunk limbs of the Chydoridae (Cladocera). Hydrobiologia 27: 337–342.

Smirnov, N. N., 1966b. Pleuroxus (Chydoridae): morphology and taxonomy. Hydrobiologia 28: 161–194.

Smirnov, N. N., 1967. O sensillach nog vetvistousich rakoobraznich (on sensillae on trunk limbs of Cladocera). Zool. Zh. 46: 286–288.

Smirnov, N. N., 1968. On comparative functional morphology of limbs of Chydoridae (Cladocera). Crustaceana 14: 76–96.

List of abbreviations

AS 4 = accessory sensillum on endite of trunk limb 4.

EJ.H. = ejector hooks of trunk limb 1.

EN = endite

EN 1:1 = first endite of trunk limb 1

EN 1:2 = second endite of trunk limb 1

EN 1:3 = third endite of trunk limb 1

EN 2 = endite of trunk limb 2

EN 3 = endite of trunk limb 3

EN 4 = endite of trunk limb 4

EN 5 = endite of trunk limb 5

EP 5 = epipodite of trunk limb 5

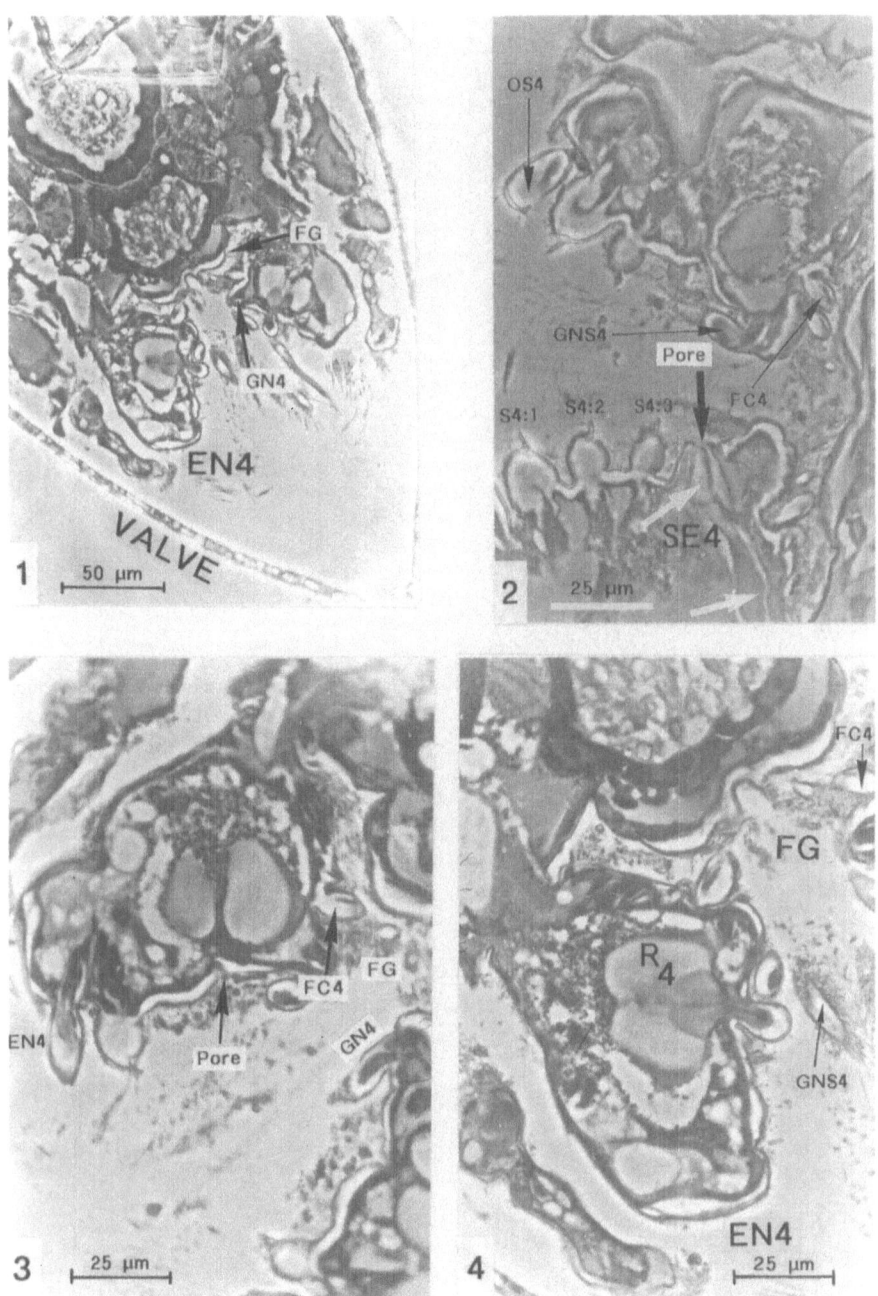

Plate 12. Camptocercus rectirostris, transverse sections through P4. 1, overview, showing the food groove, the endite, and the gnathobasic setae. The secretion reservoirs are visible in both legs, and are enlarged in 3 and 4; 2, section through the barrel-shaped sensillum, showing its aperture (pore) and a long canal leading deep into the appendage; 3, the secretion reservoir, which appears bipartitioned at this level; 4, section at the level of the connection of the reservoir (not partitioned at this site) with seta 1 of endite 4.

EX 2 = exopodite of trunk limb 2
FC 2 = internal filter comb of trunk limb 2
FC 3 = internal filter comb of trunk limb 3
FC 4 = internal filter comb of trunk limb 4

FC 5 = filter comb of trunk limb 4
FG = food groove
GN = gnathobase
GN 1 = gnathobase of trunk limb 1

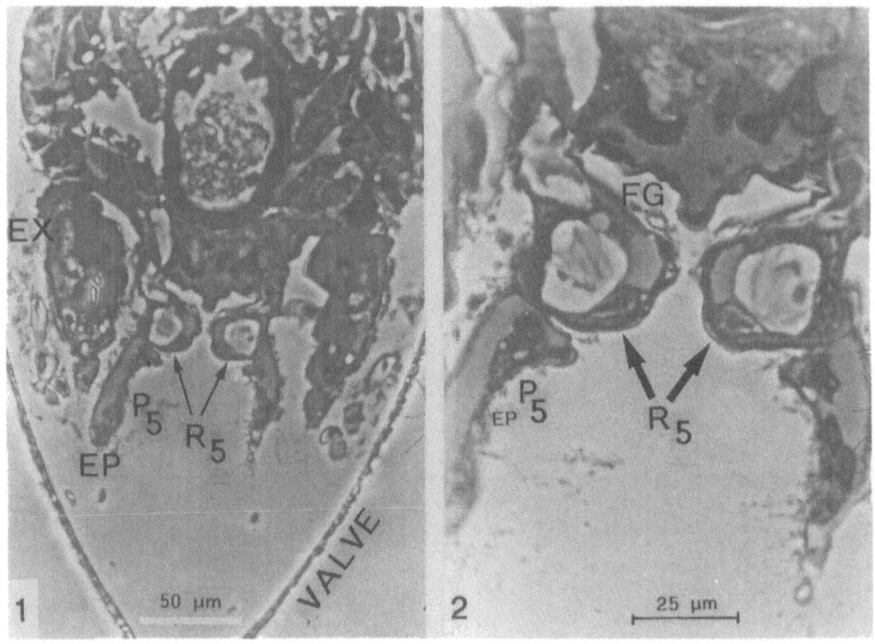

Plate 13. Camptocercus rectirostris, Transverse sections through P5 at low (1) and high (2) magnification to show the position of the secretion reservoirs inside the endites.

GN 2 = gnathobase of trunk limb 2
GN 3 = gnathobase of trunk limb 3
GB 4 = gnathobase of trunk limb 4
GNS 3 = gnathobasic 'fur' seta of trunk limb 3
GN 5 = gnathobase of trunk limb 5
GNS 4 = gnathobasic 'fur' seta of trunk limb 4
HFG = lateral horn of food groove
HS = head shield
INT = intestine
LBR = labrum
MD = mandible
MX = maxilla
OES = oesophagus
OS 3 = outer seta(e) on endite of trunk limb 3
OS 4 = outer seta on endite of trunk limb 4
P1 - P5 = trunk limbs 1 to 5
PC 3 = posterior filter comb of trunk limb 3
PC 4 = posterior filter comb of trunk limb 4

Pore 1 – 5 = pores on trunk limbs 1 to 5
PPS = papilliform sensillum
R1 – 5 = secretion reservoirs inside trunk limbs 1–5
S 4:1 = first massive seta with undulating setae on endite of trunk limb 4
S 4:2 = second massive seta with undulating setae on endite of trunk limb 4
S 4:3 = third massive seta with undulating setae on

endite of trunk limb 4
SE 3 = main perforated sensillum on endite of trunk limb 3
SE 4 = main perforated sensillum on endite of trunk limb 4.

Hydrobiologia **360:** 47–54, 1997.
A. Brancelj, L. De Meester & P. Spaak (eds), Cladocera: The Biology of Model Organisms.
©1997 *Kluwer Academic Publishers.*

Alona stochi n.sp. – the third cave-dwelling cladoceran (Crustacea: Cladocera) from the Dinaric region

Anton Brancelj

National Institute of Biology, Večna pot 111, 1000 Ljubljana, Slovenia

Key words: taxonomy, ecology, caves, Chydoridae, Slovenia

Abstract

Alona stochi n.sp., a third blind cave-dwelling cladoceran from the Dinaric region, was found on several occasions in a small cave 30 km south-east of Ljubljana (Slovenia). It is related to *Alona hercegovinae* Brancelj, 1990 and *A. sketi* Brancelj, 1992. Several characters suggest that it belongs to a primitive group within the genus *Alona*.

A. stochi was found in a 'semi-cave' environment, accompanied by a cave-associated fauna (Amphibia: *Proteus anguinus*, Decapoda: cf. *Troglocaris schmidti*, Copepoda: *Troglodiaptomus sketi*, *Diacyclops tantalus*, *D. charon*, *Elaphoidella stammeri*).

Introduction

The first record of a cladoceran from a hypogean habitat dates back to 1972, when Petkovski & Flössner (1972) described *Alona smirnovi* from the interstitial waters of lake Ohrid. This was the first cladoceran to be reported as having a reduced eye apparatus. Later, several new species from hyporheic or interstitial waters were described (Dumont, 1983, 1987; Dumont & Brancelj, 1994; Sabater, 1988), all with a vestigial eye apparatus.

The first cave-dwelling cladoceran, *Alona hercegovinae*, was described in 1990 from the cave Ljelješnica in Hercegovina (Brancelj, 1990). Two years later, *Alona sketi* was found in the Osapska Jama cave in Slovenia (Brancelj, 1992). Both show a complete reduction of the eye apparatus and a uniform comb of scrapers on P2. We can confirm Dumont & Brancelj's (1994) statement that 'the number of ground-water and cave-dwelling Cladocera has greatly expanded in recent years, ... but the impression one gets is that the subject has hardly been touched'. Most interstitial, hyporheic and cave-dwelling taxa belong to a single genus, *Alona*, although numerous other cladoceran taxa can live permanently or for some time in subterranean environments (Brancelj & Sket, 1990; Dumont & Negrea, 1996).

Material examined

3 parthenogenetic females, 7 adult females without eggs.

The specimens were collected with a hand net with a mesh-size of 60 μm, and were stored in 4% formaldehyde, later being transferred to 70% alcohol with a drop of glycerol added.

Type material

Holotype. An adult female without eggs (total length 0.47 mm, height 0.36 mm) mounted on a slide in glycerol, sealed with glyceel (collected 20 July, 1995). Held in the collection of the author (National Institute for Biology – N.SP: *Alona stochi*: 1–5).

Paratypes. Two parthenogenethic females, five adult females without eggs; two adult females completely dissected (collected between 20 July and 7 August, 1995). One adult female and one parthenogenethic female, designated as paratypes and preserved in 70% alcohol, deposited in the British Museum (Natural History), reg. no. = 1995.1665–1667. Two dissected females, one parthenogenethic female and three adult females without eggs, designated as paratypes, held in the author's collection.

Other material. One adult female without eggs and one parthenogenethic female (collected 21 July, 1993) held in the author's collection.

Male: unknown.

Juveniles: not found.

Type locality. The cave Kompoljska Jama, 30 km SE of Ljubljana (Slovenia), at the edge of a carstic polje, just beneath the ridge of the mountain Mala Gora (Little Mountain). The cave functions as a spring during heavy rains or after snow melt, and later as a sink hole, into which torrent drains the water from a temporary lake.

The entrance to the cave is at the same elevation as the bottom of the polje (450 m a.s.l.). The entrance gallery is *c.* 20 m long, slightly descending (*c.* 10°), with a bottom of stones and fine sand. At the end of the gallery is a small lake, *c.* 5 m wide, 10 m long and *c.* 1 m deep, with a bottom of stones and fine sand, mixed with clay. Beyond the lake, the gallery divides. To the right is a narrow corridor, filled with water and ending in a narrow siphon. The left fork is *c.* 30 m long, and is filled with standing water and also with a siphon. Half of the lake is in dim light, whilst the remainder lies in permanent darkness.

Normally, the water in the cave is *c.* 3 m below the level of the polje. The orientation and the slope of sand ridges in the entrance gallery indicate that there is strong water flow from the inside of the cave, while reverse flow, i.e. from the outside, is negligible. The water level was low and stable during our visits in summer (20 July–7 August). Water temperature was relatively high, 10.2 ± 0.2 °C, and oxygen concentration close to saturation. Data on other chemical and physical variables are given in Table 1.

Etymology

The new species is named after Dr F. Stoch from Trieste (Italy), who sent me the first two specimens of the new species.

Diagnosis

Completely blind, postero-ventral corner of valves smooth, their ventral margin setose; posterior margin spinose, no gap in the armature of the free margin of the valves. Tree main headpores interconnected, the two lateral headpores small. Basal spine of postabdominal claw slim, longer than in both other blind taxa. No fine striation between main striae of valves, and no reticulation on head capsule.

Description of adult female

Size: body length = 0.45 ± 0.02 mm, body height = $0.36 \pm 0,02$ mm (*n* = 6). Average length-width ratio of adult females = 1.25, range = 1.18–1.31 (*n* = 6).

Head: rostrum fairly blunt and rounded. Three main headpores interconnected and equidistant (Figure 1d). The two lateral pores small, situated at a distance of *c.* 0.8 the length of the main headpores. Compound eye and ocellus absent (Photo 1d).

Labrum: triangular, with margin undulating slightly just above its tip. Tip not uniform in shape, ranging from broadly rounded to pointed (Figure 1g, Photo 1d).

Valves: transparent or slightly yellowish. Digestive tract and appendages clearly visible. Shape of body slightly elongate or close to circular (Figure 1a; Photo 1a, 1b). Dorsal margin of valve highly arcuate, semicircular, with ventrum slightly less arcuate. Postero-ventral and anterior angles broadly rounded and smooth (Figure 1c). Valves with a series of chitinous crests running obliquely downwards from antero-drosal to postero-ventral area (Figure 1c). Cross-connections in postero-ventral region rare. No fine striation between the chitinous ridges, but a fine reticulation with hexagonal cells in the middle of valves (Photo 1c). Ventral margin of valves adorned with setae throughout, but with no zone differentiation (Figure 1c). Antero-ventral part with *c.* 24 long, fine, unilaterally feathered, equidistant seta, followed by a series of setae that decrease in length posteriorly, set equidistantly. Postero-ventral corner with a row of small denticles, slightly increasing in length dorsally (Figure 1c; Photo 1e) running up the inner surface of the valve. Posterior part with a row of small denticles, each 5-th or 6-th slightly longer than others, on inner side of valves.

Postabdomen: short, with well marked preanal corner (Figure 1b). Endclaw with inner and outer pecten relatively long and robust. Basal spine long and slim, with an additional 2 (3?) fine setae at its base. Marginal teeth relatively small, arranged in 9–10 groups (Figure 1b, Photo 1f). Their number increases from 1 in the group close to the endclaw, to 4 close to the anal

Table 1. Physical and chemical parameters measured in Kompoljska Jama cave on four different dates in 1995 (nm – not measured).

Date	N-NO$_3$ μgl^{-1}	N-NH$_4$ μgl^{-1}	totN μgl^{-1}	P-PO$_4$ μgl^{-1}	totP μgl^{-1}	COD mgl^{-1}	pH	alcalin. μeqvl^{-1}	cond. μScm^{-1}	Temp. °C	oxygen mgl^{-1}
Jan. 19	1220	45	2440	25	26.00	2.53	7.46	nm	343	6.2	7.2
Jul. 20	1170	685	1750	8.7	33.64	1.34	7.52	nm	338	9.8	11.4
Aug. 1	1183	15	2070	>1.0	20.29	1.18	7.51	nm	371	10.2	11.2
Oct. 6	741	24	1950	>1.0	18.53	2.76	7.52	4539	357	10.0	10.9

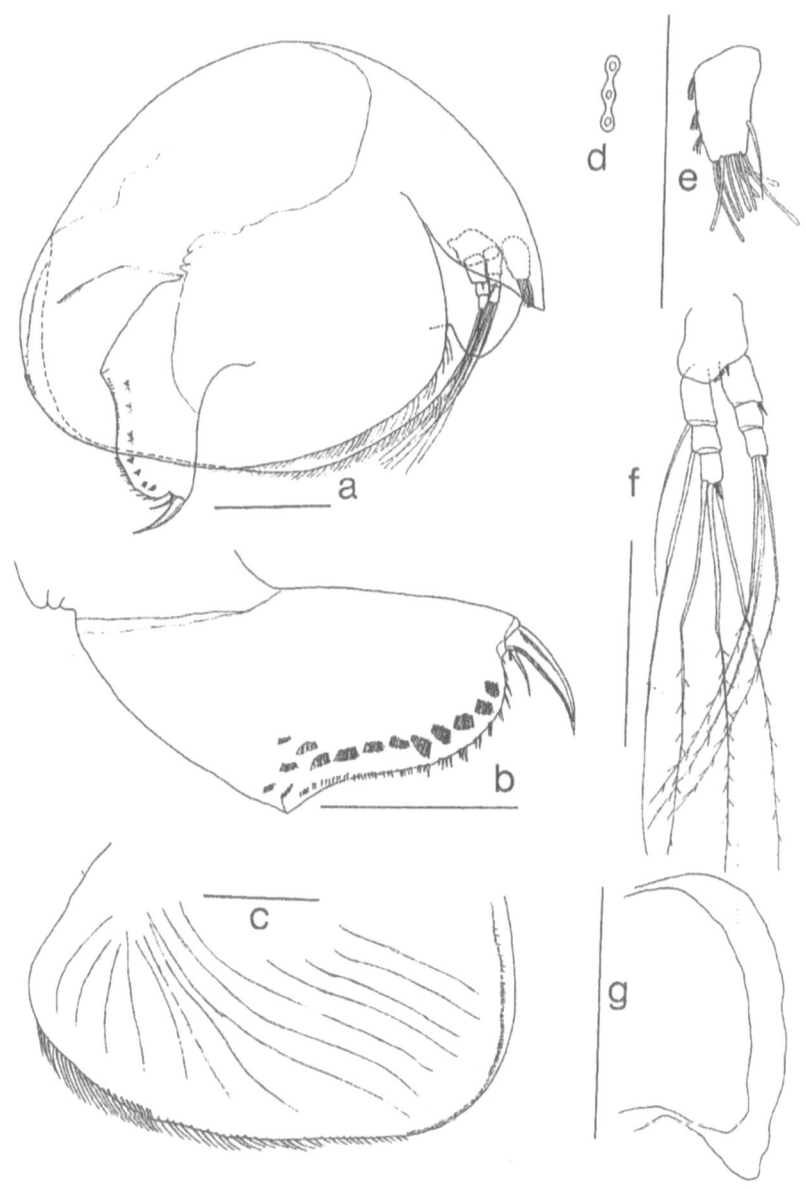

Figure 1. *Alona stochi* n.sp.; Kompoljska Jama cave (Slovenia) (all scale bars 100 μm). a: habitus of adult female (holotype); b: postabdomen; c: valve; d: headpores; e: first antenna; f: second antenna; g: labra.

zone. A large tooth, accompanied by 1–3 smaller teeth, increasing in size towards the outer margin. Anal zone with 5–6 groups of equal-sized teeth. Number of teeth in each group 2–7.

An additional series of long, slender setae (usually with more than 10 in a bundle) occurs as 5 bundles situated submarginally in the postanal zone (Figure 1b; Photo 1g). In the anal zone *c.* 6 bundles of shorter setae of uniform length. Just above the preanal corner an additional 3–4 bundles with short setae.

Antennula: not reaching apex of rostrum (Figure 1a; Photo 1d). Set with nine apical aesthetascs, three of which are longer than the others (Figure 1e). External seta attached at *c.* 2/3 from tip. On the opposite side, three bundles of small setae.

Antenna: eight natatory setae, two small apical spines and a spine on the basal segment of the exopodite (all of similar size) (Figure 1f). A further small spine on the antennal coxa between endo- and exopodite, and accompanied by a group of fine spinules. No spinules on the segments of the endo- or exopodite.

Antennal formula:

0(1) 0(0) 3(1)/1(0) 1(0) 3(1).

Mouth parts: mandibles as for the genus, and asymmetrical. Left mandible with a uniform and robust 'dental' part, the right mandible with fine, sharp teeth on the 'dental' part.

Trunk limb I: (Figure 2a). Ventral edge adorned with seven hair-like groups of setae and a few groups of small spinules. Inner distal lobe with two unilaterally feathered setae and one elongated spine. Outer distal lobe with one fine unilaterally feathered seta. Endites (End) with four, three and two setae. End3 with three setae of equal size, the fourth one shorter and more robust. End2 with two strongly unilaterally feathered setae, equal in shape and size, but with the lateral one somewhat shorter and bilateral feathered. Two setae of End1 flagellum-like and finely unilaterally feathered. Gnathobase with one seta. Epipodit circular.

Trunk limb II: (Figure 2b). Exopodite with one apical seta. Endopodite with eight scrapers, similar in shape and denticulation, increasing in size uniformly. Scraper 3 not significantly shorter than 2 or 4. Scraper 8 slightly more robust than the rest. Strong spine at the base of scraper 1. Gnathobase with a filter comb of 7 setae (6 of them equal in size), and 4 marginal spines.

Trunk limb III: (Figure 2c). Exopodite with 7 setae. Seta 3 short with very short flagellum. Seta 5 with very long flagellum. Epipodite small, round. Endopodite with an external row of 7 spines, of which the distal two scrapers are elongated and with moderately strong teeth. There is a small sensilum at the base of the distal two scrapers. The inner row with four setae. The gnathobasic plate has a filter comb of 7 uniform setae. There are two apical setae, the longest being distinctly angulate.

Trunk limb IV: (Figure 2d). Exopodite with 6 setae of subequal length. Epipodite slightly elongated. Preepipodite lobated. Endopodite with an inner row of three short setae and an outer row of four setae and a sensila. Gnathobasic filter plate with 5 setae in its filtering comb and an apical seta arching over the sensila of the endopodite.

Trunk limb V: (Figure 2e). Exopodite with 4 setae. Preepipodite lobated. Epipodite rounded with small projection. Endopodite elongated with short and robust bilateral feathered seta on its base. Gnathobase with two bilateral feathered setae and two sensilae. Gnathobasic filter plate with 5 (7?) setae.

Differential diagnosis and relationships

Alona stochi belongs, according to morphological characters, to the same group as *A. hercegovinae* Brancelj, 1990 and *A. sketi* Brancelj, 1992, described from caves in the Dinaric region. In a wider context, all three taxa are close to *A. diaphana* King, 1853, and differ from *A. protzi* Hartwig and *A. guttata* Sars groups in the shape and ornamentation of the postabdomen. Members of both complexes occur in interstitial habitats. Despite this, this 'trigeminus' of blind cave-dwelling cladocerans does not form a uniform group. They show a number of similarities: all three taxa are blind and have a circular body shape, with a highly arcuate ventrum and a widely rounded posterior valve rim. Antennal spinulation is weak, and the labrum is triangle-shaped. The main headpores are of similar shape, and the general shape of the postabdomen is similar, too. The structure of the scrapers of P2 is diagnostic, although similar within the group. The difference among cave-dwelling *Alona* can be sum-

Figure 2. Alona stochi n.sp.; Kompoljska Jama cave (Slovenia) (scale bar 100 μm). a: trunk limb I; b: trunk limb II; c: trunk limb III; d: trunk limb IV; e: trunk limb V;

marised as follows. The striation seen in *A. hercegovinae* is intensive, with cross-connections being common. There is also reticulation on the upper part of the valves, and even on the head shield. In *A. sketi* the.striation is limited to the ventral part of the valve, and cross-connections are common. In *A. stochi* the striation is similar to that of *A. sketi*, but with few cross-connections (comp. Brancelj, 1990: Photo 1; Brancelj, 1992: Photo 1 and this article).

All three taxa also differ in the ornamentation of the ventral valve rim. In *A. hercegovinae*, the row of setae is short, limited to the ventral-most part, and separated into three sections. In *A. sketi*, a row of setae runs along the entire antero-ventral and ventral margin, and is differentiated in three zones. In *A. stochi*, the row of setae is similar to that of *A. sketi*, but without differentiation into zones. The setae are situated more or less equidistantly, and their lengths decrease evenly from

52

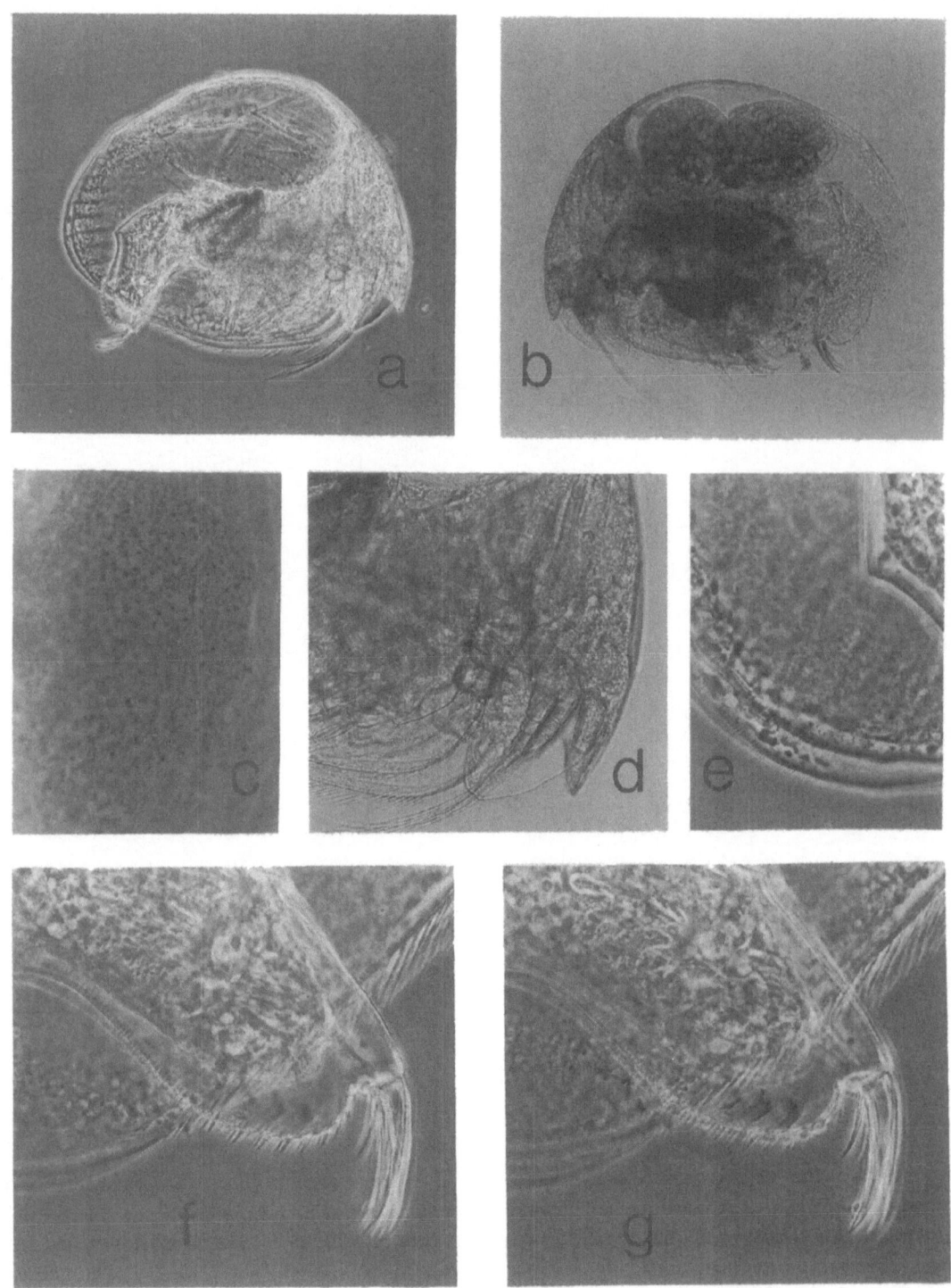

Photo 1 . Alona stochi n.sp.; Kompoljska Jama cave (Slovenia). a: habitus of adult female (holotype) – a phase-contrast photo; b: parthenogenethic female; c: polygonal cells on valve; d: head with labrum; e: posterio-ventral corner of valve; f: postabdomen with marginal teeth; g: postabdomen with submarginal teeth and basal spine.

the antero-posteriorly. In *A. sketi* and *A. stochi,* there is no gap between the marginal setae on the ventral side of the valvae and submarginal spines on the posterior part of valve.

The denticulation of the postabdomen is distinctive in all three taxa. In *A. hercegovinae* the number of marginal teeth is lowest: from 1 at the endclaw to 2–3 at the postanal corner, and arranged in 6–7 groups. In *A. sketi* 4–5 marginal teeth are arranged in 8–9 groups, whilst in *A. stochi* the marginal teeth increase from 1 at the endclaw to 4 in the anal zone, arranged in 9–10 groups.

The basal spine of *A. hercegovinae* is relatively short and slim. In *A. sketi,* it is somewhat longer, but still slim, whereas in *A. stochi* it is both long and slim.

The three blind taxa fall into two groups: *A. hercegovinae* with its specific reticulation, headshield and short row of marginal setae, and *A. sketi* and *A. stochi* with similar ornamentation and setation of valves. The most significant differences are in the armature of postabdomen, i.e. the number of teeth in a group, and in the absence of lateral headpores (?) in *A. sketi*. Both taxa have a relatively long basal spine, although it is longer in *A. stochi*.

According to the ornamentation and setation of the valve, *A. sketi* and *A. stochi* are primitive members of the genus *Alona,* whereas *A. hercegovinae,* with its reduced ventral row of setae, is the most specialised member of this complex.

Discussion

The new species of *Alona* is clearly a stygobiont, although accompanied by a fauna which is not exclusively stygobiotic. Eighteen taxa of copepods and cladocerans were found in the cave Kompoljska Jama (see Table 2). Among the cladocerans, all but the new species are members of the epigean fauna, but tolerant of some types of hypogean habitats (Brancelj & Sket, 1990). Among the 14 taxa of Copepoda, 6 are stygobionts, and most have relatively small ranges (i.e. they are endemic to Slovenia or to the Dinaric area). The remaining taxa from Table 2 belong to epigean faunas, although found frequently in subterranean environments (Brancelj, unpubl.).

Specimens of stygobiont taxa outnumbered epigean ones in a ratio of 10:1. Numerous small specimens of *Proteus anguinus* (Amphibia) and cf. *Troglocaris schmidti* (Decapoda) were found regularly in the lake. On the other hand, one can not believe that a small pool

Table 2. List of cladocerans and copepods found in Kompoljska Jama cave between July 20 and August 7, 1995. (∗ indicates taxa known as stygobitic)

Anomopoda
  Alona costata Sars, 1862
  ∗*Alona stochi* n.sp.
  Biapertura affinis (Leydig, 1860)
  Chydorus sphaericus O. F. Müller, 1785
  Leydigia leydigi (Schoedler, 1858)

Copepoda

Calanoida
  ∗*Troglodiaptomus sketi* Petkovski, 1987

Cyclopoida
  ∗*Acanthocyclops venustus stammeri* (Kiefer, 1930)
  ∗*Diacyclops charon* (Kiefer, 1931)
  ∗*Diacyclops tantalus* (Kiefer, 1937)
  Eucyclops serrulatus (Fischer, 1851)
  Megacyclops viridis (Jurine, 1820)
  Paracyclops fimbriatus (Fischer, 1853)

Harpacticoida
  Attheyella crassa (Sars, 1863)
  Bryocamptus dacicus (Chappuis, 1923)
  Bryocamptus minutus (Claus, 1863)
  ∗*Ceuthonectes serbicus* Chappuis, 1924
  Echinocamptus pilosus (van Douwe, 1910)
  ∗*Elaphoidella stammeri* (Chappuis, 1936)
  Paracamptus schmeili (Mrazek, 1893)

lying just behind the cave entrance is the true habitat of these stygobiont taxa. It is most probable that the small lake is an accidental habitat, and their real habitat lies deeper in the massif of Mala Gora. Presumably, animals are washed out during strong rains, and some of them become trapped into the lake at the cave entrance. Some can persist there for quite long periods. Conditions in the lake, i.e. basically absence of the day light, are more favourable for stygobionts than for epigean taxa. The fact that quite a large population of small specimens of *Proteus anguinus* (10–15 cm in length) inhabit the lake supports this hypothesis. Only a few specimens from the epigean taxa were observed to carry eggs.

There is no doubt that *A. sketi* and *A. stochi* are closely related. This is also supported by the relatively small distance between the type localities of the two species. The cave Kompoljska Jama (inhabited by

A. stochi) is situated *c.* 80 km from the cave Osapska Jama (inhabited by *A. sketi*). It may at first sight seem surprising that two different taxa from the genus *Alona* can be found living so close to one another. However, if the present hydrological situation is considered, *A. sketi* inhabits a water body that flows to the Adriatic Sea, whilst *A. stochi* inhabits one that drains to the Black Sea. This separation of drainage basins goes back to the Middle Pliocene, when the development of Karstic land forms (including the formation of caves) took place in the southern part of Slovenia (Bole, 1970; Sket, 1994). Today's separation of the drainage basins between epigean (and also hypogean) water bodies is the result of a series of parallel faults running in NW–SE, i.e. in a Dinaric direction.

It is probable that both *A. sketi* and *A. stochi* have a common epigean ancestor that either retreated into or penetrated into subterranean environments during the Late Pliocene or at the start of the Pleistocene (i.e. at the beginning of the glacial periods, which was accompanied by a drying-out of shallow lakes). Populations separated by hydrographic barriers developed into at least two stygobiont forms, that today represent two distinct taxa. A parallel situation can be seen in some populations of *Asellus aquaticus* (Sket, 1994).

Acknowledgments

My thanks to Dr Fabio Stoch (Trieste), who sent me the first two specimens, to my wife and daughter who assisted me during field work, to Dr Huw Griffith who made corrections to the text and to Mrs Andreja Jerebic who undertook the chemical analyses. Special thanks to Prof. Dr Henri Dumont who gave comments and suggestions for improvements of the text and figures in previous versions of the manuscript.

References

Bole, J., 1970: Subterranean Mollusca and evolution of river drainage basins (in Slovene). pp.: 247-250. Congrès Yugoslave de Spéléologie Cinquième session, 15–20 Sept. 1968. Skopje.

Brancelj, A., 1990. *Alona hercegovinae* n.sp. (Cladocera: Chydoridae), a blind cave inhabiting Cladoceran from Hercegovina (Yugoslavia). Hydrobiologia 199: 7–16.

Brancelj, A. & B. Sket, 1990. Occurrence of Cladocera (Crustacea) in subterranean waters in Yugoslavia. Hydrobiologia 199: 17–20.

Brancelj, A., 1992. *Alona sketi* sp.n. (Cladocera: Chydoridae), the second cave-inhabiting Cladoceran from former Yugoslavia. Hydrobiologia 248: 105–114.

Dumont, H. J., 1983. Discovery of groundwater-inhabiting Chydoridae (Crustacea: Cladocera) with the description of two new species. Hydrobiologia 106: 97–106.

Dumont, H. J., 1987. Groundwater Cladocera. A synopsis. Hydrobiologia 145: 169–173.

Dumont, H. J. & A. Brancelj, 1994. *Alona alsafadii* n.sp. from Yemen, a primitive, groundwater-dwelling member of the *A. karua*-group. Hydrobiologia 281: 57–64.

Dumont, H. J. & A. S. Negrea, 1996. A conspectus of the Cladocera of the subterranean waters of the world. Hydrobiologia 325: 1–30.

Petkovski, T. K. & D. Flössner, 1972. Eine neue *Alona*-art (Crustacea: Cladocera) aus dem Ohridsee. Fragm. balc. Mus. maced. Sci. Nat. 9: 97–106.

Sabater, F., 1987. On the interstitial Cladocera of the River Ter (Catalonia, NE Spain), with a description of the male of *Alona phreatica*. Hydrobiologia 144: 51–62.

Sket, B., 1994. Distribution of *Asellus aquaticus* (Crustacea: Isopoda: Asellidae) and its hypogean populations at different geographic scales, with a note on *Proasellus istrianus*. Hydrobiologia 287: 39–47.

Hydrobiologia **360**: 55–61, 1997.
A. Brancelj, L. De Meester & P. Spaak (eds), Cladocera: The Biology of Model Organisms.
©1997 *Kluwer Academic Publishers.*

Moina ephemeralis n.sp. from Central Europe

Igor Hudec

Institute of Zoology, Slovak Academy of Sciences, Löfflerova 10, SK-040 01 Košice, Slovak Republic
(e-mail: hudec@linux1.saske.sk)

Key words: Crustacea, Anomopoda, Moinidae, *Moina ephemeralis*, taxonomy, cladocerans

Abstract

Moina ephemeralis n.sp. is described from South Slovakia. It belongs to the *Moina gouldeni-lipini* group. Together with *Moina macrocopa* (Straus, 1820) it is the second species with a 2-egged ephippium from Central Europe. The species was recorded in the plankton of a highly eutrophic fish pond.

Introduction

Seven (5 native and 2 introduced) species of the genus *Moina* Baird, 1850 are currently known from Europe (*M. brachiata* (Jurine, 1820), *M. macrocopa* (Straus, 1820), *M. micrura* Kurz, 1874, *M. salina* Daday, 1888, and *M. lipini* Smirnov, 1976 – native; *M. affinis* Birge, 1893 and *M. weismanni* Ishikawa, 1896 – introduced). Of these *M. affinis* is only known from Italy (Margaritora et al., 1983).

During a two year survey (1994–1995) of zooplankton in Hrhovské carp ponds (located in the buffer zone of the Biospheric Reservation Slovak Karst), I found three moinid taxa (*Moina ephemeralis* n.sp., *M. weismanni* and *M. micrura*). *Moina ephemeralis* and *M. micrura* were found during a limited part of the year 1995 only.

Methods

Quantitative zooplankton samples were collected at bi-weekly intervals during 2 seasons (April 18–September 30 1994; April 4–October 28, 1995). Samples were taken using a No. 13 plankton net with vertical hauls taken from the bottom. Some characteristics of Small Hrhovský carp pond are given in Table 1.

Table 1. Mean values of some physical, chemical and biological parameters in open water of Small Hrhovský carp pond in production period (April–September) of 1994 and 1995.

Parameter	Unit	1994	1995
Water temperature	(°)	20.7	17.9
pH	–	7.4	7.3
Ca^{2+}	$(mg\,l^{-1})$	49.1	42.4
Mg^{2+}	$(mg\,l^{-1})$	3.7	7.6
NO_3^-	$(mg\,l^{-1})$	1.84	1.29
NO_2^+	$(mg\,l^{-1})$	0.09	0.08
NH_4^+	$(mg\,l^{-1})$	1.10	0.88
PO_4^{3-}	$(mg\,l^{-1})$	0.35	1.10
Chlorophyll-a	$(\mu g\,l^{-1})$	363.6	139.8
'cladocera'	$(N\,l^{-1})$	49.64	150.63
B. longirostris	$(N\,l^{-1})$	45.31	144.19

Moina ephemeralis **n.sp.**

Origin of name
The species was named after its short (ephemeral) life cycle in carp ponds.

Material examined
Type locality: Hrhov (S. E. Slovakia: 30° 15′ E, 30° 08′ W); small carp pond (17 ha, aproximal 10% of vegetation cover).

Records: May 1 1995 (21 parthenogenetic females); May 11 1995 (38 parthenogenetic, 22 ephip-

Table 2. The length characteristic of *M. ephemeralis* n.sp. and *M. micrura* from Small Hrhovský carp pond

Length [mm]	M. ephemeralis	M. micrura
Total production	$N = 129$	$N = 93$
mean	0.87	0.51
max	1.36	0.73
min.	0.47	0.34
Parthenogenetic		
females with eggs	$N = 38$	$N = 34$
average ± STD	1.01 ± 0.07	0.58 ± 0.06
Ephippial females	$N = 22$	
average ± STD	1.01 ± 0.04	
max	1.14	
min.	0.97	
males	$N = 12$	
average ± STD	0.90 ± 0.09	
min.	0.62	

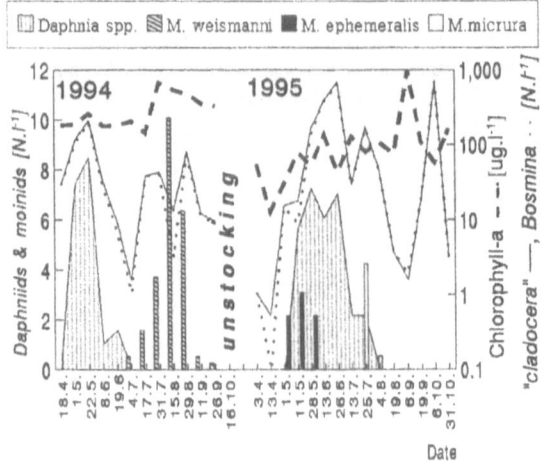

Figure 1. Seasonal dynamics and abundance of total 'cladocera', *Bosmina longirostris*, *Daphnia* spp. and moinids in Small Hrhovský carp pond during 1994–1995.

pial females and 12 males), and May 22 1995 (a few parthenogenetic, ephippial females and males).

Indication of type material and deposition
Holotype: one parthenogenetic female mounted in PVA (slide contains also one dissected specimen) deposited in the British Museum of Natural History (BMNH) in London (No. 1997.868)

Paratypes: one parthenogenetic and one ephippial females in one slide mounted in PVA (BMNH No. 1997.869); one ephippial female and one open

ephippium in one slide mounted in PVA (BMNH No. 1997.870). The series of 10 specimens (parthenogenetic and ephippial females and males in a tube (BMNH 1997.872–881).

Allotype: one adult male mounted in PVA (BMNH No. 1997.871). The slide also contains one dissected and one juvenile male.

A few individuals are in: Hungarian Natural History Museum – Budapest; Museum of Upper Nitra – Prievidza (Slovakia) and in my personal collection.

Diagnosis
Females with rounded head and supraocular depression, eye large, ocellus absent. Antennules long, slightly spaced and transversely ciliated in rows. First trunk limb with one anterior, bristled seta on penultimate segment and one on terminal segment. Postabdomen with 7–11 feathered teeth and one short bidented tooth. Terminal claw armed with two rows of setae. Two eggs in ephippium. Fine irregular cellular reticulation on ephippium surface.

Size: up to 1.36 mm.

Male with long, knee-bent antennule, sensory setae in the middle and 8 short hooks with semicircular arrangement at tip. First trunk limb with well-developed large hook on third segment; terminal segment with 2 feathered and one slightly hooked seta, exopod terminating in very long seta.

Size: up to 0.90 mm

Description
Female (Figures 2–8):
Head (Figure 2) broadly rounded, supraocular depression present; eye large, ocellus absent.

Antennules (Figure 3) long, well devolloped, slightly divergent (similar to *M. weismanni*, but longer). Sensory setae situated at middle of dorsal margin; surface densely ciliated in transverse rows.

Labrum (Figure 4) similar to that of *M. macrocopa*.

Antennae – setulation pattern of the moinid type; setae formula: 0-0-1-3/1-1-3; spine formula: 0-1-0-1/0-0-1; one spine on basipodite.

Valves (Figures 2–2d) without superficial bristles. Ventral margin with row of 30–32 short setae of similar length as the intersetal distance (Figure 2a). Behind these there is a row of minute, ungrouped setae, extending to the valve junction (Figure 2b). A pair of curved hooks, one on each valve is located near the valve junction (Figure 2c). The hooks are double (of the

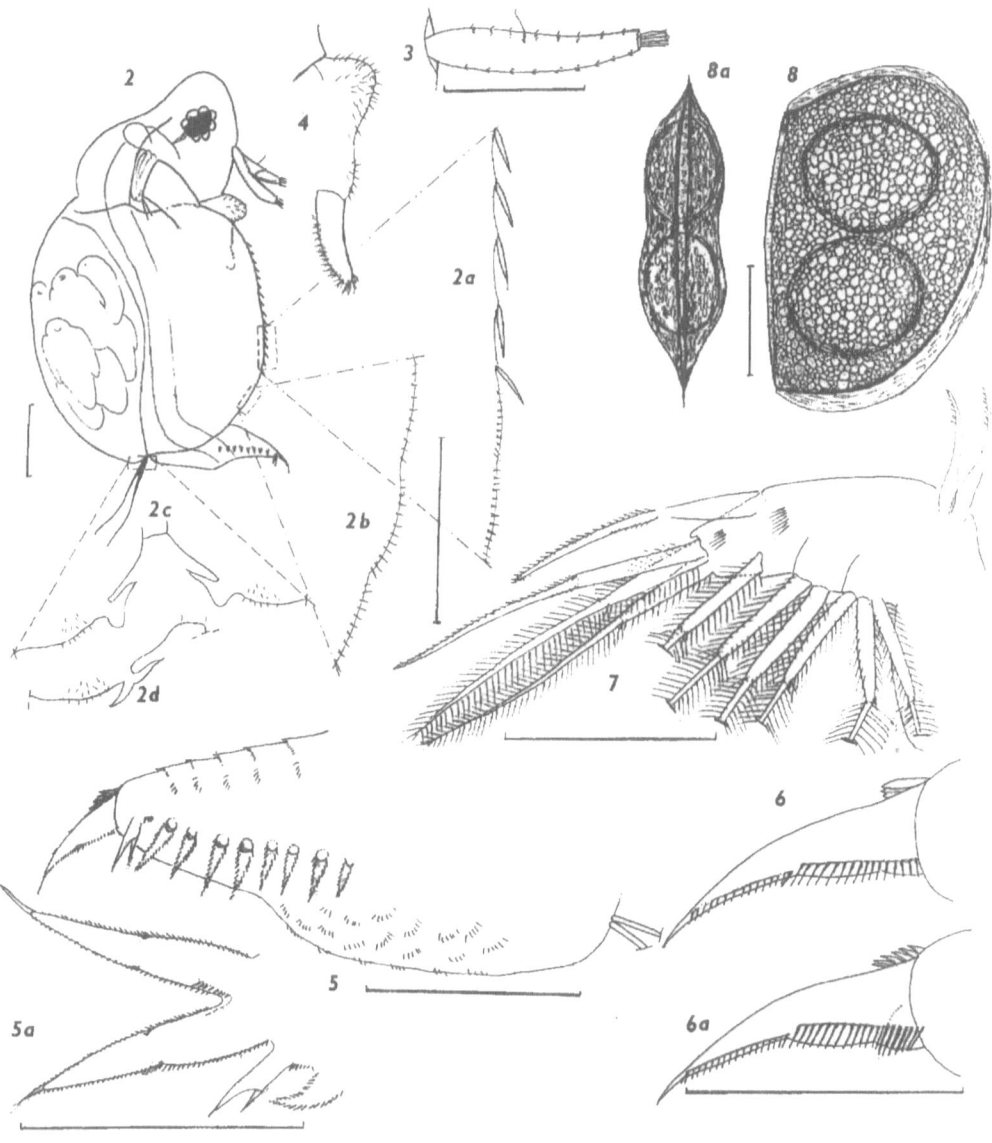

Figure 2–8. Moina ephemeralis n.sp. (female). 2: parthenogenetic female (lateral view). 2a–2b: details of valve margin. 2c–2d: valve hooks. 3: antennule. 4: labrum. 5: postabdomen (lateral view). 5a: postabdomen (distalmost paert, ventral view). 6: terminal claw (6: outer side; 6a: inner side). 7: 1-st thoracic limb. 8: ephippium (8: lateral view; 8a: dorsal view) The scale bar represents 0.1 mm.

M. macrocopa type) and positioned in opposite directions (Figure 2d). This character is also found in males (Figure 11).

Postabdomen (Figure 5) with 7–11 feathered teeth (mostly in different number on each side of postabdomen) and one short bidented tooth with unequal rami. Praeanal part with clusters of minute ciliae on the dorsal surface.

Terminal claw (Figure 6) relatively short (of the *M. macrocopa* type) with two rows of setae on the dorsal margin. The proximal group consists of slightly longer setae than the distal one. The inner side (Figure 6a) has a row of setae composed of 3 groups. The proximal group is situated submarginaly and the setae are the longest. The outer two groups of setae are very similar.

58

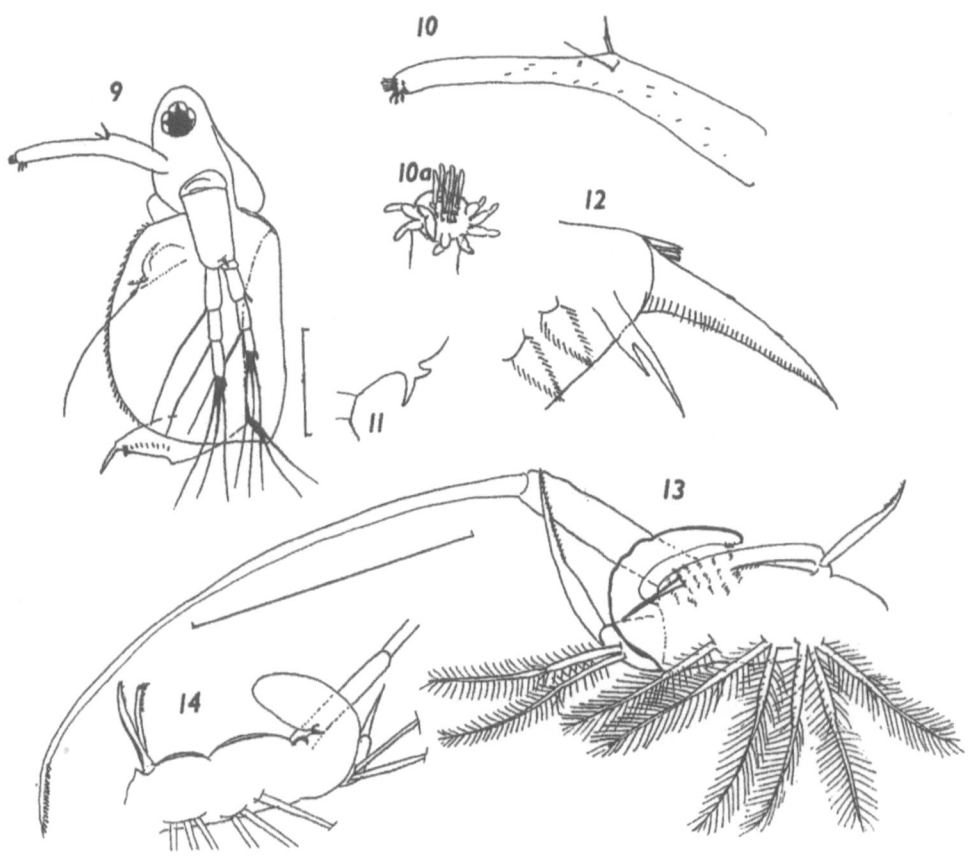

Figure 9–14. Moina ephemeralis n.sp. (male). 9: male (lateral view). 10: antennule. 10a: antennule (tip, detail). 11: valve hooks. 12: terminal claw with bidented tooth. 13: 1-st thoracic limb (adult male). 14: 1-st thoracic limb (juvenile male). The scale bar represents 0.1 mm.

First thoracic limb (Figure 7): first segment (proximal endite) with three two-segmented setae; second segment with two two-segmented setae; third segment with one two-segmented seta and one anterior short bristled seta; fourth segment with three setae of the same size (two two-segmented setae and one bristled seta).

I did not find any differences in 2-nd to 5-th thoracic limbs compared to other species of the genus.

Ephippium (Figures 8–8a): Mature ephippium with two eggs; irregular cellular reticulation on whole surface, less pigmented and sclerified than typical for moinids. Cellular reticulation not raised above the valve surface. Immature ephippium with diffuse cellular reticulation.

Male (Figures 9–13):
Head about one third of total body length, rounded, with a slight supraocular depression (Figure 9).

Antennules (Figures 10–10a) about half as long as total body length and knee-bent; two sensory setae situated at the knee (in the middle of the dorsal margin). Eight equal, short hooks at the tip composed in a semiring with the aesthetasts inside (Figure 9a).

First thoracic limb (Figure 13) thick with large hook. Terminal segement with 2 feathered setae originating from the base of a thicker and shorter, slightly bent nacked seta. The exopod terminates in a long seta, reaching the posterior valve margin.

Postabdomen as in female but setulation on terminal claw less clear (Figure 12).

A note on the biology of M. ephemeralis
A comparison of the size distribution of a population of *M. ephemeralis* and *M. micrura* from the same locality is given in Table 2, and the size frequency distribution of *M. ephemeralis* is given in Figure 15. For compar-

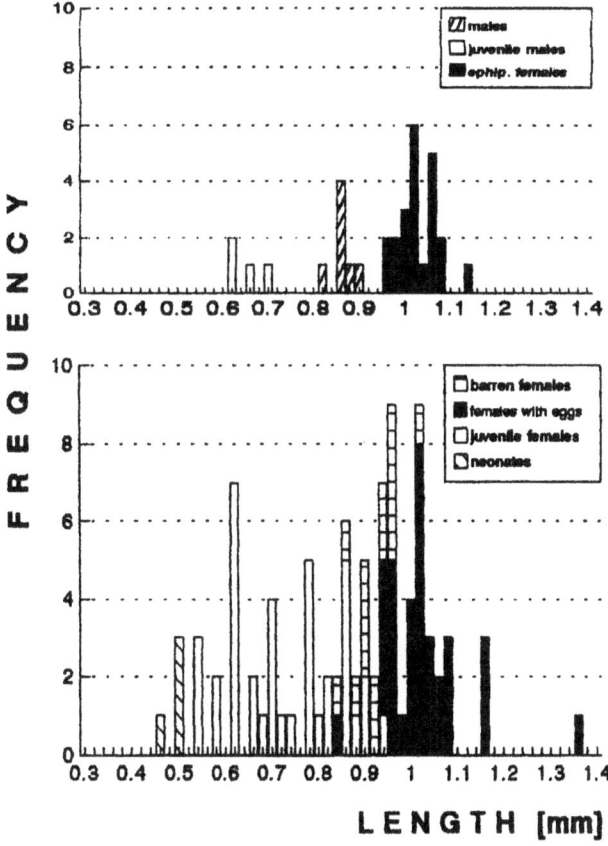

Figure 15. Size frequency distribution of sexual, reproductive stages, juveniles and parthenogenetic *M. ephemeralis* (May 11 1995).

ative measuremens of *M. microura* and *M. weismanni*, see Hudec (1990).

Moina ephemeralis is probably a limnetic species which co-occurs with *Bosmina longirostris* (O. F. Müller), *Daphnia galeata* Sars, *Ceriodaphnia pulchella* Sars. More information on this cladoceran community can be found in Timková and Hudec (1997).

Differential diagnosis
Moina ephemeralis belongs to a group of *Moina* species that has an ephippium with two eggs. Such an ephippium is known in nine moinids. Only *M. macrocopa* (Straus, 1820) and *M. lipini* Smirnov, 1976 are found in Europe.

The parthenogenetic female of *M. ephemeralis* resembles that of *M. weismanni* (Hudec, 1990) in: (1) head shape, (2) antennules, (3) setulation of the first trunk limb, and (4) the morphology of the terminal claws.

It is difficult to distinguish them without sexual females or males, as parthenogenetic females can easily be confused with *M. weismanni*.

Both species have the same supraocular depression on the head and two bristled setae on the penultimate and terminal segments of the first trunk limb.

Minute differences include:

M. ephemeralis: (1) long, slightly spaced antennules (Figure 3); (2) short setae (as long as intersetal distance) on ventral valve margin (Figure 2a); (3) long setae (Figure 6) near dorsal base of terminal claw (no spicules or teeth); (4) 3–4 fine flakes on the ventral base of terminal claw; (5) a relative short bidented tooth near the base of the ventral claw (Figure 5).

M. weismanni: (1) short stout, spaced antennules; (2) marginal valve setae shorter than intersetal distance; (3) basal pecten on terminal claw consisting of 15–21 fine long teeth; (4) 1–2 fine flakes on ventral base of the terminal claw; (5) relative long bidented tooth near the base of the ventral claw.

Table 3. Morphological differences between *Moina macrocopa* Straus , *M. ephemeralis* n.sp., *M. gouldeni* Mirabullaev and *M. weismanni* Ishikawa

Character	M. macrocopa (Goulden, 1968)	M. ephemeralis n.sp.	M. gouldeni (Mirabdullaev, 1993)	M. weismanni (Hudec,1991)
Females				
Supraocular depresion	Absent	Present	Present	Present
Superficial setules on head and valve	Present	Absent	Absent	Absent
Antennules	long spindlike	long slightly spaced	stout spindlike	stout, spaced apart and lateral
Valve Shell				
Marginal spine	Long	Short	Short	Short
spine length to interspine distance	=2	=1	≥1	<1
1st thoracic limb				
Anterior seta (3rd segment)	Toothed	Bristled	Bristled	Bristled
One terminal seta	Toothed	Bristled	Bristled	Bristled
Postabdomen				
Feathered teeth	7–10	7–11	5–8	7–9
Claws setulation	2 rows	2 rows	2 rows	1 fine pecten
Flakes on ventral base	1–2	4	3	2
Ephippium				
No. of eggs	2	2	2	1
Reticulation	Polygonal	Iregular cellular	?	Polyglonal and raised knobs in central part
Sclerotisation	Heavy	Weak	?	Heavy
Pigmentation	Brown	Brownish	?	Brown
Max size [mm]	1.5	1.36	1.47	0.95
Males				
Antennal hooks	6 short	8 short	4 longer & 2 short	4 short
1st thoracic limb terminal segment (feathered setae)	1	2	2	1

Ephippial females (number of eggs, ephippial reticulation) and males (antennular shape, shape of 1-st thoracic limb) are quite different (Table 3).

M. ephemeralis and *M. macrocopa*. Both have two eggs in the ephippium and a similar terminal claw with setulation on the ventral margin and a relatively short bidented tooth, but a quite different setular armature of the first thoracic limb (one seta on penultimate and one on the terminal segment).

Parthenogenetic female are clearly distinguishable by:

M. ephemeralis: (1) presence of supraocular depression (Figure 2); (2) absence of bristles on the head and valve surface; (3) short setae on the ventral valve margin; (4) two bristled setae on penultimate and terminal segments of first trunk limb (Figure 7).

M. macrocopa: (1) supraocular depression mostly absent; (2) presence of bristles on the head and valve surface; (3) long setae on the ventral valve margin (twice as long as intersetal distance); (4) two thick toothed setae on penultimate and terminal segments of first trunk limb (Goulden, 1968).

The two egged ephippia of both species have different surface reticulation. While *M. ephemeralis* has a small, irregular cellular reticulation (Figure 8), *M. macrocopa* has a characteristic polygonal reticulation on the ephippium surface. Morphological differences in the males are minute and are limited to the number of terminal hooks on the antennules (Table 3).

M. ephemeralis and *M. gouldeni* belongs to the same group. They have a very similar general shape (including a supraocular depression), setal armature of the 1st thoracic limbs and short setae on the ventral valve rim in females and males.

M. gouldeni has: (1) stout spindle-like antennules; (2) 3 fine flakes on the ventral base of terminal claw (Mirabdullaev, 1993).

M. lipini probably also belongs to this group, because of its similar 1-st thoracic limb and 2-egged ephippia. But its original description and drawings (Smirnov, 1976) are too brief for a detail comparision.

The ephippial reticulation cannot be compared. It was not described for *M. gouldeni* nor for *M. lipini*.

All three species occur in fish ponds and related habitats. *M. gouldeni* and *M. ephemeralis* were found in fish ponds, while *M. lipini* was found in a fish way that directly connects to fish ponds of the Moscow region (Smirnov, 1976).

Acknowledgments

I thank the Foundation IUCN Slovakia (Bratislava) who partially sponsored the Hrhovské fish ponds survey through the project 'Enviromental and economical evaluation of functions and fish pond management in Slovakia'.

References

Goulden, E., 1968. The systematics and evolution of the Moinidae. Trans. am. phil. Soc. 58, 101 pp.

Fulín, M., I. Hudec, E. Sitášová, V. Slobodník & P. Sabo, 1995. Enviromental and ecological evaluation of functions and fish pond management in Slovakia. European program IUCN. Foundation IUCN Slovakia. Komprint, Bratislava, 72 pp. (in Slovak).

Hudec, I., 1990. *Moina weismanni* Ishikawa, 1896 (Cladocera, Moinidae) in Central Europe. Hydrobiologia 190: 33–42.

Margaritora, F. G., I. Ferrari & D. Crosetti, 1983. A Far East *Moina, M. wesmanni* Ishikawa, 1896 found in an Italian ricefield. Hydrobiologia 145: 93–103.

Mirabdullaev, I., 1993. *Moina gouldeni* n.sp. (Cladocera, Moinidae) from Central Asia. Crustaceana 64: 192–196.

Smirnov, N., 1976. Macrothricidae i Moinidae fauni mira. Fauna SSSR, Leningrad 1, 3, 237 pp. (in Russ.).

Timková, J., I. Hudec (1997). Zooplankton of Hrhovské fish ponds. Natura Carpatica 38: 53–62 (in Slovak).

Hydrobiologia **360**: 63–73, 1997.
A. Brancelj, L. De Meester & P. Spaak (eds), Cladocera: The Biology of Model Organisms.
©1997 *Kluwer Academic Publishers.*

A checklist of the littoral cladocerans from Mexico, with descriptions of five taxa recently recorded from the Neovolcanic Province

Manuel Elías-Gutiérrez, Jorge Ciros-Pérez*, Martha Gutiérrez-Aguirre &
Adrián Cervantes-Martínez
*Laboratorio de Zoología. UNAM Campus Iztacala AP 314, CP 54000. Los Reyes Iztacala, Tlalnepantla,
Edo. de México, Mexico*
*Department de Microbiologia: Ecologia (Edifici D'Investigacioni) Universitat de Valencia, E-46100 Burjassot
(Valencia), Spain*

Abstract

By 1996 an inventory of Mexican cladocerans had recorded 48 species of which 22 belonged to Chydoridae, Macrothricidae and Ilyocryptidae. Unfortunately, most of the surveys were made before researchers knew that these crustaceans are not entirely cosmopolitan. For this reason and the lack of deposited type material, many of these records are doubtful and need detailed analyses. In this study, material from 18 water bodies located in the Neovolcanic Province of Mexico is analyzed and compared with the literature. Also, males of *Camptocercus dadayi* Stingelin and *Leydigia leydigi* (Schoedler) and new records *Eurycercus longirostris* Hann and *Biapertura intermedia* (Sars) are described. Half of the total number of species recorded are American endemics and represents a mixture of the North and South American fauna, suggesting that Mexico constitutes a transition between Nearctic and Neotropical zones.

Introduction

The cladocerans of Mexico are little known, and no single work has been devoted to the study of the littoral fauna belonging to the Anomopoda. A summary of most cladoceran papers published from different regions is given in Table 1. Most of the surveys have been done in the pelagic zones of lakes and reservoirs where Macrothricids, Ilyocryptids and Chydorids are caught only occasionally. By 1996, the inventory of these families recorded only 22 species. So the purpose of this review is to establish the current state of knowledge of these animals in Mexico.

Study area

Surveys were conducted in 18 water bodies including artificial reservoirs, puddles and natural lakes, located in the North (Panuco River Basin) and Central (Lerma River Basin) regions of the State of Mexico. All sampling sites are located between 98° 37′–100° 38′ W and 18° 22′–20° 17′ N, at more than 2000 m above sea level (Figure 1).

Material and methods

Samples from each water body were collected by moving a conical net (50 μm mesh size with a metal handle of 1.5 m long), through growths of macrophytes and over the substrates of the littoral zones. All samples were preserved with 4% sugar-formalin. Sampling was carried out intermittently during the period 1993–1995. The animals were analyzed and dissected under a stereomicroscope and identified using a compound microscope. Standard literature (e.g. Smirnov, 1971, 1992) and comparison with original descriptions were used. Drawings were made with a camera lucida attached to a Nikon Labophot II compound microscope. The material was deposited in the Reference Collection at Laboratorio de Zoología Campus Iztacala (Universidad Nacional Autónoma de México).

Table 1. Published data on mexican cladocerans, 1915–1996. The number into parenthesis corresponds to littoral cladocerans.

LOCALITIES	Number of species		Source
1.	3 Localities near to Mexico City	14(10)	Juday, 1915
2.	Some cenotes of Peninsula de Yucatán	10(1)	Wilson, 1936
3.	Lake of Pátzcuaro, Michoacán	4(1)	Uéno, 1939
4.	Lake of Pátzcuaro	9(3)	Rioja, 1940
5.	San Felipe Xochiltepec Lake, Puebla	7(3)	Rioja, 1942
6.	Valles, San Luis Potosí	2(1)	Osorio-Tafall, 1943
7.	10 Localities in several places	7(0)	Brehm, 1955
8.	6 Localities in Southeast Mexico	25(11)	Van de Velde et al., 1978
9.	Catemaco lake	2(1)	Suárez-M. et al., 1986
10.	Pátzcuaro lake	6(1)	Chacón et al., 1991
11.	Reserve of Sian Ka'an	12(2)	Suárez-M. & Elías-G., 1992
12.	J. A. Alzate Dam, Mexico	6(1)	Suárez-M. et al., 1993
13.	13 Reservoirs in State of Mexico	28(11)	Elías-G., 1995
14.	17 Localities in North Mexico	26(12)	Rodríguez-A. & Leija-T., 1995
15.	Several localities in State of Mexico and Aguascalientes	1(1)	Ciros-P. et al., 1996
16.	10 Localities in State of Mexico	16(10)	Ciros-P. & Elías-G., 1996
17.	19 Reservoirs and ponds in State of Aguascalientes	31(9)	Dodson & Silva-Briano, 1996

Results and discussion

An updated list of the littoral cladocera known from Mexico is given in Table 2. The total number of taxa found is 48. The probability of occurrence of about 15% of these taxa in Mexican waters is remote according to recent standard literature (see Table 2). Therefore, they need a revision to establish their identity and affinities with conspecific taxa. For example, *Dunhevedia setigera* is recorded in two old publications. One of the records consisted of a single specimen (Rioja, 1940) and there is no material available anymore. Actually, this record has been considered a synonymy of *D. crassa* King, 1853 by Smirnov (1971). *Kurzia longirostris* (Daday, 1898) was recorded by Van de Velde et al. (1978), but there is no material available to check its status. The occurrence of this taxon in America is unlikely if the non-cosmopolitanism idea of Frey (1995) is taken in account. In total, about 51% of the taxa from Mexico should be reconfirmed by revision of the extensive samples and comparison of the material with type material and original descriptions.

Table 3 shows the results of all the species recorded in the present survey. Five taxa of special interest were found. They are described and discussed below.

Ilyocryptus sp.
(Figures 2–8)

Material examined: 35 parthenogenetic females from Laguna El Sol collected on June 4, 1994.

Length: 0.56–0.72 mm. Body subquadrangular with prominent postero-ventral corner. Granulate valve surface. Posterior margin almost straight. Incomplete molting with 5 to 7 previous carapaces (Figure 2). Ventral setae feathered and posterior ones bifurcated (Figures 6–7). First antenna bisegmented with a short distal segment ornamented with 4 circumspherical rows of spines. Second antenna robust, with spines 0-1-0-1/0-0-1 and setae 0-0-0-3/1-1-3. Seta from the second segment of the endopod with long and thin setules on the proximal segment (Figure 5). Postabdomen large, with the ventral margin almost straight and armed with some distal spinules. Dorsal margin curved, with one lobe, and with strong and long marginal spines diminishing in length proximally. Lateral spines followed by a row of small spinules. Anus opening in the distal

Table 2. Littoral species known from Mexico until 1996. x = well documented species (good descriptions and deposited material available); ? = uncertain species (needs revision, lack of abundant deposited material) and 0 = species unlikely to be found in Mexico (needs revision, no material available, species restricted to other region according to the geographical distribution given in the literature). * = found in State of Mexico. Source is according to Table 1.

Taxa	Source	Taxonomic Status
Family Macrothricidae		
Macrothrix hirsuticornis Norman & Brady, 1867	14	?
*M. laticornis (Fischer, 1851)	7, 13, 14, 17	0
*M. mexicanus Ciros, Silva & Elías, 1996	15, 13, 17	x
M. rosea (Jurine, 1820)	1, 14, 11	?
*M. triserialis (Brady, 1866)	16	?
*Macrothrix n. sp. Ciros & Elías 1996		?
Family Ilyocryptidae		
*Ilyocryptus agilis Kurz, 1878	13	?
*I. spinifer (Brady, 1886)	1, 8, 14	x
Family Chydoridae		
Pleuroxus sp.	17	?
*Pleuroxus denticulatus Birge, 1879	1, 8	x
P. aduncus (Jurine, 1820)	13	?
*Alonella excisa (Fischer, 1854)	16	x
*Dadaya macrops (Daday, 1898)	13	x
*Disparalona hamata (Birge, 1879)	14, 16, 17	x
*Chydorus brevilabris Frey, 1980	16	x
*C. cf eurynotus	16	?
C. eurynotus Sars, 1901	8	?
C. sphaericus O. F. Müller, 1785	1, 4, 5, 8, 10, 14	0
C. cf sphaericus	13	?
Chydorus. sp.	14, 17	?
*Pseudochydorus globosus (Baird, 1893)	1, 17	x
*Ephemeroporus acanthodes Frey, 1982	16	x
*E. hybridus Daday 1905	16	x
Dunhevedia crassa King, 1853	17	x
Dunhevedia setigera King, 1853	1, 4	0
Alona circumfimbriata Megard, 1967	14	?
*A. cf rectangula	13	?
A. costata Sars, 1901	1, 4, 5	?
*A. diaphana King, 1853	8	x
A. eximia Kiser, 1948	8	?
A. guttata Sars, 1872	8	?
A. monocantha Sars, 1901	8	?

Table 2. Continued

Alona sp.	6, 17	?
*Biapertura affinis (Leydig, 1860)	16	x
*B. cf. pseudoverrucosa	13	0
B. karua (King, 1853)	14	?
Oxyurella tenuicaudis (Sars, 1862)	2, 11	?
*Acroperus harpae (Baird, 1834)	13	?
Eurycercus lamellatus (O. F. Müller, 1785)	1	0
Euryalona orientalis (Daday, 1898)	8	?
*Camptocercus dadayi Stingelin, 1913	16	x
*C. cf rectrirostris	13	0
Graptoleberis testudinaria (Fischer, 1848)	1	?
*Leydigia acanthocercoides (Schoedler, 1862)	16, 14	x
*L. leydigi (Schoedler, 1862)	13	x
L. quadrangularis (O. F. Müller, 1785)	14	?
*Kurzia latissima (Kurz, 1874)	1, 14, 17	x
K. longirostris (Daday, 1898)	8	0

region of postabdomen, between the first and fifth marginal spines. Postabdominal claw slightly curved, the concave margin with two long basal spines, subequal in length, and a row of three groups of very small setules. The convex margin with a proximal spine, about half the length of the basal spines.

On the basis of these features, this taxon strongly resembles *I. gouldeni* Williams, 1978. After a detailed analysis, however, it was found to differ in the ornamentation of the natatorial setae of the antenna, the spines on the basipod, the absence of special spinulation surrounding the anal aperture, the morphology of the postabdominal claw and its armature and the armature of the posterior region of shell. Considering the position of the anus and associated spinulation (Stifter, 1991), this taxon could be related to the *I. acutifrons* species complex.

Eurycercus longirostris Hann, 1982
(Figures 9–13)

Material examined: 15 juvenile females from Ignacio Ramírez reservoir collected on February 26, 1994.

Total length: 0.64 to 0.94 mm. The maximum height is 55 to 65% of the total length. Dorsal margin smooth, without keel. Head pore large (about 25 μm) and located on a bulb-like projection. Ventral margin almost straight, with about 100 short and plumosae setae. Posterior margin of the valve with short setules. Shell with no noticeable ornamentation. Rostrum long

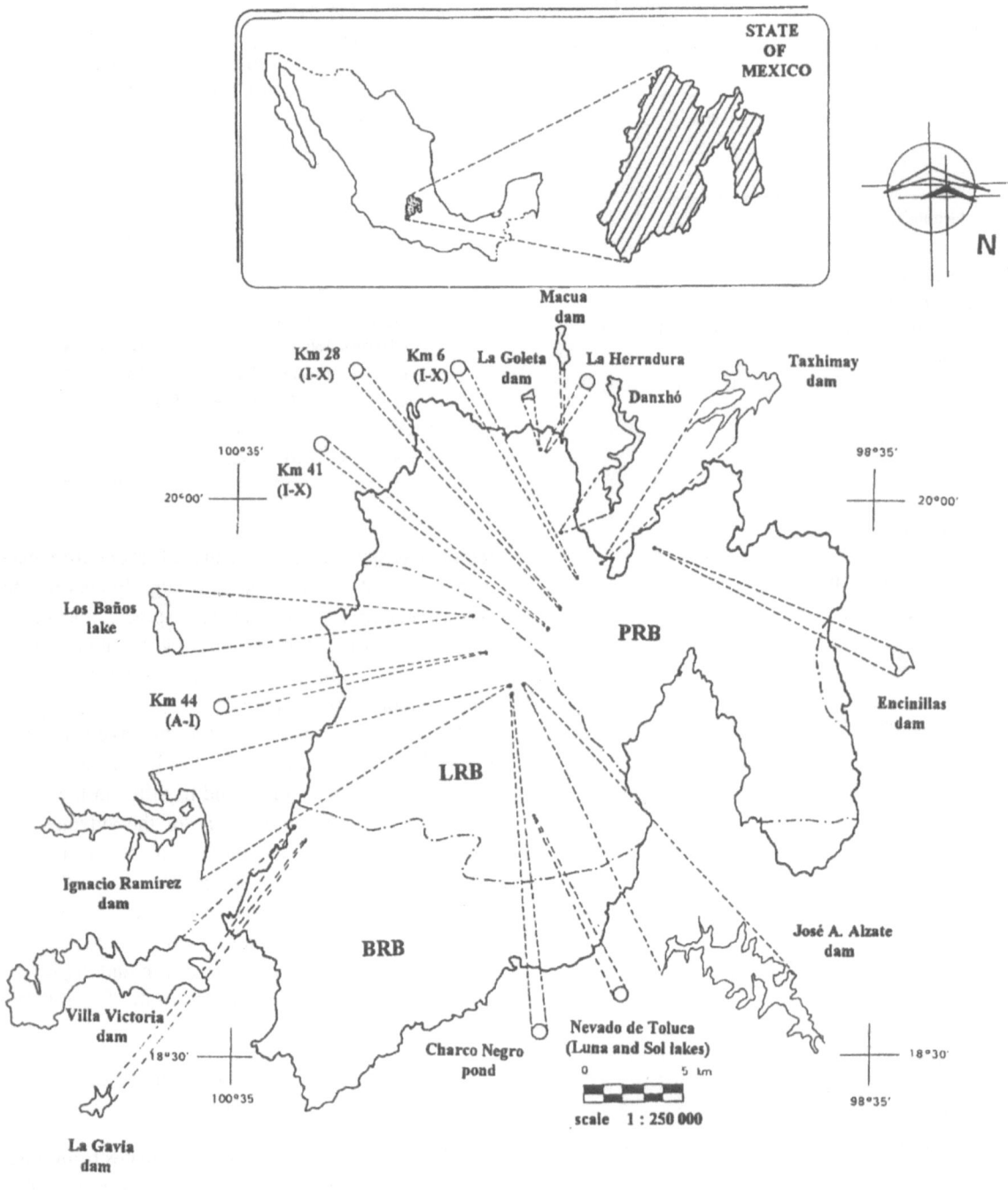

Figure 1. Location of sampling sites in the State of Mexico. PRB = Panuco River Basin. LRB = Lerma River Basin. BRB = Balsas River Basin. I-X = Ixtlahvaca-Jilotepec highway. A-I = Atlacomulco-Ixtlahvaca highway. Scale bar applies to large reservoirs Taxhimay, Macua, La Goleta, Los Baños, Ignacio Ramírez, Villa Victoria, La Gavia, José A. Alzate. Remaining systems are not at scale. State of Mexico is at 1: 1 600 000 scale.

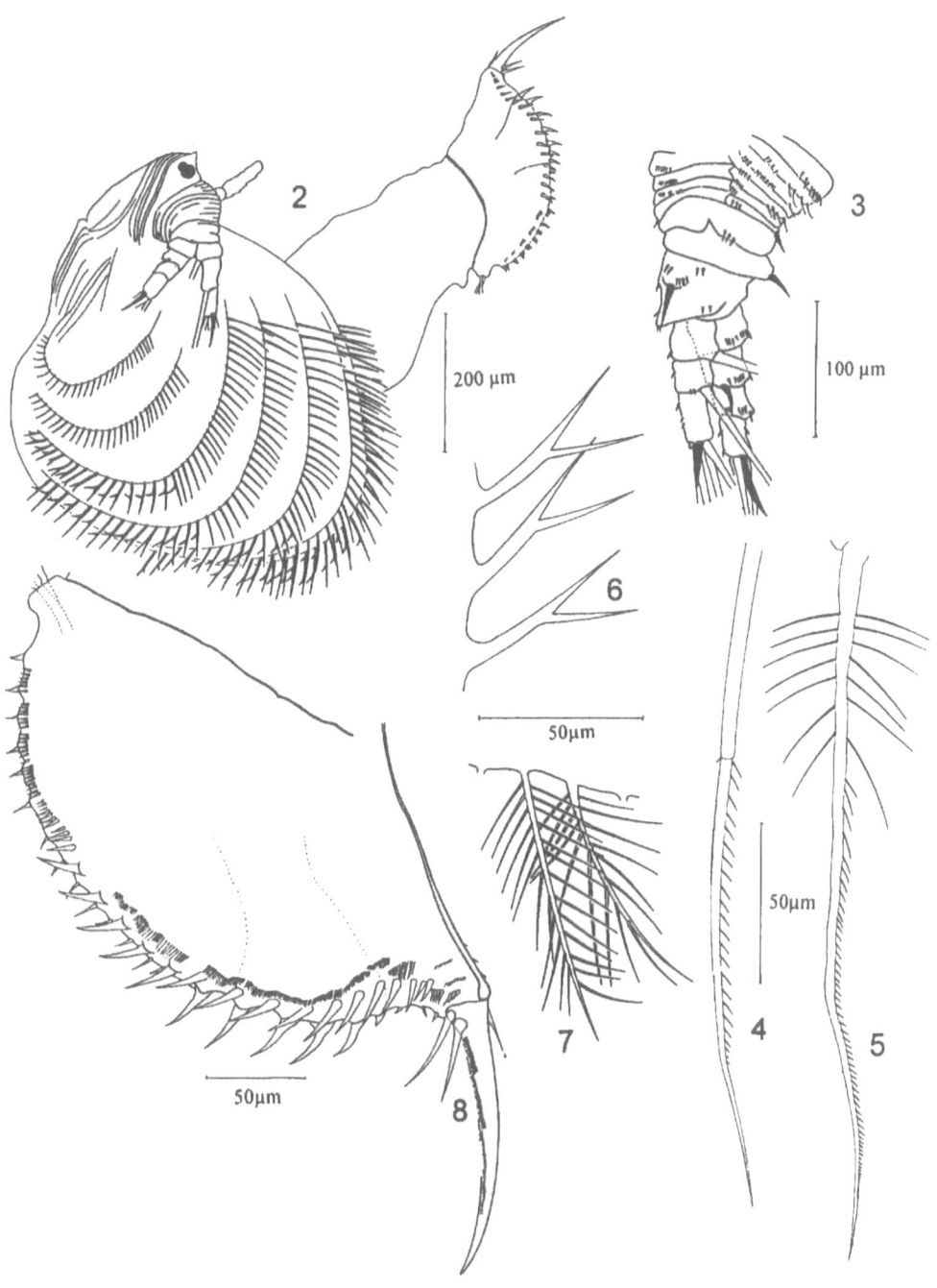

Figures 2–8. Ilyocryptus sp. parthenogenetic female. 2: Habit, lateral view. 3: Antenna. 4: Natatorial seta from second segment of the exopod. 5: Natatorial seta of the first segment of the exopod. 6: Setae from posterior margin of the valve. 7: Setae from the ventral margin of the valve. 8: Postabdomen, lateral view.

and attenuated. Total cephalic length 370–490 μm. Postabdomen with a row of 76–86 marginal teeth, about 215 to 310 μm in length. Postabdominal claw with two basal spines, concave margin armed with 21 to 31 teeth. Total length of claw 45–90 μm. Inner distal lobe (IDL) of trunklimb I with 3 clasping hooks, the middle one strongly chitinized and fused with the IDL. The setules of this lobe are fused in the distal region

Table 3. Species found in fresh water bodies in the State of México where, A = Encinillas dam, B = Taxhimay dam, C = Danxhó dam, D = La Herradura pond, E = Macua dam, F = La Goleta dam, G = km 6 (I–X) pond, H = km 28 (I–X) pond, I = km 41 (I–X) pond (Panuco River Basin); J = José A. Alzate dam, K = Ignacio Ramírez dam, L = Los Baños lake, M = Luna lake, N = Sol lake, O = Charco Negro pond, P = km 44 m (A–I) pond (Lerma River Basin); Q = Villa Victoria dam and R = La Gavia dam (Balsas River Basin).

Taxa	A	B	C	D	E	F	G	H	I	J	K	L	M	N	O	P	Q	R
Family Macrothricidae																		
Macrothrix laticornis (Fischer, 1851)	+		+	+														
M. mexicanus Ciros, Silva & Elías, 1996	+		+							+	+							
M. triserialis (Brady, 1866)							+									+		
M. n.sp. Ciros & Elías, 1996								+										+
Family Ilyocryptidae																		
Ilyocryptus agilis Kurz, 1878		+			+			+										
I. spinifer (Brady, 1886)	+	+						+										
Ilyocryptus sp.														+				
Family Chydoridae																		
Eurycercus longirostris Hann, 1982											+							
Pleuroxus denticulatus Birge,1879		+					+	+	+		+					+		+
Alonella excisa (Fischer, 1854)							+	+										
Disparalona hamata (Birge,1879)				+														
Chydorus brevilabris Frey,1980		+		+			+	+	+		+					+		
C. cf. eurynotus																+		+
Pseudochydorus globosus (Baird,1893)					+			+	+									
Ephemeroporus acanthodes Frey, 1982							+									+		
Alona cf. *rectangula*	+	+	+	+	+			+			+	+				+		
A. diaphana King, 1853								+				+				+		
Alona cf. *setulosa*		+		+				+	+		+	+	+			+		
Biapertura affinis (Leydig, 1860)								+					+					
B. intermedia Sars, 1862														+				
Acroperus harpae (Baird, 1834)					+													
Camptocercus dadayi Stingelin, 1913										+	+					+		
Leydigia acanthocercoides (Schoedler, 1862)				+														
L. leydigi (Schoedler, 1862)	+		+		+						+	+						
Kurzia latissima (Kurz, 1874)																+		

forming a type of cutting membrane as described by Hann (1982); median surface of IDL carrying 4 groups of clasping setules, the distal and proximal groups with 3–5 members, the marginal group with 2–3 and the basal group with 5–7. On the second trunk limb, the teeth on the scratching setae are sharply pointed. The gnatobase has three terminal setae, the middle one short, wide and armed with 6 denticles; the two lateral ones are larger and feathered distally. Adjacent to these setae, there is a short, stout spine narrowing distally.

E. (Bullatifrons) longirostris has previously been recorded only in eastern USA (Hann, 1982). Although only juveniles were found in the present survey, the original description took the whole ontogeny of the species into account. All the characters of the juveniles found by us match with the original description. The only species of this genus recorded near Mexico City by Juday (1915) was *E. (Eurycercus) lamellatus*, a taxon geographically restricted to Europe (Frey, 1975). So, we suspect that Juday's specimens could belong to the species described here.

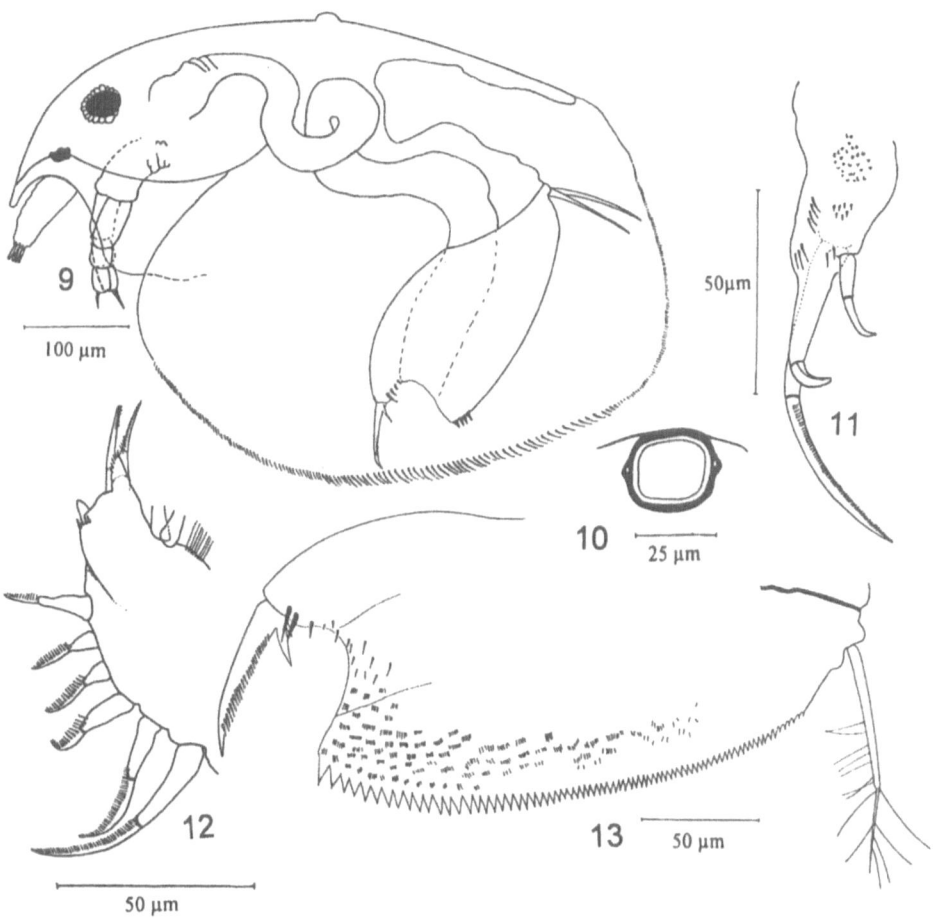

Figures 9–13. Eurycercus longirostris, female. 9: Habit, lateral view. 10: Head pore. 11: Trunk limb I, inner distal lobe (IDL). 12: Part of Trunk limb II. 13: Postabdomen, lateral view.

Camptocercus dadayi Stingelin, 1913
(Figures 14–18)

Material examined: 10 males, 4 parthenogenetic females and 10 ephippial females from km 44 Toluca-Atlacomulco highway collected on January 28 and July 7, 1994.

Male: Total length: 0.50–0.55 mm. Body elongated but smaller than that of the parthenogenetic female. Ventral margin slightly convex or almost straight, with 60–70 feathered setae followed by a row of small submarginal setae running from the distal third to the posterior margin. Valves with parallel grooves or furrows on dorsal and ventral areas. Rostrum truncated and wide. First antenna long, with a sensory seta inserted near the middle, and with an accessory, long subapical seta. Postabdomen long (about 50% of total body length), narrowing distally. Postanal region long

(postanal length/maximum height: 3.5), with 14–15 groups of marginal setae, each with 5 to 9 setae; distal group of 3 small setules. Eleven lateral fascicles of setae, similar in arrangement to the marginal ones. First trunk limb with a copulatory hook bearing two small crescent ridges near the tip. Genital pore opening in a ventral notch, close to the distal portion of postabdomen. Claw with 10 denticles on the concave margin, diminishing in length proximally, followed distally by a row of fine setules. Basal spine on claw often serrated on one or both sides, sometimes smooth.

Parthenogenetic females were previously recorded and re-described by Ciros-Pérez & Elías-Gutiérrez (1996) from an adjacent site to the pond surveyed in the present study.

Ephippial female: Total body length 0.76 mm. Ephippium elongated, its length being almost twice

70

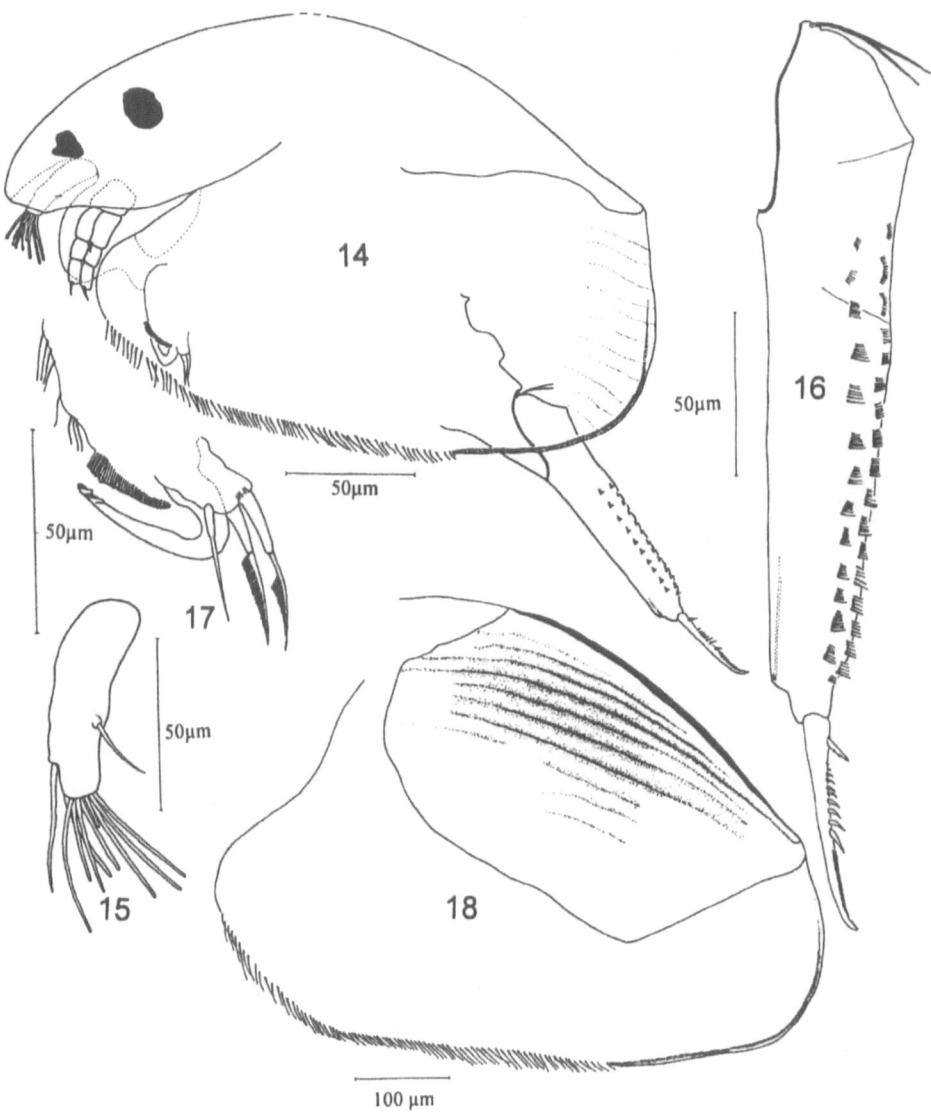

Figures 14–18. Camptocercus dadayi, male and ephippium. 14: Habit, lateral view. 15: Antennule, lateral view. 16: Postabdomen, lateral view. 17: Part of trunk limb I. 18: Ephippium.

its width. Bearing longitudinal and parallel grooves, deeper on the dorsal region.

This species was first described from South America as *C. australis dadayi* Stingelin, 1913. On the basis of two parthenogenetic females, Smirnov (1971) considered it to be a variety of *C. lilljeborgi*, also with a South American distribution. Finally Rey & Vasquez (1986) placed it as a true species and re-described the parthenogenetic female. Males were unknown until the present study.

Leydigia leydigi (Schödler, 1862)
(Figures 19–22)

Material examined: 20 males and 20 ephippial females from Laguna Los Baños, collected on January 28, 1994.

Male: Total length 0.45–0.55 mm. General shape of body similar to that of parthenogenetic female. Posterodorsal corner almost at level of maximum body height. Inner distal lobe of trunk limb I with 3 setae, subequal in length. Distal seta hook-like, stronger than the others. Copulatory hook long and robust with 3

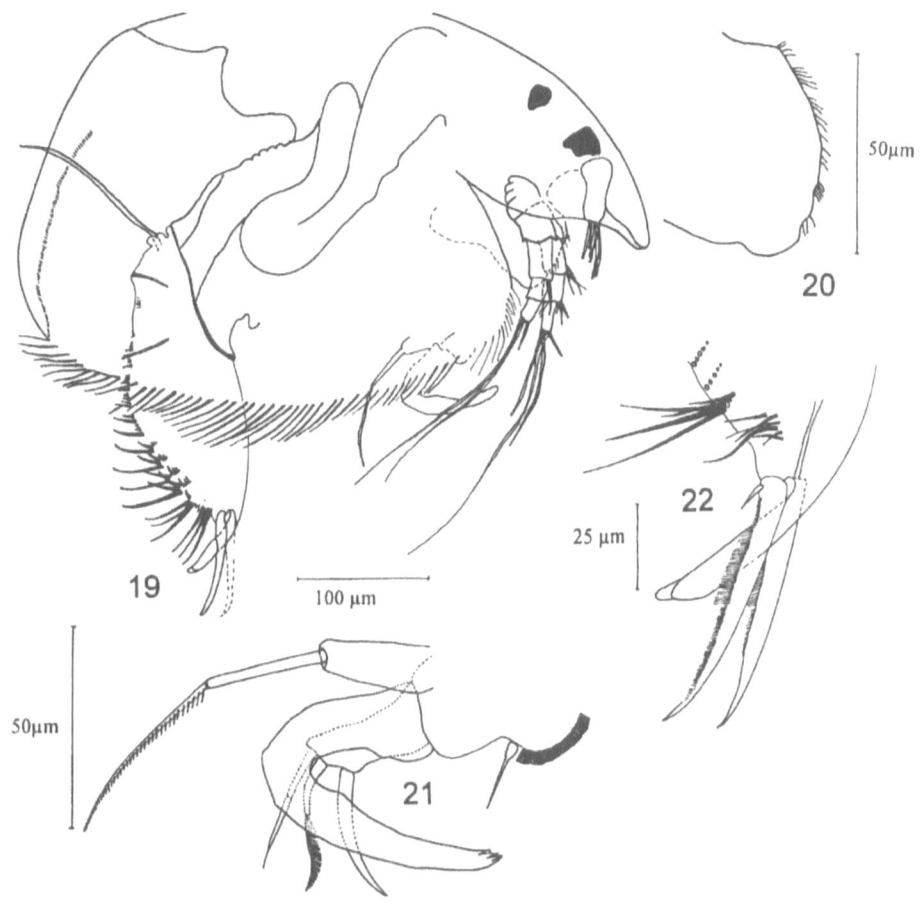

Figures 19–22. Leydigia leydigi, male. 19: Habit, lateral view. 20: Labral plate. 21: Part of trunk limb I, IDL and ODL. 22: Distal part of postabdomen and claw, lateral view.

crescent ridges near the tip. Male seta with an enlarged base and tapering distally. Postabdomen with a projection between claws resembling a penis, the vas deferens flows through it.

Smirnov (1971) pointed out that *L. leydigi* is holarctic and tropical in distribution, with records in North and South America. The first record of the species in Mexico was from a nearby reservoir (Elías-Gutiérrez, 1995, 1996), but only parthenogenetic females were described.

Biapertura intermedia (Sars, 1862)
(Figures 23–29)

Material examined: Numerous parthenogenetic females from lake El Sol, collected on June 4, 1994.

Total length from 0.48 to 0.5 mm. Body oval with rounded posterior margin. Maximum height is 65–72%

of total length. Ventral margin with 40 to 45 feathered setae, the first 5 to 6 longer than the remaining ones. Posterior margin of the head shield rounded with two main pores connected with a channel. Labral keel large and smooth, with a pointed apex. Antennule with nine aestethascs and a sensory seta arising from the distal third. Postabdomen rounded, widening distally; ventral side strongly chitinized, slightly convex, with three groups of minute denticles; preanal region somewhat longer than the postanal; marginal spines relatively small. Lateral surface with about 15 lateral fascicles, with the distalmost one or two projecting beyond the postabdominal margin. Claw with setules on concave side; basal spine 1/4 length of the claw, with a group of basal spinules decreasing in size proximally. Inner distal lobe with three setae, two of them similar in length and distally setulated, the remaining one short

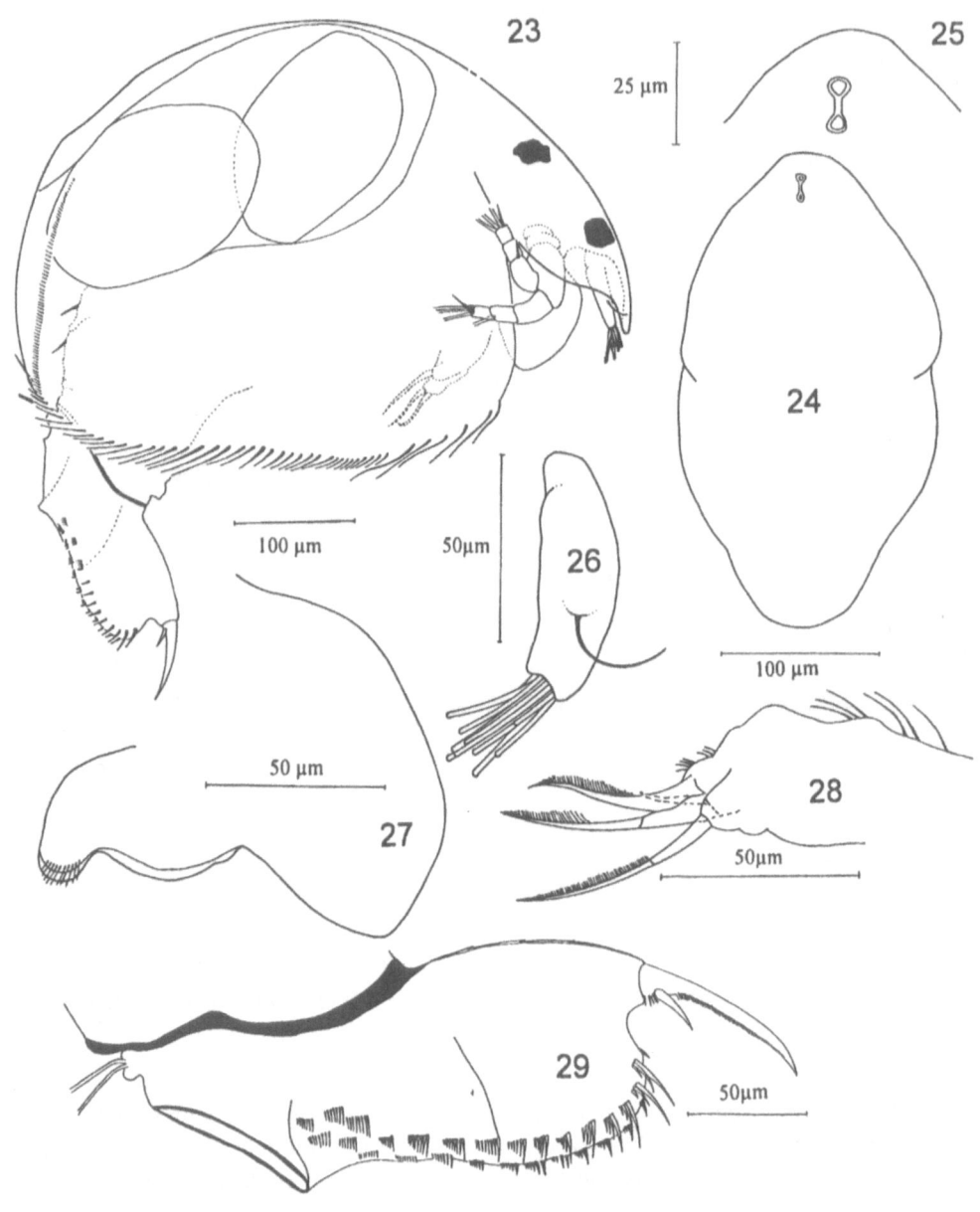

Figures 23–29. *Biapertura intermedia* parthenogenetic female. 23: Habit, lateral view. 24: Head shield, dorsal view. 25: Head pores. 26: Antennula, lateral view. 27: Labrum. 28: Part of trunk limb I, IDL and ODL. 29: Postabdomen, lateral view.

and hook-like. Exopod with a bisegmented seta, setulated on one side of the distal margin.

This species has been found in South America (Sars, 1901) and North America (Smirnov, 1971), and is here reported for the first time in Mexico.

Conclusions

There is a need to increase studies on littoral cladocerans from Mexico. Almost half of the already known records require reconfirmation based on detailed analyses of large gamogenetic populations (with parthenogenetic, sexual females and males in all possible developmental stages). The present study shows that the mexican cladoceran fauna is a mixture of North Amer-

73

ican and South American taxa, which emphasizes the transitional biogeographical location of the country. Males, bearing specific characters, are described for the first time in the fauna of this region.

Acknowledgments

We thank the Comisión Nacional para el Estudio y Conocimiento de la Biodiversidad en México (CONABIO) for partially supporting this study (Grant H-112).

References

Brehm, V., 1955. Mexikanische Entomostraken Öster Zool. Zeitschr 6: 412–420.

Ciros-Pérez, J. & M. Elías-Gutiérrez., 1996. Nuevos registros de Cladoceros (Crustacea: Anomopoda) en México. Rev. Biol. Trop. 44: 297–304.

Ciros-Pérez, J., M. Silva-Briano & M. Elías-Gutiérrez., 1996. A new species of Macrothrix (Anomopoda: Macrothricidae) from Central Mexico. Hydrobiologia 319: 159–166.

Chacón, T. A, M. R. Pérez & I. E. Musquiz., 1991. Biología Acuática 1. Síntesis Limnológica del Lago de Pátzcuaro, Michoacán, México. Univ. Mich. San Nicolás de Hidalgo, México: 23–26.

Dodson, S. I. & M. Silva-Briano. 1996. Crustacean zooplankton richness and associations in reservoirs and ponds of Aguascalientes State, Mexico. Hydrobiologia 325: 163–172.

Elías-Gutiérrez, M., 1995. Notas sobre los Cladoceros de embalses a gran altitud en el Estado de México, México. An. Esc. Cienc. Biol., Méx. 40: 184–197.

Elías-Gutiérrez, M., 1996. Taxonomía y algunos aspectos ambientales de los grupos Ctenopoda y Anomopoda (antes Cladocera en parte) de la provincia Neovolcanense, Subprovincia Meridional de la República Mexicana. Tesis Doctoral (in spanish). Instituto Politécnico Nacional. México, 185 pp.

Frey, D. G., 1975. Subgeneric differentiation within Eurycercus (Cladocera,Chydoridae) and a new species from northern Sweden. Hydrobiologia 46: 263–300.

Frey, D. G., 1995. Changing attitudes toward chydorid anomopods since 1769. Hydrobiologia 307: 43–55.

Hann, B. J., 1982. Two new species of Eurycercus (Bullatifrons) from Eastern North America. (Chydoridae, Cladocera) Taxonomy, Onthogeny and Biology. Int. Revue. ges. Hydrobiol. 67: 585–610.

Juday, C., 1915. Limnological studies on some lakes in Central America. Trans. Wis. Acad. Sci. Arts Lett. 18: 214–250.

Osorio-Tafall, B. F., 1943. Observaciones sobre la fauna acuática de las cuevas de la región de Valles, San Luis Potosí (México). Rev. Soc. mex. Hist. nat. IV: 43–71.

Rey, J. & E. Vásquez, 1986. Note Taxonomique sur Camptocercus dadayi Stingelin, 1913, comb. nov. (Crustacea, Cladocera). Ann. Limnol. 22: 177–180.

Rioja, E., 1940. Notas acerca de los crustáceos del Lago de Pátzcuaro. An. Inst. Biol. Mex. 11: 469–475.

Rioja, E., 1942. Observaciones del Plancton de San Felipe Xochiltepec, Puebla, México. Analyt. Inst. Biol. Mex. 23: 503–526.

Rodríguez-Almaraz, G. & A. Leija-Tristan, 1995. Cladocerans (Branchiopoda: Anomopoda; Ctenopoda) of the Nuevo Leon State, México, S. West Nat. 40: 322–350.

Sars, G. O., 1901. Contribution to the knowledge of the Freshwater Entomostraca of South America. Part I. Cladocera. Arch. Math. Naturv. 23: 1–102.

Smirnov, N. N., 1971. Chydoridae of the World fauna. Fauna of the USSR. Crustacea. 1. New Series No. 101. Leningrad: 531 pp (in russian). (English Transl. A. Mercado. Israel Prog. Sci. Trans. Jerusalem. 1974), 643 pp.

Smirnov, N. N., 1992. The Macrothricidae of the World. Guides to the identification of the microinvertebrates of the Continental waters of the World 1. SPB Acad. Pub. The Hague, 143 pp.

Stifter, P., 1991. A review of the genus Ilyocryptus (Crustacea: Anomopoda) from Europe. Hydrobiologia 225: 1–8.

Suárez-Morales, E., L. Segura & M. Fernández, 1986. Diversidad y abundancia del zooplancton en la Laguna de Catemaco, Veracruz durante un ciclo anual. An. Inst. Cienc. del Mar y Limnol. México 13: 313–316.

Suárez-Morales, E & M. Elías-Gutiérrez, 1992. Cladóceros (Crustacea: Branchiopoda) de la reserva de la biósfera de Sian Ka'an, Quintana Roo y Zonas adyacentes. In Navarro, D. & E. Suárez-Morales (eds), Diversidad Biológica en la Reserva de la Biósfera de Sian Ka'an, Quintana Roo, México. Vol II CIQRO/SEDESOL, México: 145–154.

Suárez-Morales, E., A. Vázquez-Mazy & E. Solís, 1993. On the Zooplankton community of a Mexican eutrophic reservoir, a seasonal survey. Hidrobiológica. 3: 71–80.

Uéno, M., 1939. Zooplancton of Lago de Pátzcuaro, México. Annot. Zool. Japon. 18: 105–114.

Van de Velde, I., H. Dumont & P. Grootaert, 1978. Report on a collection of Cladocera from Mexico and Guatemala. Arch. Hydrobiol. 83: 391–404.

Wilson, C. B., 1936. Copepods from the cenotes and caves of Yucatán Peninsula with notes on cladocerans. In Pearce, A. S., E. P. Creaser & F. G. Hall (eds), The Cenotes of Yucatan, a Zoological and Hidrografic Survey. Carnegie Inst. Wash. Publ.: 77–88

Hydrobiologia **360**: 75–78, 1997.
A. Brancelj, L. De Meester & P. Spaak (eds), Cladocera: The Biology of Model Organisms.
©1997 *Kluwer Academic Publishers.*

The Cladocera of the Godthåbfjord area, SW Greenland

Ulrik Røen[†]

Zoological Museum, Universitetsparken 15, DK-2100 Copenhagen Ø, Denmark

Introduction

In July and August 1973, I made a number of collections in the Godthåb Fjord (now Nuuk) area, partly around the bottom of the fjord, partly in the land area which separates the fjord from the open sea (see Figure 1). The branchiopod fauna of Godthåb area has previously been treated by Wesenberg-Lund (1894), but he recorded only 10 species from the area, a number of which were unidentified, and there is no information on the location or the type of freshwater from where the specimens were taken.

Localities

The area around the head of Godthåbsfjord has a continental, subarctic climate. It is the only area in Greenland where the Atlantic salmon (*Salmo salar*) spawns. Eelgrass (*Zostera marina*) is found only here in Greenland and at the head of a fjord just south of Godthåbfjord. On land, a number of more southern phanerogams occur. The coastal area has an oceanic, low arctic climate. The vegetation on land consists mostly of low copses of willows, areas with grass, lichens and mosses, or barren ground.

The freshwaters investigated consisted of 28 larger or smaller lakes, 74 ponds, and 13 temporary pools. No lotic water was investigated. Most of the freshwaters around the head of the fjord had an abundant vegetation, but the water at the coastal area had poor or no vegetation.

[†] Dr Ulrik Røen died on February 9th 1997. This manuscript was accepted in a letter of 20 May 1997 from the editor after changes according to the suggestions of the referees. These changes were made by G. Høpner Petersen, a collegue from the Zoological Museum, Copenhagen.

Methods

In total, 115 freshwater bodies were investigated: 60 around the head of fjord, and 55 in the coastal area. From most of the sites, water samples were taken in 'Pyrex' bottles and kept as cool as possible until they could be analysed by the Chemical Department of the Geological Survey of Denmark. The samples of Entomostraca were collected with a plankton net of 25 cm in diameter, and a mesh of 80 μm. All the samples were made as horizontal hauls at *c.* 12 m.

Results

The conductivity of the freshwaters was nearly the same in the two areas, varying in both from 30 to 120 μohm, but the dominating ions in the coastal area were Cl^- and Na^+, while in the continental area HCO_3^- and Mg^{++} dominated. The water temperature was on average about 2 °C higher around the head of the fjord than in the coastal area.

23 species of Cladocera were found (Table 1). All these species were found around the head of the fjord, but only 21 species were found in the coastal area. The species found belong to 3 of the 5 faunal elements found in Greenland (Røen, 1962, 1994): (1) the element found throughout Greenland. (2) the southern element and (3) the southern most element.

Only two species, *Daphnia pulex* and *Chydorus arcticus,* are found throughout Greenland and they are both very common in the coastal part and at the head of the fjord.

The southern element comprises 18 species, all of which have a northern limit in Greenland and occur in both areas. Generally, these species are not so common in the coastal area as in the area around the head, but *Eurycercus glacialis, Acroperus harpae, Bosmina longispina obtusirostris, Alonella excisa* and *Polyphemus pediculus*, which are all rather common north of

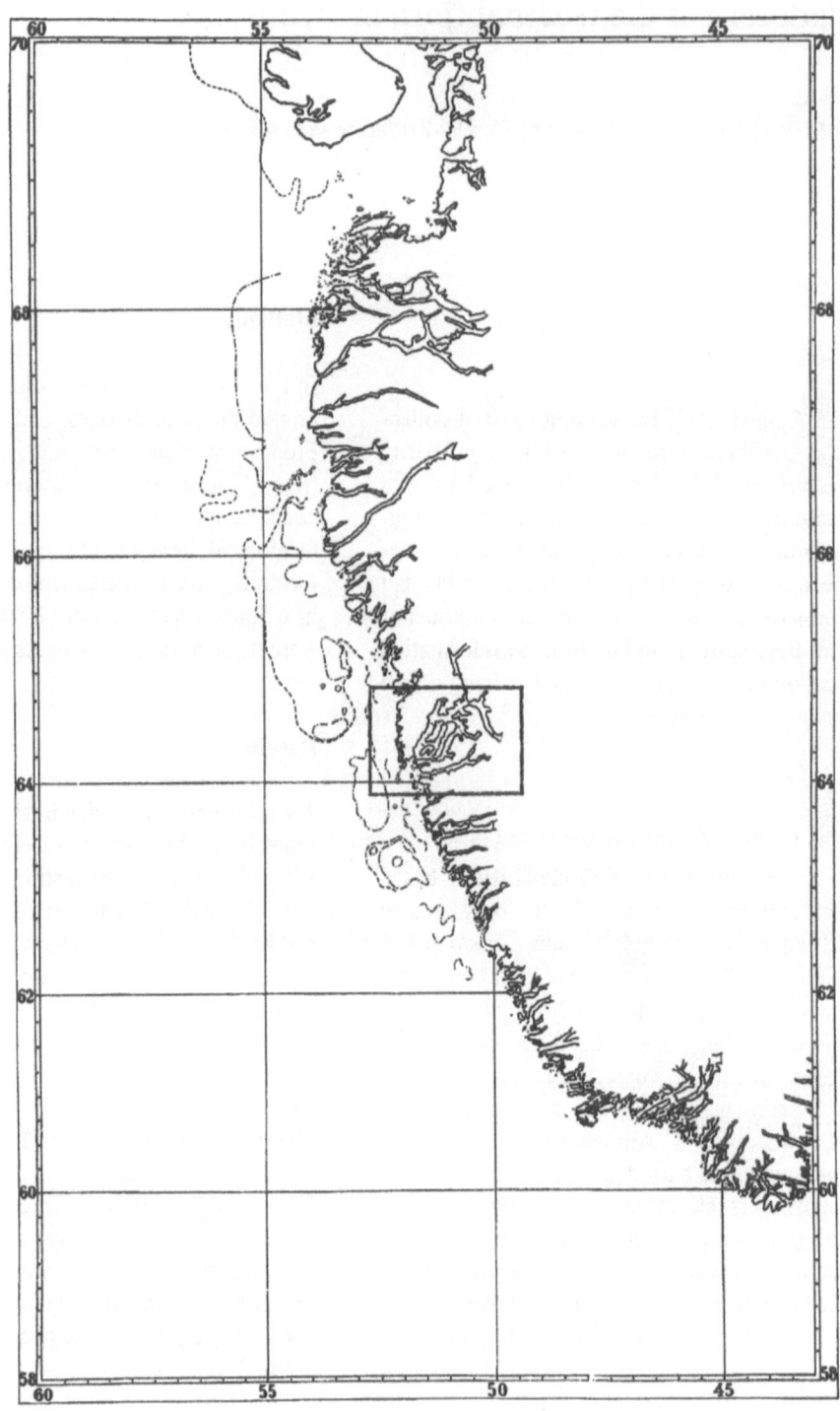

Figure 1. Map of West Greenland with the Godthåb (= Nuuk) Fjord area indicated.

Table 1. List of cladocerans found in 115 water bodies in the Godthåbfjord area, SW Greenland, during July and August 1973.

	Head of Fjord 55 freshwaters	Coastal area 60 freshwaters
Latona setifera (O. F. Müller)	29	18
Holopedium gibberum Zaddach	10	12
Daphnia pulex Leydig	23	43
Simocephalus	6	21
S. serrulatus (Koch)	3	4
Scapholeberis mucronata (O. F. Müller)	22	28
Ceriodaphnia quadrangula O. F. Müller	12	26
Bosmina (Eubosmina) longispina obtusirostris G. O. Sars	43	36
Streblocerus serricaudatus (Fischer)	15	18
Eurycercus glacialis Lilljeborg	33	22
Acroperus harpae (Baird)	33	23
Alona guttata G. O. Sars	3	6
A. fabricii Rren		1
A. intermedia G. O. Sars	5	5
A. rectangula G. O. Sars	11	12
A. quadrangularis (O. F. Müller)	2	4
A. affinis (Leydig)	26	30
Graptoleberis testudinaria (Fischer)	3	6
Alonella excisa (Fischer)	13	12
A. nana (Baird)	11	14
Pleuroxus truncatus (O. F. Müller)		1
Chydorus arcticus Rren	50	58
Polyphemus pediculus (L.)	46	33
Number of species	21	23
Average no. of species per water body	7.41	7.20

Table 2. Cladoceran species comprising more than 50% of the total number of individuals at a given water body in Godthåbfjord area, SW Greenland.

	Coastal area oceanic	Head of Fjord Continental
Latona setifera	1	
Holopedium gibberum	2	1
Daphnia pulex	1	7
Simocephalus vetulus		1
Simocephalus serrulatus		1
Scapholeberis mucronata		2
Bosmina longirostris obtusirostris	18	12
Eurycercus glacialis		1
Acroperus harpae	1	1
Alona affinis	1	
Chydorus arcticus	8	14
Polyphemus pediculus	3	1
Total number of water bodies with dominant species	35	41

The average number of species in the two areas is nearly the same, 7.4 in the coastal area and 7.2 in the continental. However, the number of specimens is smaller in the coastal area than in the continental. This is propably due to the lower water temperature in the coastal area.

Zoogeographical remark

In arctic environments, one or a few species often dominate, while in subarctic or temperate areas the number of specimens of a single species does not dominate. In the present investigation, one species is represented in more than 50% of the total number of specimens in 35 freshwaters in the subarctic, continental area (Table 2). The two species found throughout Greenland, *Daphnia pulex* and *Chydorus arcticus*, dominate in 21 freshwaters in the continental area, but only in 9 freshwaters in the coastal area. 7 species of southern group dominate in the continental area. Of these *Latona setifera* and *Holopedium gibberum* are of greatest interest, as they have their northern limit just north of Godthåbfjord. However, they nevertheless also dominate in the coastal area in one and two freshwaters. Of the southernmost element only one species, *Simocephalus serrulatus*, is dominating in the continental area.

Godthåbfjord, occur more often in the coastal area than in the area around the head.

The southern most element consists of only 3 species, of which *Simocephelus serrulatus* occurs in both areas, while *Alona fabricii* and *Pleuroxus truncatus* are found only in the area around the head. Of these, *S. serrulatus* and *A. fabricii* are known from the Kap Farvel area (Røen, 1994), while *P. truncatus* is new to Greenland and has never been found outside the palaearctic region before.

Thus there is no evidence supporting a separation of low arctic and subarctic areas on basis of cladocerans.

References

Røen, U., 1962. Studies on freshwater Entomostraca in Greenland, II. Localities, ecology and geographical distribution of the species. Meddel. Grønland 170: 1–240.

Røen, U., 1994. Studies on freshwater Entomostraca in Greenland, VI. The Entomostraca of the Kap Farvel area, southern most Greenland. Meddel. Grønland. Bioscience 41: 1–21.

Wesenberg-Lund, C., 1894. Grønlands Ferskvandsentomostraca. I. Pyhllopoda: Branchiopoda at Cladocera. Vid. Meddel. Dansk Nat. Foren. 46: 82–193.

Hydrobiologia **360**: 79–88, 1997.
A. Brancelj, L. De Meester & P. Spaak (eds), Cladocera: The Biology of Model Organisms.
©1997 *Kluwer Academic Publishers.*

Offspring size in *Daphnia*: does it pay to be overweight?

Maarten Boersma
Max-Planck-Institut für Limnologie, Postfach 165, D-24302 Plön, Germany
(e-mail: boersma@mpil-ploen.mpg.de)

Key words: Life history, food

Abstract

Variation in offspring size and number has been described for a wide range of organisms. In this study I investigated the relationship between resource level of the mother and size of her offspring in the cladoceran *Daphnia magna*, in order to assess whether offspring produced at different food levels are optimal in size for these food levels. Optimal offspring size was defined as the size of offspring that yields the highest parental fitness (i.e. offspring of optimal size have the highest juvenile fitness per unit maternal **effort** invested in them). I observed that especially at the higher food levels, daphnids produced offspring that are **larger** than the computed optimal offspring size at these food levels. I interpret this as a mechanism to avoid starvation of neonates in the case of suddenly deteriorating food conditions.

Introduction

One of the general axioms in the theories of life-history evolution is that selection should favour those parental strategies which maximize parental fitness (e.g. in the determination of the number and size of offspring produced, Lack (1947)). According to the classic paper of Smith & Fretwell (1974), maximizing parental fitness will lead to a single optimal investment per offspring for any given environment, provided that (1) a trade-off exists between size and number of individual offspring, i.e. the amount of energy invested in reproduction per breeding attempt is fixed; and (2) that as energy expended on individual offspring increases, fitness of individual offspring increases. The optimal investment per offspring is characterised by the highest fitness of individual juveniles per unit effort put into these animals. Changes in either total resource acquisition, or in the proportion of resources which is allocated to reproduction, should not change this optimal investment per offspring, as the total effort per breeding attempt does not influence the parental fitness-function (Smith & Fretwell, 1974). In planktonic cladocerans such as *Daphnia*, the trade-off between size and number of offspring has been repeatedly observed (e.g.

Ebert, 1993; Boersma, 1997), but is often masked by differences in total available resources for reproduction (see van Noordwijk & de Jong, 1986). Moreover, as daphnids live in changing environments with respect to food conditions, the resource acquisition of the adults may be used as an estimate of environmental circumstances which their offspring will encounter after birth. Adults and juveniles generally show a large overlap in their resources, and the mothers adjust the size of their offspring according to their own resource availability, resulting in a relationship between resource level and egg size in cladocerans, characterized by larger offspring produced at lower food levels (e.g. Tessier & Consolatti, 1991; Glazier, 1992; Guisande & Gliwicz, 1992; Ebert, 1994). This relationship is easy to understand intuitively, as offspring produced by females grown at low food levels should have more reserves to withstand the concurrent food conditions than when food is in ample supply and juveniles can start feeding immediately. Thus far, however, to my knowledge only one study exists which tries to link the investment in individual offspring with their successive fitness in a more formal way, in order to study whether the offspring produced at different food levels are optimal for the ambient environmental conditions (Tessier & Consolatti, 1989).These authors, however, only described a theoretical framework, and made no attempt to quantify optimal offspring sizes. In the present study, I set out to perform this quantification. I collected neonates with different initial weights produced by first adult instar females (Boersma, 1997), and cultured them at four different food levels to assess the optimal offspring weight under these conditions. These values of optimal offspring weight were then compared to observed offspring weights.

Materials and methods

Offspring fitness

Fitness, usually defined as the average number of offspring produced by individuals with a certain genotype, relative to the number produced by individuals with other genotypes (e.g. Ridley, 1993), is in the current literature often represented by the intrinsic rate of population increase, r (e.g. Stearns, 1992). Most laboratory studies on cladocerans use the Euler-Lotka equation to estimate r iteratively (e.g. Ebert & Jacobs, 1991; Spitze, 1992; Weider, 1993; De Meester, 1994; Spaak & Hoekstra, 1995; Lampert & Trubetskova, 1996). Often, these values of the intrinsic rate of population increase are computed using individual animals (Weider, 1993; De Meester, 1994; Spaak & Hoekstra, 1995). This causes problems, as doing so makes it impossible to find negative values of r, except for the infinite negative, when an animal does not reproduce at all. The production of one single egg in the animal's life-time will lead to an r-value equal or larger than zero. Moreover, even if r equals zero, this does not mean that fitness equals zero, as the individual is still capable of replacing itself, leading to a stable population density. It could therefore be argued that the intrinsic rate of population increase is not suitable as a fitness measure, but that rather e^r (λ) would be the appropriate fitness measure, as this value equals zero when r approaches negative infinity. However, this fitness measure essentially only rescales the observations, and hence does not solve the above mentioned problems with r, as now values between 0 and 1 will be absent. This interval of $0 < e^r < 1$ is of lesser interest when assessing optimality of offspring size, because when population growth rates are less then 1, the population will die out. Therefore, the domain of the optimization problem for realistic purposes has a lower limit of $e^r = 0$, which will result in a stable pop-

ulation. Moreover, given the nature of r, with values usually small and close to zero, the distinction between r and e^r is rather subtle, as e^r shows a linear relationship with r for small intervals. To allow comparison with other studies, and given the considerations above r, was used as a measure of fitness in this study.

Laboratory derived r-values as a measure of fitness do not incorporate mortality occurring in the field. Evolution of egg size obviously has occured with mortality sources, such as predation, present. Size differences between offspring are, however, usually not large, and thus far it has been very difficult to properly estimate differences in mortality in the field between animals with such subtle size differences (e.g. Boersma et al., 1996). As these differences is size-specific mortality are unknown, and are likely to change within a growing season, it is difficult to incorporate field-mortality in the computation of optimal offspring size. One way to circumvent this is to estimate fitness under different assumptions of juvenile size-selective mortality. However, as very little is known about this size-selective juvenile mortality, the power of this analysis would be limited. Alternatively, optimal offspring sizes can be computed under the assumption that mortality differences are non-existent. Then, from the computed optimal offspring sizes and the measured actual offspring sizes, the size-specific mortality needed to make the observed offspring size the optimal offspring size can be computed. The validity of this set of size-specific mortality assumptions could then be assessed by field observations and experiments.

A significant correlation has been found between juvenile growth rate (somatic growth between birth and maturity) and the intrinsic rate of population increase, r (Lampert & Trubetskova, 1996). As juvenile growth rates are easier and quicker to establish, I collected one individual per experimental vessel when the animals were five days old, and again upon reaching maturity. These animals were used to establish dry-weights in order to estimate growth rates. In order to estimate the exact relation between juvenile growth and r under the conditions of the present study, I also established r-values for animals placed in 20 experimental chambers per series; these animals were cultured until they released their second broods. Development times and number of juveniles produced were used to estimate r by solving the Euler-Lotka equation iteratively. The relationship between growth rates and r-values computed for these individuals was then used to establish r-values for all experimental animals.

Offspring weight offspring fitness

The *Daphnia magna* clone used in this study has been kept in the laboratory for many years, and was originally collected from a pond in Frankfurt, Germany. From a stock culture, juvenile animals were collected randomly. These animals were placed individually in 120 ml flow-through chambers, with a flow rate of 1 litre d^{-1}, and fed a *Scenedesmus acutus* suspension with an algal carbon content of 0.8 mg C l^{-1} at 20 °C, under continuous light conditions. The algae were grown in 3-litre chemostats in Chu-12 medium (Lampert et al., 1988). Food suspensions were prepared daily by adding *Scenedesmus* to 0.45 μm filtered lake water. Algal concentrations were measured spectrophotometrically. The only way to obtain a large set of offspring with different initial weights produced by first adult instar females is to harvest these neonates from animals cultured at different food levels. Therefore, first brood neonates of the animals cultured at 0.8 mg C l^{-1} were collected and these neonates were placed in flow-through chambers at four different food levels: 0.1; 0.2; 0.4 and 1.0 mg C l^{-1}. Each chamber contained 5–10 individuals. Upon reaching maturity, these animals were measured, the length of the neonates produced by these animals was established, and two or three neonates from every chamber were weighed to establish dry weight of individual offspring. If possible, five other neonates from each chamber were collected and placed in flow-through chambers to assess their fitness. Neonates produced at the different food levels were divided evenly across all four food levels mentioned above.

The assumption behind the Smith & Fretwell (1974) model is that offspring fitness increases monotonically with offspring weight up to an asymptote (i.e. increasing per-offspring investment results in diminishing fitness gains). Hence, I iteratively fitted the three parameter model proposed by Tessier & Consolatti (1991) to describe the dependence of offspring fitness (f) on offspring investment (n):

$$f = f_m[1 - e^{-k(n-n_o)}],$$

where n_o = the minimum viable neonate mass, f_m = the maximum fitness level, and k the rate of rise of fitness with neonate mass (see also Winkler & Wallin, 1987). As pointed out by Tessier & Consolatti, f_m will be most sensitive to changes in food concentration in the environment, with lower food levels leading to lower values of f_m. However, the optimal solution in the offspring weight-offspring number trade-off is independent of the value of f_m (Parker & Begon, 1986), and therefore f_m was estimated separately, being the average fitness of all animals with a birth-weight larger than 10 μg.

Egg quality

The premise for the approach taken in this paper to be valid is that, apart from the differences in size between the offspring, no quality differences between offspring produced at different food levels should exist. This was tested for the four different food levels, in an experimental set-up identical to the one described above. Experimental animals were collected, and placed in flow-through chambers at the four food levels. Each chamber contained 10 individuals, with 22 replicate chambers per food level. Once these animals reached maturity they were harvested, the eggs produced by these animals were separated from the mothers, and analysed for volume, dry weight, carbon content and fat content. Observations were made every 12 hours, so the average age of these eggs was six hours. The analysis of carbon content and fat content are mutually exclusive so two different sets of samples had to be prepared, one to be analysed further for ash free dry weight and carbon content of the eggs, and the other one to be used in the analysis of total fat content of the eggs. A minimum of five eggs per sample were collected, if possible from the same mother. For the individuals cultured at the lower food levels, however, this was not always possible. *Daphnia* eggs are nearly spherical in early development, and hence one measured diameter per egg sufficed for the estimation of egg volume (Lampert, 1993). Egg diameter of three eggs per brood was established to the nearest 0.01 mm. For the analysis of dry weight and carbon content, eggs were collected in pre-combusted small silver weighing boats, and dried for 24 hours at 60 °C. These samples were stored in a desiccator, weighed to the nearest 0.1 μg using an electronic microbalance, and subsequently analysed for carbon content.

For the analysis of fat (triglyceride) content, eggs were collected and stored at –20 °C in 100 μl phosphate buffered saline solution (Sigma; P–4417). They were homogenized, and the fat content was established using triglyceride (GPO-Trinder) reagent (Sigma; Catalogue number 337), which is normally used for the quantitative enzymatic determination of triglycerides in serum or plasma (see also Stibor, 1995).

Table 1. Quantity differences between eggs produced at different food levels. Four different food levels were tested (0.1, 0.2, 0.4, and 1.0 mg C l^{-1}), a total of 70 clutches were analysed. *P*-values of analyses of variance are shown

Trait	*P*-value
Volume (mm³)	< 0.001
Carbon content (μg)	< 0.001
Ash-free dry weight (μg)	< 0.001
Fat (μg)	0.20

Table 2. Summary table of the iterative curve fitting of the relationship between offspring fitness and initial offspring weight (standard error of estimate). The following function was fitted: $f = f_m\{1 - e^{-k(n-n_0)}\}$. Total explained variance and significance of the whole model are shown. Values for f_m were determined by averaging the realised fitness of animals with a initial dry weight of more than 10 μg

Food level	f_m	k	n_0	r^2	N	P
(mg C l^{-1})	(d^{-1})	(μg^{-1})	(μg)			
0.1	no feasible solution					
0.2	0.19	0.48	3.46	0.24	62	< 0.001
		(0.25)	(1.49)			
0.4	0.26	0.54	2.79	0.22	72	< 0.001
		(0.42)	(2.10)			
1.0	0.45	0.36	1.92	0.31	82	< 0.001
		(0.11)	(1.27)			

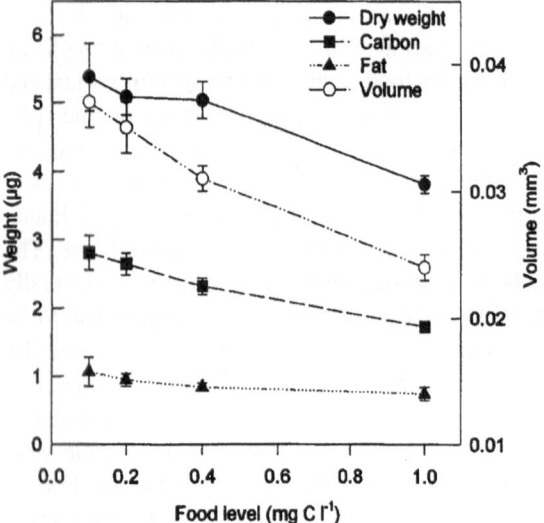

Figure 1. Quality differences between eggs produced at different food levels. Four different food levels were tested (0.1, 0.2, 0.4, and 1.0 mg C l^{-1}), a total of 70 clutches were analysed. Mean values and standard errors of carbon content (μg), ash-free dry weight (μg), fat (μg) and volume (mm³) are shown.

Results

Egg quality

Differences in volume, carbon content and ash-free dry weight of eggs produced at the four experimental food levels were highly significant (Table 1), whereas I observed no signifcant differences in fat content of the eggs. The pattern was similar for all four measures: highest values were found for the eggs produced by females grown at the lowest food level, and lowest values were observed for eggs produced by females cultured at 1.0 mg C l^{-1} (Figure 1). A two-way ANOVA with food concentration and trait as the grouping variables, and the log-transformed measurements as

the dependent ones showed a non-significant interaction ($F_{9,280} = 0.205$; $P = 0.99$). This means that the relative differences between food levels are similar for the different traits, and implies that no quality differences exist between the food levels.

Offspring weight offspring fitness

Figure 2 shows the relation between initial offspring weight and the computed values for r from the daily growth increments. Unexpectedly, most data points seem to fall within the saturated part of the curves relating offspring fitness with offspring investment, making it difficult to fit curves of the shape described by Smith & Fretwell (1974). Especially for animals cultured at the lowest food level, no solution could be found. As a result of the distribution of the data, even the curves fitted through the data of the animals cultured at the higher food levels explain only a small (but significant) portion of the variation (Table 2; Figure 2). With the help of these curves, using the marginal value solution proposed by Smith & Fretwell (1974) it was possible to estimate the optimal offspring weight at the different food levels (Figure 2). For the food levels 0.2, 0.4, and 1.0 mg C l^{-1} the optimal offspring weights were computed to be 6.4, 5.3, and 4.6 μg dry weight, respectively. The method of estimating optimal birth weight from the fitness profile is in essence an inverse regression problem, and the techniques to assign confidence intervals for the linear case are well-developed. (Sokal & Rohlf, 1981). In the non-linear case, however, these techniques are less readily available. Since k

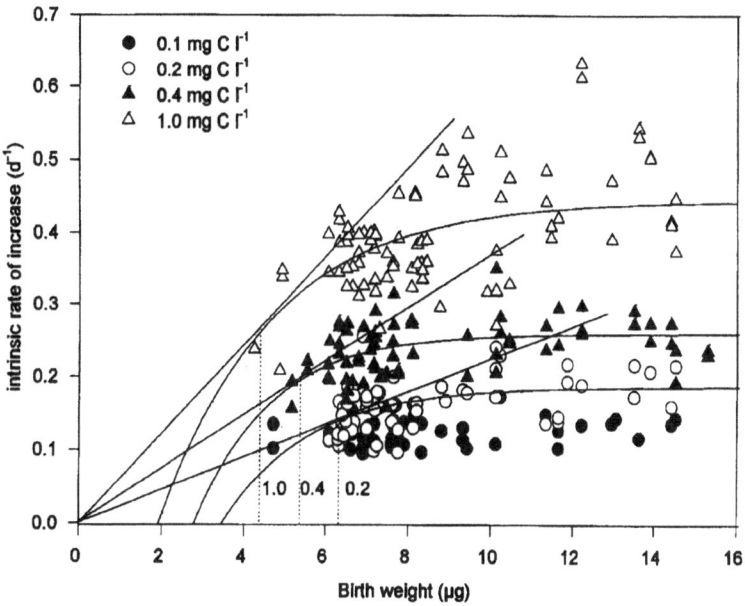

Figure 2. Initial offspring weight (μg) and offspring fitness (r, d^{-1}) as computed from the daily growth increments. For the highest three food levels the following functions were fitted: $f = f_m\{1 - e^{-k(n-n_0)}\}$ (see also Table 2) in order to estimate the optimal offspring weight at these food levels (indicated on the abscissa).

essentially is the regression coefficient in the current study, I assessed the variability of the estimates of optimal offspring weight by computing the optimal weight at different food levels, with $k \pm$ one standard error. These 'confidence-intervals', obviously asymmetrical around the estimated values, were found to be 5.8–7.7 (avg = 6.4), 4.4–7.1 (5.3) and 4.4–5.3 (4.6) for the different food levels. The comparison of the computed values for optimal offspring weights at the different food levels with the weights of the offspring actually produced by these animals revealed that the actual offspring weight, even for the first adult instar mothers used in this study, was larger than the computed optimal offspring weights for these food levels, especially for the animals at the higher food levels (Figure 3). At the lower end of the food concentration spectrum, the differences between the computed optimal and actual offspring weights were smaller.

Mortality

Fitness, r, as established from abbreviated life-tables inadequately describes fitness under field conditions. Mortality in the field is obviously much higher than in the laboratory, as mortality by predation is completely ruled out in the laboratory. This implies that the measurements of r in the laboratory are only related to dif-

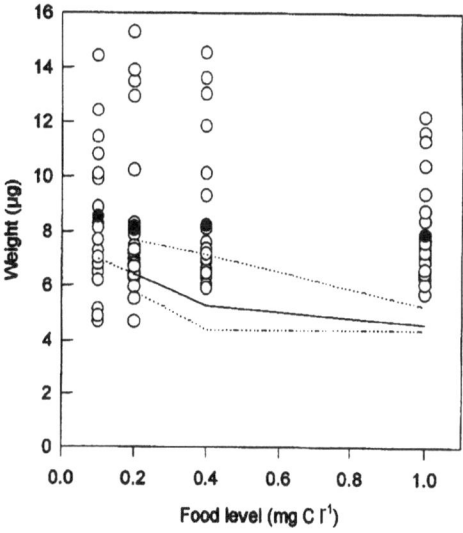

Figure 3. Optimal offspring weight (μg) (line) and actual offspring weight (μg) (symbols) at different food concentrations (dotted line indicates 'confidence interval' for optimal offspring weight, filled symbols represent average offspring weight at different concentrations).

ferences in food concentration. Size-selective mortality in the field would change the relative r values of the different size classes, which could change the conclusions of the present study considerably. I assessed the possible influence of size-selective mortality as follows: first I assumed that the observed average offspring weight produced at the different food levels is actually the optimal offspring weight under these food conditions under natural circumstances. This implies that the curve relating juvenile fitness to offspring weight should be redrawn in such a way that the actual average weight becomes the point on the abscissa at which the line passing through the origin is tangent to the curve (Figure 4). As the value of the maximal attainable fitness, f_m, does not influence this point, f_m values were kept constant. The values for n_0 are most likely only influenced by concurrent food conditions, and not by predatory effects. Hence, the curves were redrawn, with only k fitted, and f_m and n_0 kept constant. Under the assumption that daily mortality is constant, mortality (m) can be entered in the Euler equation by substituting l_x, the probability of survival to time x, by e^{xm} (Stearns, 1992). This implies that m and r scale linearly, and hence that the mortality needed to decrease the observed r-values in such a way that the observed average weights produced at the different food levels are optimal for these food levels can be computed. Hence, the adjusted r-values fall on the line described by the old f_m and n_0 , and the recomputed k (Figure 5). It should be kept in mind that, since f_m was not changed, the absolute value of the mortalities is not of interest. It is the change in mortality rate with changes in initial body mass that is relevant. The calculated decrease in mortality with the increase in birth weight equals 0.006 d^{-1} μg^{-1} ($r = -0.19$; $P < 0.01$; $N = 211$), which implies that individuals with a birth-weight of 6 μg should have a mortality in the field, which is 0.054 d^{-1} higher when compared with animals which are 15 μg at birth.

Discussion

In this study I set out to investigate the relationship between initial weight of neonates of *Daphnia magna* and their fitness. The main conclusion of this paper should be that this is not an easy task. First of all, as was already indicated in the Materials and Methods section, a measure of juvenile fitness, given the individual culturing set-up as *the* method of choice in life-history experiments, is difficult to find. Although the vast majority of papers do take the intrinsic rate of population increase, r, as a fitness measurement it should be reiterated that fitness equals zero when r equals negative infinity, i.e when the animal dies before

reproducing. Values between negative infinity and 0 (when an animal produces exactly one offspring during its life-time) are not possible in such experimental systems. e^r (λ) would be a better fitness measurement, as this value equals zero when the animal does not reproduce, but then again as a result of the culturing conditions, no values between 0 and 1 (r equals 0) will be found, and all values will be larger than one. This would not make the curve-fitting procedure, as carried out in this paper, any easier. Total reproductive output, R_0, does not take into account the time needed to reproduce, and hence is not suitable as a measurement of fitness either. Fitness is a relative quantity, describing the relative contribution of a certain individual or clone to the next generation, and in most studies values of r are used to compare the reaction of identical individuals to different experimental conditions (e.g. Weider, 1993; Spaak & Hoekstra, 1995), or to compare different animals under similar conditions (e.g. Ebert & Jacobs, 1991; Spitze, 1992). Although it should be kept in mind that r does not scale linearly with λ, and hence that taking r as a fitness measure could bias the results, the relative differences between both estimates will be small, because r is usually close to zero. As argued in the Materials and Methods section, an optimal weight of offspring yielding a value of r smaller then zero would result in a population going to extinction. A computed optimum offspring size resulting in a negative r is therefore outside the domain, and cannot occur. The optimum offspring size should thus be in the range where $r > 0$.

Tessier & Consolatti (1991) reported differences in quality between offspring produced at different food levels: they observed that neonates born at the lowest food level (≈ 0.05 mg C l^{-1}) had a higher relative nitrogen content than neonates produced at higher food levels. In contrast to these findings, I did not observe quality differences between offspring produced at different food concentrations. Although food quality has been reported to affect egg quality (Müller-Navarra, 1993), my results indicate that this is not the case for differences in food quantity, at least for the rather crude measures of offspring quality used here. Further research should clarify whether indeed differences between offspring produced by females grown at different food levels are only quantitative in nature, or whether more sophisticated measures (e.g. fatty acid spectrum) would reveal also qualitative differences.

The explained variance of the curves fitted through the data relating offspring weight with offspring fitness was low, but significant (Table 2). Generally, it has

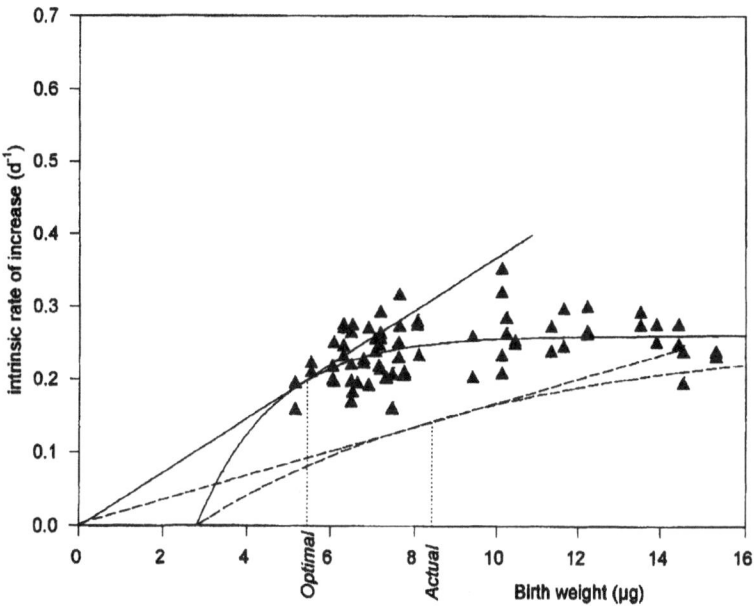

Figure 4. An example of a refit of the function: $f = f_{\mathrm{m}}\{1 - e^{-k(n-n_0)}\}$, for animals grown at 0.4 mg C l^{-1}, with the parameters k and n_0 changed in such a way that the observed average offspring weight becomes the optimal offspring weight (dotted line).

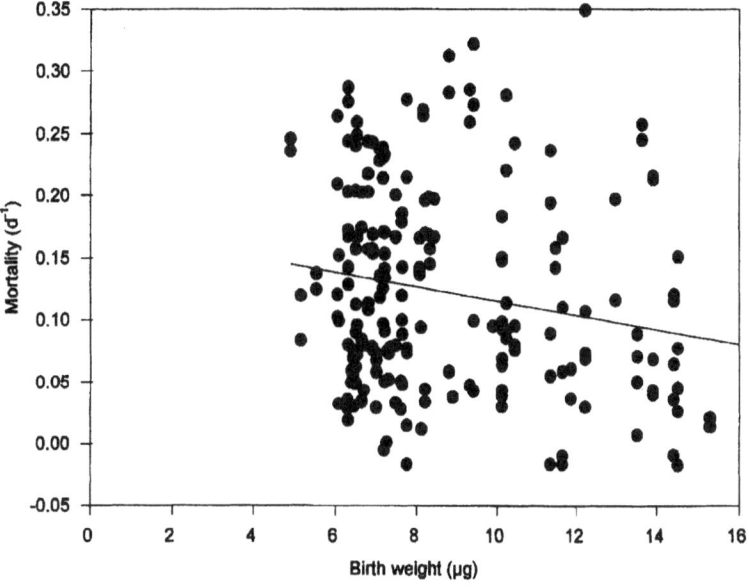

Figure 5. Mortality (d^{-1}) needed to explain the discrepancy between optimal offspring weight and observed offspring weight.

been found that the smallest offspring are produced at the highest food levels, while larger offspring are produced at lower food levels (e.g. Tessier & Consolatti, 1991; Glazier, 1992; Guisande & Gliwicz, 1992). Hence, it was to be expected that fitting the curves for animals cultured at the highest and at the lowest food levels would prove to be most difficult: at the high-

est food level even the smallest neonates were not 'too small' for this food level, while at the lowest food level there were no individual newborn that were 'too large'. Even at the two intermediate food levels variation was large, and also at these food levels animals that were too small for these food concentrations seemed to be lacking. It could be argued that, given the large variation in the observed r-values, fitting a three parameter model through the data is not valid, and that instead the more parsimonious solution, i.e. linear regression, should have been chosen. This was not done, as fitness of the juveniles cannot increase indefinitely, but rather should show some maximum value. The main problem, however, was the large variation in the data, and the observation that very few offspring seemed to be too small for their environment. The causes of this are unclear. On the one hand, it could have been the case that the mothers used in these experiments were badly fed, leading to the production of large offspring. However, Lampert (1993) reported that eggs with a diameter of 0.22 mm always failed to hatch. Using the relationships between volume and dry weight established in the first experiment, this would translate in a dry weight of around 4 μg. Figure 2 shows that a number of the offspring produced in this study were close to this value, so most likely it was not the case that all offspring used in this experiment were too large, but rather it seems to be that in this study differences in initial weight of the offspring are rather unimportant, and that only the actual food level plays a role in the determination of r, an observation contradicted by a substantial body of literature, showing maternal effects in daphnids (e.g. Lynch & Ennis, 1983; Lampert, 1993).

The values computed for n_0 were smaller than the value for the minimum viable offspring reported by Lampert (1993) of 4 μg. Bell (1983) also reported for *Daphnia pulex* that eggs smaller than a certain volume never hatched. This would imply that my estimates for n_{rmo} are too high, which would imply that the computed values for optimal offspring weight are underestimated. Alternatively, the value of minimum viable egg weight as estimated by Lampert (1993) might be too small as these experiments were done *in vitro*, with a total hatching success of only 50%. This value is certainly lower than *in vivo*, and it could be the case that the smaller eggs would have hatched in the broodpouch.

As only a few studies have focused on size-selective mortality in the field (Vijverberg & Richter, 1982; Boersma et al., 1996) it is difficult to judge whether my estimates of the differences in mortality needed to make the actual average birth weight the optimal birth weight are realistic. This would involve a very detailed study of the fate of newborn daphnids with size differences which are rather small. Using the paper of Vijverberg et al. (1993), I estimated how large the differences in mortality would be between the 15 smallest individu-

als produced in this study and the 15 largest individuals (ranked according to their lengths). In summer, these differences are in the order of 0.1 d^{-1}, a value comparable to the difference in mortality I computed, and hence it is conceivable that these differences in natural mortality could explain the discrepancy between the observed average weight and the computed optimal weight. This difference in mortality would then comprise of either selective predation on smaller individuals (e.g. by invertebrate predators) or higher nonpredation mortality rates of smaller individuals.

It seems to be the case that especially at the higher food levels, neonates are heavier than the computed optimal offspring weights. The question of size of the offspring produced by daphnids cultured on different food levels has been addressed by a number of workers (e.g. Tessier & Consolatti, 1989, 1991; Glazier, 1992; Guisande & Gliwicz, 1992; Ebert, 1993, 1994; Boersma, 1995, 1997), but thus far only a few studies have investigated the relationship between initial offspring size and fitness (e.g. Lynch & Ennis, 1983). Only Tessier & Consolatti (1991) tried to link these traits in a more formal way. In an earlier study (Boersma, 1997), I analysed the effect of maternal age on offspring weight and fitness, and observed that, although primiparous females produce the smallest offspring, these neonates had the highest fitness per unit effort. Therefore I concluded that the youngest females produced the optimal offspring weight. The results found in the current study, however, would implicate that even these small offspring produced by the first adult instar females are larger than the optimal weight, and that producing even smaller offspring would yield a higher parental fitness. It is difficult to envisage that daphnids, which frequently experience periods of high food abundance, are suboptimally adapted to their environment, as is suggested by my finding that especially at higher food levels they produce offspring larger than the computed optimal offspring weight. The mean observed offspring weight of the animals cultured at carbon concentrations of 0.2 mg l^{-1} was similar to the computed optimal weight, whereas at the highest food level the difference amounted to a factor two. This might be a result of an incomplete trade-off between size and number of offspring at these food levels. Indeed, Ebert (1993) reported that the negative correlation between size and number of offspring is strongest at lower food levels, and close to zero when food conditions are good. Although Ebert & Yampolsky (1993) suggested that oocytes are originally produced in excess, so that the actual egg number might be adjusted downward, it is unknown as yet how many oocytes are actually produced. It is conceivable that especially at higher food levels, the number of oocytes, rather than the amount of energy available is limiting the number of eggs produced. This means that the animals have excess energy, which might be distributed to the available oocytes. This 'luxus-nourishment' of the eggs would then result in offspring with a size larger than the optimal size.

Environmental conditions for daphnids in the field are rather unstable. High levels of edible algae at one moment in time can be followed by much lower levels only a few days later, caused by the rapid depletion of food as a result of high densities of zooplankton (e.g. Jeppesen et al., 1990). Moreover, it has become increasing clear that aquatic environments are not as homogenous as previously assumed, but that many zooplankton and phytoplankton populations show a high degree of patchiness in their distributions (George & Edwards, 1976; Harris, 1980), with, as was reported by Malone & McQueen (1983) differences of upto a factor three in chlorophyll content within a distance of a few meters. This implies that a slight horizontal or vertical movement of the daphnids could lead to different feeding conditions rather rapidly. As a result, it may not always be the best strategy to adjust completely to concurrent food conditions, as these may change within a few days. Egg weight and number are determined around 0.6 developmental units (instar durations) before they are actually deposited into the broodchamber (e.g. Bradley et al., 1991; Ebert & Yampolsky, 1993; Stibor, 1995). This means that the timelag between the determination of the weight of the offspring and their release from the broodchamber is 1.6 instar durations, typically around 5.5 days at 20 °C. Larger offspring have higher starvation resistance (Threlkeld, 1976; Tessier et al., 1983), and as a result of the shape of the curves relating parental fitness with effort per offspring (see also Smith & Fretwell, 1974), the parental fitness loss of producing offspring which are slightly heavier than would be optimal is much lower than the fitness loss associated with the production of offspring which are slightly lighter than the optimal weight. As the likelihood of deteriorating food conditions is obviously greater at high than at low food levels, the advantage of producing larger-than-optimal-sized offspring is also larger under high food conditions. At low food levels, energy is limiting and producing overweight offspring might imply producing one offspring less, which would obviously decrease parental fitness substantially. In addition, it was suggested that offspring size at low food levels might be limited by other factors, such as the existence of some maximal offspring size (e.g. Ebert, 1994; Boersma, 1995), which would make it impossible to invest extra energy in individual offspring for animals grown under low food conditions.

In conclusion, in this study I observed that especially at higher food levels the difference between observed offspring weights and the calculated optimal offspring weight is large. The most likely explanation for this observation is that daphnids experiencing high food conditions produce these slightly obese offspring to avoid possible starvation in the case of degenerating food conditions. The experiments in this paper were carried out with one laboratory clone of *Daphnia magna*. Obviously, clones with different backgrounds may react differently, and hence the results obtained here should be corroborated by experiments with a range of clones coming from different locations. It would be of interest to investigate whether clones from more instable habitats produce offspring that are overweight to a higher degree than clones from more stable habitats.

Acknowledgements

I thank Herwig Stibor for his help with the analysis of the fat content of the *Daphnia* eggs and Hinnerk Boriss, Dieter Ebert, Winfried Lampert, Klaus Plath, Larry Weider and two anonymous referees for their comments on the manuscript.

References

Bell, G., 1983. Measuring the cost of reproduction. 3. The correlation structure of the early life history of *Daphnia pulex*. Oecologia 60: 378–383.

Boersma, M., 1995. The allocation of resources to reproduction in *Daphnia galeata*: Against the odds? Ecology 76: 1251–1261.

Boersma, M., 1997. Offspring size and parental fitness in *Daphnia magna*. Evol. Ecol.: 11: 439–450.

Boersma, M., O. F. R. van Tongeren & W. M. Mooij, 1996. Seasonal patterns in the mortality of *Daphnia* species in a shallow lake. Can. J. Fish. aquat. Sci. 53: 18–28.

Bradley, M. C., D. J. Baird & P. Calow, 1991. Mechanisms of energy allocation to reproduction in the cladoceran *Daphnia magna* Straus. Biol. J. Linn. Soc. 44: 325–333.

De Meester, L., 1994. Life histories and habitat selection in *Daphnia* – divergent life histories of *D. magna* clones differing in phototactic behaviour. Oecologia 97: 333–341.

Ebert, D., 1993. The trade-off between offspring size and number in *Daphnia magna* – the influence of genetic, environmental and maternal effects. Arch. Hydrobiol. Suppl. 90: 453–473.

Ebert, D., 1994. Fractional resource allocation into few eggs – *Daphnia* as an example. Ecology 75: 568–571.

Ebert, D. & J. Jacobs, 1991. Differences in life-history and aging in 2 clonal groups of *Daphnia cucullata* Sars (Crustacea, Cladocera). Hydrobiologia 225: 245–253.

Ebert, D. & L. Y. Yampolsky, 1992. Family planning in *Daphnia*: when is clutch size determined? Russian J. aquat. Ecol. 2: 143–148.

George, D. G. & R. W. Edwards, 1976. The effect of wind on the distribution of chlorophyll a and crustacean plankton in a shallow eutrophic reservoir. J. appl. Ecol. 13: 667–692.

Glazier, D. S., 1992. Effects of food, genotype, and maternal size and age on offspring investment in *Daphnia magna*. Ecology 73: 910–926.

Guisande, C. & Z. M. Gliwicz, 1992. Egg size and clutch size in 2 *Daphnia* species grown at different food levels. J. Plankton Res. 14: 997–1007.

Harris, G. P., 1980. Temporal and spatial scales in phytoplankton ecology. Mechanisms, methods, models, and management. Can. J. Fish. aquat. Sci. 37: 877–900.

Jeppesen, E., M. Søndergaard, O. Sortkjær, E. Mortensen & P. Kristensen, 1990. Interactions between phytoplankton, zooplankton and fish in a shallow, hypertrophic lake: a study of phytoplankton collapses in Lake Søbygård, Denmark. Hydrobiologia 191: 149–164.

Lack, D., 1947. The significance of clutch size. Ibis 89: 309–352.

Lampert, W., 1993. Phenotypic plasticity of the size at first reproduction in *Daphnia*: the importance of maternal size. Ecology 74: 1455–1466.

Lampert, W. & I. Trubetskova, 1996. Juvenile growth rate as a measure of fitness in *Daphnia*. Funct. Ecol. 10: 631–635.

Lampert, W., R. D. Schmitt & P. Muck, 1988. Vertical migration of freshwater zooplankton: test of some hypotheses predicting a metabolic advantage. Bull. Mar. Sci. 43: 620–640.

Lynch, M. & R. Ennis, 1983. Resource availability, maternal effects, and longevity. Exper. Geront. 18: 147–165.

Malone, B. J. & D. J. McQueen, 1983. Horizontal patchiness in zooplankton populations in two Ontario kettle lakes. Hydrobiologia 99: 101–124.

Müller-Navarra, D., 1993. Quantifizierung von Nahrungsqualität für herbivores Zooplankton. PhD Thesis, University of Kiel, 137 pp.

Parker, G. A. & M. Begon, 1986. Optimal egg size and clutch size: effects of environment and maternal phenotype. Am. Nat. 128: 573–592.

Ridley, M., 1993. Evolution. Blackwell. Boston, 670 pp.

Smith C. C. & S. D. Fretwell, 1974. The optimal balance between size and number of offspring. Am. Nat. 108: 499–506.

Sokal, R. R. & F. J. Rohlf, 1981. Biometry. Freeman and Company, San Francisco.

Spaak, P. & J. R. Hoekstra, 1995. Life history variation and the coexistence of a *Daphnia* hybrid with its parental species. Ecology 76: 553–564.

Spitze, K., 1992. Predator-mediated plasticity of prey life history and morphology – *Chaoborus americanus* predation on *Daphnia pulex*. Am. Nat. 139: 229–247.

Stearns, S. C., 1992. The Evolution of Life Histories. Oxford University Press, Oxford, 249 pp.

Stibor, H., 1995. Chemische Informationen in limnischen Räuber-Beute Systemen: Der Effekt von Räubersignalen auf den Lebenszyklus von *Daphnia* spp. (Crustacea: Cladocera). PhD Thesis, University of Kiel, 150 pp.

Tessier, A. J. & N. L. Consolatti, 1989. Variation in offspring size in *Daphnia* and consequences for individual fitness. Oikos 56: 269–276.

Tessier, A. J. & N. L. Consolatti, 1991. Resource quantity and offspring quality in *Daphnia*. Ecology 72: 468–478.

Tessier, A. J., L. L. Henry, C. E. Goulden & M. W. Durand, 1983. Starvation in *Daphnia*: Energy reserves and reproductive allocation. Limnol. Oceanogr. 28: 667–676.

Threlkeld, S. T., 1976. Starvation and the size structure of zooplankton communities. Freshwat. Biology 6: 489–496.

van Noordwijk, A. J. & G. de Jong, 1986. Acquisition and allocation of resources: their influence on variation in life history tactics. Am. Nat. 128: 137–142.

Vijverberg, J. & A. F. Richter, 1982. Population dynamics and production of *Daphnia hyalina* Leydig and *Daphnia cucullata* Sars in Tjeukemeer. Hydrobiologia 95: 235–259.

Vijverberg, J., R. D. Gulati & W. M. Mooij, 1993. Food-web studies in shallow eutrophic lakes by the Netherlands Institute of Ecology: Main results, knowledge gaps and new perspectives. Neth. J. aquat. Ecol. 27: 35–49.

Weider, L. J., 1993. Niche breadth and life history variation in a hybrid *Daphnia* complex. Ecology 74: 935–943.

Winkler, D. W. & K. Wallin, 1987. Offspring size and number: a life history model linking effort per offspring and total effort. Am. Nat. 129: 708–720.

Hydrobiologia **360**: 89–99, 1997.
A. Brancelj, L. De Meester & P. Spaak (eds), Cladocera: The Biology of Model Organisms.
©1997 *Kluwer Academic Publishers.*

Phenotypic plasticity of *Daphnia* life history traits in response to predator kairomones: genetic variability and evolutionary potential

Anke Weber[1] & Steven Declerck[2]
Publication number 2334 of The Netherlands Institute of Ecology, Centre for Limnology
[1] *Netherlands Institute for Ecology, Rijksstraatweg 6, 3631 AC Nieuwersluis, The Netherlands*
(e-mail: weber@cl.nioo.knaw.nl)
[2] *Laboratory of Animal Ecology, Ledeganckstraat 35, 9000 Gent, Belgium (e-mail: steven.declerck@rug.ac.be)*

Key words: broad-sense heritability, *Chaoborus*, *Perca*, kairomone mixture

Abstract

Cladoceran populations can respond to changing predation regimes by a phenotypical response as well as by shifts in genotype frequencies. In this study, we investigated the phenotypic plasticity exhibited by life history traits of *D. galeata* in response to the presence of predator kairomones, as well as the extent to which natural selection may act on these traits and their phenotypic plasticity. In a life-table experiment, seven clones of a natural *D. galeata* population were subjected to kairomones from fish (*Perca*), from an invertebrate predator (*Chaoborus*) or a mixture of both. Life history traits were affected by the kairomones of both predators, but effects of *Chaoborus* were neutralised by *Perca* in the kairomone mix. No apparent trade-off was found between growth- and reproduction related traits: although daphnids from the *Chaoborus* treatment grew faster than daphnids from the other treatments, no reduction in the reproductive output was observed. Broad-sense heritabilities were found to be relatively high for some life history traits (size at maturity, neonate size, number of neonates) as well as for the phenotypic plasticity response of these traits. This reflects the evolutionary potential of life history traits and their phenotypic response to predator kairomones in the *D. galeata* population.

Introduction

Cladocerans exhibit various life-history strategies which are considered to be responses to strong selection pressures imposed by predators, and predation is thought to act as potential mechanism influencing the evolution of cladocera life histories (review by Lynch, 1980; Hanazato & Yasuno, 1989; Brett, 1992). For example, seasonal changes in morphology in *Daphnia* have received much attention during the last decades (review by Jacobs, 1987), and both proximate (environmental) as well as ultimate (fitness) aspects of cyclomorphosis are well-studied (Sorensen & Sterner, 1992; Weider & Pijanowska, 1993; Spaak, 1995).

Recently, it has been found that organisms react to dissolved chemicals (kairomones) released by potential predators (review by Havel, 1987; Larsson & Dodson, 1993). These induced defences are of a temporary nature; they are only produced when the kairomone concentration is sufficiently high. The phenotypic responses to maximise fitness under predation can be changes in morphology, e.g. spines and neck teeth (Krüger & Dodson, 1981), behaviour, such as vertical migration in the presence of fish (Dawidowicz et al., 1990; Ringelberg, 1991) and life-history changes (Spitze, 1992, 1995; Stibor, 1991, 1992). The evolutionary significance of various life history strategies has been examined to cast more light on the trade-offs between somatic growth and reproduction (Stearns, 1992; Weider & Pijanowska, 1993; Tollrian, 1995).

Inducible defences as a strategy to reduce predation are generally assumed to impose costs (Stemberger & Gilbert, 1988). Costs can result, for example, from synthesis of extra body tissue or defence chemicals, or from escape to suboptimal environments. The key idea for the explanation of the evolution of inducible rather than constitutive defences is that costs associated with the maintenance of the defence are saved during times when the protection is not needed (Havel et al., 1990). Thus, phenotypic plasticity might offer great advantages to an organism that experiences fluctuating predation regimes during its life cycle.

Recent work (Taylor & Gabriel, 1993) suggests contrasting shifts in optimal resource allocation patterns in cladocerans exposed to either vertebrate or invertebrate predators. Vertebrate predators were shown to induce an increase in reproductive investment. Typical responses of daphnids exposed to fish kairomones are a relatively early maturation, a small size at first reproduction and larger clutch size (Reede, 1997). In contrast, invertebrate predators such as *Chaoborus* are known to induce a greater investment in growth, at the cost of reproduction: exposure to *Chaoborus* kairomones was found to result in a later maturation at a larger body size, often with fewer eggs in the first clutch (Brett, 1992; Black, 1993; Stibor & Lüning, 1994; Lüning, 1995). These contrasting life history responses are mostly considered to be adaptive (Larsson & Dodson, 1993) and are explained by differences in prey size selectivity between vertebrate and invertebrate predators. Whereas optimal prey size of invertebrate predators is restricted to daphnids smaller than approximately 0.9 mm (Pastorok, 1981; Herzig & Auer, 1990), vertebrate predators hunt visually and select larger prey (review by O'Brien et al., 1980). *Daphnia* clones reproducing early and at a less conspicuous size will be favoured under fish predation, while animals mainly investing in growth during the juvenile period and growing beyond the size of high risk before investing in egg production will be more successful under high *Chaoborus* predation.

Although the phenotypic response of life history traits and morphological traits of *Daphnia* to predator kairomones has already been extensively investigated, most studies have focused on the effects of kairomones of single predators. Under natural conditions, however, several predators, often with differing prey size preferences, may occur simultaneously. In such a case, prey are faced with an important dilemma: a phenotypic response maximising survival to one predator may increase vulnerability to the other predator. To date, only two studies are available to illustrate the effect of a combination of predator kairomones (Lüning, 1995; Black, 1993) both using *Chaoborus* and *Notonecta* cues.

Most studies that focus on the phenotypic response of *Daphnia* traits to kairomones are based on one or a few clones. As a result, information on the genetic component of such phenotypic plasticity in populations is scarce. *Daphnia* populations consist of a number of clones that may reproduce by ameiotic parthenogenesis for many generations, especially in permanent environments such as lakes. Clones may differ in mean values as well as in phenotypic plasiticity of traits. The genetic variance in a population determines the ability of that population to respond to changing selection regimes. Natural selection can modify the clonal composition of a population by acting on the mean value of a trait in a given environment, as well as on their phenotypic plasticity.

In the present study, we investigated the phenotypic plasticity of life history traits in a population of *Daphnia galeata* exposed to predator kairomones. In a life table experiment, seven clones of *D. galeata* from a lake with high fish predation were tested on their reaction to water containing kairomones a vertebrate predator (*Perca*) and an invertebrate predator (*Chaoborus*) in a full factorial design. Genetic variability and evolutionary potential of several life history traits and their phenotypic plasticity were quantified through the estimation of broad-sense heritabilities.

Material and methods

Experimental design

To unravel the effect of kairomones of contrasting predatory regimes on *Daphnia* life history, a full factorial design was chosen. Seven *D. galeata* clones, collected from one population (Lake Tjeukemeer, The Netherlands), were cultured either in kairomone-free water (control) or in water enriched with *Chaoborus* kairomone, *Perca* kairomone or a mixture of both. In total, the experimental set-up consisted of 112 individual cultures for seven clones under four treatments, with four replicates per clone and per treatment. The mothers of the experimental animals were raised under the experimental conditions in order to minimize interference from maternal effects (Lampert, 1993). Neonates originating from the third brood of the mother were raised as experimental animals. Initially, two of them were incubated per experimental tube (100 ml). Another sister neonate was measured and discarded. After three days, one randomly chosen animal of each tube was discarded so that one *Daphnia* remained in each tube. Animals were examined every 12 hours. The age at maturity (time of deposition of the first clutch in the brood pouch), size at maturity, timing of subsequent clutches, number of neonates per clutch, and neonate sizes were recorded. This paper is based on data of the first four adult instars except for mortality estimates which were made after the daphnids reached the sixth adult instar.

Media

Water of Lake Maarsseveen served as basic medium. Prior to use, the water was aged for at least one day in a sandfilter to ensure the bacterial break-down of naturally occurring kairomones. The predators (*Perca fluviatilis, Chaoborus americanus*) were kept in two separate aquaria, each of them containing 10 litres of water. In one of the aquaria, two perch, equal in size and with a total biomass of 7.1 g, were daily fed fresh oligochaete worms (*Tubifex*). In the other aquarium, 200 fourth instar *Chaoborus* larvae were daily fed *Daphnia* (approximately five individuals per litre). To prevent possible confounding effects of the presence of *Daphnia*-chemicals on the *Chaoborus* treatment, an equal amount of daphnids was daily inoculated in the fish-containing aquaria as well as in the control medium. These *Daphnia* densities of 5 ind. l^{-1} in the control medium are far beyond the *Daphnia* densities known to lead to crowding effects (40 ind. l^{-1}; Goser & Ratte, 1994). In contrast to the control, feeding of *Chaoborus* and *Perca* on the *Daphnia* has probably resulted in an amount of alarm substances of *Daphnia* in all of the predator treatments. It should therefore be kept in mind that the outcome of the experiment should be considered as the result of the combined effect of both kairomones and alarm substances, as it normally occurs in the field. From each predator aquarium, seven litres were daily taken and replaced by sandfiltered water. This water was further used to prepare four treatments: (1) a control, without any predator kairomone, solely consisting of sandfiltered water; (2) a *Chaoborus* treatment, a 1:1 mixture of control water and water inhabited by *Chaoborus*, (3) a mixed treatment, a 1:1 mixture of water inhabited by *Chaoborus* and water inhabited by *Perca*, (4) a *Perca* treatment, a 1:1 mixture of control water with water inhabited by *Perca*. Media for each treatment were filtered through 0.45 μm membrane filters prior to use, to eliminate bacteria, faeces from the predators and single surviving *D. pulex*. The green alga *Scenedesmus acutus* was added as food source in a concentration of

1 mg C l^{-1}, which is well above incipient limiting level (Lampert, 1977). The algae were cultured at high nutrient levels in a continuous culture with a growth rate of 1.3 d^{-1}. The concentration was established photometrically. The media were refreshed daily to keep food and kairomone concentrations sufficiently high. The photoperiod consisted of a 16 hours light to 8 hours dark cycle. Ambient temperatures were kept constant at 17.5 °C.

Daphnia

The experiment was performed with seven randomly taken clones (labelled: clone 1–7) of *D. galeata* from Tjeukemeer, a shallow eutrophic lake in the northern part of the Netherlands (Beattie et al., 1979). Predation pressure by young planktivorous fish (0+ smelt, bream, perch, pikeperch and roach) is generally high in Tjeukemeer (Lammens et al., 1985; Vijverberg et al., 1990) while invertebrate predation (mainly by *Leptodora*) is probably of less importance (Vijverberg, pers. comm.). Twenty clones were collected in November 1995 and cultured for three months in 1 l jars. Seven of them were randomly selected for the experiment. Genetic analyses by RAPD (random amplified polymorphic DNA: Williams et al., 1990; Hadrys et al., 1992) confirmed that all clones belonged to the taxon *D. galeata* but differed in genotype.

Data analysis

Somatic growth rate of an individual (mm d^{-1}) was estimated as the body length increase per unit of time during the juvenile period. This estimate may, however, be confounded by changes in animal morphology. To account for this, additional measurements were done on body width. The intrinsic rate of increase (r) was calculated according to the Euler-equation:

$$1 = \sum e^{-rx} \times l_x \times m_x,$$

where r is the per capita rate of increase for the population (d^{-1}), x the age class (0, 1...N), l_x the probability of surviving to age x, and m_x the fecundity at age x. The estimates of the intrinsic rate of population increase r were used as a measure of fitness.

Prior to analysis, the life history data were tested for normality with the Wilk-Shapiro procedure after checking for equality of variance with Levene's test. Mortality, estimated for each clone as the percentage of animals that died before reaching the sixth adult instar, was transformed to the arcsinus of its square root. The effects of the independent variables *'Chaoborus'* and *'Perca'* and of the random variable 'clone' were tested by a three-way mixed-model analysis of variance (ANOVA). Sequential Bonferroni corrections were applied to minimise the occurrence of type 1 errors (Rice, 1989). Significant clone effects or interaction effects between treatments were further explored with the Tukey HSD test for unequal N (Sokal & Rohlf, 1981). When significant clone × treatment interactions were found, one- or two-way ANOVAs, followed by a sequential Bonferroni correction, were performed for each clone separately in order to determine which clones differed in their response to the treatments.

Broad-sense heritabilities (H^2) were calculated for each trait in each treatment through a clonal repeatability analysis (Lessels & Boag, 1987). We also consider the plasticity of the phenotypic response of each trait to the presence of a predator kairomone as a trait. Therefore, the broad-sense heritability of the phenotypic response of each trait to the presence of *Perca* kairomone, *Chaoborus* kairomone and the mixture of both was calculated.

Results

Reproduction-related traits

For age at maturity a strong interaction between *Perca* and *Chaoborus* kairomone was noticed (Table 1; Figure 1a). A posthoc test (Tukey HSD) revealed a significantly lower age at maturity ($p<0.01$) in the *Chaoborus* treatment than in the control (Table 1). The fact that no differences could be shown between the control and the mixed treatment indicates that the presence of *Perca* kairomone in the mixed treatment hampered the effect of the *Chaoborus* kairomones.

Although no main effects were found for the total number of neonates produced during the first four adult instars (Figure 1b) clones responded differently to the presence of *Perca* kairomone (Table 1). One way-ANOVAs performed on data of each separate clone showed that only clone 4 produced significantly less neonates in the *Perca* (Table 1).

No effects of neither clone nor treatment were found for the intrinsic rate of increase (r) (Table 1; Figure 1c).

Size-related traits

The presence of *Perca* kairomones had a significant effect on the size at maturity (Table 1; Figure 1d). The daphnids matured at a smaller size in treatments containing the *Perca* kairomone (Figure 1d).

A strong interaction between *Chaoborus* and *Perca* kairomone was noticed (Table 1, Figure 1e). Tukey HSD revealed a significantly higher somatic growth rate in the *Chaoborus* treatment (Table 1) than in all other treatments ($p<0.01$).

Neonates were significantly smaller in the *Perca* and mixed treatment than in the control and the *Chaoborus* treatment (Figure 1f). Moreover, significant differences in neonate size were found between clones and clones were affected differently by the interplay of kairomones (clone × *Perca* × *Chaoborus* interaction). The results of two way-ANOVAs done on data of the separate clones show that three out of the seven clones changed their neonate size in the presence of kairomones (clones 4, 6 and 7), each showing a different pattern (Figure 2). Clone 7 was only affected by the *Perca* kairomone and produced significantly smaller neonates in the presence of this kairomone ($p<0.01$; Figure 2c). Clone 6 responded to both the *Perca* and the *Chaoborus* kairomone with a decrease in neonate size ($p<0.01$, Figure 2b). The response of neonate size in the mixed treatment was intermediate to that in the *Perca* and *Chaoborus* treatment. For clone 4, a main effect for *Chaoborus* kairomone was found ($p<0.05$) as well as an interaction between both kairomones ($p<0.001$, Figure 2a). Neonates were larger in the *Chaoborus* treatment than in the control (Tukey HSD: $p<0.0001$), but this effect was neutralised by the presence of *Perca* kairomones in the mixed treatment.

Mortality

A tendency towards an interaction effect between *Perca* and *Chaoborus* was observed (Table 1, Figure 1g). Mortality was higher in the *Chaoborus* than in the control treatment (Tukey HSD: $p<0.05$), but this effect was not significant on the table-wide level.

Broad-sense heritability

Significant broad-sense heritabilities were found for the traits number of neonates in the *Perca* treatment, size at maturity and somatic growth rate in the mixed treatment, and for neonate size in the control (Table 2). In addition, a significant broad-sense heri-

Table 1. Results of three-way ANOVAs (with clone as random factor). On mortality estimates a two-way ANOVA was performed. Traits: (a) age at maturity, (b) size at maturity, (c) somatic growth rate, (d) total number of neonates, (e) neonate size, (f) r and (g) mortality. Data were analyzed for the effect of Chaoborus kairomone (Ch), Perca kairomone (P) and clone (C). *: p<0.05, **: p<0.01, ***: p<0.001. — * Indicates significance using the sequential Bonferroni test with an initial critical probability value of p<0.007 (seven non-independent traits tested for clone, Chaoborus and Perca or 0.008 (six non-independent traits tested for clone × Chaoborus, clone × Perca, Chaoborus × Perca and clone × Chaoborus × Perca.

Source	df	df error	a) age at maturity MS	MS error	F	b) total number of neonates MS	MS error	F	c) r MS	MS error	F	d) size at maturity MS	MS error	F	e) somatic growth rate MS	MS error	F	f) neonate size MS	MS error	F
clone (C)	6	72	0.0033	0.0021	1.62	143.688	80.0139	1.80	0.0014	0.0008	1.76	0.0223	0.0073	3.04*	0.0496	0.0234	2.12	0.0135	0.0028	4.77***
Chaoborus	1	6	0.0094	0.0029	3.22	201.671	140.978	1.43	0.0065	0.0006	9.39	0.0008	0.0092	0.09	0.0675	0.0221	3.05	0.0004	0.0053	0.09
Perca (P)	1	6	0.0001	0.0033	0.03	1065.681	282.484	3.77	0.0006	0.0005	1.25	1.2724	0.0216	58.82***	0.8423	0.0693	12.15*	0.1268	0.0041	31.08**
C × Ch	6	72	0.0029	0.0021	1.44	140.977	80.0138	1.76	0.0007	0.0008	0.84	0.0092	0.0073	1.26	0.0221	0.0234	0.95	0.0053	0.0028	1.89
C × P	6	72	0.0033	0.0021	1.64	282.483	80.0138	3.53**	0.0005	0.0008	0.64	0.0216	0.0073	2.95	0.0693	0.0234	2.95*	0.0041	0.0028	1.44
Ch × P	1	6	0.0157	0.0003	39.28**	118.836	74.3007	1.60	0.0011	0.0003	12.19*	0.0076	0.0131	0.58	0.1248	0.0077	16.08**	0.0076	0.0158	0.48
C × Ch × P	6	72	0.0003	0.0021	0.20	74.301	80.0139	0.93	0.0003	0.0008	0.44	0.0131	0.0073	1.78	0.0077	0.0234	0.33	0.0158	0.0028	5.57***

g) mortality

source	df	MS	df error	MS error	F
C	1	0.0106	24	0.0029	3.58
Ch	1	0.0011	24	0.0029	0.35
P	1	0.0191	24	0.0029	6.38*

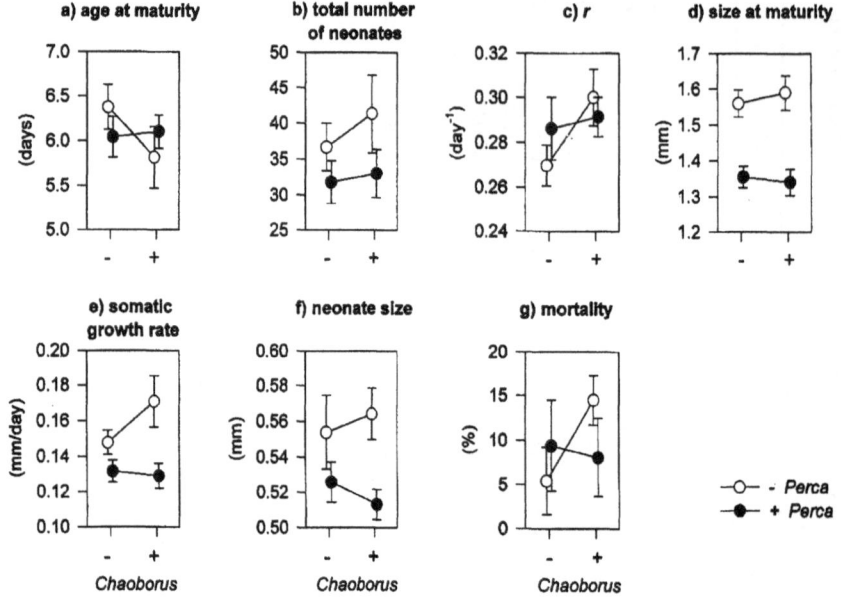

Figure 1. Reaction norm plots of life history traits in response to kairomone treatments (means ± 2 SE).

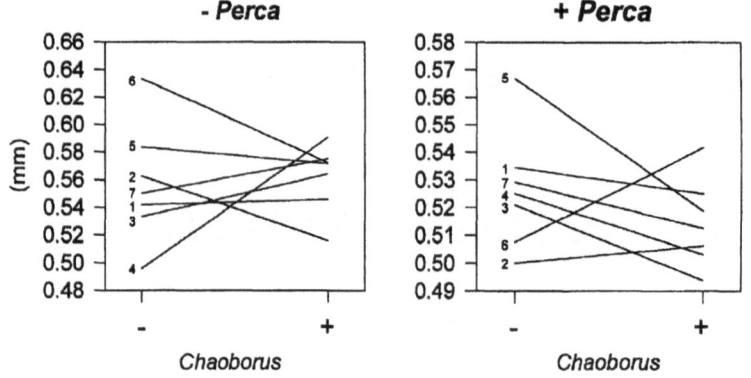

Figure 2. Reaction norm plots of neonate size in response to kairomone treatments. Numbers refer to clones 1–7.

tability for the phenotypic response was calculated for number of neonates in the presence of *Perca* kairomone (Table 3), and for neonate size to the interaction of *Chaoborus* × *Perca* kairomone (Table 3).

Discussion

Treatment effects

The response of the *D. galeata* individuals to fish chemicals was in accordance with the results of former studies (Reede, 1995, 1997). The presence of fish kairomone resulted in a reduction of the somatic growth rate and in a smaller size at first reproduc-

tion (Figure 1). This apparent reduction of allocation towards growth, however, did not result in an increased reproductive output. No increase in the number of neonates was observed and the mean neonate size was found to be lower than in the other treatments.

In contrast to *Perca*, no main effects of *Chaoborus* were found for any of the traits studied. However, some life history traits were affected by interactions between the *Perca* and *Chaoborus* kairomone (Table 1). Size at maturity was the same in the control and *Chaoborus* treatment, but this size was reached at an earlier age in the *Chaoborus* treatment than in the control due to an enhanced somatic growth rate (Figure 1). The effect of *Chaoborus* was completely neutralised, however, in the mixed treatment, i.e. in the presence of

Table 2. Broad-sense heritabilities (H^2) calculated per trait and per treatment.*: $p<0.05$; **: $p<0.01$; ***: $p<0.001$. −* Indicates significance using the sequential Bonferroni test with an initial critical probability value of $p<0.0125$ (four non-independent traits tested).

Trait	*Chaoborus*	Mix	*Perca*	Control	Mean
a) age at maturity	−0.02	−0.10	0.25	0.38*	0.13
b) total number of neonates	0.19	0.35*	0.65***	−0.11	0.27
c) r	0.04	0.04	−0.08	−0.07	0.02
d) size at maturity	0.16	0.71**	0.08	0.18	0.28
e) somatic growth rate	0.04	0.58**	0.35*	0.10	0.27
f) neonate size	0.26	0.28	0.33	0.53**	0.35

Perca kairomone. Similar patterns were found for neonate number, neonate size and percentage mortality. Mean values of these traits were higher in the *Chaoborus* treatment than in the control, in contrast to the mixed treatment. The interaction between *Perca* and *Chaoborus* kairomone was found to be highly clone-dependent for neonate size (Table 1).

The differences found between the *Chaoborus* treatment and the control in our study are not in agreement with previous studies. Exposure of *Daphnia* to *Chaoborus* is most often found to result in an increase of the age at maturation (Black, 1993; Stibor & Lüning, 1994; Lüning, 1995) and size at maturity (Black, 1993). Moreover, *Chaoborus*-induced phenotypic plasticity has been found to be associated with a reduced rate of body growth (Lüning, 1995) and a reduced allocation of energy to reproduction (Stibor & Lüning, 1994). Most of these studies, however, were done on *D. pulex*. The difference in results between our study and previous work may indicate differences in strategies between different taxa, on the species as well as clonal level, to the presence of a predator. Nevertheless, in a life-history evolution study on *D. pulex* under *Chaoborus* predation, Spitze (1991) found trends similar to those of our study. Under *Chaoborus* predation, populations evolved toward larger body size and an increased and earlier reproduction. An increase in reproduction was also the outcome of individual-based simulations of size-selective predation on unconditioned *D. galeata* (Mooij & Boersma, 1996). This might imply, that for larger *Daphnia* species such as *D. pulex* the response to kairomones from *Chaoborus*, a predator selecting for small prey, differs significantly from the response to predation by this predator.

Under the assumption of a fixed energy budget for growth and reproduction, a trade-off between energy allocated to growth versus energy allocated to reproduction is expected. Such a trade-off is not appar-

Table 3. Broad-sense heritabilities (H^2) calculated for the phenotypic plasticity of traits in response to the presence of *Chaoborus* and *Perca* kairomone. *: $p<0.05$; **: $p<0.01$; ***: $p<0.001$. −* Indicates significance using the sequential Bonferroni test with an initial critical probability value of $p<0.008$ (six non-independent traits tested).

Trait	Clone × Chaoborus	Clone × Perca	Clone × Chaoborus × Perca
a) age at maturity	0.08	0.10	−0.13
b) total number of neonates	0.10	0.26**	−0.01
c) r	−0.02	−0.05	−0.09
d) size at maturity	0.04	0.21	0.10
e) somatic growth rate	0.02	0.16	−0.09
f) neonate size	0.12	0.07	0.40***

ent from our results. The relatively high somatic growth rate during the juvenile period of animals in the *Chaoborus* treatment suggests a high allocation of energy to growth. Such high investment of energy in growth did not, however, result in a reduced reproductive output. Animals had a higher intrinsic rate of increase and tended to produce more and larger neonates in the presence of *Chaoborus* than in the control. Although not significant, a tendency towards an increased mortality in the *Chaoborus* treatment may indicate a trade-off between early versus late energy investment, rather than a trade-off between growth and reproduction. The fact that no increased mortality was found in the mixed treatment indicates that the enhanced mortality of animals in the *Chaoborus* treatment can not have been due to toxicity associated with the *Chaoborus* kairomone. In the *Perca* treatment, no trade-off was found at all. A reduced somatic growth rate in the juvenile period was not compensated by an increased survival or an increased reproductive effort: neonate size and total number of neonates were smaller in the *Perca* treatment than in the control. Trade-offs may become less apparent when the total available energy budget for growth and reproduction is affected by the treatment. Ingestion rates of *D. galeata* are modified by kairomones (Weber, unpublished results), while behavioural responses of *Daphnia* to predator kairomones, such as an increased alertness or a modification of the vertical migration behaviour (Dodson & Ramcharan, 1991; De Meester, 1996) have been described. Weber & Declerck (in prep.) have observed reduced swimming activity of *D. galeata* in the presence of *Chaoborus*. Therefore, the total amount of energy available for growth and reproduction may have been higher in the *Chaoborus* treatment than in the control due to a reduction of the amount of energy invested in locomotary activity. The lack of trade-offs in the *Perca* treatment may be explained by costs associated with a small body size. Fish are size selective feeders because of their visual hunting mode. Therefore, the observed response towards smaller neonate size and a smaller size at maturity in *Daphnia* may be considered to be adaptive (Larsson & Dodson, 1993). At a small size, the number and size of neonates is restricted by the volume of the brood pouch. The reduced somatic growth rate may have been due to metabolic costs associated with a small body size, such as a reduced filtration rate and a higher respiration rate per unit of body mass as proposed by the size efficiency hypothesis (Hall et al., 1976).

In our study, an interaction between *Perca* and *Chaoborus* kairomones was found for several traits, such as age at maturity, somatic growth rate and neonate size (Table 1, Figure 1). For all these traits, the effects induced by the *Chaoborus* cue were neutralised by fish kairomone in the mixed treatment. As a matter of fact, similar patterns were noticed to a variable extent for most other characteristics (Figure 1). The combined effect of kairomones on the life history of *Daphnia* was analysed in two former studies (Lüning, 1992; Black, 1993). In contrast to our work, both studies investigated the effect of *Chaoborus* and *Notonecta* on *D. pulex*. There is, however, one important similarity between our work and the work of Lüning (1992) and Black (1993). In all cases, the combined effects of two predators with contrasting prey size preference were investigated: *Chaoborus* select for small or young *Daphnia* individuals, whereas both *Perca* and *Notonecta* select for large, adult animals. Though the response on *Chaoborus* was different in our study from the one described by Lüning (1992), similar interaction effects

were found between predators selecting large prey and predators selecting small prey.

The phenotypic response of a *Daphnia* clone to predator kairomones is expected to be influenced by the history of selection regimes in the lake from where the clone originates. Lüning (1992) used only two clones and there is not much information available about the ecological conditions in the lakes from which her clones were isolated. The clones used in our experiment were isolated from Tjeukemeer. The densities of the *Chaoborus* population in Tjeukemeer have been very low during the last decades (Vijverberg, pers. comm.). Adaptation to *Chaoborus* predation in Tjeukemeer can have evolved in the past (no paleolimnological data available) or, more probably, Tjeukemeer may originally have been colonised by clones originating from habitats where *Chaoborus* was an important predator. The fact that Tjeukemeer clones still have the ability to respond to the presence of *Chaoborus* indicates that the cost associated to phenotypic plasticity is probably low or non-existent. The interaction effect of the kairomones on the *Daphnia* clones in our experiment may be understood when present predation regimes in Tjeukemeer are considered. Fish has been shown to be the most important predator in this lake (Lammens et al., 1985; Vijverberg et al., 1990). Therefore, a strong selection may be expected against clones that do not properly respond to the presence of fish (e.g. by a reduction of size at maturity and neonate size). Though the ability to respond to *Chaoborus* kairomones has been maintained in the population, clones responding to the occasional presence of *Chaoborus* in the presence of fish have probably been wiped out. Due to a lack of a way to quantify kairomone concentrations precisely, differences in kairomone concentrations between treatments could not be taken into account. Therefore, more investigations should be done where animals are exposed to dilution series of different kairomones. In order to determine in how far *Daphnia* is able to 'estimate' the relative threat of each predator based on kairomone concentrations, experiments should be done where *Daphnia* is exposed to kairomones of different predators at combinations of varying concentrations.

Genetic variability and evolutionary potential of life history traits

The significant clone effect and the two clone by treatment interactions revealed by the ANOVAs for several traits indicate genetic variation for life history in the

D. galeata population of Tjeukemeer (Table 1). Broad-sense heritability (H^2) of a trait is defined as the proportion of the total phenotypic variance explained by the total genetic variance. In populations that reproduce by ameiotic parthenogenesis, broad-sense heritability can be correctly inserted in the breeder's equation (Spitze, 1995): $R = H^2 S$ which relates evolutionary response (R) to selection (S). Therefore, in the parthenogenetically reproducing *D. galeata* population of Tjeukemeer, the broad-sense heritability estimates of a given trait reflect the evolutionary potential of this trait. Though broad-sense heritability is a basic tool for the understanding of the evolutionary dynamics of parthenogenetically reproducing populations, estimates are scarce (review by Spitze, 1995). Most H^2-estimates have been made for large *Daphnia* species, typically inhabiting temporary ponds. H^2-estimates were done for some life history traits in *D. pulex*, *D. obtusa* or *D. magna* populations (Lynch, 1984, 1994) and for phototactic behaviour in *D. magna* populations (De Meester, 1996). However, very few estimates are available for smaller *Daphnia* species like *D. galeata*, which typically occur in permanent lakes. Spitze (1995) calculated a mean broad-sense heritability for size at maturity, age at maturity and clutch size across 19 populations of 6 *Daphnia* species, based on a compilation of published life history data. Mean values were rather high and ranged between 0.31 and 0.34. In our study, similar mean broad-sense heritabilities were found for life history traits such as somatic growth rate and neonate size, as well as for the phenotypic response of several life history traits to the presence of predator kairomones. Variation between traits was high, however. Mean H^2-values (Table 2) were very low for the intrinsic rate of population increase (0.02) and age at maturity (0.13), while relatively high mean H^2-values were found for neonate size (0.35), size at maturity (0.28), neonate number (0.27) and somatic growth rate (0.27). In addition, a significant H^2-value of 0.26 was found for the phenotypic response of the number of neonates to the presence of *Perca* kairomone and the response of neonate size to the interaction of both the *Chaoborus* and *Perca* kairomone (H^2-value: 0.40; Table 3). Due to genotype-by-treatment interactions, broad-sense heritabilities of some traits were found to vary between treatments (Table 2).

Life history traits strongly determine the fitness of an individual in a given environment. Constant directional selection will lead to low genetic variance for fitness-associated traits. Low broad-sense heritabilities should therefore be expected for life history traits of parthenogenetically reproducing *Daphnia* populations, as long as there are no alternative strategies with similar fitness (Spitze, 1995). High levels of expressed genetic variance may be reached following a bout of sexual reproduction, but a strong decrease of this variance should be expected after a limited number of generations, as has been documented for a *D. pulex* population (Lynch, 1984). In our study, the relatively high broad-sense heritabilities found for some life history traits in some treatments were remarkable. This might be explained by the existence of different strategies with similar fitness (De Meester, 1994), though our data do not provide any evidence for alternative life history strategies: trends for which significant H^2-values were calculated do not appear to be in mutual trade-off (Table 2). The *D. galeata* clones used in our experiment were originally isolated during fall, a period during which hatching of sexual eggs is generally expected to be low. Apparently, expressed genetic variance was maintained throughout the growing season, possibly due to fluctuations in selection regimes. Lake Tjeukemeer contains both vertebrate as well as invertebrate predators (*Leptodora*, cyclopoid copepods, *Notonecta*). Predation intensity exerted by these predators on *D. galeata* and other environmental factors change in time and space and may favour the coexistence of clones with differing phenotypes and with differing phenotypic plasiticities.

The relatively high heritabilities calculated for some life history traits (neonate size, size at maturity, somatic growth rate and number of neonates; Tables 2, 3) reveal the evolutionary potential with respect to these traits in the *D. galeata* population of Tjeukemeer. Selection on these traits can result in a shift in genotype frequencies. Moreover, the genetic variation found for phenotypic plasticity of these traits indicates a potential for evolution of phenotypic plasticity for life history traits. This supports the idea that phenotypic plasticity is a fundamental component of evolutionary change rather than constraining it (Thompson, 1991). As a consequence of genotype-dependent phenotypic plasticity, a high variation in heritabilities of traits was found between treatments in our study. Such genotype-by-treatment interaction indicates that monitoring of the heritability dynamics of natural populations should be done under conditions resembling the conditions in the field as much as possible.

Acknowledgments

We would like to thank Piet Spaak, Luc De Meester, Maarten Boersma, Bill DeMott, Ramesh Gulati, Wolf Mooij, Arie van Noordwijk, Koos Vijverberg and Henri Dumont for their valuable suggestions and constructive comments on earlier versions of the manuscript. We also thank Peter Mac Gillavry, Guus Postema and Klaas Siewertsen for practical assistance. S. Declerck acknowledges a scholarship provided by the Flemisch Institute for the stimulation of Scientific and Technological Research in the Industry (I.W.T., Belgium) and logistic support by the Centre of Limnology (NIOO, The Netherlands) and by project 01103595 of the University of Ghent (Belgium).

References

Beattie, D. M., H. L. Golterman & J. Vijverberg, 1979. An introduction to the limnology of the Friesian lakes. Hydrobiologia 58: 49–64.

Black, A. R., 1993. Predator-induced phenotypic plasticity in *Daphnia pulex*: Life history and morphological responses to *Notonecta* and *Chaoborus*. Limnol. Oceanogr. 38: 986–996.

Brett, M. T., 1992. *Chaoborus* and fish-mediated influences on *Daphnia longispina* population structure, dynamics and life history strategies. Oecologia 89: 69–77.

Davidowicz, P., J. Pijanowska & K. Ciechomski, 1990. Vertical migration of *Chaoborus* larvae is induced by the presence of fish. Limnol. Oceanogr. 35: 1631–1637.

De Meester, L., 1994. Life histories and habitat selection in *Daphnia*: divergent life histories of *D. magna* clones differing in phototactic behaviour. Oecologia 97: 333–341.

De Meester, L., 1996. Evolutionary potential and local genetic differentiation in a phenotypically plastic trait of a cyclical parthenogen, *Daphnia magna*. Evolution 50: 1293–1298.

Dodson, S. & C. Ramcharan, 1991. Size-specific swimming behaviour of *Daphnia pulex*. J. Plankton Res. 13: 1367–1379.

Goser, B. & H. T. Ratte, 1994. Experimental evidence of negative interference in *Daphnia magna*. Oecologia 98: 354–361.

Hadrys, H., M. Ballick & B. Schierwater, 1992. Applications of random amplified polymorphic DNA (RAPD) in molecular ecology. Mol. Ecol.: 55–63.

Hall, D. J., S. T. Threlkeld, D. C. W. Burns & P. H. Crowley, 1976. The size-efficiency hypothesis and the size structure of zooplankton communities. Ann. Rev. Ecol. Syst. 7: 177–208.

Hanazato, T. & M. Yasuno, 1989. Zooplankton community structure driven by vertebrate and invertebrate predators. Oecologia 81: 450–458.

Havel, J. E., 1987. Predator-induced defenses: A review. In W. C. Kerfoot and A. Sih (eds), Predation: Direct and Indirect Impacts on Aquatic Communities. New England: 263–278.

Havel, J. E., P. D. N. Hebert & L. D. Delorme, 1990. Genetics of sexual Ostracoda from a low arctic site. J. evol. Biol. 3: 65–84.

Herzig, A., & B. Auer, 1990. The feeding behaviour of *Leptodora kindtii* and its impact on the zooplankton community of Neusiedler See (Austria). Hydrobiologia 198: 107–117.

Jacobs, J., 1987. Cyclomorphosis in *Daphnia*. Memoire dell'Instituto Italiano di Idrobiologica Dott. Marco de Marchi 45: 325–352.

Krüger, D. A. & S. Dodson, 1981. Embryological induction and predation ecology in *Daphnia pulex*. Limnol. Oceanogr. 26: 219–223.

Lammens, E. H. R. R., H. W. De Nie, J. Vijverberg & W. L. T. Van Densen, 1985. Resource partitioning and niche shifts of bream (*Abramis brama*) and eel (*Anguilla anguilla*) in Tjeukemeer. Can. J. Fish. aquat. Sci. 42: 1342–1351.

Lampert, W., 1977. Studies on the carbon balance of *Daphnia pulex* De Geer as related to environmental conditions II. The dependence of carbon assimilation on animal size, temperature, food concentration and diet species. Arch. Hydrobiol. Suppl. 48: 310–335.

Lampert, W., 1993. Phenotypic plasticity of the size at 1st reproduction in *Daphnia* – the importance of maternal size. Ecology 74: 1455–1466.

Larsson, P. & S. Dodson, 1993. Invited review – chemical communication in planktonic animals. Arch. Hydrobiol. 129: 129–155.

Lessels, C. M. & P. T. Boag, 1987. Unreparable repeatabilities: a common mistake. The Auk 104: 116–121.

Lüning, J., 1992. Phenotypic plasticity of *Daphnia pulex* in the presence of invertebrate predators – morphological and life history responses. Oecologia 92: 383–390.

Lüning, J., 1995. Life-history responses to *Chaoborus* of spined and unspined *Daphnia pulex*. J. Plankton Res. 17: 71–84.

Lynch, M., 1980. The evolution of cladoceran life histories. Quart. Rev. Biol. 55: 23–42.

Lynch, M., 1984. The limits to life history evolution in *Daphnia*. Evolution 38: 465–482.

Lynch, M., 1994. Evolutionary genetics of *Daphnia*. In Real, L. (ed.), Ecological Genetics. Princeton University Press, Princeton, NY.: 109–128.

Mooij, W. M. & M. Boersma, 1996. An object-orientated simulation framework for individual-based simulations (OSIRIS): *Daphnia* population dynamics as an example. Ecol. Modell. 93: 139–153.

O'Brien, W. J., D. Kettle, H. Riessen, D. Schmidt & D. Wright, 1980. Dimorphic *Daphnia longireis*: Predation and competitive interactions between two morphs. In Kerfoot, W. C. (ed.), Evolution and Ecology of Zooplankton Communities. Univ. Press New England, Hanover: 497–505.

Pastorok, R. A., 1981. Prey vulnerability and size selection by *Chaoborus* larvae. Ecology 62: 1311–1324.

Reede, T., 1995. Life history shifts in response to different levels of fish kairomones in *Daphnia*. J. Plankton Res.17: 1661–1633.

Reede, T., 1997. Effects of neonate size and food concentration on the life history responses of a clone of the hybrid *Daphnia hyalina* × *galeata* to fish kairomones. Feshwat. Biol. 37: 389–396.

Rice, W. R., 1987. Analyzing tables of statistical tests. Evolution 43: 223–225.

Ringelberg, J., 1991. A mechanism of predator-mediated induction of diel vertical migration in *Daphnia hyalina*. J. Plankton Res. 13: 83–89.

Sokal, R. R. & F. J. Rohlf, 1981. Biometry. Freeman and Company, San Francisco, 859 pp.

Sorensen, K. H. & R. W. Sterner, 1992. Extreme cyclomorphosis in *Daphnia lumholtzi*. Freshwat. Biol. 28: 257–262.

Spaak, P., 1995. Cyclomomorphosis as a factor explaining success of a *Daphnia* hybrid in Tjeukemeer. Hydrobiologia 307: 283–289.

Spitze, K., 1991. *Chaoborus* predation and life-history evolution in *Daphnia pulex*: temporal pattern of population diversity, fitness, and mean life history. Evolution 45: 82–92.

Spitze, K., 1992. Predator-mediated plasticity of prey life history and morphology – *Chaoborus americanus* predation on *Daphnia pulex*. Am. Nat. 139: 229–247.

Spitze, K., 1995. Quantitative genetics of zooplankton life histories. Experientia 51: 454–465.

Stearns, S., 1992. The Evolutionary Life Histories. Oxford University Press, New York, USA: 123–179.

Stemberger, R. S. & J. J. Gilbert, 1987. Defenses of planktonic rotifers against predators. In Kerfoot, W. C. & A. Sih (eds), Predation: Direct and Indirect Impacts on Aquatic Communities. New England: 227–239.

Stibor, H., 1991. Größenvariabilität von *Daphnia* spp. bei der ersten Reproduktion. Diplom–thesis, Univ. Kiel. Germany.

Stibor, H., 1992. Predator-induced life-history shifts in a freshwater cladoceran. Oecologica 92: 162–165.

Stibor, H. & J. Lüning, 1994. Predator-induced phenotypic variation in the pattern of growth and reproduction in *Daphnia hyalina* (Crustacea: Cladocera). Funct. Ecol. 8: 97–101.

Taylor, B. E. & W. Gabriel, 1993. Optimal adult growth of *Daphnia* in a seasonal environment. Funct. Ecol.: 513–521.

Thompson, J. D., 1991. Phenotypic plasticity as a component of evolutionary change. Trends Ecol. Evol. 6: 246–249.

Tollrian, R., 1995. Predator-induced morphological defences: Costs, life history shifts, and maternal effects in *Daphnia pulex*. Ecology 76: 1691–1705.

Vijverberg, J., M. Boersma, W. L. T. van Densen, W. Hoogenboezem, E. H. R. R. Lammens & W. M. Mooij, 1990. Seasonal variation in the interactions between piscivorous fish, planktivorous fish and zooplankton in a shallow eutrophic lake. Hydrobiologia 207: 279–286.

Weider, L. J. & J. Pijanowska, 1993. Plasticity of *Daphnia* life histories in response to chemical cues from predators. Oikos 67: 385–392.

Williams, J. G. K., A. R. Kubelik, K. J. Livak, J. A. Rafalski & S. V. Tingey, 1990. DNA polymorphisms amplified by arbitrary primers are useful as genetic markers. Nucl. Acids Res. 18: 6531–6535.

Hydrobiologia **360**: 101–108, 1997.
A. Brancelj, L. De Meester & P. Spaak (eds), Cladocera: The Biology of Model Organisms.
©1997 *Kluwer Academic Publishers.*

Life tables of *Moina macrocopa* (Straus) in successive generations under food and temperature adaptation

Eugeny S. Burak
Institute of Zoology of Academy of Sciences of Belarus, 27 Scorina St., 220072 Minsk, Belarus

Key words: Moina macrocopa, life history, longevity, food concentration, reproduction, size

Abstract

Life tables of *Moina macrocopa* (Straus) cultured at seven food concentrations (FC) (*Scenedesmus sp.*, 1.49–1490 mg wet weight l^{-1}, $10^4 - 10^6$ cell ml^{-1}) were investigated for animals of the first generation ('nonadapted animals') and for animals of the third generation ('adapted animals') cultivated at these FC. Adapted animals showed a 'trophic preferendum', i.e. a narrow FC-range at which maximal R_0 values were observed in comparison with nonadapted animals. In adapted animals, the maximal R_0 was 115.3 individual, observed at 20 °C and a FC of 74.5 mg wet weight l^{-1}.

Abbreviations:

FC – food concentration
R_0 – net reproduction
T – cohort generation time
L_x – maximal individual longevity
AA – adapted animals
NA – nonadapted animals

Introduction

Moina macrocopa is an eurythermic polycyclic cladoceran, which lives predominantly in temporary hypereutrophic ponds, where it often constitutes about 90% of the zooplankton biomass. It is widely distributed geographically (Manuylova, 1964; Smirnov, 1976) and is frequently used as living food for mass cultivation of fish larvae (Maksimova, 1968).

Several features of the biology and ecology of *M. macrocopa* are known quite well (Maksimova, 1968; Ivleva, 1969). There are also data on the growth and reproduction of *M. macrocopa* as a function of temperature and trophic conditions (Sushchenya et al., 1990; Martinez-Jeronimo & Gutierrez-Valdivia, 1991). Nevertheless, the dependence of the growth and reproduction of *M. macrocopa* on the food amount is studied inadequately. A food concentration (FC) of

5 mg dry weight l^{-1} has been found to be better for growth and reproduction of *Moina* than lower concentrations (Martinez-Jeronimo & Gutierrez-Valdivia, 1991). However, it cannot be claimed that this concentration is optimal with respect to the reproductive output of *M. macrocopa*, since there are no data obtained at higher food concentration. It is also interesting to investigate the reproduction strategy of *M. macrocopa* in successive parthenogenetic generations under constant and variable trophic conditions.

The goal of the present work is to investigate the main parameters of life tables of *Moina macrocopa* in successive generations under food and temperature adaptation. Life table stadies are important both for the optimization of conditions for the extensive cultivation of *Moina* and for exploring the possibility to use *Moina* in toxicological research.

Material and methods

M. macrocopa macrocopa from prolonged extensive laboratory culture were used for the experiments. Food concentration in the stock culture varied from 1.49 to 1490 mg wet weight l^{-1}.

Settled tap water was used as a medium. The unicellular alga *Scenedesmus sp.* was used as food (cell volume is $149.2 \pm 81.59 \, \mu m^3$). The alga was cultivated in Tamya medium (Vasser, S. P., N. V. Kondratyeva, N. P. Masyuk and other, 1989) and cultures were harvested every five days. The FC was determined by microscope count. There are fourteen treatments: two temperatures (20 °C and 25 °C) were combined with seven food concentrations (FC: 1.49; 7.45; 14.9; 74.5; 149; 745; 1490 mg wet weight l^{-1}).

Adult parthenogenetic females from the laboratory stock culture with embryos in stage 8 (Green, 1956) were placed into Petri dishes. Neonates aged ± 2 hours were used in the first experimental series. Each neonate was cultured in a separate tube of 10 ml at a certain FC and temperature during its entire life time. Ten replicate individuals were studied for each treatment. These individuals were assumed to be 'non-adapted' (NA) because they represent the first generation cultured on the particular experemental conditions. The second experimental series was carried out with offspring of the third clutch from females of the first series. The animals in the second experimental series were cultured under identical conditions as these mothers, and were considered to be 'adapted' (AA). (Galkovskaya & Morozov, 1981). Offsprings of females that were cultivated at a certain FC were grown at the same concentration during the whole life time. All experiments were carried out under dim continuous light.

Every day the animals were transferred into tubes with fresh medium and the appropriate alga concentration. All offspring were reared, counted and removed. The body length of the animals was measured daily with a microscope (resolution 0.025 mm). The age at maturity (in days), longevity (in days), the number and and of the neonates produced, the size of the neonates and the size at maturity were determined for each individual. The data on fecundity and mortality of the nonadapted and adapted animals were used to calculate R_0:

$$R_0 = \sum l_x m_x,$$

where l_x is the proportion of individuals that reached age x and m_x is the average number of female offspring produced at the age x.

Results

For NA and at 20 °C the age at maturity was four days for a wide FC range in food concentration (7.45–745 mg wet weight l^{-1} (Table 1)). A decrease in the FC to 1.49 mg wet weight l^{-1} resulted in an increase in the age at maturity to seven days, whereas and an increase in the FC to 1490 mg wet weight l^{-1} led to a decrease age at maturity to two days. At a FC of 1.49 mg wet weight l^{-1} an increase age at maturity and a reduction in the size of the clutch were found at both temperatures in both the adapted and non-adapted animals (Table 1).

The highest values for the maximal clutch size were observed in the range of FC 74.5–745 mg wet weight l^{-1}, at both temperatures and in the adapted as well as the non-adapted animals. In all series minimal values for clutch sizes were recorded at FC 1.49 mg wet weight l^{-1}.

The relationship between the net reproduction (R_0) and the FC is shown in Figure 1. The plot of R_0 versus the FC is a dome-shaped curve. The maximal R_0 was 115.3 ind. at 20 °C and 74.5 mg l^{-1} FC for the animals of the second generation ('adapted' animals). For animals of the first generatia ('non-adapted' animals), the range of maximal R_0 was wide (at 20 °C from 14.9 to 745 mg l^{-1}; at 25 °C from 74.5 to 745 mg l^{-1}). The AA thus showed a narrow 'trophic preferendum', i.e. a narrow range of FC, at which maximal R_0 were observed (Figure 1) in comparison with NA.

At the lowest FC tested, the difference between the non-adapted and the adapted animals depended on the temperature. At 20 °C, R_0 was almost equal for the adapted and non-adapted animals; whereas at 25 °C, the average R_0 of the AA was eight times higher than that of the non-adapted individuals. Probably, the increase in temperature is accompanied by a decrease in the trophic threshold of reproduction.

The maximal individual longevity and the cohort generation time tended to decrease as the FC increase (Table 1). This trend is the most pronounced at 20 °C. The maximal Lx (32 days) was recorded at 20 °C and a FC of 14.9 mg wet weight l^{-1}, in NA. At 25 °C, the maximal Lx was 20 days, observed at a FC of 149 mg wet weight l^{-1} in NA. For the NA, the maximal individual longevity and the generatia time at the higher food concentration (1490 mg wet weight l^{-1}) were 2.77 and 2.16 times smaller than they are at the lowest FC, respectively; for the AA, the values are 2.4 and 2.6, respectively.

Table 2 contains sizes of adult females, the number of neonates of the first clutch and the size of the neonates. At 20 °C, adaptation to trophic conditions resulted in a lower neonate number of the first clutch at all FC. The size of the neonate of the first clutch is found to increase as the size of the adult female increases (Figure 2), resulting in positive values of

Table 1. The main parameters of *Moina macrocopa*'s life tables

20 °C

	NONADAPTED							ADAPTED						
Food concentration, mg wet weight l^{-1} (FC)	1.49	7.45	14.9	74.5	149	745	1490	1.49	7.45	14.9	74.5	149	745	1490
Food concentration, cell 10^6 ml^{-1}	0.01	0.05	0.1	0.5	1	5	10	0.01	0.05	0.1	0.5	1	5	10
Time of first clutch, in day	7	4	4	4	4	4	2	16	6	4	4	5	5	4
Maximal clutch size, ind (± sd)	5 ± 0.01	9.2 ± 3.02	10.3 ± 0.99	21.3 ± 1.69	23 ± 1.29	24.5 ± 1.21	18.9 ± 1.5	4.5 ± 0.5	5 ± 0.5	12 ± 0.71	22.8 ± 0.71	19.5 ± 1.11	10 ± 0.1	17 ± 0.23
Maximal individual longevity, in day (Lx)	25	21	32	21	15	16	9	24	25	19	18	18	10	10
Net reproduction, ind (R_0)	8	27.7	86.4	66	76.1	79.7	45.4	7.75	42.4	57	115.3	75	32.2	27.6
Cohort generation time, day (T)	12.2	9.7	14.8	10.4	7.8	8.45	5.63	19.54	13.8	10.5	10.8	10	7.29	7.5

25 °C

	NONADAPTED							ADAPTED						
Food concentration, mg wet weight l^{-1} (FC)	1.49	7.45	14.9	74.5	149	745	1490	1.49	7.45	14.9	74.5	149	745	1490
Food concentration, cell 10^6 ml^{-1}	0.01	0.05	0.1	0.5	1	5	10	0.01	0.05	0.1	0.5	1	5	10
Time of first clutch, in day	10	3	3	2	3	3	3	4	4	3	3	2	2	2
Maximal clutch size, ind (± sd)	3 ± 0.01	6 ± 0.01	22 ± 0.12	22 ± 0.19	21 ± 0.21	24 ± 0.01	17 ± 2.9	6 ± 1.01	0.25 ± 1.0	12.5 ± 1.09	24 ± 1.5	27.75 ± 2.86	24 ± 1.2	25.3 ± 0.71
Maximal individual longevity, in day (Lx)	13	7	15	12	20	8	7	7	12	12	13	11	11	12
Net reproduction, ind (R_0)	1.2	18	55.952	69.253	68.18	69.25	28	9.5	33.05	48.35	57.25	93.8	87	45.58
Cohort generation time, day (T)	11	5	8.53	6.64	9.16	5.91	5.51	5.63	9.005	7.24	7.16	6.8	6.84	6.16

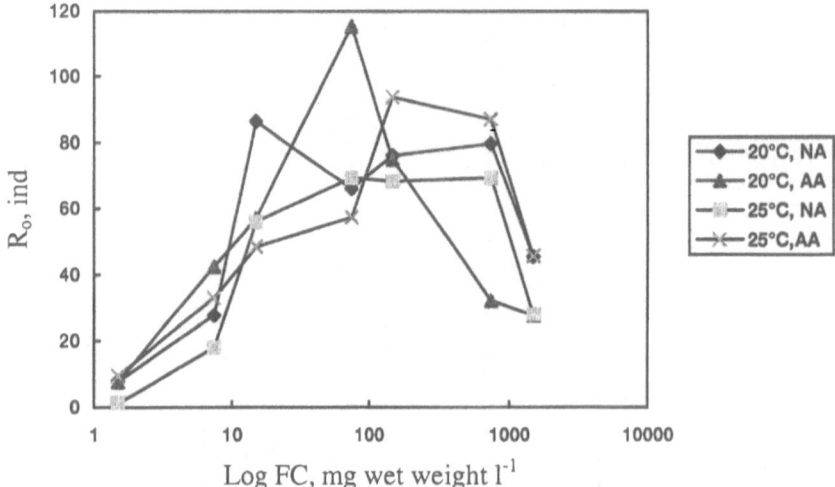

Figure 1. The relationship between the net reproduction (R_0) (ind.) and food concentration (FC) (mg wet weight l^{-1}).

Table 2. Size of maturity of *Moina macrocopa* adult female (body length, *L*, mm), the number of neonates in the first clutch (*N*, ind.) and the average size of the neonates (body length, L_n, mm)

FC, mg l^{-1}	FC, cell 10^6 ml^{-1}	Non-adapted			Adapted		
		N	*L*	L_n	*N*	*L*	L_n
		20 °C					
1.49	0.01	2.0 ± 1.12	1.25 ± 0.01	0.46 ± 0.01	1.5 ± 0.52	1.23 ± 0.03	0.45 ± 0.02
7.45	0.05	19.2 ± 0.45	1.38 ± 0.1	0.55 ± 0.1	7.0 ± 1.55	1.43 ± 0.06	0.55 ± 0.01
14.9	0.1	12.0 ± 2.6	1.65 ± 0.1	0.75 ± 0.1	16.0 ± 0.95	1.68 ± 0.01	0.55 ± 0.01
74.5	0.5	10.3 ± 1.05	1.8 ± 0.1	0.63 ± 0.1	7.8 ± 1.51	1.78 ± 0.02	0.6 ± 0.01
149	1	21.0 ± 2.31	1.75 ± 0.1	0.65 ± 0.1	14.8 ± 1.9	1.7 ± 0.03	0.68 ± 0.01
745	5	18.75 ± 1.02	1.9 ± 0.01	0.58 ± 0.03	6.5 ± 0.85	1.63 ± 0.02	0.58 ± 0.02
1490	10	6.7 ± 1.38	1.63 ± 0.01	0.6 ± 0.03	6.0 ± 0.7	1.5 ± 0.01	0.6 ± 0.02
		25 °C					
1.49	0.01	3.0 ± 0.25	0.85 ± 0.1	0.45 ± 0.1	3.5 ± 0.5	1.03 ± 0.09	0.43 ± 0.01
7.45	0.05	6.0 ± 1.62	0.8 ± 0.05	0.45 ± 0.1	4.5 ± 1.16	1.0 ± 0.09	0.64 ± 0.01
14.9	0.1	11.75 ± 1.58	0.88 ± 0.01	0.45 ± 0.1	12.5 ± 0.69	1.28 ± 0.04	0.57 ± 0.03
74.5	0.5	4.75 ± 1.51	1.1 ± 0.04	0.6 ± 0.1	7.0 ± 1.47	1.13 ± 0.04	0.65 ± 0.02
149	1	11.3 ± 0.84	1.13 ± 0.01	0.6 ± 0.1	6.6 ± 0.22	1.2 ± 0.01	0.73 ± 0.02
745	5	12.75 ± 0.71	0.9 ± 0.01	0.55 ± 0.1	9.0 ± 0.7	1.18 ± 0.04	0.65 ± 0.02
1490	10	9.75 ± 0.92	1.18 ± 0.04	0.58 ± 0.01	11.2 ± 0.85	1.08 ± 0.03	0.58 ± 0.02

the coefficiens of correlation between the sizes of the neonates and the sizes of maturity (Table 3). Body size of the neonates and adult females was substantially lower at 25 °C compared to 20 °C. The number of neonates in the first clutch decrease almost at all FC following adaptation (Table 2). The higest average body size at maturity (1.9 mm) was found at 20 °C and a FC of 745 mg wet weight l^{-1} (Table 2). As the size of adult females increase, R_0 also increase, which is

confirmed by high coefficients of correlations between these quantities (Table 3).

Discussion

Among Cladocera, members of the genus *Daphnia* have been studied most intensively with respect to the effect of food abundance and its fluctuations on the

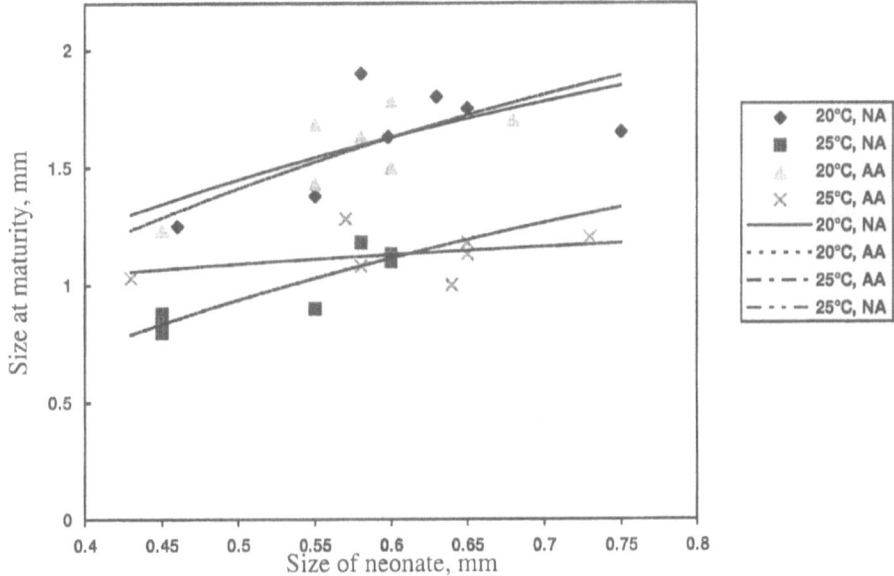

Figure 2. The dependence of the size of neonate of the first clutch (mm) from the size of adult female (mm) (nonadapted (NA) and adapted (AA) animals) at the 20 and 25 °C.

Table 3. The coefficients of correlation between net reproduction (R_0), size of *Moina macrocopa* adult female (L_n) and size of neonate (L_n)

	20 °C			
	NA		AA	
	L	L_n	L	L_n
L	1		1	
L_n	0.5964068	1	0.7629716	1
R_0	0.8878195	0.8530921	0.9523984	0.7298428

	25 °C			
	NA		AA	
	L	L_n	L	L_n
L	1		1	
L_n	0.8917919	1	0.355137	1
R_0	0.4471957	0.5346807	0.6363971	0.8448591

Table 4. The coefficients of regression between size of neonate (L_n) and FC, and L_n and size *Moina macrocopa* adult female (L)

	20 °C	
	NA	AA
FC	−3.91E-05	3.43E-05
L	0.2605659	0.2815144

	25 °C	
	NA	AA
FC	−1.06E-06	1.038E-05
L	0.4230646	0.3407471

the first clutch is related to the selection process under different environmental conditions (Lynch, 1980). It has also been shown that large neonates maturate at large sizes than small neonates (Lampert, 1993).

Growth and reproduction of *M. macrocopa* has been studied inadequately. Most of the available literature is based on studies of the growth and reproduction parameters in a relatively narrow range of food concentrations that are low in comparison with the concentrations that are considered to be optimal for the

growth and reproduction parameters, a proved good model organisms in various kinds of environmental studies (Hrbackova, 1974; Lampert, 1977; Paloheimo et al., 1982; Cowgill et al., 1985a; Sterner, 1993). It has been shown in *Daphnia* that the size of the neonates of cultivation of *M. macrocopa* (Kokova, 1982; Maksimova, 1968).

From literature data it is known that an increase in food concentration results in a decrease of the duration of the juvenile period. For example, Semenchenko (1990) showed that an increase in the concentration of *Chlorella sp.* from 0.072 to 1.44 mg dry weight l^{-1} resulted in a decrease in the duration of the juvenile period of *M. macrocopa* from 3 to 2.1 days at a temperature of 20 °C. An increase in the temperature from 20 to 35 °C cause the same effect (Sushchenya et al., 1990).

Our results indicate that the age at maturity increases substantially as the FC decreases to 1.49 mg wet weight l^{-1}, but remains almost identical for a wide range of higher FC (see Table 1). This may result from food shortage; a similar fact was reported earlier for other Cladocera (Romanovsky, 1984). At 20 °C and for non-adapted animals, age of maturity was observed at a very high FC (10^6 cell ml^{-1}, 1490 mg wet weight l^{-1}). Our data suggest that the higher food concentration tested does not provide optimal conditions. In non-adapted animals, the maximal clutch size is found to decrease at both temperatures at a food concentration above $5 \cdot 10^6$ cell ml^{-1} (745 mg wet weight l^{-1}). A food concentration lower than 74.5 mg wet weight l^{-1} also resulted in a sharp decrease of the maximal clutch size. At 20 °C, males were present in the clutches at food concentrations lower than 14.9 mg l^{-1}. The appearance of males may be a result of the combination of food shortage and low temperature, but this requires farther studies. There are Zadeev & Gubanov (1995) have reported an increase in the number of females with sexual eggs at a concentration of $5 \cdot 10^4$ *Chlorella* cell ml^{-1} compared to a concentration of $1 \cdot 10^5$ *Chlorella* ml^{-1}. Adaptation causes a decrease in the food concentrations at which the maximal clutch size is highest. While in the non-adapted animals the maximal clutch size is rather high at a wide renge of food concentrations of (74.5–745 mg wet weight l^{-1} and 14.9–745 mg wet weight l^{-1} at 20 and 25 °C, respectively), the range of food concentrations at which the highest maximal clutch size is observed is more narrow in the adapted animals, especially which is at 20 °C. Several earlier studies have investigated the clutch size of *M. macrocopa* in relation to food concentration (range: 0.072–5.7 mg dry weight l^{-1}; 10^4–10^6 cell ml^{-1}) and temperature (20, 25, 30 and 35 °C; Sushchenya et al., 1990; Semenchenko, 1989; Ruslutskii, 1995). Maximal clutch size (18 neonates) were recorded at 25 °C and at a food concentration of 5.7 mg dry weight l^{-1}.

Gladyshev et al. (1997) cultivated *M. macrocopa* on *Chlorella* in a wider range of higher food concentrations. The experiments were carried out at 20 °C. The specific growth rate of *M. macrocopa* was observed to be maximal at a *Chlorella* concentration higher than 10 mg dry weight l^{-1} (about 0.02 h^{-1}) and did not decrease until 20 mg dry weight l^{-1}. In flow-through cultures at a specific flow velocity of 0.39 h^{-1}, a maximal biomass gain of 0.223 mg dry weight in 60 h was recorded at *Chlorella* concentrations maintained at the level 10.6 mg dry weight l^{-1}. A decrease in the food concentration was accompanied by an abrupt drop in the biomass gain, while increasing the food concentration to 42.3 mg dry weight l^{-1} did not result in a strong reduction of the bioman gain.

Martinez-Jeronimo & Gutierez-Valdiria (1991) compared the survivorship and fecundity of *Moina macrocopa* cultivated on the three algae species, *Ankistodesmus convolutus*, *Scenedesmus incrassatulus* and *Chlorella vulgaris* (at 27 °C). Irrespective of food quality, survivorship and longevity were found to be maximal at the highest of the food concentrations tested, i.e. 5 mg dry weight l^{-1} or 10^6 cell ml^{-1}.

With regard to the size of adult females and neonates, it is known from the literature that the size of adult *Moina* females increases with the increasing food concentration. Sushchenya et al. (1990) showed that within a food concentration range of 0.44 to 5.7 mg dry weight l^{-1} of *Chlorella* and at 20, 30 and 35 °C, the size at maturity of *Moina* females increased within increasing food concentrations. Semenchenko (1989) obtained similar result. However, at lower food concentrations the size of adult females was found to decrease with increasing food concentration (Razlutskii, 1995). This author reports a maximal size of neonates at a *Chlorella* concentration $2 \cdot 10^3$ cell ml^{-1}. We have not obtained a clearent relation between size of maturity and food concentration. In both adapted and nonadapted individuals, we found a smaller size of the neonate and size at maturity at 25 °C compared to 20 °C (Table 2). The number of neonatesin the first clutch was also lower at 25 °C compared to 20 °C at almost all food concentrations. It is known that embryos of *M. macrocopa* are fed by a placenta (Smirnov, 1976; Goulden & Hornig, 1980). Therefore, the food supply of embryos is closely related to the energy reserver of the female. Survival of the neonates is also dependent upon the level of reserves which remain after completion of the embryonic development (Smirnov, 1976; Goulden & Hornig, 1980). An increase in temperature obviously causes higher energy consumption by both adults and embryos. Probably, the neonate number and neonate size of *M. macrocopa* show a decrease under unfavourable food and temperature conditions. Malhorta & Lander Seema (1993) observed that the size of *M. macrocopa* neonates was smaller at 32 °C than at 10 °C. Moreover, a decrease in temperature from 32 to 18 °C resulted in an increase in the number of neonates produced by the females.

It is known that unfavourable trophic conditions results in production of larger eggs Cladocera, which is favourable for the survival of the neonates under food stress (Piatakov, 1956; Green, 1966). For example, Trubetskova & Lampert (1995) reported that the diameter of *Daphnia magna* eggs produced at a constantly low concentration of *Scenedesmus acutus* (0.1 mg C l^{-1}) was larger than those produced at a constantly high food concentration (1.5 mg C l^{-1}). Our data suggest that at low food concentrations, the neonate size in the first clutch depends to a greater extent on the size of mother than on the food concentration.

We conclude that our life table experiment with *Moina macrocopa*, carried out ower a wide range of food concentrations (10^4–10^6 cell ml^{-1}; 1.49–1490 mg wet weight l^{-1}) allowed us to specify the range of optimal food concentrations with respect to a number of the life history parameters and to estimate the check of adaptation to food conditions and temperature on net reproduction of this species. The maximal net reproduction (115.3 ind.) was recorded for animals adapted at 20 °C and food concentration of 10^3 *Scenedesmus* cell ml^{-1} (74.5 mg wet weight l^{-1}).

Acknowledgements

I am deeply indebted to Prof. Galkovskaya, Dr Mityanina and these anonymous reviewers for valuable suggestions on earlier versions of the manuskript. I also thank Mrs Kortneva for help in preparing the English text.

References

Cowgill, U. M., K. I. Keating & I. T. Takahashi, 1985a. Fecundity and longevity of *Ceriodaphnia dubia/affinis* in relation to diet at two different temperatures. J. Crust. Biol. 5: 420–427.

Galkovskaya, G. A & A. M. Morozov, 1981. The formation of temperature adaptations in *Daphnia*. J. Ob. Biol. 42: 113–117. (in Russian)

Gladyshev, M. I., A. P. Tolomeev, T. A. Temerova & A. G. Degrermenji, 1997. Dependence of growth rate of *Moina macrocopa* on the concentration of *Chlorella vulgaris* in differentially flowing cultivators. DAN 352: 281–283. (in Russian)

Goulden, C. E. & L. L. Hornig, 1980. Population oscillations and energy reserves in planktonic Cladocera and their consequences to competition. Proc. Nat. Acad. Sci. 77: 1716–1720.

Green, J., 1956. Growth, size and reproduction in *Daphnia* (Crustacea; Cladocera). Proc. zool. Soc. Lond. 126: 173–204.

Green, J., 1966. Seasonal variation in egg production by *Cladocera*. J. anim. Ecol. 35: 77–104.

Hrbackova, M., 1974. The size of primiparae and neonates of *Daphnia hyalina* Leydig (Crustacea: Cladocera) under natural and enriched food conditions. Vestn. Cs. spolec. zool. 38: 98–105.

Ivleva, I. V., 1969. Biological bases and methods of mass cultivation of food invertebrates. Moscow: 'Nauka', 170 pp. (in Russian)

Kokova, V. E., 1982. Continuous Cultivation of Invertebrates. Nauka, Novosibirsk, 167 pp.

Lampert, W., 1977. Studies on the carbon balance of *Daphnia pulex* de Geer as related to environmental conditions. II. The dependence of carbon assimilation on animal size, temperature, food concentration and diet species. Arch. Hydrobiol. 48: 314–317.

Lampert, W., 1993. Phenotypic plasticity of the size at first reproduction in *Daphnia*: the importance of maternal size. Ecology 74: 1455–1466.

Lynch, M., 1980. The evolution of cladoceran life histories. Quart. Rev. Biol. 55: 23–42.

Manuylova, E. F., 1964. Cladocera of fauna of USSR. Moskva, Leningrad: 'Nauka', 327 pp. (in Russian)

Maksimova, L. P., 1968. Biology of *Moina* and rotifers and their growing as an alive food for *Coregonus* larvae. Izv. Gos NIORCh. 67: 107–135. (in Russian)

Malhorta, Y. R. & L. Seema, 1993. Effect of temperature on growth and fecundity of selected species of Cladocera. J. Ind. Inst. Sci.73: 335–345.

Martinez-Jeronimo, F. & A. Gutierrez-Valdivia, 1991. Fecundity, reproduction, and growth of *Moina macrocopa* fed different algae. Hydrobiologia 222: 49–55.

Matveev, V. F., 1978. The structure of zooplankton community of lake Glubokoe. Avtoref. Dis.... Kand. Biol. Nauk. M., 24 pp. (in Russian)

Paloheimo J. E., S. J. Crabtree & W. D. Taylor, 1982. Growth model of *Daphnia*. Can. J. Fish. aquat. Sci. 39: 598–606.

Piatakov, M. L., 1956. Concerning seasonal change of fecundity of Cladocera. Zool. J. 35: 1814–1819. (in Russian)

Razlutskii, V. I., 1995. The effect of trophic conditions on the duration of embryonic development of Cladocera. DAN 39: 77–80.

Romanovsky, Y. E., 1984. Prolongation of postembryonic development in experimental and natural cladoceran populations. Int. Rev. ges. Hydrobiol. 69: 613–632.

Semenchenko, V. P., 1989. Variation in the fecundity and egg size of *Moina macrocopa* (Crustacea, Cladocera) during its life cycle. Zool. j. 68: 135–138. (in Russian)

Semenchenko, V. P., 1990. The comparative analysis of reproduction strategy in Cladocera at different food concentrations. J. ob. biol. 51: 828–835. (in Russian)

Smirnov, N. N., 1976. Fauna of the world. Crustacea. V.1. Leningrad: 'Nauka': 174 180. (in Russian)

Sterner, R. W., 1993. *Daphnia* growth on varying quality of *Scenedesmus*: mineral limitation of zooplankton. Ecology 74: 2351–2360.

Sushchenya, L. Ì., V. P. Semenchenko, G. A. Semenyuk & I. L. Trubeckova, 1990. Production of planktonic crustacean and factors of environment. Navuka i technika, Minsk: 133–141. (in Russian)

Trubetskova, I. & W. Lampert, 1995. Egg size and egg mass of *Daphnia magna*: response to food availability. Hydrobiologia 307: 139–145.

Vasser, S. P., N. V. Kondratyeva, N. P. Masyuk and other, 1989. Algae. Reference book. Navukova dumka, Kiev. 605 pp.

Zadeev, E. S. & V. G. Gubanov, 1995. The effect of *Moina macrocopa* (Cladocera) population density and food availability on the change of the method of reproduction of *Moina macrocopa*. Ecologia: 409–410. (in Russian)

Hydrobiologia **360**: 109–115, 1997.
A. Brancelj, L. De Meester & P. Spaak (eds), Cladocera: The Biology of Model Organisms.
©1997 *Kluwer Academic Publishers.*

Preliminary experiments on resource competition between a migrating and a non-migrating clone of the hybrid *D. galeata* × *hyalina*

Tineke Reede
University of Amsterdam, Department of Aquatic Ecology, Kruislaan 320, 1098 SM Amsterdam, The Netherlands

Key words: Anomopoda, cladocerans, diel vertical migration, intraspecific competition

Abstract

Competition experiments between two clones of the hybrid *D. galeata* × *hyalina* differing in vertical migration behaviour were carried out in the laboratory, in a set-up which separated the clones spatially. The influence of fish kairomones, artificial predation and the imitation of diel vertical migration circumstances were investigated. In the presence of fish kairomones, the non-migrating clone outnumbered the migrating one. The introduction of artificial predation reduced this effect in such a way that the migrating clone had a competitive advantage over the non-migrating clone. It was concluded that competition results do not contradict life history studies performed with the same two clones, since the presence of fish kairomones enhances the reproduction of the non-migrating clone more than that of the migrating one.

Introduction

Competition between species can be of very strong influence on the composition of freshwater zooplankton (e.g. Seitz, 1980; DeMott & Kerfoot, 1982; Romanovsky & Feniova, 1985; MacIsaac & Gilbert, 1991; Milbrink & Bengtsson, 1991; Matveev & Gabriel, 1994). Several hypotheses exist about competitive dominance (see Bengtsson 1987 for review) and they often contradict each other. Most studies on this subject deal with interspecific competition while only few studies examine the role of intraspecific competition (Loaring & Hebert, 1981; Perrin et al., 1992; Weider, 1992; De Meester et al., 1995; Enserink et al., 1996). In the case of *Daphnia*, many clones can co-exist in an ecosystem, competing for the same food (Mort & Wolf, 1985; Spaak & Hoekstra, 1993; Spaak, 1996). The high genetic diversity in *Daphnia* populations might indicate that competitive exclusion of clones is rare. De Meester (1994) conducted competition experiments with three *D. magna* clones differing in phototactic behaviour and stated that coexistence is probably partly mediated by habitat selection. A study by Spaak & Hoekstra (1995) has revealed that hybrid clones regularly outnumber their parental species. This

is the case in Lake Maarsseveen (The Netherlands), where many clones of the hybrid *D. galeata* × *hyalina* coexist throughout the year (Spaak & Hoekstra, 1993) but where the parental species are rare or absent. Most of the year the hybrids occupy the same habitat. However, during a few weeks from the end of May until the beginning of July, habitat partitioning occurs during the daytime since part of the population migrates to a greater depth while the other part remains in the upper water layers. The start of this diel vertical migration (DVM) behaviour coincides with the appearance of large shoals of voracious juvenile perch (*Perca fluviatilis*) preying upon *Daphnia* (Ringelberg, 1991). DVM is generally believed to be a way of avoiding visual predators (Lampert, 1989). Performing DVM, however, can be disadvantageous: low temperatures at greater depths prolong egg development times and because less food is available a lower egg production is realised. Although food availability is not less at greater depth in all lakes (Williamson et al., 1996) it is usually the case in Lake Maarsseveen. Therefore, the advantages of migration might be outweighed by a lower fecundity. However, DVM is probably the favourable strategy when predators are abundant (Flik et al., 1997). This is reflected by the fact that after a few weeks of migration in Lake Maarsseveen, in most years few or no daphnids are found in the upper water layers during the day (Flik & Ringelberg, 1993) suggesting that the non-migrating clones have been eaten.

The appearance of juvenile perch is one of the factors triggering DVM (van Gool & Ringelberg, 1995). Like many other predatory fish, perch apparently excrete an as yet unknown chemical substance known as fish kairomone. This substance has been shown to induce life history changes enhancing reproductive chances in the presence of fish (Stibor, 1992; Tátrai & De Bernardi, 1992; Macháček, 1993; Weider & Pijanowska, 1993). Daphnids tend to mature at an earlier age while producing more offspring in the presence of kairomones (Macháček, 1991; Stibor, 1992). The instar at which the exposure to kairomones occurs (Macháček, 1995) and the concentration (Reede, 1995) determines the extent of the changes.

The purpose of this study is to investigate whether competition between two clones derived from the same lake that differ in life history characteristics and also in phototactic behaviour will lead to dominance or exclusion of one clone. Starting-point of the competition experiments are the life history experiments by Reede & Ringelberg (1995) on a migrating and a non-migrating hybrid *D. galeata* × *hyalina* clone from Lake Maarsseveen. The supposedly non-migrating clone showed a higher intrinsic rate of increase when fish kairomones were present, and it also reduced its size at maturity as compared to the migrating clone. When no kairomones were present, no difference in intrinsic rate of increase was found. Therefore, I expect that the non-migrating clone will dominate only when fish kairomones are present in the medium. Also, when the migrating clone is kept under migrating circumstances (low temperature and no food during the day), it will probably be present in smaller numbers than the non-migrating clone unless artificial predation changes this dominance pattern. To test these hypotheses, I performed competition experiments in the laboratory.

Material and methods

The experiments were carried out with two clones of the hybrid *D. galeata* × *hyalina* which were also used in the life history experiments by Reede and Ringelberg (1995). One clone was caught near the surface around noon at the beginning of a DVM period in Lake Maarsseveen I, while the other was caught at a depth of 20–25 metres in the same period. Therefore, the first is considered to be a non-migrating clone while the latter is considered to be a migrating one.

The water used in the experiments was taken from Lake Maarsseveen. The water was filtered continuously through a sand filter to facilitate bacterial breakdown of any kairomones present. Before use in the experiments, the water was filtered through a 0.2 μm membrane filter. Algae – *Scenedesmus acuminatus* – were grown in Woods Hole Medium (Guillard, 1975) in continuous culture. Only log-phase cells were used as food in the experiments.

The competition experiments were carried out in a culture system in which the two clonal populations were spatially separated and food medium was circulated between them. Each clonal population was kept in a 700 ml glass jar. Algal suspension was pumped into each of the connected jars with a peristaltic pump from a 500 ml Erlenmeyer. Both jars received food at the same speed to enable competition. At the same time, water from the lower part of each culture jar was pumped back into the Erlenmeyer, where the suspension was mixed continuously by gentle aeration.

Table 1 describes the experimental circumstances. Six different experiments were carried out. In experiment 1 and 2 only intraclonal competition took place since both population jars contained the same clone. These experiments were used as reference. In all other experiments, one jar contained the migrating clone and the other the non-migrating clone. At the beginning of the experiments ($t = 0$), the populations in the jars consisted each of 25 adults with eggs, 25 adults without eggs and 25 juveniles from stock cultures. Twice a week, all individuals in the culture jars were counted. At the same time, jars were cleaned to remove precipitated algae and dead daphnids. The water in the jars was replaced with filtered Maarsseveen water with no added food. There was no indication of damage to the animals due to handling. The experiments were continued for about 40 days. When shorter time-spans occurred this was due to technical failures. The algal suspensions in the Erlenmeyers were renewed every 48 hours. At the beginning of each 48 h period, food concentration in the Erlenmeyers was 1.4 μg C. ml^{-1} and the pumping speed was 20 ml h^{-1}. In a preliminary experiment with two populations, each consisting of 80 adults, food concentration in the culture jars and Erlenmeyers changed as shown in Figure 1. After 48 hours hardly any food was left in the Erlenmeyers or in the jars containing the populations. Food concentration in the culture jars reached values above the incipient limiting level (ILL) (Lampert, 1987) for only a few

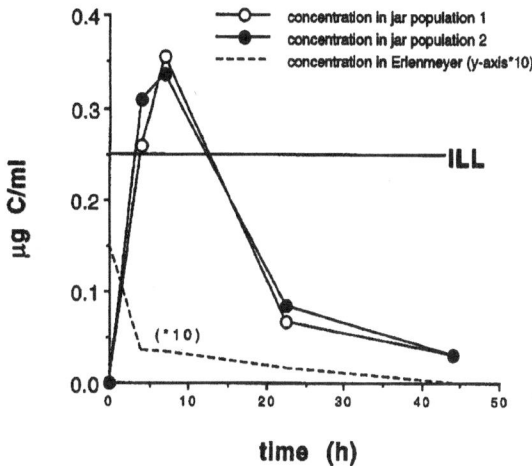

Figure 1. Changes in food concentration during a 48-hour period in a food Erlenmeyer and two connected population jars. ILL indicates the incipient limiting level for *D. longispina* (Lampert 1987).

Table 1. Competition experiments. Pop = population; *n* = number of replicate experiments; C = competition simulated between clones; Kai = kairomones present in food medium; P = artificial predation; DVM = diel vertical migration simulated for the migrating clone; m = migrating clone; nm = non-migrating clone.

Treatment			*n*	C	Kai	P	DVM	Results
Pop1	*	Pop2						
1 nm	*	nm	2	–	–	–	–	text
2 m	*	m	2	–	–	–	–	text
3 nm	*	m	4	+	–	–	–	Figure 2A
4 nm	*	m	2	+	+	–	–	Figure 2B
5 nm	*	m	2	+	+	–	+	Figure 2C before day 45
6 nm	*	m	4	+	+	+	+	Figure 2C after day 45

consisted of two parts hypolimnion water to one part water from the aquarium containing the perch.

In all experiments, the jars containing the daphnids were kept at a constant temperature of 20 ± 0.5 °C, except for the migrating clone in experiments 5 and 6. In these experiments a temperature regime was imposed on the migrating clone to imitate migration circumstances. The non-migrating clone was kept at 20 °C and food was pumped into the jar throughout the day, while the jars containing the migrating clone were kept in a water-bath which for 15 hours, during the light period, remained at a temperature of 10 °C and for 7 hours, during the dark period, at 20 °C. This time scheme allowed for one hour warming up and one hour cooling down between these periods. In addition, the migrating clone received food only during the warm (dark) period. Since kairomones were added to the food medium, these were also only supplied to the migrating clone during the dark period.

Predation was simulated in experiment 6 (removal of animals with a pipette during counting) Only gravid females were taken out since they are the most visible and therefore stand a greater chance of being eaten. We estimated that non-migrating daphnids are exposed to predation during the entire light period of 16 hours while the migrating animals are eaten only during dusk and dawn which we established to be 4 hours per day. Therefore, the number of non-migrating gravid females which were removed was four times as high as the number of migrating females. This resulted in the artificial predation of two thirds of the gravid non-migrating females and one sixth of the gravid migrating females twice a week. Lengths of the daphnids which were taken out to imitate predation were measured.

All experiments were performed twice, except for experiment 3 and 6 which were performed four times (Table 1). Technical failures caused some cultures to break down prematurely. To present the results of the competition experiments, relative abundances were calculated. The total numbers of both populations connected to one food Erlenmeyer was made to be 100%, and both population sizes were expressed as a percentage of this total at each point in time. Then the percentage of the migrating clone was deducted from that of the non-migrating clone, so that a positive result indicates that the population size of the non-migrating clone is larger than that of the migrating one, whereas a negative result indicates that the population size of the migrating clone is larger than that of the migrating clone. At point 0, the populations have equal sizes. A repeated measures ANOVA (Zar, 1996) was car-

hours during the 48-hour period. In all experiments a light:dark cycle of 16:8 h was maintained.

In experiments 4, 5 and 6, fish kairomones were added to the food suspension (Table 1). Water containing fish kairomones was taken from an aquarium filled with 20 l of water from Lake Maarsseveen housing ten 0+ perch (*Perca fluviatilis*) of about 4 cm long. This water was filtered through a 0.2 μm membrane filter before use in the experiments. One third of the water in the aquarium was refreshed every day and at the same time faeces were removed. The perch were caught in Lake Maarsseveen and were fed *Chironomus* larvae. In experiments 4, 5 and 6, water in the Erlenmeyers

Table 2. Results of repeated measures ANOVA for the population sizes during the competition experiments. Clone was used as factor and time as repeated measures factor. Absolute numbers were used in these analyses. In experiments 1 and 2, 'clone' stands for population as both populations consisted of the same clone.

Experiment	Factor	df	MS	F
1/2	clone	1	413.44	0.308
	time	8	3237.8	4.739**
	clone × time	8	275.132	0.403
	error	16	683.22	
3	clone	1	1624.5	0.365
	time	8	1443.2	1.230
	clone × time	8	414.06	0.353
	error	48	1172.9	
4	clone	1	32321.5	21.382*
	time	7	765.75	4.866**
	clone × time	7	3086.6	19.614***
	error	14	157.37	
5	clone	1	22002.8	27.983*
	time	8	2517.4	6.532***
	clone × time	8	583.6	1.514
	error	16	385.40	
6	clone	1	7708.02	2.384
	time	6	392.5	2.534*
	clone × time	6	1142	7.372***
	error	36	154.9	

*$P < 0.05$, **$P < 0.01$, ***$P < 0.0001$

ried out using Statistica (Statsoft, ver. 5.0) to compare population sizes in time. For this procedure, absolute numbers were used. A period of 14 days was taken into account for populations to stabilise; the data of the first two weeks are therefore not included in the analyses.

Results

The relative abundances of the non-migrating and migrating clones in experiment 1 and 2 (the control experiments) oscillated around point 0, indicating that none of the populations could suppress the other. A repeated measures ANOVA (Table 2) shows that population size differences between both jars and between both clones are not significant although there is a significant effect of time. No significant time × population interaction was found.

Figure 2A shows the results of the competition experiment (#3) without kairomones. When no kairomones are present in the medium, population sizes are equal for the migrating and non-migrating clone during the whole experiment. Both clones are equal competitors under these circumstances. No significant time effect or time × population interaction was found (Table 2).

In experiment 4 where competition in the presence of kairomones is simulated, the non-migrating clone outnumbered the migrating clone during the course of the entire experiment (Figure 2B). The difference in population size between the two populations was significant (ANOVA Table 2). The population sizes also changed significantly over time. A highly significant interaction shows that population numbers changed differently in both jars.

The results of experiments 5 and 6, which simulated DVM circumstances, are depicted in Figure 2C. From day 45, indicated by the vertical line, predation is imposed on the populations. In the first half of the experiment, the densities of the non-migrating clone were largest. This difference was significant (Table 2). Also, the population densities changed significantly during the course of the experiment. There was no significant time × population interaction present. However, after the introduction of artificial predation the migrating clone became dominant. The difference in population size between the two clones during the second part of the experiment was not significant but there was a significant effect of time. The highly significant interaction effect shows that both populations react differently to the changed conditions. The egg carrying adults of the migrating clone were significantly larger than those of the non-migrating one (sizes and SE: 1.72 ± 0.04 mm versus 1.58 ± 0.02 mm, Student's t: $p < 0.05$).

Discussion

My results show that the non-migrating clone obtains a significantly higher density than the migrating clone only when fish kairomones are present in the water. In all other cases, both clones are equally strong competitors. The ANOVA results (Table 2) show that a significant difference between the population sizes of the two clones exists only in experiment 4 and 5. However, an interaction effect is found in experiments 4 and 6. In all but experiment 3, there is a time effect, indicating that population sizes changed significantly during the course of the experiment. For instance, in experiment 6 one population increased from 65 indi-

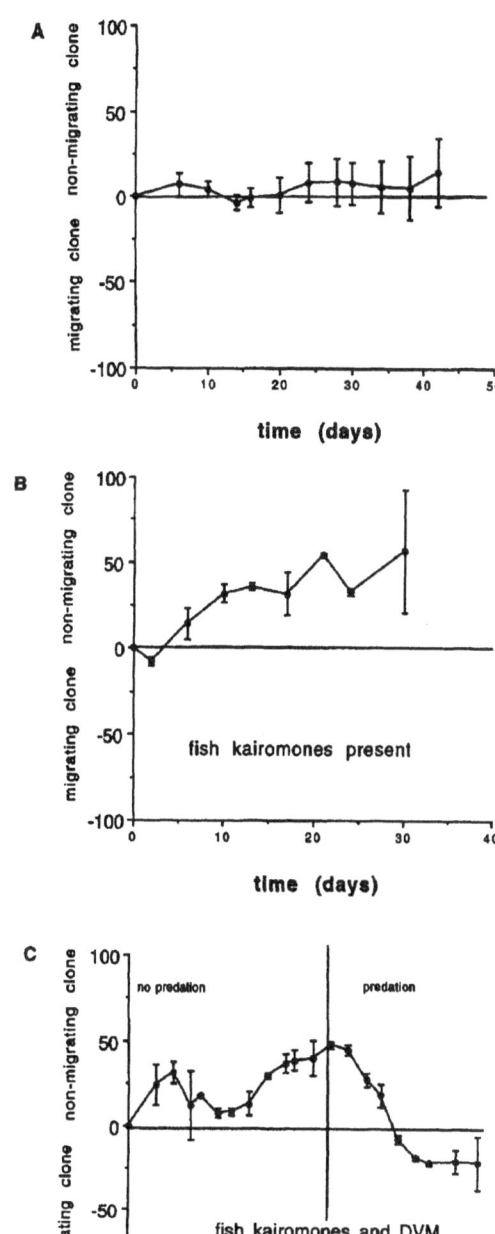

Figure 2. Relative abundance of two populations connected to one food Erlenmeyer. A: competition between a migrating and a non-migrating clonal population in the absence of fish kairomones. B: competition between a migrating and a non-migrating clonal population in the presence of fish kairomones. C: competition between a migrating and a non-migrating clone while DVM circumstances are simulated (the migrating clone received low temperatures and no food during the day, high temperatures and food during the night). From day 45, fish kairomones were added to the medium and artificial predation was imposed on the clones (predation for the non-migrating clone four times higher than predation for the migrating clone, see M & M). Error bars represent 1 Standard Error.

viduals to 111 and then decreased to 45, while another in experiment 1 went from 179 to 73 individuals.

Life table experiments which were performed earlier by Reede & Ringelberg (1995) with the same two clones showed that the non-migrating clone realised a higher intrinsic rate of increase in the presence of kairomones. Therefore, the competition results are in accordance with the r-max hypothesis (Goulden et al., 1978), which predicts that the organism with the higher intrinsic rate of increase wins in a competition situation. The competition results do not comply with the size-efficiency hypothesis which predicts that large animals are better competitors than small ones (see for instance Jacobs, 1978) since the life table experiments by Reede & Ringelberg (1995) showed that the non-migrating clone is generally smaller than the migrating one. De Meester et al. (1995), who performed migration experiments in plankton towers also showed that the largest clones were generally found at a greater day depth in the presence of chemicals or fish than the smaller ones.

If we presume that all non-migrating clones in Lake Maarsseveen have a higher intrinsic rate of increase than migrating clones, one would expect the non-migrating clones to be dominant over the migrating clones during the DVM period. However, this is not the case: towards the end of the migration period no daphnids are found in the upper water layers during the day (Flik & Ringelberg, 1993). This is either due to the fact that not all non-migrating clones have a higher intrinsic rate of increase than migrating clones or to predation by juvenile perch. As is shown in Figure 2C, (artificial) predation can change clonal dominance. In my predation experiments I applied differential predation rates with predation pressure solely being determined by the time that either migrating or non-migrating clones are usually exposed to predation (see Materials and methods). These data are derived from field experiments in Lake Maarsseveen (Ringelberg et al., 1991). In a natural environment, the intensity of predation changes over time. A higher or lower predation pressure than the one established in the experiments might lead to different results. This was also concluded by De Meester et al. (1995). Therefore, more experiments with several predation levels would give additional information on the competitive abilities of the clones.

Food limitation restricts migration in large animals (Johnsen & Jakobsen, 1987). In Lake Maarsseveen food limitation led to a delayed migration period in 1991 (Flik & Ringelberg, 1993). Migration was started when the juvenile perch were already present in the open water zone for a month. Such a delay can have important consequences for the clonal composition of the daphnid populations. When they have not moved to greater depth the large clones are more vulnerable to visual predation by fish than smaller ones. This may lead to a relatively higher relative abundance of non-migrating animals. However, when more food is available and migration is possible, the migrating animals can increase in relative abundance again. In the present experiments, only one food regime was applied. Competition experiments by Goulden et al. (1982) have shown that food concentration and changes in food concentration influence the relative competitive ability of *D. magna*, *D. galeata* and *Bosmina longirostris*. It is known that food concentration influences the effect of fish kairomones on the non-migrating clone (Reede, 1997). When food concentrations are low, small neonates are produced. Small neonates showed a greater difference in age at maturity and number of eggs in the first clutch between a kairomone and a control treatment than large neonates. Large neonates which were born when food was abundant differed more in size at maturity (Reede, 1997). Therefore, it is very likely that with different food regimes, the competition outcome between the clones will differ. In the present experiments, factors such as the pulsed feeding regime might have facilitated the co-occurrence of the clones. Schulze et al. (1995) concluded from their competition experiments between *Diaptomus* and *Daphnia* that subtleties of culture conditions (such as resource supply) may be critical to competitive ability. MacIsaac & Gilbert (1991) also found different competition results between *Keratella* and *Daphnia* when different feeding regimes were applied.

Cladocerans cultured in continuous cultures vary in population density through time (see for instance Schulze et al.,1995). From time to time, many juveniles are present which will increase competition within a population. The higher densities then result in changes of life history characteristics (Guisande, 1993). Competition between different cohorts will occur and the outcome depends on the food regime. Enserink et al. (1996) found that in competition experiments between different cohorts of *D. magna* the larger daphnids performed better at constant low food levels while the smaller ones were better competitors at fluctuating food concentrations.

Food concentration and fish kairomones are not the only factors influencing competition. Allan (1977) showed that temperature may play a key role in the dynamics of a mixed population of cladocerans. Dur-

ing the DVM period in Lake Maarsseveen, temperature is certainly of importance because of the habitat segregation of migrating and non-migrating animals. In the hypolimnion, temperatures can be up to 10 °C lower than in the upper water layers. As was stated in the introduction, this can have serious consequences for egg development times.

The present study suggests that results from life history experiments (as the ones in Reede & Ringelberg 1995) are a valid basis for making more general conclusions at the population level. Life history studies are indispensable for investigating the effects of various environmental factors on populations. Nevertheless, putting them into a larger perspective by performing experiments on a larger scale provide additional insight.

Acknowledgments

I wish to thank Erik van Gool, Joop Ringelberg, Piet Spaak, Luc de Meester and one anonymous reviewer for critical comments which substantially improved the manuscript, Inge Somhorst for doing much of the practical work involved, and Hans Dekker for correcting the English.

References

Allan, J. D., 1977. An analysis of seasonal dynamics of a mixed population of *Daphnia*, and the associated cladoceran community. Freshwat. Biol. 7: 505–512.

Bengtsson, J., 1987. Competitive dominance among Cladocera: Are single factor explanations enough? An examination of the experimental evidence. Hydrobiologia 145: 245–257.

De Meester, L., 1994. Habitat partitioning in *Daphnia*: coexistence of *Daphnia magna* clones differing in phototactic behaviour. In Beaumont, A. R. (ed.), Genetics and Evolution of Aquatic Organisms. Chapman & Hall London: 323–335.

De Meester, L., L. J. Weider & R. Tollrian, 1995. Alternative antipredator defences and genetic polymorphism in a pelagic predator-prey system. Nature 378: 483–485.

DeMott, W. R. & W. C. Kerfoot, 1982. Competition among cladocerans: nature of the interaction between *Bosmina* and *Daphnia*. Ecology 63: 1949–1966.

Enserink, E. L., N. Van der Hoeven, M. Smith, C. M. van der Klis & M. A. van der Gaag,1996. Competition between cohorts of juvenile *Daphnia magna*: a new experimental model. Arch. Hydrobiol. 136: 433–454.

Flik, B. J. G., D. K. Aanen & J. Ringelberg, 1997. The extent of predation by juvenile perch during diel vertical migration of *Daphnia*. Arch. Hydrobiol. Ergebn. Limnol. (in press).

Flik, B. J. G. & J. Ringelberg, 1993. Influence of food availability on the initiation of diel vertical migration (DVM) in Lake Maarsseveen. Arch. Hydrobiol. Ergebn. Limnol. 39: 57–65.

Guisande, C., 1993. Reproductive strategy as population density varies in *Daphnia magna* (Cladocera). Freshwat. Biol. 29: 463–467.

Goulden, C. E. E., E. L. Hornig & C. Wilson, 1978. Why do large zooplankton species dominate? Verh. int. Ver. Limnol. 20: 2457–2460.

Goulden, C. E. E., L. L. Henry & A. J. Tessier, 1982. Body size, energy reserves and competitive ability in three species of cladocera. Ecology 63: 1780–1789.

Guillard, R. R. L., 1975. Culture of phytoplankton for feeding marine invertebrates. In Smith, W. L. & M. H. Chanley (eds), Culture of Marine Invertebrate Animals. Plenum New York: 29–60.

Jacobs, J., 1978. Coexistence of similar zooplankton species by differential adaptation to reproduction and escape in an environment with fluctuating food and enemy densities. III. Laboratory experiments. Oecologia 35: 35–54.

Johnsen, G. H. & P. J. Jakobsen, 1987. The effect of food limitation on vertical migration in *Daphnia longispina*. Limnol. Oceanogr. 32: 873–880.

Lampert, W., 1987. Feeding and nutrition in Daphnia. In Peters, R. H. & R. De Bernardi (eds), *Daphnia*. Mem. Inst. ital. Idrobiol. 45: 143–192.

Lampert, W., 1989. The adaptive significance of diel vertical migration of zooplankton. Funct. Ecol. 3: 21–27.

Loaring, J. M. & P. D. N. Hebert, 1981. Ecological differences among clones of *Daphnia pulex* Leydig. Oecologia 51: 162–168.

Macháček, J., 1991. Indirect effect of planktivorous fish on the growth and reproduction of *Daphnia galeata*. Hydrobiologia 225: 193–197.

Macháček, J., 1993. Comparison of the response of *Daphnia galeata* and *Daphnia obtusa* to fish-produced chemical substance. Limnol. Oceanogr. 38: 1550–1554.

Macháček, J., 1995. Inducibility of life history changes by fish kairomone in various developmental stages of *Daphnia*. J. Plankton Res. 17: 1513–1520.

MacIsaac, H. J. & J. J. Gilbert, 1991. Competition between *Keratella cochlearis* and *Daphnia ambigua*: effects of temporal patterns of food supply. Freshwat. Biol. 25: 189–198.

Matveev, V. & W. Gabriel, 1994. Competitive exclusion in Cladocera through elevated mortality of adults. J. Plankton Res. 16: 1083–1094.

Milbrink, G. & J. Bengtsson, 1991. The impact of size-selective predation on competition between two *Daphnia* species: a laboratory study. J. anim. Ecol. 60: 1009–1028.

Mort, M. A. & H. G. Wolf, 1985. Enzyme variability in large-lake *Daphnia* populations. Heredity 55: 27–36.

Perrin, N., D. J. Baird & P. Calow, 1992. Resource allocation, population dynamics and fitness: some experiments with *Daphnia magna* Straus. Arch. Hydrobiol. 123: 431–449.

Reede, T., 1995. Life history shifts in response to different levels of fish kairomones in *Daphnia*. J. Plankton Res. 17: 1661–1667.

Reede, T., 1997 Effects of neonate size and food concentration on the life history responses of a clone of the hybrid *Daphnia hyalina × galeata* to fish kairomones. Freshwat. Biol. 37: 389–396.

Reede T. & J. Ringelberg, 1995 . The influence of a fish exudate on two clones of the hybrid *Daphnia galeata × hyalina*. Hydrobiologia 307: 207–212.

Ringelberg, J., 1991. A mechanism of predator-mediated induction of diel vertical migration in *Daphnia hyalina*. J. Plankton Res. 13: 83–89.

Ringelberg, J., B. J. G. Flik, D. Lindenaar & K. Royackers, 1991. Diel vertical migration of *Daphnia hyalina* (sensu latiori) in Lake Maarsseveen: Part 2. Aspects of population dynamics. Hydrobiologia 122: 385–401.

Romanovsky, Y. E. & I. Y. Feniova, 1985. Competition among Cladocera: effect of different levels of food supply. Oikos 44: 243–252.

Schulze, P. C., H. E. Zagarese & C. E. Williamson, 1995. Competition between crustacean zooplankton in continuous cultures. Limnol. Oceanogr. 40: 33–45.

Seitz, A., 1980. The coexistence of three species of *Daphnia* in the Klostersee. Oecologia 47: 333–339.

Spaak, P., 1996. Temporal changes in the genetic structure of the *Daphnia* species complex in Tjeukemeer, with evidence for backcrossing. Heredity 76: 539–548.

Spaak, P. & J. R. Hoekstra, 1993. Clonal structure of the *Daphnia* population in Lake Maarsseveen: its implications for diel vertical migration. Arch. Hydrobiol. beih. Ergebn. Limnol. 39: 157–165.

Spaak, P. & J. R. Hoekstra, 1995. Life history variation and the coexistence of a *Daphnia* hybrid with its parental species. Ecology 76: 553–564.

Statistica for Windows, version 5.0, 1995. Statsoft, inc. 2325 East 13th Street, Tulsa, OK 74104, USA.

Stibor, H., 1992. Predator induced life-history shifts in a freshwater cladoceran. Oecologia 92: 162–165.

Tátrai, I. & R. de Bernardi, 1992. Indirect impact of cyprinid fish fry on the growth and fecundity of *Daphnia obtusa*. Arch. Hydrobiol. 125: 371–381.

Van Gool, E. & J. Ringelberg, 1995. Swimming of *Daphnia galeata × hyalina* in response to changing light intensities: influence of food availability and predator kairomone. Mar. Fresh. Behav. Physiol. 26: 259–265.

Weider, L. J., 1992. Disturbance, competition and the maintenance of clonal diversity in *Daphnia pulex*. J. Evol. Biol. 5: 505–522.

Weider, L. J. & J. Pijanowska, 1993. Plasticity of *Daphnia* life histories in response to chemical cues from predators. Oikos 67: 385–392.

Williamson, C. E., R. W. Sanders, R. E. Moeller & P. L. Stutzman, 1996. Utilization of subsurface food resources for zooplankton reproduction: implications for diel vertical migration theory. Limnol. Oceanogr. 41: 224–233.

Zar, J. H., 1996. Biostatistical Analysis. Prentice Hall, Upper Saddle River, New Jersey 07458, 662 pp.

Hydrobiologia **360**: 117–125, 1997.
A. Brancelj, L. De Meester & P. Spaak (eds), Cladocera: The Biology of Model Organisms.
©1997 *Kluwer Academic Publishers.*

Electron transport system (ETS) activity and respiration rate in five *Daphnia* species at different temperatures

Tatjana Simčič & Anton Brancelj
National Institute of Biology, Večna pot 111, 1000 Ljubljana, Slovenia

Key words: Cladocera, metabolic potential, oxygen consumption, habitats, ecology, physiology

Abstract

Electron transport system activity (ETS) and respiratory rate (R) were measured in five *Daphnia* species at different temperatures in the laboratory. The animals were collected from different habitats: *Daphnia hyalina* Leydig from an eutrophic subalpine lake, *D. pulicaria* Forbes from an oligotrophic high-mountain lake, *D. pulex* Leydig from a temporary lowland pool, and *D. obtusa* Kurz from a temporary highland pool. *D. magna* Straus was obtained from laboratory cultures.

ETS activities and respiratory rates were studied in juveniles and adult females without eggs. The rates were measured at 5, 10, 15, and 20 °C. For adult females Q_{10} of ETS-activity ranged from 1.9 for *D. pulicaria* to 2.2 for *D. obtusa*. The values for juvenile individuals showed greater variations, ranging from 1.6 for *D. pulicaria* to 3.4 for *D. hyalina*. Q_{10} values for respiration were from 1.3 to 1.7 for adults, and from 1.4 to 2.1 for juveniles. The accelerating effect of temperature (μ) ranged from 18.5 to 93.4 kJ mol^{-1}. Average ETS/R ratios for single species for two life stages ranged from 1.0 to 2.5.

Introduction

The overall use of chemical energy by animals to carry out their various functions is often referred to as their energy metabolism (Schmidt-Nielsen, 1979). The activity of the respiratory electron transport system (ETS) is a biochemical measure of the potential metabolic activity (Lampert, 1984). The ETS is localised in the mitochondrial inner membrane and acts as bridge between the oxidising organic matter and O_2. This multi-enzyme complex contains flavoproteins, metallic proteins and cytochromes which are arranged in a complete biochemical redox system for transporting electrons from the coenzymes NADH, NADPH and succinate, arriving from the Krebs' cycle, to the terminal electron acceptor – O_2. Formazan production in ETS assays has been found to be in close correlation with the O_2 consumption of the organisms (G.-Tóth et al., 1995).

The method for measuring formazan production is simple and extremely sensitive. After it was first proposed by Packard (1971), the ETS assay has been modified and improved by various authors (Kenner & Ahmed, 1975a; Owens & King, 1975; G.-Tóth, 1993). Its use to measure metabolic activity has been extended from specific components of marine plankton (Kenner & Ahmed, 1975b; Bamstedt, 1980) and freshwater plankton (Borgmann, 1978; James, 1987) to plankton community respiration in oceans (Packard et al., 1975; Vosjan & Olanczuk-Neyman, 1991) and lakes (Giorgio, 1992), benthic organisms (Cammen et al., 1990), sediment respiration (G.-Tóth et al., 1994) and even biofilms (Blenkinsopp et al., 1991).

The rate of oxygen consumption by *Daphnia* has been reported in many studies (Vollenweider & Ravera, 1958; Urabe & Watanabe, 1990), but there are not many published data on the ETS activity of *Daphnia* (Borgmann, 1977, 1978).

Daphnia are mainly eurythermal, because they are exposed to considerable changes in temperatures during daily cycles of heating and cooling in small ponds and during vertical migration in stratified lakes (Peters, 1987). Since the body temperature of most invertebrates is determined by immediate environment, their

ETS must function under a wide range of temperatures. Thus adaptation to environmental temperatures might be expected. When comparing ETS activities of invertebrate species from different environments, some information on the effect of temperature on ETS activity is required (Borgmann, 1978). The functional response of respiration rate may also differ among species (Urabe & Watanabe, 1990). Goss & Bunting (1980) state that respiration rate functions are correlated with habitat characteristics, spatial distribution and upper temperature tolerance levels. Vosjan & Olanczuk-Neyman (1991) found that organisms which live in relatively stable temperature environment are not adapted to wide temperature fluctuations. In this case respiratory ETS activity is much more sensitive to temperature variations.

The aim of our research was to examine the ETS activity (e.g. metabolic potential) and the oxygen consumption in five *Daphnia* species. We tested the differences in ETS activity and respiration among species as well as between juveniles and adults. The ratio between ETS activity and respiration rate (ETS/R) was calculated to obtain an idea of the degree to which the potential of the metabolism is exploited. We looked for differences in the ETS/R ratios between pond and lake species and tested the hypothesis that *Daphnia* species which inhabit mountain habitats (*D. obtusa* and *D. pulicaria*) are less sensitive to decreasing temperature than *Daphnia* species from lowland habitats. To test this hypothesis, we pre-adapted specimens of all five species to 'standard' temperature of 20 °C, and subsequently exposed them for a short time to lower temperatures. We expected that differences in respiration rate as well as ETS activity among species would remain, even after one month of pre-adaptation.

Material and methods

Cultures and handling before measurements

ETS activity and respiratory rate were measured in five *Daphnia* species, which were brought into the laboratory from different Slovenian habitats: *D. hyalina* Leydig from the eutrophic subalpine lake Bled, *D. pulicaria* Forbes from the oligotrophic mountain lake Srednje Kriško Jezero, *D. pulex* Leydig from a lowland pool by lake Cerkniško Jezero and *D. obtusa* Kurz from a highland pool in Dol pod Studorjem. *D. magna* Straus was obtained from a permanent laboratory culture, which has been kept in a laboratory for ten years (the original source of this culture is unknown). It was used as a reference. All species were cultured for at least one month in the laboratory to get a steady state population. They were kept in 10-l aquaria and some hundred specimens were there all the time. The animals were fed every second day with a suspension of *Scenedesmus* sp. and yeast. After feeding the food level was $1-2 \times 10^5$ cells ml^{-1} for *Scenedesmus* sp. and $5-6 \times 10^6$ cells ml^{-1} for yeast. We selected only healthy-looking adult females without eggs to reduce errors due to an 'uncontrolled mass' of eggs with different ETS and R. Animals were considered to be adults when they were bigger than the smallest ovigerous female. The water temperature was 21.0 ± 2.5 °C, oxygen concentration 8.3 ± 1.0 mg l^{-1} and pH 7.9 ± 0.4. Dry weight of each species was determined by measuring seven subsamples (100 juveniles or 50 adults) on an electrobalance (resolution 10 μg). Subsamples were dried for 24 h at 60 °C (Table 1). Body length was measured from the top of the helmet to the base of the spine. Thirty juveniles and thirty adults were measured from each species. ETS activities and respiration rates were measured on two different groups of animals, taken simultaneously from the same population, approximately 24 hours after feeding. The animals were pre-sorted to ensure that they were all of the same size. Before the experiment started, specimens were pre-adapted to the experimental temperatures for 3 hours (temperature of 15 °C), 4 hours (10 °C) or 5 hours (5 °C).

ETS activity

ETS activity was measured using the method developed by Owens & King (1975) and improved by G.-Tóth (1993). The buffer for homogenisation, substrate solution, reagent solution and stopping solution were prepared just before the experiments to avoid substrate decomposition and bacterial contamination. Twenty (*D. magna* and *D. pulicaria*) or fifty adults without eggs (*D. hyalina*, *D. obtusa*, *D. pulex*) or one hundred juveniles (all species) were concentrated with a plankton net and homogenised in 4 ml final volume of ice-cold homogenisation buffer in a homogeniser (potter) for 3 min at 500 rpm. The homogenate was sonicated by an ultrasonic homogeniser (Cole-Parmer) for 25 s at 40 W and stored subsequently on ice. Within 10 min, 0.5 ml of the homogenate (in triplicate) was incubated in 1.5 ml substrate solution with 0.5 ml reagent solution for 40 min at 5, 10, 15, and 20 °C. Electron transfer assays were carried out in glass tubes. The reaction was stopped by adding 0.5 ml stopping-

Table 1. Length (mm) and dry weight of the specimens (μg) of different *Daphnia* species, that were used in the experiments. Mean $\pm$ SD ($N = 30$).

	Length (mm)		Individual dry weight (μg)	
	Juveniles	Adults	Juveniles	Adults
D. hyalina	0.91 ± 0.10	1.84 ± 0.13	2.41 ± 0.87	21.93 ± 4.46
D. magna	1.30 ± 0.13	3.07 ± 0.15	12.66 ± 5.68	179.25 ± 18.02
D. obtusa	0.91 ± 0.09	1.78 ± 0.14	2.97 ± 1.13	22.87 ± 3.89
D. pulex	0.86 ± 0.09	2.04 ± 0.22	3.77 ± 1.23	39.78 ± 5.08
D. pulicaria	1.09 ± 0.14	2.57 ± 0.18	8.17 ± 2.45	92.11 ± 16.67

solution. Blanks (1.5 ml substrate solution and 0.5 ml reagent solution) were incubated and stopped as the samples, while 0.5 ml of homogenate was added after the stopping. The formazan production was determined spectrophotometrically from the absorption of the sample at 490 nm wavelength against the blank within 10 min after stopping the reaction. ETS activity was calculated according to Kenner & Ahmed (1975a):

$$\text{ETS activity } (\mu\text{l O}_2 \text{ s}^{-1} \text{ h}^{-1}) = \frac{\text{Abs}^{490\text{nm}} \times V_r \times V_h \times 60}{V_a \times S \times t \times 1.42},$$

where:
$\text{Abs}^{490\text{nm}}$ = absorption of the sample
V_r = final volume of the reaction mixture (3 ml)
V_h = volume of the original homogenate (4 ml)
V_a = volume of the aliquot of the homogenate (0.5 ml)
S = sample size (mg)
t = incubation time (min)
1.42 = factor for conversion to volume O_2

Respiration

The respiration rate was estimated by the closed bottle method (Lampert, 1984). Ground glass stoppered bottles (300 ml) were filled with filtered (glass microfibrile filter Whatman GF/C – particle retention approximately 1 μm) and aerated tap water from a common container. Ten bottles received animals (100 juveniles or 50 adults of each species), while two bottles with no animals served as final controls. Bottles were stoppered and kept at incubation temperatures (5, 10, 15, and 20°C). After 24 h the concentration of dissolved oxygen in the experimental and control bottles was measured by the Winkler method. The difference between the oxygen concentration of the experimental and that of the control bottles was taken as the amount of oxygen consumed by animals. The amount of oxygen consumed was then converted to a dry weight-specific respiration rate. At least six replicates were measured

for each of five species and two life stages (juveniles, adults).

Calculations

Respiratory data are expressed as μl O$_2$ mg^{-1} dw h^{-1}. ETS activity was measured as a rate of tetrazolium dye reduction which was converted to equivalent oxygen utilised mg^{-1} dw h^{-1} (μl O$_2$ mg^{-1} dw h^{-1}) as described by Kenner & Ahmed (1975a). ETS/R ratios were calculated as the ratio of the calculated maximum rate of oxygen consumption as measured by *in vitro* enzymatic rates to the rate of respiration (R) *in vivo*. Single classification ANOVA were carried out to test for differences among species in ETS activity and respiration rate for two development stages and four different temperatures. Significance levels were corrected using the sequential Bonferroni procedure (Rice, 1989). The temperature coefficient, Q_{10}, was used to characterise the increase of ETS activity and respiration associated with a temperature increase for 10 °C. Thermodynamic responses of both processes were shown by Arrhenius activation energy (μ). The numerical value of μ is a 'temperature characteristic' which describes the accelerating influence of temperature on the metabolic rate. When the graph is plotted as the logarithm of the respiratory rate on the ordinate and the inverse of the absolute temperature on the abscissa the data points form a straight line (Robinson & Williams, 1993). The regression line was calculated by the least squares method. The Arrhenius activation energy (μ) was then calculated from the slope of the regression line by the equation:

$$\mu = -RS$$

where μ is expressed in kJ mol^{-1}, R is the gas constant (8,31 J mol^{-1} degree^{-1}) and S is the slope of the Arrhenius plot.

Table 2. Results of analysis of variance (Single classification) applied to ETS activity of juveniles and adults at 5, 10, 15 and 20 °C for five *Daphnia* species. An * indicates significance using the sequential Bonferroni test with an initial critical probability value of $P<0.0063$ (eight non independent series of data tested).

	d.f. Between groups	d.f. Within groups	F	P-value
Juveniles				
5 °C	4	25	2.14	0.105
10 °C	4	25	2.39	0.078
15 °C	4	25	2.58	0.062
20 °C	4	26	10.74	0.000*
Adults				
5 °C	4	24	3.09	0.035
10 °C	4	24	4.30	0.009*
15 °C	4	25	11.44	0.000*
20 °C	4	26	21.51	0.000*

Table 3. Results of analysis of variance (Single classification) applied to respiration of juveniles and adults at 5, 10, 15 and 20 °C for five *Daphnia* species. An * indicates significance using the sequential Bonferroni test with an initial critical probability value of $P<0.0063$ (eight non independent series of data tested).

	d.f. Between groups	d.f. Within groups	F	P-value
Juveniles				
5 °C	4	27	1.43	0.250
10 °C	4	25	6.33	0.001*
15 °C	4	27	3.04	0.034
20 °C	4	31	7.92	0.000*
Adults				
5°C	4	25	1.99	0.127
10°C	4	28	7.81	0.000*
15°C	4	28	10.87	0.000*
20°C	4	31	6.05	0.001*

Results

ETS activity

ETS activity were higher in juvenile than in adults for all species and at all temperatures, except for *D. hyalina* at 5 °C and *D. obtusa* at 10 °C (Figure 1). ETS activity increased with increasing temperature for all species in juveniles as well as in adults. ETS activity did not differ significantly among juveniles of the different species when kept at 5, 10 or 15 °C and among adults of the different species when kept at 5 °C (Table 2). *D. magna* had the lowest ETS activity of all species at all temperatures. The highest ETS activity depended both on species and temperature (Figure 1).

Respiration

Weight-specific respiration rates in juveniles were higher than in adult females at all temperatures (Figure 1). The rates significantly differed among species for both juveniles and adults at 10 and 20 °C, but not at 5 and 15 °C (Table 3). Juveniles of *D. hyalina* and *D. obtusa* had similar body sizes (Table 1), but their respiration rates differ significantly at 20 °C (Table 3). *D. magna* had the lowest respiration rates, in juveniles as well as in adults, at all temperatures. *D. obtusa* had the highest respiration rates at all temperatures for juveniles and at 5, 10 and 15 °C for adults.

The ETS/R ratio

The ETS/R ratio varied among species as well as within species, among temperatures (Figure 2). Mean ETS/R ($\pm$ SD) ratios, averaged over juveniles and adults, ($n=8$) were 1.2 ± 0.4 for *D. obtusa*, 2.0 ± 0.4 for *D. pulex*, 2.1 ± 0.5 for *D. hyalina*, 2.1 ± 0.6 for *D. magna* and 2.5 ± 0.5 for *D. pulicaria*. Pond species (*D. obtusa* and *D. pulex*) have significantly lower mean ETS/R ratios (2.28 ± 0.54) than lake species (*D. hyalina* and *D. pulicaria*: 1.49 ± 0.53).

Influence of temperature

For adult females, the Q_{10} of ETS activity ranged from 1.9 for *D. pulicaria* to 2.2 for *D. obtusa* (Table 4). ETS activity ranged more widely in juveniles: from 1.6 for *D. pulicaria* to 3.4 for *D. hyalina*. Q_{10} values for respiration ranged from 1.3 to 1.7 for adults, and from 1.4 to 2.1 for juveniles. Q_{10} values were higher for ETS than for respiration in all species treatments combinations.

Arrhenius activation energy differed among all five species of *Daphnia* (Table 5). Adults of *D. pulicaria* and *D. obtusa* had significantly lower μ values (20.15 ± 2.33) for respiration than the other three species (41.73 ± 6.08) (*t*-test, $t=5,56$, d.f.$=2.7$), while μ values for ETS activity did not differ significantly. For both species, the coefficient μ for ETS activity of

ETS activity　　　　　　　Respiration

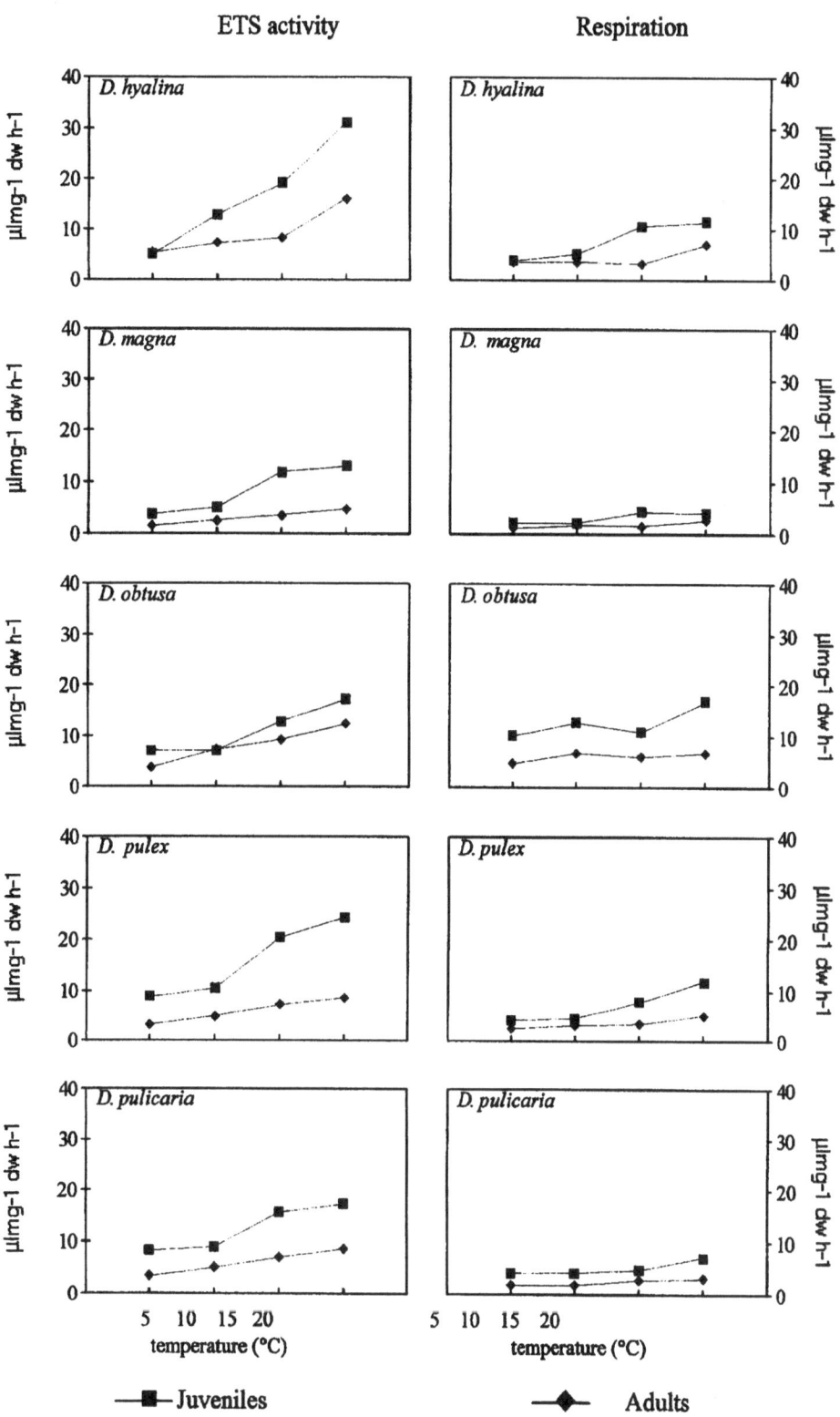

Figure 1. ETS activity and respiration rate of *Daphnia* species (μl O$_2$ mg^{-1} dw h^{-1}).

122

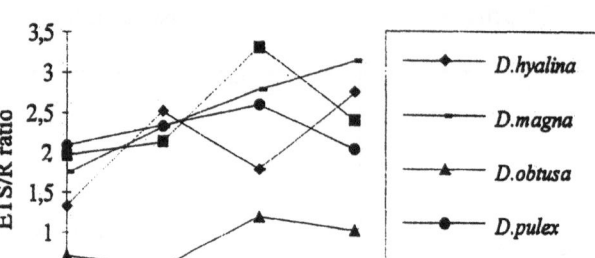

Juveniles

ETS/R ratio vs temperature (°C)

- D.hyalina
- D.magna
- D.obtusa
- D.pulex
- D.pulicaria

Adults

ETS/R ratio vs temperature (°C)

- D.hyalina
- D.magna
- D.obtusa
- D.pulex
- D.pulicaria

Figure 2. ETS/R ratio for *Daphnia* species incubated at 5, 10, 15 and 20 °C.

Table 4. Q_{10} values of ETS activity and respiration rate for five *Daphnia* species (Q_{10} were calculated from means at 5 and 20 °C).

	ETS activity		Respiration	
	Juveniles	Adults	Juveniles	Adults
D. hyalina	3.4	2.1	2.1	1.6
D. magna	2.3	2.1	1.5	1.7
D. obtusa	1.8	2.2	1.4	1.3
D. pulex	2.0	1.9	2.0	1.6
D. pulicaria	1.6	1.9	1.4	1.4

Table 5. Arrhenius activation energy μ (kJ mol^{-1}), calculated from the slope of the regression line. Each value was calculated from $n = 25$.

	ETS activity		Respiration	
	Juveniles	Adults	Juveniles	Adults
D. hyalina	87.3	53.5	93.4	35.3
D. magna	66.5	50.9	42.2	47.4
D. obtusa	50.0	51.6	54.0	21.8
D. pulex	54.8	49.2	35.7	42.5
D. pulicaria	44.0	43.5	41.4	18.5

juveniles and adults was similar, but μ for respiration was much lower in adults than in juveniles. Juveniles of *D. hyalina*, *D. obtusa* and *D. pulex* are similar in body length, but theirs μ values differed significantly. Juveniles of *D. hyalina* had the highest μ. Adult *D. pulex* and *D. pulicaria* showed similar changes in ETS activ-

ity with increasing temperature, but the response of respiration was different. *D. pulicaria* had much lower μ for respiration than *D. pulex*.

Discussion

ETS activity and respiration rate differed among species as well as between juveniles and adults (Figure 1). *Daphnia* species vary greatly in body size and thus characters related to body size are also expected to vary between species (Peters, 1987). Bamstedt (1988) suggests that ETS activity and respiration rate of *Acartia clausii, Pseudocalanus* sp. and *Calanus finmarchicusis* scale similarly. Therefore, we expected that ETS activity and respiration would also be influenced by the different size of our study species. Armitage & Lei (1979) report that the weight-specific respiration rate of *Daphnia ambigua* is inversely related to body size at different temperatures. As expected, ETS activities and weight-specific respiration rates were higher for juveniles than for adult females at all temperatures. The reason for this might be a higher ratio between metabolic active versus metabolic inactive tissue in juveniles. The proportion of body mass that is metabolically inert increases as the animal grows, and this component might have contributed to the negative allometry of respiration rate relative to body mass (Glazier, 1991).

Urabe & Watanabe (1990) stated that respiration rate is basically similar among species if the body weight is similar. Our results have shown that juveniles of *D. hyalina, D. obtusa* and *D. pulex,* also they have a similar body length (Table 1), have a significantly different respiration rate at 20 °C. The reason for this is differences in body mass among species.

In our experiments, respiration rates were significantly different among species for both stages on all temperatures, except at 5 °C (see Table 3). At low temperature, animals have a minimal metabolic rate, necessary for survival, which might be similar for all species.

The values of ETS/*R* ratios varied within species at different temperatures (Figure 2). Our ETS/R ratios are similar to the values found for marine zooplankton (Owens & King, 1975; Bamsted, 1980) and *Ceriodaphnia dubia* Richard (James, 1987). Since this ratio is a measure of the efficiency or degree of control of the enzymes and electron carriers in the ETS (Kenner & Ahmed, 1975b), we assume that lake species control their metabolism in a different way than pond species. Pond species live in habitats with a higher food concentration than lake species, so we expected higher metabolic demands in pond species. In our experiments, we measured the basal metabolism and expenditure on locomotion. The expenditure on filtration and specific dynamic action (SDA) were minimal,

because the animals were not fed just prior or during the experiments. The ETS/*R* ratio reflects that fraction of the maximum respiratory capacity that the organism is effectively using (Martinez, 1992). The pond species *D. obtusa* and *D. pulex* exploited 67% of the potential of metabolism for basal metabolism and locomotion, while *D. hyalina* and *D. pulicaria* used 44% of the metabolic potential. This could mean that pond species have higher energy demands for basal metabolism and locomotion than lake species. Hrbáček (1987) reports that *D. pulex* grows better at high concentrations of food than does *D. pulicaria*.

Increased temperature may cause: (a) a shock reaction producing an unrealistically high metabolic rate, (b) a stabilised state where metabolism is maintained at an elevated rate or (c) an acclimatised state where compensatory mechanisms reduce metabolism to a rate similar to that prior to the temperature shift (Robinson & Williams, 1993). We kept the animals 3 h before the experiments on experimental temperatures. This time should be long enough to avoid shock reaction but too short for complete acclimation or changes in ETS activity (Bamsted, 1980). We only worked with temperature reductions. Q_{10} values for respiration ranged from 1.3 to 1.7 for adults, and from 1.4 to 2.1 for juveniles (Table 2). Vollenweider & Ravera (1958) reported that Q_{10} for respiration mostly ranges from 2 to 3. Peters (1987) states that a Q_{10} of 2 is a reasonable first approximation for *Daphnia*. Q_{10} values were equal as higher for ETS than for respiration during all experiments. ETS activity is a direct enzymatic process, dependent upon enzyme concentration and temperature, whereas respiration is a complex physiological process. This means that when temperature changes, respiration is also influenced by changes in substrate concentration, the intracellular environment, structure of the lipid membranes, etc. (Withers, 1992). Q_{10} values for ETS activity in adults were nearly 2 and not significantly different among species. In juveniles, however, values differed among species. *D. pulicaria* had low Q_{10} value for both ETS activity and respiration while *D. obtusa* had low Q_{10} values for respiration in both stages. *D. obtusa* and *D. pulicaria* had significantly lower Q_{10} values for respiration than for ETS activity. Since both species were obtained from highland water bodies and have a relatively high metabolism at 5 °C (60–70% of maximum respiration rate), we hypothesize that the low Q_{10} values for respiration are an adaptation to the lower temperatures which prevail commonly in alpine water bodies.

Arrhenius activation energy is another way to illustrate the influence of temperature on organisms. Adults of *D. pulicaria* and *D. obtusa* also had significantly lower μ values for respiration than did the other three species Juveniles of *D. hyalina* had the highest μ for respiration among species. This high value of μ means that *D. hyalina* is metabolically more sensitive to increasing temperature. Brancelj & Blejec (1994) found that juveniles of *D. hyalina* live in relatively warmer lake strata. In general, the animals obtained from lowland habitats appear to be more sensitive to increasing temperature than those from mountain habitats. Packard et al. (1975) state that marine organisms can develop adaptive mechanisms that increase metabolic efficiency at low temperature and decrease it at high temperature, thus achieving homeostatic control over their enzyme reactions. Goss & Bunting (1980) found that cladocerans and probably many other aquatic invertebrates with narrow geographical distribution and from particular habitats with limited temperature changes will be metabolically (*sensu* respiration) more sensitive to increasing temperatures.

Adult *D. pulex* and *D. pulicaria* had similar responses of ETS activity to increasing temperature, but the respiration response differed. Thus, metabolic potential is similar between these two species, but the oxygen consumption differs. As temperature increases, *D. pulicaria* showed a more stable response than *D. pulex*. The value of μ for *D. pulex* was 2.3 times as high than that for *D. pulicaria*. It is reasonable to assume that, in addition to the metabolic potential, there were other factors and/or mechanisms that influenced respiration rates at different temperatures.

We can conclude that there are differences in metabolism between pond and lake species as well as between species from lowland and mountain water bodies. If we had done experiments on animals that were isolated directly from the field without a period of acclimatization in laboratory, the differences among species would probably have been much more obvious. A lot of work on this topic has to be done.

Acknowledgments

We thank D. S. Glazier, L. G.-Tóth and P. Spaak for constructive and helpful comments on the previous versions of the manuscript.

References

Armitage, K. B. & C. Lei, 1979. Temperature acclimatization in the filtering rates and oxygen consumption of *Daphnia ambigua* Scourfield. Comp. Biochem. Physiol. 62A: 807–813.

Bamstedt, U., 1980. ETS activity as an estimator of respiratory rate of zooplankton populations. The significance of variations in environmental factors. J. exp. mar. Biol. Ecol. 42: 267–283.

Bamstedt, U., 1988. Ecological significance of individual variability in copepod bioenergetics. Hydrobiologia 167/168: 43–59.

Blenkinsopp, S. A., P. A. Gabbott, C. Freeman & M. A. Lock, 1991. Seasonal trends in river biofilm storage products and electron transport system activity. Freshwat. Biol. 26: 21–34.

Borgmann, U., 1977. Electron transport system activity in *Daphnia* and crayfish. Can. J. Zool. 55: 847–854.

Borgmann, U., 1978. The effect of temperature and body size on electron transport system activity in freshwater zooplankton. Can. J. Zool. 56: 634–642.

Brancelj, A. & A. Blejec, 1994. Diurnal vertical migration of *D. hyalina* Leydig, 1860 (Crustacea: Cladocera) in Lake Bled (Slovenia) in relation to temperature and predation. Hydrobiologia 284: 125–136.

Cammen, L. M., S. Corwin & J. P. Christensen, 1990. Electron transport system (ETS) activity as a measure of benthic macrofaunal metabolism. Mar. Ecol. Prog. Ser. 65: 171–182.

Giorgio, P. A., 1992. The relationship between ETS (electron transport system) activity and oxygen consumption in lake plankton: a cross-system calibration. J. Plankton Res. 14: 1723–1741.

Glazier, D. S., 1991. Separating the respiration rates of embryos and brooding females of *Daphnia magna*: Implications for the cost of brooding and allometry of metabolic rate. Limnol. Oceanogr. 36: 354–362.

Goss, L. B. & D. L. Bunting, 1980. Temperature effects on zooplankton respiration. Comp. Biochem. Physiol. 66A: 651–658.

G.-Tóth, L., 1993. Measurement of the terminal electron system (ETS)-activity of plankton and sediment. Proceedings of the ILEC/UNEP International Training Course, 149–152.

G.-Tóth, L., Zs. Langó, J. Padisák & E. Varga, 1994. Terminal electron transport system (ETS)-activity in the sediment of Lake Balaton, Hungary. Hydrobiologia 281: 129–139.

G.-Tóth, L., M. Szabo & D. Webb, 1995. Adaptation of the tetrazolium reduction test for the measurement of the electron transport system (ETS) activity during embryonic development of medaka. J. Fish. Biol. 46: 835–844.

Hrbáček, J., 1987. Systematics and biogeography of *Daphnia* species in the Northern temperate region. In Peters, R. H. & R. Bernardi (eds), '*Daphnia*', Mem. Ist. ital. Idrobiol. 45: 37–76.

James, M. R., 1987. Respiratory rates in cladoceran *Ceriodaphnia dubia* in lake Rotiongaio, a monomictic lake. J. Plankton Res. 9: 573–578.

Kenner, R. A. & S. I. Ahmed, 1975a. Measurements of electron transport activities in marine phytoplankton. Mar. Biol. 33: 119–127.

Kenner, R. A. & S. I. Ahmed, 1975b. Correlation between oxygen utilization and electron transport activity in marine phytoplankton. Mar. Biol. 33: 129–133.

Lampert, W., 1984. The measurement of respiration. In Downing, J. A. & F. H. Rigler. A Manual on Methods for the Assessment of Secondary Productivity in Fresh Water. IPB Handbook 17, 2nd edn., Blackwell Sci. Publ. 413–468.

Martinez, R., 1992. Respiration and respiratory electron transport activity in marine phytoplankton: growth rate dependence and light enhancement. J. Plankon Res. 14: 789–797.

Owens, T. G. & F. D. King, 1975, The measurement of respiratory electron transport system activity in marine zooplankton. Mar. Biol. 30: 27–36.

Packard, T. T., 1971. The measurement of respiratory electron-transport activity in marine phytoplankton. J. mar. Res. 29: 235–244.

Packard, T. T., A. H. Devol & F. D. King, 1975. The effect of temperature on the respiratory electron transport system in marine plankton. Deep-Sea Res. 22: 237–249.

Peters, R. H., 1987. Metabolism in *Daphnia*. In Peters, R. h. & R. de Bernardi (eds) '*Daphnia*'. Mem. Ist. ital. Idrobiol. 45: 193–243.

Rice, W. R., 1989. Analyzing tables of statistical tests. Evolution 43: 223–225.

Robinson, C. & P. J. B. Williams, 1993. Temperature and antarctic plankton community respiration. J. Plankton Res. 15: 1035–1051.

Schmidt-Nielsen, I., 1979. Animal Physiology: Adaptation and Environment. 2nd edn. Cambridge University Press, 560 pp.

Urabe, J. & Y. Watanabe, 1990. Influence of food density on respiration rate of two crustacean plankters, *Daphnia galeata* and *Bosmina longirostris*. Oecologia 82: 362–368.

Vollenweider, R. A. & O. Ravera, 1958. Preliminary observations on the oxygen uptake by some freshwater zooplankters. Verh. int. Ver. Limnol. 12: 369–380.

Vosjan, J. H. & K. M. Olanczuk-Neyman, 1991. Influence of temperature on respiratory ETS-activity of micro-organisms from Admiralty Bay, King George Island, Antarctica. Neth. J. Sea Res. 28: 221–225.

Withers, P. C., 1992. Comparative Animal Physiology. Saunders Collage Publishing: 82–187.

Hydrobiologia **360**: 127–133, 1997.
A. Brancelj, L. De Meester & P. Spaak (eds), Cladocera: The Biology of Model Organisms.
©1997 *Kluwer Academic Publishers.*

Hybridization in the *Daphnia galeata* complex: are hybrids locally produced?

Piet Spaak
Max-Planck-Institut für Limnologie, Postfach 165, D-24302 Plön, Germany and EAWAG / ETH, Department of Limnology, Ueberlandstrasse 133, 8600 Dübendorf, Switzerland (Permanent address)
(e-mail spaak@eawag.ch)

Key words: genetic distances, cladocera, allozymes, genetic diversity

Abstract

Within the species complex of *Daphnia galeata*, *D. cucullata* and *D. hyalina* various combinations of hybrids and parental taxa occur in lakes throughout Europe. Since daphnids are cyclic parthenogens and mostly reproduce asexually, hybrid populations can be maintained by asexual reproduction and without recurrent hybridization events. Therefore, it is possible that hybridization events have been rare, with range expansion occurring by dispersal of hybrids.

Allozyme data from seven European populations were used to compare genetic variation within and between hybrid and parental taxa. An UPGMA cluster analysis of genetic distances showed that *D. cucullata* × *galeata* hybrids from different lakes grouped in different clusters according to the lake from which they were isolated, suggesting multiple hybridization events. Clonal diversity within hybrid taxa was comparable to parental taxa. Furthermore, evidence was found for introgression of the *Pgi-S* allele from *D. cucullata* to *D. galeata* in three lakes. These results indicate that multiple hybridization events within this species complex are likely, and that hybrid taxa can reproduce sexually.

Introduction

Interspecific hybridization is a common phenomenon in *Daphnia*. Hybrids occur in several species complexes in North America (Colbourne & Hebert, 1996; Hebert & Finston, 1996), Australia (Hebert & Wilson, 1994) and Europe (Schwenk & Spaak, 1995). The wide distribution of *Daphnia* hybrids may be caused by their reproductive mode. Since *Daphnia* reproduce through cyclic parthenogenesis, only one or a few hybridization events are needed to establish a population of hybrids. Since these hybrids can reproduce asexually, no further hybridization events are necessary to maintain a hybrid population; presuming that environmental conditions are such that hybrids may remain present throughout the year. However, hybrid taxa also produce males and sexual females (Carvalho & Wolf, 1989; Spaak, 1995). This creates the possibility of producing a myriad of backcrosses with parental and other hybrid taxa. Recent studies (Taylor & Hebert, 1993; Spaak, 1996)

have shown that backcrossing may be a more common process than previously believed (Wolf, 1987; Hebert et al., 1989).

Several genetic studies have shown clearly that the production of *Daphnia* hybrids is a continuing process (Taylor & Hebert, 1992; Müller & Seitz, 1995; Spaak, 1996). One argument to support this notion is that genetic variation within hybrids is comparable, or even higher, than within parental species. This would not be expected in cases where hybrids were formed a long time ago and populations persist only as parthenogenetic females. In that case, the only source for genetic variation would be the initially available variation supplemented by the variation added via mutations. As mutations also add variation to parental species, they offer an unlikely explanation for the relatively high genetic variability within hybrid populations. Furthermore, at least one hybrid was found to hatch from ephippia collected from the bottom of the Shösee in Northern Germany (Carvalho & Wolf, 1989), and

Schwenk (pers. comm.) recently succeeded in producing hybrids from laboratory crosses. However, these observations do not exclude the possibility that a substantial part of the hybrid lineages may be old.

One of the questions concerning hybridization in *Daphnia* is whether hybridizing taxa are separate species. Based on the 'Biological species concept' (Mayr, 1942) of genetically isolated taxa, one could argue that they are not if hybrids are regularly produced and if back-crossing occurs. On the other hand, several genetic studies (Taylor & Hebert, 1992; Spaak, 1996; Schwenk, 1997) have shown that *Daphnia* species that produce interspecific hybrids have unique combinations of alleles, suggesting some reproductive isolation among these groups. Furthermore, parental species are morphologically distinct from each other (Flößner, 1993) and not all possible intermediate forms are observed as expected in a randomly mating 'hybrid swarm'.

More evidence for local production of interspecific hybrids in *Daphnia* can be obtained through comparisons of the genetic structure of the hybrid populations from isolated lakes. Let us assume that hybrids are evolutionary old, dispersed from one or a few hybrid 'sources', and that no additional hybrids or backcrosses were produced. In this scenario, one would expect hybrid populations to be characterised by: (i) a lower number of clones compared to parental taxa; (ii) a lower genetic differentiation among populations of different lakes than among hybrids and parental taxa within lakes. If hybridization is a local phenomenon, however, one would expect relative large genetic distances among hybrid taxa from different lakes. In the present study, I use two allozyme loci to compare *Daphnia* populations of the *Daphnia galeata* species complex (*D. galeata*, *D. cucullata*, *D. hyalina* and all three of their possible hybrids) from seven European lakes. Genetic distances among taxa within and between lakes were compared using the unweighed pair-group (UPGMA) clustering algorithm. I tested the hypothesis that hybrid populations from different lakes will cluster with the parental taxa from the same lakes rather than with the hybrids of other lakes. Support for this hypothesis indicates local and likely recurrent hybridization events in the genus *Daphnia*.

Materials and methods

Data on the genetic structure of seven *Daphnia* populations were used in this study (Table 1). Tjeukemeer is a shallow eutrophic lake in the northern part of The Netherlands, Maarsseveen is a mesotrophic lake in the centre of The Netherlands, Ringsjön is a eutrophic shallow lake situated in southern Sweden, Plußsee, Belauersee and Höftsee are situated in northern Germany, and the Bodensee is a large, deep mesotrophic lake in southern Germany. From all lakes except Ringsjön only September and October samples were used: Lake Tjeukemeer was sampled in 1989, 1990 and 1991, Maarsseveen in 1989 and 1990, Belauersee and Höftsee in 1994 and Plußsee in 1994 and 1995. Ringsjön was sampled on 16 June and 14 October 1994 by Eva Bergstrand. Data for the Bodensee were taken from Weider & Stich (1992); I limited the analysis of this data set to only September and October sampling dates. For sampling methods and dates in Tjeukemeer see Spaak (1994; 1996); the other lakes were sampled using vertical plankton net tows (see Spaak & Hoekstra, 1993).

Individual adult females were picked randomly and assayed using cellulose acetate electrophoresis following standard methods (Hebert & Beaton, 1989). Generally, 100 animals per sampling date were assayed for three enzyme loci: phosphoglucomutase (*Pgm*, EC 5.4.2.2), phosphoglucose isomerase (*Pgi*, EC 5.3.1.9) and glutamate oxaloacetate transaminase (*Got*, EC 2.6.1.1). *Got* is a diagnostic marker for discrimination between *D. galeata*, *D. cucullata* and *D. hyalina* (Wolf & Mort, 1986). *D. galeata* is fixed for the *F* (fast) allele, *D. cucullata* for the S^- (very slow) allele, and *D. hyalina* for the *S* (slow) allele. Interspecific hybrids can be identified as heterozygotes (S^-F, SF, S^-S) at *Got* (*D. galeata* × *hyalina* [*G* × *H*]: *SF*; *D. cucullata* × *hyalina* [*C* × *H*]: S^-S; *D. cucullata* × *galeata* [*C* × *G*]: S^-F). Variation at the *Pgi* and *Pgm* loci made it possible to identify a number of two-locus genotypes within each taxon.

Clonal diversity

Clonal diversity was calculated for all taxa in all populations. As a measure of clonal diversity, I used Simpson's (1949) index of concentration, $C = \Sigma p_i^2$, where p_i is the frequency of the *i*th clone (two-locus genotype) in the sample, and the summation is over all clones. Clonal diversity *D* was calculated as $D = -\log C$ (Pielou, 1975). This diversity index is a composite of abundance and evenness: it exhibits low values when a single clone is overwhelmingly predominant and high values when the number of clones is high and when clones tend to be equally abundant.

Table 1. Average frequencies of the different taxa of the *Daphnia galeata* species complex for the seven studied lakes. N is to the total number of *Daphnia* individuals analysed

Lake	Country	N	*D. cucullata*	*D. cuc. × gal.*	*D. galeata*	*D. gal. × hyl.*	*D. hyalina*	*D. cuc. × hyl.*
Tjeukemeer	Netherlands	1184	0.30	0.38	0.32			
Maarsseveen	Netherlands	316			0.14	0.86		
Plußsee	North Germany	743	0.36	0.20	0.15	0.16		0.13
Belauersee	North Germany	229	0.82	0.04			0.14	
Höftsee	North Germany	111		0.10	0.90			
Ringsjön	Sweden	275	0.61	0.11	0.28			
Bodensee[1]	South Germany	486			0.49	0.20	0.31	

[1] Data from Weider & Stich (1992)

Genetic distances

Unbiased genetic distances (Nei, 1978) among samples of the different taxa from each lake were calculated from genotype frequencies and the relatedness of the different samples was assessed using the unweighed pair-group (UPGMA) clustering program of BIOSYS-1 (Swofford & Selander, 1981). Because the number of samples was unequal for several lakes, analyses of genetic distances were based on mean frequency data per taxon per lake combining all available sampling dates. Analyses were restricted to *Pgi* and *Pgm* (*Got* genotypes were used to differentiate between the taxa). Höftsee was excluded from the analysis, since only 11 *C × G* individuals were found, leaving only one taxon for the analysis. For the same reason, *C × G* from Belauersee was excluded from the analysis, but the remaining two taxa in this lake (*D. cucullata* and *D. hyalina*) were included.

I also calculated mean genetic distances between several categories of populations to make comparisons among several classes of taxa. A population in this analysis was defined as the individuals that belong to the same taxon (hybrid or parental) in a specific lake. Comparisons were made among parental taxa (different lakes); within parental taxa (among lakes); within hybrid taxa (among lakes); and among hybrid taxa and their parental taxa in the same lake.

Results

Taxon frequencies

Based on their *Got* genotypes individuals were assigned to *D. galeata* (*FF*), *D. cucullata* (*S⁻S⁻*), *D. hyalina* (*SS*) or the interspecific hybrids: In all seven studied lakes, two or more *Daphnia* taxa were found (Table 1). *D. cucullata*, *D. galeata* and their interspecific hybrid were the most common taxa. In three lakes, both *D. cucullata* and *D. galeata* were found together with their hybrid. In Bodensee, *D. galeata* and *D. hyalina* were found together with their interspecific hybrid. The *G × H* hybrid was also found in two other lakes, co-occurring with only one of the parental species (*D. galeata*). The *C × H* hybrid was found only in the Plußsee, along with one of the parental species (*D. cucullata*).

Allele frequencies

The most common *Pgi* allele found was *M* (Table 2) with a mean frequency of 72%. Both *D. galeata* and *D. hyalina* were almost fixed for this allele (frequency > 90%) in all but one population (the *F* allele had a frequency of 16% in the *D. galeata* population in the Höftsee). In *D. cucullata*, the *M* and *S* alleles were found at comparable frequencies. For this species, the Ringsjön population in Sweden was an exception whith the *S⁻* allele rather than the *M* allele having a frequency of greater then 30%. As expected, the hybrid taxa showed intermediate allele frequencies, except for *C × G* in the Höftsee where no *S* allele was found. In three *D. galeata* populations (Maarsseveen, Höftsee, Bodensee), no *Pgi-S* allele was found. These were also the populations were *D. cucullata* was absent.

Four alleles were found at the *Pgm* locus (Table 3). The most common *M* and *F* alleles had an equal mean overall frequency of 40%. Allele frequencies of *Pgm* were less taxon-specific than for *Pgi* (Table 3). The *S* allele had the highest frequency in *D. cucullata*, whereas the *F* allele seemed to be more frequent in *D. galeata* and *D. hyalina*. No *Pgm-F⁺* allele was found in *D. cucullata*.

Table 2. Mean allele frequencies of *Pgi* averaged over all sampling dates for *D. cucullata, D. galeata, D. hyalina* and their interspecific hybrids in seven European lakes

Lake	D. cucullata				D. cuc. × gal.				D. galeata				D. gal. × hyl.				D. hyalina				D. cuc. × hyl.			
	S⁻	S	M	F	S⁻	S	M	F	S⁻	S	M	F	S⁻	S	M	F	S⁻	S	M	F	S⁻	S	M	F
Tjeukemeer	.04	.52	.42	.02	.00	.18	.80	.02	.00	.04	.95	.01												
Maarsseveen									.00	.00	1.0	.00	.00	.00	1.0	.00								
Plußsee	.13	.59	.28	.00	.17	.31	.49	.03	.01	.03	.92	.05	.01	.05	.94	.00					.04	.50	.44	.01
Belauersee	.00	.45	.55	.00	.00	.45	.55	.00									.00	.02	.98	.00				
Höftsee					.00	.00	.96	.04	.00	.00	.84	.16												
Ringsjön	.37	.56	.06	.01	.19	.33	.48	.00	.00	.01	.98	.01												
Bodensee[1]									.00	.00	.94	.06	.00	.00	1.0	.00	.00	.00	1.0	.00				

[1] Data from Weider & Stich (1992)

Table 3. Mean allele frequencies of *Pgm* averaged over all sampling dates for *D. cucullata, D. galeata, D. hyalina* and their interspecific hybrids in seven European lakes

Lake	D. cucullata				D. cuc. × gal.				D. galeata				D. gal. × hyl.				D. hyalina				D. cuc. × hyl.			
	S	M	F	F⁺	S	M	F	F⁺	S	M	F	F⁺	S	M	F	F⁺	S	M	F	F⁺	S	M	F	F⁺
Tjeukemeer	.24	.49	.27	.00	.10	.28	.62	.00	.08	.31	.61	.00												
Maarsseveen									.18	.33	.49	.00	.06	.33	.61	.00								
Plußsee	.38	.51	.11	.00	.18	.53	.25	.04	.01	.30	.62	.08	.18	.53	.25	.04					.05	.52	.06	.37
Belauersee	.06	.89	.05	.00	.00	.80	.20	.00									.00	.07	.82	.11				
Höftsee					.41	.49	.05	.05	.18	.42	.36	.04												
Ringsjön	.49	.46	.05	.00	.14	.72	.14	.00	.04	.64	.30	.02												
Bodensee[1]									.04	.38	.53	.05	.01	.02	.93	.04	.00	.00	.95	.05				

[1] Data from Weider & Stich (1992)

Clonal diversities

Clonal diversity for several lakes did not differ much for *D. cucullata, D. galeata* and the hybrid *C × G* (Table 4). Mean values for these taxa were all above 0.60. Clonal diversities were much lower in the *D. hyalina, C × H* and *G × H* populations (mean values 0.30 or less).

Genetic distances

Mean genetic distances were highest among populations of different 'parental species' (Figure 1) and lowest within these species (0.08). An average genetic distance of 0.13 was found for different populations of a given hybrid taxon. The mean genetic distance between hybrid populations and their parental taxa from the same lake was 0.12 and not significantly different from the value obtained in the species or within hybrids comparison.

The UPGMA cluster analysis of Nei's genetic distances separated the 20 taxon-lake combinations into two main groups (Figure 2). These two groups, which I will call the *D. galeata* and the *D. cucullata* group,

were separated by a distance of more than 0.67. In the *D. galeata* group, *D. hyalina* and *G × H* from Bodensee and *D. hyalina* from Belauersee were separated from the rest of the group by a distance of 0.09. The Tjeukemeer population of *C × G* was the only '*D. cucullata* like' hybrid that clustered in the *D. galeata* group. The *C × H* from Plußsee and the two populations of *C × G* clustered within the *D. cucullata* group. The Ringsjön population of *D. cucullata* was quite distinct from the rest of the *D. cucullata* group (Figure 2).

Discussion

Individual *Daphnia* taxa from the *D. galeata* species complex, sampled from geographic distant locations in Europe, showed large similarities in allele frequencies. Within species, populations from different localities showed only minor genetic differentiation among them, with the one exception being the *D. cucullata* population from Lake Ringsjön, Sweden (Figure 2). This population differed from the other *D. cucullata* populations by showing a high frequency (0.37) of the *Pgi-S⁻* and by the occurrence of a *Pgm-S* allele. Nev-

Table 4. Clonal diversity D, calculated as the negative logarithm of Simpson's index of concentration (see text) and averaged over all sampling dates, for all taxa of the *D. galeata* species complex in seven European lakes

Taxon	Bodensee[1]	Maarsseveen	Plußsee	Ringsjön	Tjeukemeer	Belauersee
D. cucullata			1.15	0.95	1.09	0.70
D. cuc. × *gal.*			0.54	0.60	0.79	
D. galeata	0.61	0.65	0.61	0.51	0.65	
D. gal. × *hyl.*	0.10	0.33	0.46			
D. hyalina	0.08					0.24
D. cuc. × *hyl.*				0.27		

[1] Data from Weider & Stich (1992)

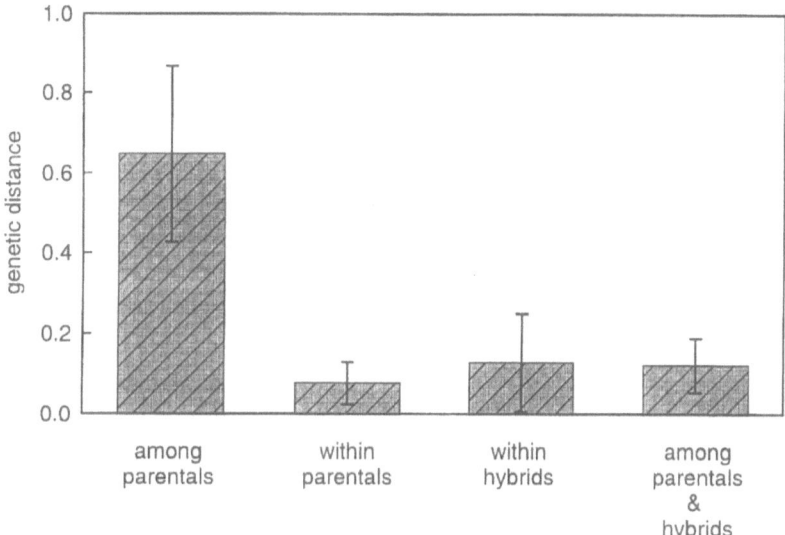

Figure 1. Mean values of Nei's genetic distance among populations of different or the same taxa. Four categories where made: among parental species (different populations), within parental species (different populations), within hybrid taxa (different populations, all hybrid taxa averaged), among parentals and hybrid taxa (within the same lake). Error bars are 95% confidence limits of the mean.

ertheless, Ringsjön *D. cucullata* still clustered within the *D. cucullata* group.

Recently, several studies have shown that backcrossing and introgression are important processes in *Daphnia* hybrid species complexes (Taylor & Hebert, 1992; Spaak, 1996; Gießler, 1997; Schwenk, 1997). Spaak (1996) found that the population of hybrid $C \times G$ in Tjeukemeer probably consisted of a large number of backcrosses with *D. galeata*, and showed that $C \times G$ hybrids from Tjeukemeer clustered close to *D. galeata*. The data also suggested that the *Pgi-S* allele may be introgressed from *D. cucullata* to *D. galeata* through the $C \times G$ hybrid. The present study shows that the same process probably happened in Ringsjön and Plußsee. In both these lakes, *D. galeata* co-occurs

with *D. cucullata* and their interspecific hybrid, and the *Pgi-S* allele was found in *D. galeata* of both lakes in low frequencies (0.01 and 0.03 respectively; Table 2), whereas this allele was not found in the three other lakes (Maarsseveen, Höftsee, Bodensee) where *D. galeata* occurred, but did not co-occur with *D. cucullata*. Despite this suggestion for backcrossing and reticulated evolution, the Plußsee and Ringsjön $C \times G$ populations clustered in the *D. cucullata* group (Figure 2). This suggests that other processes (e.g. differences in sexual reproduction, selection, different ages of hybrid populations) play a more important role in causing the genetic distinction between populations than the occurrence of backcrosses.

132

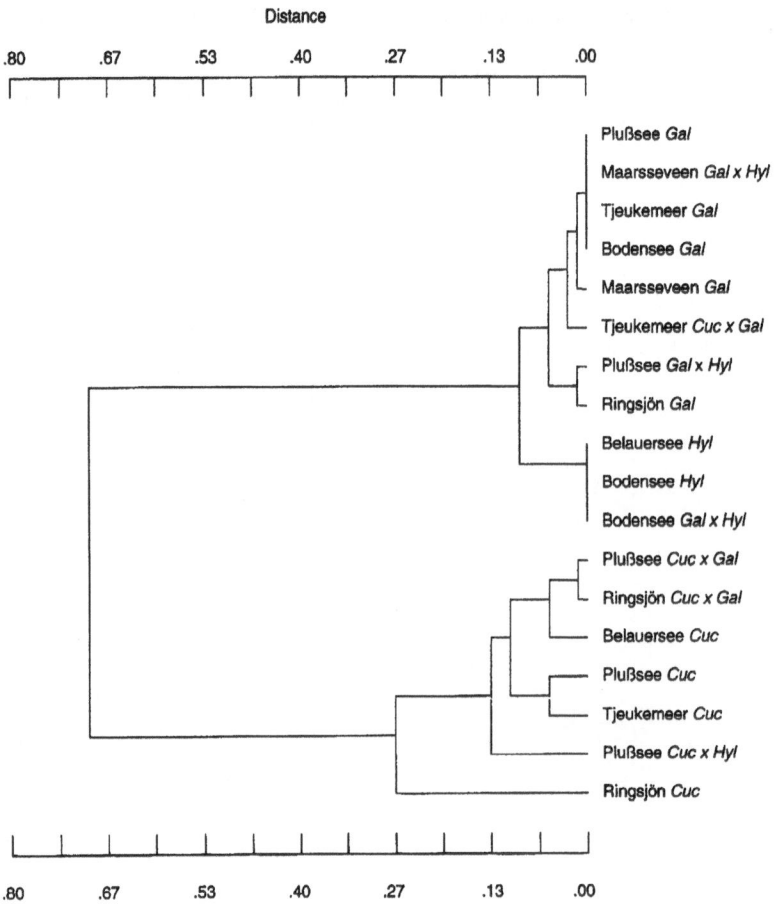

Figure 2. UPGMA clustering of Nei's genetic distance calculated for *Pgi* and *Pgm* among pooled samples of *Daphnia* taxa. Samples originated from six different lakes (see text for explanation).

Of the $C \times G$ hybrids, three populations were investigated. One (Tjeukemeer) clustered within the 'galeata' group, whereas the other two populations (Plußsee and Ringsjön) clustered together in the 'cucullata' group. Also, the $G \times H$ hybrids showed strong differences between populations. The Bodensee population clustered very close to *D. hyalina*, whereas the Maarsseveen and Plußsee populations clustered closer to *D. galeata* than to *D. hyalina*. These observations support the hypothesis of multiple hybridization events within the *D. galeata* species complex. If hybridization would have occurred once with the hybrids afterwards being dispersed over different European lakes, one would expect a clustering comparable to the parental species. Furthermore, if hybrids were relatively old and reproduced only parthenogenetically, one would expect a relatively low clonal diversity within hybrid populations. The observed clonal diversity, however, is in the range of the parental species. This

also support the hypothesis that hybridization events still occur. Further support for this hypothesis can be derived from the mean genetic distances between taxa. In the case of 'old dispersing hybrids' one would expect a much lower differentiation within hybrid taxa compared to hybrids and parental taxa from the same lake. In fact no difference was found (Figure 1).

$C \times G$ hybrids showed a high clonal diversity in all three studied populations (Table 4), similar to the *D. cucullata* and *D. galeata* populations from the same lake. From my data it seems that sexual reproduction and hybridization events are more frequent in these taxa than in *D. hyalina* populations. This may be related with the type of habitats in which these taxa live (see Hrbáček, 1987). *D. cucullata* is a typical species of eutrophic lakes which displays high abundances in summer and autumn and disappears during winters, whereas *D. galeata* is frequent in lakes that freeze during winter. In contrast, *D. hyalina* is found in deep lakes where they can over-winter more easily as partheno-genetic females. This may explain the low clonal diversity within *D. hyalina* and their hybrids, although the low clonal diversity may also be caused by the low numbers of '*hyalina* like' taxa found in the present study. My observations are in agreement with earlier genetic studies (Taylor & Hebert, 1993; Müller & Seitz, 1995) which have shown that the production of *Daphnia* hybrids within this species complex is most likely an ongoing process. Clearly, further experimental and field studies are needed to solve the question why hybrid and parental taxa are still morphologically and genetically distinct.

Acknowledgements

The samples of lake Ringsjön were provided by Eva Bergstrand. I thank Maarten Boersma, Luc De Meester, Jacob Müller, Klaus Schwenk, James Ward, Larry Weider, and two anonymous reviewers for their critical comments on earlier versions of this manuscript. Chris Robinson is acknowledged for correcting the English. During the German part of this study I was supported by a research fellowship from the Max Planck Society.

References

Carvalho, G. R. & H. G. Wolf, 1989. Resting eggs of lake-*Daphnia*. I. Distribution, abundance and hatching of eggs collected from various depths in lake sediments. Freshwat. Biol. 22: 459–470.

Colbourne, J. K. & P. D. N. Hebert, 1996. The systematics of North-American *Daphnia* (crustacea, anomopoda) – a molecular phylogenetic approach. Phil. Trans. r. Soc. Lond. B 351: 349–360.

Flößner, D., 1993. Zur Kenntnis einiger *Daphnia*-Hybriden. Limnologica 23: 71–79.

Gießler, S., 1997. Analysis of reticulate relationships within the *Daphnia longispina* species complex. Allozyme phenotype and morphology. J. Evol. Biol. 10: 87–105.

Hebert, P. D. N. & M. J. Beaton, 1989. Methodologies for allozyme analysis using cellulose acetate electrophoresis. Helena laboratories Beaumont, Texas, 32 pp.

Hebert, P. D. N. & T. L. Finston, 1996. A taxonomic reevaluation of North-American *Daphnia* (Crustacea, Cladocera). 2. New species in the *Daphnia-pulex* group from the South-Central United-States and Mexico. Can. J. Zool. 74: 632–653.

Hebert, P. D. N., S. S. Schwartz & J. Hrbáček, 1989. Patterns of genotypic diversity in Czechoslovakian *Daphnia*. Heredity 62: 207–216.

Hebert, P. D. N. & C. C. Wilson, 1994. Provincialism in plankton: endemism and allopatric speciation in Australian *Daphnia*. Evolution 48: 1333–1349.

Hrbáček, J., 1987. Systematics and biogeography of *Daphnia* species in the northern temperate regions. In R. H. Peters & R. de Bernardi (eds), *Daphnia*, Pallanza, Mem. Inst. ital. Idrobiol. 45: 37–76.

Mayr, E., 1942. Systematics and the origin of species. Columbia University Press, New York, 332 pp.

Müller, J. & A. Seitz, 1995. Differences in genetic structure and ecological diversity between parental forms and hybrids in a *Daphnia* species complex. Hydrobiologia 307: 25–32.

Nei, M., 1978. Estimation of average heterozygosity and genetic distance from a small number of individuals. Genetics 89: 583–590.

Pielou, E. C., 1975. Ecological Diversity. Wiley-Interscience, New York, 165 pp.

Schwenk, K., 1997. Evolutionary genetics of *Daphnia* species complexes – hybridism in syntopy. Ph.D Thesis, University of Utrecht, The Netherlands, 141 pp.

Schwenk, K. & P. Spaak, 1995. Evolutionary and ecological consequences of interspecific hybridization in cladocerans. Experientia 51: 465–481.

Simpson, E. H., 1949. Measurement of diversity. Nature 163: 688.

Spaak, P., 1994. Genetical ecology of a coexisting *Daphnia* hybrid species complex. Ph.D Thesis, University of Utrecht, The Netherlands, 125 pp.

Spaak, P., 1995. Sexual reproduction in *Daphnia*: interspecific differences in a hybrid species complex. Oecologia 104: 501–507.

Spaak, P., 1996. Temporal changes in the genetic structure of the *Daphnia* species complex in Tjeukemeer, with evidence for backcrossing. Heredity 76: 539–548.

Spaak, P. & J. R. Hoekstra, 1993. Clonal structure of the *Daphnia* population in Lake Maarsseveen: its implications for diel vertical migration. Arch. Hydrobiol. Beih. Ergebn. Limnol. 39: 157–165.

Swofford, D. L. & R. B. Selander, 1981. BIOSYS-1: a fortran program for the comprehensive analysis of electrophoretic data in population genetics and systematics. J. Hered. 72: 281–283.

Taylor, D. J. & P. D. N. Hebert, 1992. *Daphnia galeata mendotae* as a cryptic species complex with interspecific hybrids. Limnol. Oceanogr. 37: 658–665.

Taylor, D. J. & P. D. N. Hebert, 1993. Habitat-dependent hybrid parentage and differential introgression between neighboringly sympatric *Daphnia* species. Proc. Natl. Acad. Sci. USA. 90: 7079–7083.

Weider, L. J. & H. B. Stich, 1992. Spatial and temporal heterogeneity of *Daphnia* in Lake Constance; intra- and interspecific comparisons. Limnol. Oceanogr. 37: 1327–1334.

Wolf, H. G., 1987. Interspecific hybridization between *Daphnia hyalina*, *D. galeata* and *D. cucullata* and seasonal abundance of these species and their hybrids. Hydrobiologia 145: 213–217.

Wolf, H. G. & M. A. Mort, 1986. Interspecific hybridization underlies phenotypic variability in *Daphnia* populations. Oecologia 68: 507–511.

Hydrobiologia **360**: 135–142, 1997.
A. Brancelj, L. De Meester & P. Spaak (eds), Cladocera: The Biology of Model Organisms.
©1997 *Kluwer Academic Publishers.*

Among-populational genetic differentiation in the cyclical parthenogen *Daphnia magna* (Crustacea, Anomopoda) and its relation to geographic distance and clonal diversity

J. Vanoverbeke & L. De Meester
Laboratory of Ecology and Aquaculture, Katholieke Universiteit Leuven, Naamsestraat 59, 3000 Leuven, Belgium

Key words: Daphnia, genetic differentiation, geographic distance, clonal diversity, gene flow, genetic drift

Abstract

Using allozyme data based on four polymorphic enzyme loci, we present an analysis of genetic differentiation among eight *Daphnia magna* populations, separated by less than 100 m to more than 500 km from each other. In spite of the large range of geographic distances, there was only a slight tendency for an increase in genetic differentiation with increasing geographic distance between populations, and the relation was not significant. This was mainly due to the fact that neighbouring populations were already highly genetically differentiated. Our results suggest that in populations in which only a few abundant clones are present after a period of strong clonal selection, among-populational genetic differentiation as revealed by allozyme markers is inflated as a result of stochasticity involving chance associations of alleles with specific abundant genotypes. Indices quantifying genetic differentiation were much higher among populations with a low clonal diversity than among populations with a high clonal diversity.

Introduction

During the last decades, thanks to the development of allozyme electrophoresis and methods to study DNA polymorphisms, much information has been gathered on the genetics of *Daphnia* populations (Hebert, 1974, 1987; Carvalho, 1994; Lynch & Spitze, 1994). Studies have focused both on genetic variation within populations and on genetic differentiation between populations. With respect to interpopulational genetic differentiation, some remarkable results have been reported. *Daphnia* species produce propagules (ephippia containing resting eggs) that can withstand adverse conditions such as drying and freezing, and therefore can be passively dispersed by, for instance, waterfowl (Proctor & Malone, 1965). Nevertheless, one expects a reduction in genetic similarity with increasing geographic distance between populations, due to a lower exchange of propagules between distant populations compared to neighbouring populations. Innes (1991), however, reported only a slightly positive relationship between geographic and genetic distance in a

study of 11 pond populations of cyclic parthenogenetic *D. pulex*. Lynch & Spitze (1994) obtained similar results when reviewing the data of several studies involving different *Daphnia* species. From their data, it is clear that the lack of a pronounced pattern is due to the fact that genetic differentiation among nearby populations is already high, rather than to a low genetic differentiation among populations that are further apart from each other. Indeed, several studies involving pond-dwelling *Daphnia* have revealed that neighbouring populations can have very different allele frequencies (Hebert, 1974; Hebert & Moran, 1980; Korpelainen 1984; Innes, 1991). In this respect, Mort & Wolf (1986) and Wolf (1988) pointed to a difference between pond and lake populations. In lake-dwelling *Daphnia*, genetic differentiation between neighbouring populations is relatively low, and there is a clear pattern of increasing genetic differentiation with increasing geographic distance (Mort & Wolf, 1986; Wolf, 1988; Jacobs, 1990). One possible explanation for this difference between pond- and lake-dwelling *Daphnia* is that dispersal among lakes is more common than among ponds. Ephippia of pond-dwelling *Daphnia magna* sink to the bottom, whereas ephippia from lake-dwelling species (e.g. the *D. longispina* complex) float and may therefore be more prone to dispersal. However, Weider (1989) observed a low among-populational genetic differentiation in lake-dwelling *Polyphemus pediculus*, even though this species produces no ephippial envelopes to protect the resting eggs. In addition, there is a difference in the degree of genetic differentiation between intermittent and permanent pond populations of *D. magna*, with permanent pond populations showing a higher among-populational genetic differentiation than intermittent pond populations (Innes, 1991; De Meester, 1996). This is unlikely to be the result of differences in dispersal capacity, since the observations are based on data from the same species. We hypothesize that the differences in genetic differentiation among lake populations, intermittent pond populations and permanent pond populations are largely due to differences in clonal structure among those types of populations, *in casu* clonal diversity. To test this hypothesis we sampled eight *D. magna* populations from habitats that widely differ in size (<10m² to >70 ha), clonal diversity (1 to >20 detectable clones in a sample of 40–50 animals), and geographic distance among each other (<100 m to >500 km). We analysed the relationship between genetic differentiation and geographic distance in this set of populations, and investigated the impact of clonal diversity on this relationship.

Material and methods

We studied eight populations of the cladoceran *Daphnia magna* Straus (Crustacea: Anomopoda) (Table 1). *D. magna* inhabits a wide range of mostly eutrophic habitats, from small ponds and canals to (less frequently) shallow lakes. Four of the eight populations studied were situated in or near Gent (Belgium), three near Woumen (Belgium), about 60 kilometres from Gent, and one population was situated in Lebrader (Germany), 560 kilometres from Gent (Table 2).

All populations were sampled once, during a phase of sexual reproduction. Lake Blankaart is an exception, since no ephippial females or males occurred at the time of sampling. In this lake, traditional sampling techniques did not reveal the presence of *D. magna*. After the establishment of fish-free enclosures of 3000 l content (see Declerck et al., 1997), however, exponentially growing *D. magna* populations appeared, indicat-

ing that *D. magna* is present in extremely low densities in this lake. We sampled the *D. magna* population of one of the enclosures, ten days after they were established. No further data are available on the *D. magna* population of Lake Blankaart, but all other populations are predominantly intermittent in nature (Vanoverbeke & De Meester, pers. obs.), in that they engage periodically in sexual reproduction and that massive hatching from resting eggs is observed at the beginning of the growing season. The probability remains, however, that in some populations (e.g. the Driehoeksvijver and the Citadelpark Small Pond), clones may survive until the next growing season. In the Citadelpark Small Pond, for instance, the population crashed during the 1992–1993 winter season, but a few individuals survived and contributed to the population of the following spring, which consisted largely of new hatchlings (Vanoverbeke & De Meester, unpubl. data). Though the different populations were sampled at different times (Table 1), all populations except that of Lake Blankaart were sampled at an ecologically comparable period (i.e. during a sexual phase, after a period of clonal selection). Clonal selection is expected to have eroded clonal diversity in the different populations, the resulting clonal diversity being determined by the intensity of clonal selection, the length of the growing season and the number of clones at the beginning of the growing season.

The whole water column was sampled with a 200 μm dipnet and the animals were brought alive to the laboratory. Adult animals were isolated in an aselective way within a few hours after sampling, and stored in liquid nitrogen. In some populations, genotype frequencies differed significantly between parthenogenetic females, sexual females, and males (Vanoverbeke & De Meester, unpubl. data). However, parthenogenetic females always made up the largest part of the population (60–100%) and we only used data on the population of parthenogenetic females in our analyses. From each population, 40–65 parthenogenetic females were analysed (see Table 1).

We used cellulose acetate electrophoresis (following Hebert & Beaton, 1989) to screen for allelic variation at four different loci: GPI (EC 5.3.1.9), MPI (EC 5.3.1.8), AAT (EC 2.6.1.1) and MDH (EC 1.1.1.37). The loci were chosen because they are polymorphic in many *D. magna* populations and showed to be polymorphic in a preliminary survey of some of the populations studied here. As the loci are chosen for their polymorphic character, they give no good representation of the general level of polymor-

Table 1. *D. magna* populations studied, showing name with abbreviation, location (village or town, province or Bundesland, country), surface area, sampling date, number of individuals analysed per population (N), number of multilocus genotypes detected in the sample (C) and clonal diversity (CD).

Name	Location	Surface area (ha)	Date	N	C	CD
Citadelpark Small Pond (CSP)	Gent East-Flanders Belgium	0.03	20/10/92	64	3	1.48
Citadelpark Big Pond (CBP)	Gent East-Flanders Belgium	0.15	06/07/92	48	6	1.65
Pond Botanic Garden (PB)	Gent East-Flanders Belgium	0.0004	03/07/92	45	1	1
Driehoeksvijver (DV)	Heusden East-Flanders Belgium	0.75	30/07/92	42	25	16.02
Lake Blankaart (LB)	Woumen West-Flanders Belgium	30	25/05/94	41	19	15.21
Eendekooi (EK)	Woumen West-Flanders Belgium	0.4	03/08/92	47	12	5.95
Fish Pond (FP)	Woumen West-Flanders Belgium	0.55	03/08/92	42	13	8.09
Lebrader Teich (LT)	Lebrader Schleswig-Holstein Germany	75	20/08/94	40	10	7.63

Table 2. Distance (km) between the different regions in which *D. magna* populations were sampled. The number of populations sampled within each region is given in parentheses; * mean distance between populations within a region.

	Ghent	Heusden	Woumen	Lebrader
Ghent (3)	0.23*			
Heusden (1)	5	–		
Woumen (3)	60	65	0.3*	
Lebrader (1)	560	560	600	–

phism in the analysed populations. We constructed multilocus genotypes by combining the genotypes of the different loci in one individual. In clonal organisms, multilocus genotypes are an appropriate way to characterize clonal groups. Although multilocus genotypes may harbour additional genetic variation, they will further on be called clones.

Clonal diversity (CD) was calculated using the Shannon-Wiener index: $H = -\sum_i p_i \log_e p_i$, with p_i being the relative frequency of the i-th clone (Peet, 1974). This index not only takes into account the number of clones present in a sample, but also the evenness in relative abundance of the different clones. The index is expressed as $e^{(H)}$, giving a value ranging from one (if only one clone is detected) to N (= number of clones, if all clones have equal frequencies).

For all analysed loci, fixation indices ($F = (h_s - h_o)/h_s$; with h_s being the expected heterozygosity under Hardy-Weinberg expectations, and h_o being the observed heterozygosity) were calculated in the different populations. Deviations of genotype frequencies from Hardy-Weinberg expectations were analysed using χ^2-tests. As a measure of genetic differentiation, we used G_{ST} values calculated from allele frequencies (Nei, 1973). The correlation between G_{ST} and geographic distance or clonal diversity was analysed using a Mantel test for dependent variables (Sokal & Rohlf, 1995). Analyses were done using the computer programs Biosys-1 (Swofford & Selander, 1981) and Genepop (Raymond & Rousset, 1995).

Results

Genetic structure of the populations

For both GPI and AAT, a total of two alleles were observed. MPI and MDH showed a higher allelic variation. The Belgian populations were found to have two MPI variants (called S and F). In Lebrader Teich the S allele was not found, but another allele (F^+) was present. Although there were indications that a fourth allele (M) was present at the MPI locus, the resolution was not good enough to make a clear distinction between the F and the M allele in all individuals, and we considered both as being F. In most populations, the MDH locus revealed three allelic variants (S, M and F). A substitution of the S allele by the S^- allele was observed in Lebrader Teich. All populations except two were polymorphic at all four loci. Lebrader Teich was monomorphic at the GPI locus, and the small pond in the Botanic Garden, which contained only one multilocus genotype, was polymorphic at the AAT locus only.

The number of clones and the clonal diversity observed in each population are given in Table 1. The highest clonal diversity was observed for the Driehoeksvijver (CD = 16.02). Some populations revealed a very low number of detectable genotypes, resulting in clonal diversity values of less than two. Table 3 reports fixation indices and the results of tests for deviations from Hardy-Weinberg equilibrium, executed for all loci and all populations. In six of the eight populations, at least one of the four analysed loci is characterized by allele frequencies deviating significantly from Hardy-Weinberg equilibrium at the table-wide level.

Genetic differentiation between populations

G_{ST} values calculated over the total set of populations are given in Table 4. The highest value of G_{ST} is observed for MPI ($G_{ST} = 0.418$), reflecting the substitution of the S allele for the F^+ allele in Lebrader Teich. Excluding Lebrader Teich results in a decrease of the G_{ST} value for MPI ($G_{ST} = 0.349$). Although Lebrader

Table 3. Fixation indices (F; values <0: excess of heterozygotes; values >0 deficiency of heterozygotes) in the different populations of *D. magna* are given for all loci analysed. Significance levels for deviations from Hardy-Weinberg equilibrium calculated using χ^2-tests are marked with *: $p<0.05$; **: $p<0.01$; +: $p<0.05$ at the table-wide level, after sequential Bonferroni correction (Rice, 1989); for population abbreviations see Table 1.

	GPI	MPI	AAT	MDH
CSP	−1.000**+	−0.792**+	−0.058	0.417**+
CBP	0.478**+	1.000**+	−0.833***+	−0.043
PB	–	–	−1.000***+	–
DV	0.041	0.438**	0.344*	−0.073
LB	0.061	0.754***+	−0.428**	−0.120
EK	−0.035	0.850***+	−0.106	−0.367*
FP	−0.018	−0.224	−0.477***+	0.343**+
LT	–	0.329*	−0.429**	−0.133

Table 4. Genetic differentiation (G_{ST}) between the analysed *D. magna* populations. Values are based on A: all eight populations; B: all populations except Lebrader Teich.

Locus	A	B
GPI	0.297	0.258
MPI	0.418	0.349
AAT	0.202	0.216
MDH	0.256	0.265
mean	0.299	0.273

Teich also harbours an unique allele at the MDH locus, the value of G_{ST} for this locus does not change much when Lebrader Teich is excluded from the analysis. Probably the S^- allele was found in too low frequency ($n = 2$) in the electrophoretic analysis to be of any importance with respect to genetic differentiation. The mean value of G_{ST} over all four loci is 0.299 (0.273 excluding Lebrader Teich). This indicates that genetic differentiation between the analysed populations is on average quite strong.

No significant correlation could be found between G_{ST} values for each pair of populations and the logarithm of the geographic distance between the populations, using a Mantel test for dependent variables (8 populations; 10 000 permutations). Indeed, at most a slight tendency for an increase in genetic differentiation with increased geographic distance could be observed (Figure 1a). To verify whether clonal diversity of the populations affected genetic differentiation between populations, we tested if a significant cor-

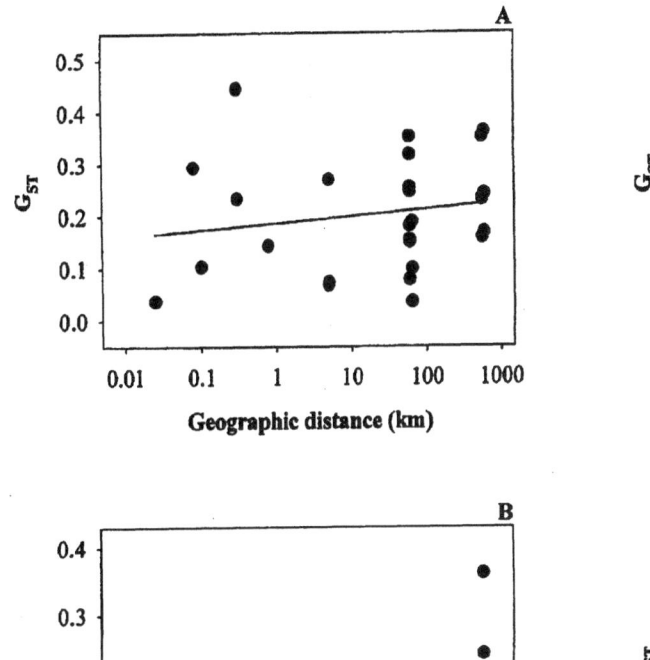

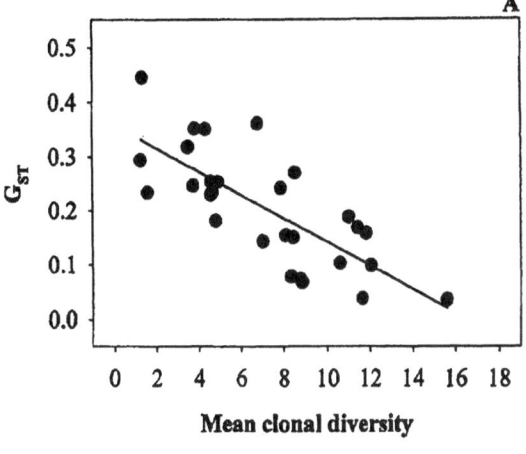

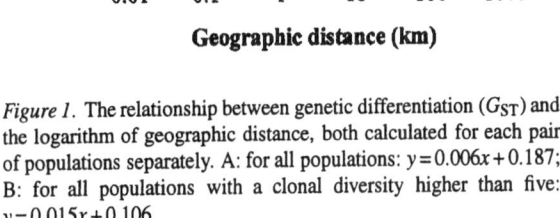

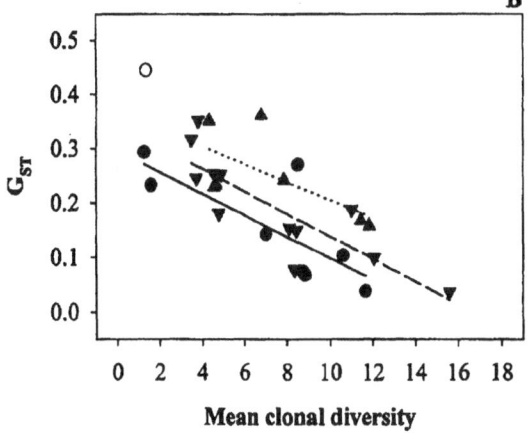

Figure 1. The relationship between genetic differentiation (G_{ST}) and the logarithm of geographic distance, both calculated for each pair of populations separately. A: for all populations: $y = 0.006x + 0.187$; B: for all populations with a clonal diversity higher than five: $y = 0.015x + 0.106$.

Figure 2. The relationship between genetic differentiation (G_{ST}) and mean clonal diversity (CD) of the populations, both calculated for each pair of populations separately. A: for all populations ($y = -0.022x + 0.359$); B: for each of three subgroups of populations; ● and full line: populations situated five or less kilometers from each other ($y = -0.019x + 0.296$; ○ excluded from the analysis); ▼ and dashed line: populations situated about 50–60 kilometers from each other ($y = -0.020x + 0.346$); ▲ and dotted line: populations situated about 600 kilometers from each other ($y = -0.017x + 0.371$).

relation could be found between G_{ST} values and the mean clonal diversity for each pair of populations, using a Mantel test. The correlation was highly significant ($p < 0.01$; 8 populations; 10 000 permutations), indicating that populations with a low clonal diversity are more genetically differentiated with respect to allozyme markers than populations with a high clonal diversity (Figure 2a). When we exclude the populations with a clonal diversity of less than five (less than 10 clones; see Table 1) from our analysis, the correlation between G_{ST} and geographic distance for each pair of populations becomes significant ($p < 0.05$; 5 populations; 120 permutations; Figure 1b). This suggests that an influence of geographic distance on genetic differentiation is present, but that the pattern is blurred by the

influence of clonal diversity, which is much stronger. This is also illustrated by Figure 2b, in which the relationship between G_{ST} and the mean clonal diversity is plotted for three subgroups of population pairs: one subgroup containing populations located five or less kilometers from each other, a second subgroup containing populations located at about 60–65 kilometers from each other, and a third subgroup containing populations located about 600 kilometers from each other (= comparisons with Lebrader Teich). We calculated regressions for each of these subgroups separately, and the slopes are remarkably parallel (note that for clarity one value was excluded for the subgroup of the smallest geographic range; including this value gives the same pattern, but the slope of the regression for this subgroup is somewhat steeper). Figure 2b shows that the geographic distance between the populations does influence the position of the regression line, but the influence of geographic distance on G_{ST} is small compared to the effect of clonal diversity on the level of genetic differentiation between populations.

Discussion

Only a slight tendency for a higher genetic differentiation with increasing geographic distance between *D. magna* populations could be observed. The lack of a strong relationship is due to the fact that populations in close vicinity of each other are strongly genetically differentiated rather than to more remote populations being not differentiated. The finding that neighbouring populations can be highly genetically differentiated is in agreement with the results found in other studies (Hebert, 1974; Hebert & Moran, 1980), as is the occurrence of only a weak relationship between genetic and geographic distance (Innes, 1991; Lynch & Spitze, 1994). Similar to our observations, several studies have reported that genetic differentiation among neighbouring populations is due to differences in allele frequencies of shared alleles, while populations at a larger geographic distance may also reveal differentiation through substitution of alleles (e.g. Hebert, 1974; Hebert & Moran, 1980; Korpelainen, 1984).

The high genetic differentiation between neighbouring populations is generally ascribed to founder effects as a result of colonisation of habitats by only a few individuals, combined with a low gene flow between populations (e.g. Hebert, 1987). Boileau et al. (1992) provide evidence for the fact that founder effects can persist over many generations, even if there is a considerable amount of gene flow between populations. In a recent review, De Meester (1996) argues, however, that effective gene flow rather than dispersal is low among populations, effective gene flow being reduced when establishment of a second colonist is hampered by competition with resident genotypes. Finally, Lynch (1987) shows that fluctuating selection can also give rise to the same pattern of strong genetic differentiation between nearby populations.

Korpelainen (1984) suggests that genetic differentiation may be enhanced by a yearly recurrent founder effect if populations start each growing season with only a few hatchlings. This, however, is likely to be important in very small populations only, e.g. in habitats of a few m^2 at most. Our data suggest that stochasticity may interfere at another level. Indeed, in many populations in which a large number of clones hatch from ephippia at the beginning of the growing season, the population is entirely dominated by only a few clones at the end of the growing season. This may be due to strong clonal selection, or to an extended phase of clonal selection (i.e. a long growing season). A considerable reduction in clonal diversity, however, may cause important changes in allozyme frequencies compared to the beginning of the growing season, as the allele frequencies will be largely determined by the occurrence of the alleles in the few clones which dominate the population (see also Lynch & Spitze, 1994). A chance effect is indeed involved, to the extent that the allozyme alleles are supposed to be neutral variants. The presence of a particular allele, therefore, is not directly associated with the fitness of the genotype, and alleles will simply be hitch-hiking with successful genotypes to which they got fortuitously associated. Due to the above, frequency shifts between subsequent growing seasons may be enhanced in populations with a low clonal diversity at the end of the growing season, as associations are broken down during each phase of sexual reproduction. Moreover, these enhanced shifts in allele frequencies may enhance genetic differentiation between populations because frequencies will drift independently in the different populations. The phenomenon described here can indeed be considered similar to genetic drift. The sampling effect, however, is exhibited at the level of the genotype rather than at the level of the individuals. Rhomberg & Singh (1989) have pointed to this chance effect in their study on the rose aphid *Macrosiphum rosae*. They observed that one infestation (e.g. the aphids on one shrub) is largely dominated by a few abundant clones, and that allozyme frequencies may differ markedly among infestations. In most aphids, however, the sexuals or the sexuparae (i.e. the generation prior to the sexuals) are winged. Migration among infestations therefore rules out the effect of low clonal diversity within infestations on genetic differentiation at the populational level (i.e. among groups of infestations, see Rhomberg & Singh, 1989). This is in contrast to the situation in *Daphnia*, where low clonal diversity is observed at the level of the population. It should be pointed out that, although gene flow among *Daphnia* populations is lower than among local infestations of aphids, G_{ST} values may be misleading with respect to the amount of gene flow between populations with a low clonal diversity. Indeed, preliminary results of computer simulations revealed that even with a considerable amount of gene flow, genetic differentiation among cyclical partheno-genetic populations increases when the mean clonal diversity decreases (Vanoverbeke, unpubl. data).

The division often made in literature between inter-mittent pond populations, permanent pond populations and lake populations (see reviews in Hebert, 1978, 1987; Mort, 1991; De Meester, 1996) is not completely accurate to describe differences in genetic structure and genetic differentiation within and among *Daphnia* populations. As mentioned in the results section, most of our populations are characterized by significant devi-ations from Hardy-Weinberg equilibrium. Moreover, most of them showed strong indications for genotyp-ic linkage disequilibrium between the analysed loci (Vanoverbeke & De Meester, unpubl. data). These are characteristics normally ascribed to permanent pond populations (Hebert, 1987; Lynch & Spitze, 1994). In general, however, our populations do reveal more similarity with intermittent pond populations as they are characterized by massive hatching of resting eggs at the beginning of the growing season and mass pro-duction of ephippia at the end of the growing season. Furthermore, the observations on the genetic structure of lake-dwelling *Daphnia* are at odds with the 'classic' distinction between intermittent and permanent pond populations (see Mort & Wolf, 1986; Hebert, 1987; Jacobs, 1990). Whereas there is undoubtedly a tenden-cy for clonal diversity to be associated with perma-nency of the habitat, there is a confounding effect of habitat size and of clonal selection that makes the dis-tinction somewhat artificial. Clonal selection in a small intermittent population may be so strong as to reduce the clonal diversity to very low levels, such as in the Citadelpark Pond populations studied by us. On the other hand, semi-permanent but large populations may harbour a high clonal diversity, as is the case in typical large-lake populations. We feel that the main distinc-tion to be made is between populations of high and low clonal diversity. Our data indeed indicate that the observed genetic differentiation among *Daphnia* pop-ulations is to a large extent a function of clonal diver-sity of the populations. The data summarized by Innes (1991; his Table 8) support this notion: the intermittent *D. magna* populations studied by Hebert (1974) were less genetically differentiated than populations classi-fied as permanent populations (Hebert, 1974), but the small intermittent populations studied by Korpelainen

(1984) showed strong among-populational genetic dif-ferentiation. Large-lake *Daphnia* populations are char-acterized by a very large number of clones, and reveal a low genetic differentiation (Mort & Wolf, 1986; Wolf, 1988; Weider, 1989).

In conclusion, our data suggest that the degree of among-populational genetic differentiation observed in one-locus markers is to a large extent determined by the clonal diversity of the populations, as chance associations between alleles and certain clones may inflate genetic differentiation in populations with a low number of coexisting clones. If this complicat-ing factor is taken into account, a positive relation-ship between genetic differentiation and geographic distance is observed.

Acknowledgments

Samples of Lebrader Teich were collected in co-operation with Suzanne Mitchell; data on Lake Blankaart were collected in co-operation with Steven Declerck and Christophe Cousyn. We thank Thier-ry Backeljau, Konjev Desender, Jean-Pierre Maelfait and three anonymous reviewers for comments on an earlier version of the manuscript. JV is a fellow of the Flemish Institute for the Promotion of Scientific-Technological Research in Industry (IWT). LDM is a postdoctoral researcher with the Fund for Scientif-ic Research, Flanders. The research was supported by grants n° 2.0128.94 and n° 1.5.012.96N of the Fund for Scientific Research, Flanders and grant OT/96/13 of the Katholieke Universiteit Leuven.

References

Boileau, M. G., P. D. N. Hebert & S. S. Schwartz, 1992. Non-equilibrium gene frequency divergence: persistent founder effects in natural populations. J. evol. Biol. 5: 25–39.

Carvalho, G. R., 1994. Genetics of aquatic clonal organisms. In Beaumont, A. R. (ed.), Genetics and Evolution of Aquatic Organ-isms. Chapman & Hall, London: 291–323.

Declerck, S., L. De Meester, N. Podoor & J. M. Conde-Porcuna, 1997. The relevance of size efficiency to biomanipulation theory: a field test under hypertrophic conditions. Hydrobiologia 360: 265–275.

De Meester, L., 1996. Local genetic differentiation and adaptation in freshwater zooplankton populations: patterns and processes. Ecoscience 3: 385–399.

Hebert, P. D. N., 1974. Enzyme variability in natural populations of *Daphnia magna* I. Population structure in East Anglia. Evolution 28: 546–556.

Hebert, P. D. N., 1978. The population biology of *Daphnia* (Crustacea, Daphnidae). Biol. Rev. 53: 387–426.

Hebert, P. D. N., 1987. Genetics of *Daphnia*. In Peters, R. H. & R. De Bernardi (eds), *Daphnia*. Mem. Ist. ital. Idrobiol., Pallanza: 439–460.

Hebert, P. D. N. & M. J. Beaton, 1989. Methodologies for allozyme analysis using cellulose acetate electrophoresis. Helena Laboratories, Beaumont.

Hebert, P. D. N. & C. Moran, 1980. Enzyme variability in natural populations of *Daphnia carinata* King. Heredity 45: 313–321.

Innes, D. J., 1991. Geographic patterns of genetic differentiation among sexual populations of *Daphnia pulex*. Can. J. Zool. 69: 995–1003.

Jacobs, J., 1990. Microevolution in predominantly clonal populations of pelagic *Daphnia* (Crustacea: Phyllopoda): selection, exchange, and sex. J. evol. Biol. 3: 257–282.

Korpelainen, H., 1984. Genic differentiation of *Daphnia magna* populations. Hereditas 101: 209–216.

Lynch, M., 1987. The consequences of fluctuating selection for isoenzyme polymorphisms in *Daphnia*. Genetics 115: 657–669.

Lynch, M. & K. Spitze, 1994. Evolutionary Genetics of *Daphnia*. In Real, L. A. (ed.), Ecological Genetics. Princeton Univ. Press, Princeton: 109–128.

Mort, M. A., 1991. Bridging the gap between ecology and genetics: the case of freshwater zooplankton. Trends Ecol. Evolution 6: 41–45.

Mort, M. A. & H. G. Wolf, 1986. The genetic structure of large-lake *Daphnia* populations. Evolution 40: 756–766.

Nei, M., 1973. Analysis of gene diversity in subdivided populations. Proc. natn. Acad. Sci. USA 70: 3321–3323.

Peet, R. K., 1974. The measurement of species diversity. Ann. Rev. Ecol. Syst. 5: 285–307.

Proctor, V. W. & C. Malone, 1965. Further evidence of the passive dispersal of small aquatic organisms via the intestinal tract of birds. Ecology 46: 728–729.

Raymond, M. & F. Rousset, 1995. GENEPOP (version 1.2): population genetics software for exact tests and ecumenicism. J. Hered. 86: 248–249.

Rhomberg, L. R. & R. S. Singh, 1989. Evidence for a link between local and seasonal cycles in gene frequencies and latitudinal gene clines in a cyclic parthenogen. Genetica 78: 73–79.

Rice, W. R., 1989. Analysing tables of statistic tests. Evolution 43: 223–225.

Sokal, R. R. & F. J. Rohlf, 1995. Biometry, 3rd edn. W. H. Freeman and Company, New York.

Swofford, D. L. & R. B. Selander, 1981. Biosys-1: a FORTRAN program for the comprehensive analysis for electrophoretic data in population genetics and systematics. J. Hered. 72: 281–283.

Weider, L. J., 1989. Population genetics of *Polyphemus pediculus* (Cladocera: Polyphemidae). Heredity 62: 1–10.

Wolf, H. G., 1988. Differences in the genetic structure of pond-dwelling and lake dwelling *Daphnia*. Verh. int. Ver. Limnol. 23: 2056–2059.

Hydrobiologia **360**: 143–152, 1997.
A. Brancelj, L. De Meester & P. Spaak (eds), Cladocera: The Biology of Model Organisms.
©1997 *Kluwer Academic Publishers.*

Ideal free distribution in *Daphnia*? Are daphnids able to consider both the food patch quality and the position of competitors?

Petter Larsson
Department of Zoology, University of Bergen, Allégt. 41, N-5007 Bergen, Norway
(e-mail: petter. larsson@zoo.uib.no)

Key words: Zooplankton, *Daphnia pulex*, behaviour, food search, aggregation

Abstract

The distribution of the planktonic crustacean *Daphnia pulex* was tested in a ring-formed flow-through chamber divided into eleven sections. The distribution of the animals under both homogeneous food conditions and in a food gradient were studied. The distribution of repeated registrations of single animals was randomly distributed in a homogeneous, low food environment. Single daphnids exposed to a food gradient tended to reside during 88% of the time at the highest or next highest food concentration, suggesting that daphnids can detect food gradients. When a group of approximately 100 co-occurring animals were given low homogeneous food conditions, they also tended to be randomly distributed. In one out of three cases, however, they were slightly less aggregated than expected from a random distribution. Exposed to a food gradient (0.5–0.0015 mg C l^{-1}), the distribution of the daphnids approximated the ideal free distribution. A very high maximum density in the food gradient (2 mg C l^{-1}) resulted in less strong aggregations than expected from the ideal free distribution.

Introduction

Daphnia are opportunistic animals that are able to utilize temporary blooms of planktonic algae. Strong selection for efficient food gathering is thus expected. The ideal free distribution principle refers to a distribution in an environmental gradient in which equal competitors achieve the same amount of food resources. This principle assumes that the animals are able to assess both the food density and the impact of competitors. The term 'ideal free distribution' was introduced by Fretwell & Lucas (1970) to describe the distribution of animals in a patchy habitat. If animals were 'free' to choose without any constrains or restrictions, they would move to the patch in the habitat were they would gain most of the available resources. This will often be the patch where the animals can have the highest food intake. Such patches, however, also attract other animals feeding on the same resource. When more animals move in, the value of an area will decline. It may then pay to move to an area with a lower food concentration but with fewer competitors. If all the animals are equally strong competitors, they will distribute in a pattern that reflects the distribution of the resources in a habitat. More specifically, the animals will distribute in such a way that all of them will have the same amount of available resources.

If the distribution and the renewal rate of food items in a habitat are known, one can predict the distribution of the animals under ideal free conditions. Ideal free distributions have been found in various animals such as ducks (Harper, 1982), cichlid fishes (Godin & Keenleyside, 1984) and sticklebacks (Milinski, 1979), and the principle seems applicable to many groups of animals.

Experiments done by Jakobsen & Johnsen (1987) on *Daphnia pulex* indicate that the ideal free distribution might also be applicable to animal groups without an advanced brain system. Daphnids were given a choice between two different food concentrations in a set of two connected flow-through chambers. The results showed that daphnids occurred in significantly higher densities in the chamber with the highest food concentration. Not all the animals, however, moved to the highest food concentration, and it looked as if there was a tendency for an ideal free distribution. Jakobsen & Johnsen (1987) found that the absolute concentration of food algae was important. At densities above the incipient limiting level, the daphnids seemed to ignore differences in food densities. Neary et al. (1994) tested similar situations in a long flow-through chamber and Cuddington & McCauley (1994) and Larsson & Kleiven (1997) tested similar situations in a ring-formed chamber. All these studies reported a clear food-search behaviour with the daphnids swimming towards the highest densities of algae.

Although it has been shown in laboratory experiments that daphnids, given a choice, occur in higher densities at high food concentrations than at low, it has not been tested whether they disperse according to the ideal free distribution principle. It could be that the presence of animals in other places than at the highest food concentration could be the result of an imperfect ability to locate the highest food density. However, if the animals are able to detect the food concentration precisely, single animals and groups of animals should react differently when placed in a food gradient. The objective of the present study was to investigate whether or not this is the case.

If a single *Daphnia* is given the chance to choose where to move in a food gradient, it is expected to search for the area with the highest food density. As there are no competitors, there is no reason for searching areas of less high food concentration, unless the maximum food concentration were above the incipient limiting level. In case several food concentrations in the gradient are above incipient limiting level, the animal is expected to choose the area with food concentrations above the incipient limiting level, without showing any preference for a given concentration within that area. In a food gradient with multiple other animals, competition is assumed where the maximum food density is below the incipient limiting level, and an ideal free distribution is expected.

In this paper, the distribution of repeated registrations of single animals was compared with the distribution of several animals together in a ring-formed, flow-through chamber. The food was distributed in a concentration gradient in the chamber, or uniformly low throughout the entire chamber.

According to the ideal free distribution theory, the distribution of solitary animals is expected to be more concentrated around the area of maximum food density than the distribution of groups. In the latter situation, if the competitors are equal, a proportional relationship is expected between the amount of available food algae and the density of the daphnids.

It is, however, still unknown how the *Daphnia* individuals can detect the algae resources in their environment. Are they able to asses the concentration of algae directly, or do they respond to their own food intake? If the distribution of *Daphnia* individuals is in agreement with ideal free expectations, there are two possibilities with respect to the mechanisms by which the ideal free distribution is attained:

(1) The animals can perceive food concentration and the decrease in food intake directly. In a group of animals, the decrease in food intake will be faster than for solitary animals, and the animal will search an area where the ingestion rate is highest. They do not react directly to the presence of other *Daphnia*, but to the actual food intake.

(2) The food concentration is perceived, but they do not necessarily react to a decrease in food concentration. Instead they are able to judge the amount of neighbour competitors, and react to the absolute food level combined with the density of neighbours.

In flow-through chambers the food concentrations can be maintained at nearly constant level (Larsson & Kleiven, 1996), and the animals should not recognize any reduction in food concentration over time. If *Daphnia* still distribute in a way that resembles an ideal free distribution, the second explanation given above is more likely. If the first explanation is correct, one expects a stronger concentration of animals around the maximum food density in a flow-through system than predicted for a more static situation.

Methods

The experimental animals belonged to a *Daphnia pulex* clone isolated from Lake Myravann in Bergen, Norway. Only adult females with eggs were used in the experiments. *Scenedesmus acutus* was used as food algae.

The ring-shaped flow-through chamber (Figure 1) was made of plexiglas. The ring has a diameter of 49.1 cm for the outer wall and 45.9 cm for the inner wall, giving a chamber width of 1.6 cm. The chamber was filled to about 14.0 cm from the bottom, giving a total volume of approximately 3 l. The chamber was divided into 11 sections. The sections were separated by vertical ribs reaching about 4 mm into the chamber from the inner wall. These ribs reduced mixing currents in the chamber without hindering the animals from swimming from one section to the next. In each section, glass tubes at the surface and at the bottom functioned as the inlet and outlet, respectively. The in- and out-flow were regulated by a peristaltic pump yielding a flow rate of about 5 ml min^{-1}. The concentration of food algae in the inlet medium was determind photometrically.

The daphnids were monitored with a black and white video camera (1:1.4, 6 mm lens) sensitive to infrared light. The light sources were a 20 W, 12 V (ran at 9 V) halogen lamp producing visible light and a ring with 154 diodes (1.5 V) producing infrared light with a wavelength of 910 nm. The halogen lamp was placed in the top and center of the covering cowl and provided homogeneous illumination to all parts of the chamber. The light intensity at the surface was moderate (160–170 μE s^{-1} m^{-2}) and equivalent to about 10% of the midday light on a partially clouded winter day in northern Germany. A ring of infrared diodes was placed below the ring-chamber to increase the light conditions for the video-camera. Infrared light of 910 nm can not be seen by the daphnids (Smith & Macagno, 1990).

A covering cowl of stainless steel was placed above the chamber, and the video camera was mounted on a rotating arm. The video camera could be moved around the periphery of the chamber by an electric motor. The camera was held at a constant distance of 17 cm from the chamber, sufficient to give a picture of slightly more than one of the sections of the ring chamber.

The ring chamber with the camera system was placed in a separate room kept in total darkness except for the light used in the experiments. The video monitor, the peristaltic pumps and the flasks for the in- and outflow were kept outside the experimental room. The movement of the video camera, the light intensity and the flow rate of medium were controlled from outside.

Two types of experiments were carried out. The first type was done with approximately 100 daphnids placed together in the ring chamber. The animals were introduced in one randomly selected section and there was a homogeneous food distribution of 0.03 mg C^{-1} in all the sections. After two hours the distribution of the animals was recorded. If the animals were randomly distributed, the ratio between the mean and the variance among the sections should be close to unity, and a deviation from the expectation can be tested by an F-test (Sokal & Rohlf, 1981: 187). After the distribution of the animals was recorded, the distribution of food levels was altered by a differentiated food input.

The section with highest food concentration was randomly selected for each replicate. After one hour, the outflows had a food gradient reasonably close to the inflows, and after two hours the distribution of animals was registered. Food gradients with dilution factors of 2 and 4 were used for each type of food distribution. With a dilution factor of two, the food concentration was halved from one section to the next on both sides of maximum density, resulting in a density gradient from 0.5 to 0.015 mg C l^{-1}. In case the distribution of animals is consistent with an ideal free distribution, the density of animals in each section should be proportional to the food concentration (except for sections with a food concentration above ILL). The distribution was tested by a G-test (Sokal & Rolhf, 1981: 705) against an ideal free distribution and a homogeneous distribution. In the experiment with dilution factor of 4 the maximum density was raised to 2.0 mg C l^{-1} and the lowest reduced to 0.002 mg C l^{-1}.

The second type of experiments used single *D. pulex* individuals. Animals were randomly placed in one of the sections in the chamber. The position of the animal was registered after two hours, and subsequently at five minutes intervals for one hour. Since *D. pulex* from lake Myravann has been found to swim in low food concentrations with an average speed of about 16 mm s^{-1} (Fullwood & Larsson, unpublished), an animal would require approximately 100 s to swim around the whole ring-chamber (about 1.5 m). If the animal preferred any part of the chamber, it could easily reach it within five minutes. In the experiments with single animals, the daphnids were exposed to low, homogeneous food conditions (0.015 mg C l^{-1}) and to a food gradient in which the dilution factor between neighbour sections was two. The section with highest food concentration was randomly selected before each replicate. Five different animals were observed singly in both treatments.

The sum of 12 observations from five replicates yielded 60 observations of single animals distributed among the 11 sections, or an average number of of 5.45 observations per section. Under homogeneous low food condition, the single animals were expected to be randomly distributed between the sections. This could be tested by an F-test. Five other animals were monitored in a food gradient. If the animals have a strong capacity to recognize food concentrations, they should be found frequently at the highest density, whereas they are expected to disperse more in case they have a weak capacity to determine food concentrations.

146

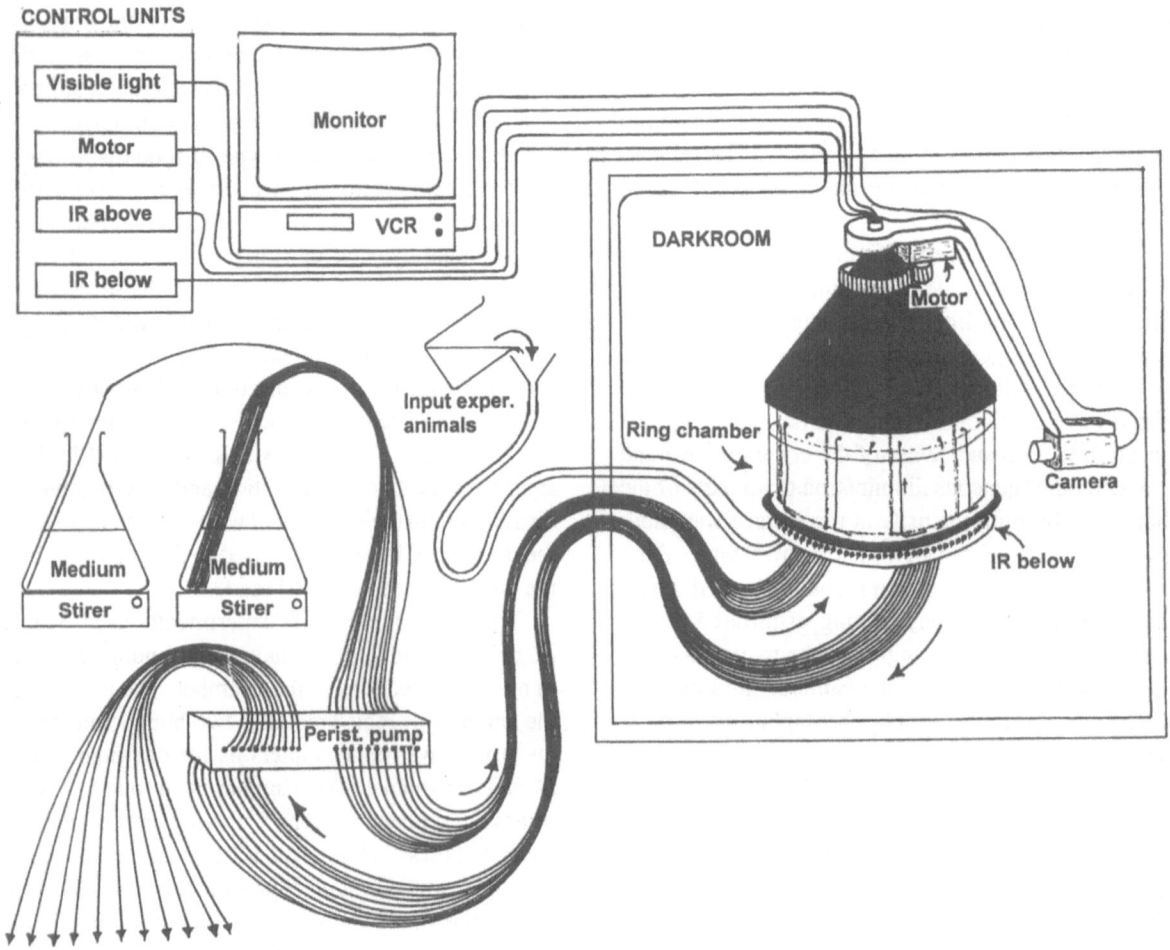

Figure 1. A schematic diagram of the experimental set-up, showing the ring-formed flow-through chamber, the video system and the control units. (Larsson & Kleiven, 1996)

In a food gradient, single animals were expected to be more frequently found at the highest food concentration than expected when they occur in groups. The two types of experiments present above differed in many respects, but were useful to indicate differences between the preferences of animals alone and in groups. If the frequency distribution of single animals in a homogeneous low food concentration is similar to the distribution of a group of animals in the same situation, than this may indicate that comparisons are not unrealistic. Since low homogeous food concentration should not give preference to any of the sections, random distribution should be expected in both cases. In a food gradient, the ideal free distribution is assumed for groups of animals, whereas single animals should more frequently choose the highest food concentration than expected from the ideal free distribution of groups.

In principle, divergence from the ideal free distribution principle has only realism when the distribution of several animals is considered. If the dispersal of animals is a result of unprecise localizing, however, it is of interest to compare the frequency distribution of single animals with an imaginary ideal free distribution. This was done by G-test for goodness of fit.

Results

In the experiments with single animals in homogeneous low food, the daphnids were distributed randomly in the ring chamber. The ratio $s^2/\bar{x}$ was 0.932 and not significantly different from 1 (Figure 2, Table 1). When 100 animals were together in the ring chamber at homogeneous low food concentration, they showed in one of three replicates a tendency to be more regularly distributed than random ($p<0.05$). The other two replicates, however, did not deviate significantly from random distribution (Figure 3, Table 1). When exposed to a food gradient, the single animals showed a clear tendency to stay in the section with the highest food density and were found there or in the adjacent chamber in 88% of the cases. The pooled observations showed a distribution significantly more aggregated around the highest density than in an imaginary ideal free distribution assuming that all observations would represent individuals in a co-occurring group of animals (Figure 4, Table 2). Thus, without competitors, the animals preformed better finding the highest food density than expected from the ideal free distribution principle. When occurring in groups, *D. pulex* showed a distribution close to the ideal free distribution (Figure 5a–c) when the maximum food density was 0.5 mg C l^{-1} and the dilution factor was two. The distribution of the animals in the three replications was very similar and did not significantly differ from an ideal free distribution. The distributions were, however, strongly different from a homogeneous distribution (Table 2). When the maximum food density was made higher, close to an assumed incipient limiting level of 2 mg C l^{-1} and with a dilution factor increased to four, the picture became different. *D. pulex* did not distribute proportionally to the food density in any of the replicates. They did, however, differ significantly from a homogeneous distribution (Figures 10 & 11, Table 2). The animal density at the highest food concentration was much lower than expected from differences in food concentration among compartments, and the whole distribution differed strongly from the expected ideal free distribution (Table 2).

Discussion

Single animals in a homogeneous low-food environment showed a random distribution and did not have any recognizable preference for special parts of the ring chamber. In Larsson & Kleiven (1996), the swimming speed of *D. magna* in the ring chamber was measured at various food concentrations. Under low food conditions, the swimming speed was found to be high, about 16 mm s^{-1}. *D. pulex* has a similar swimming speed (Fullwood & Larsson, unpublished). The animals probably swim fast as part of a food searching behaviour. When food particles are scarce, they seem to engage in random food searching behaviour,

Table 1. Statistics for the distribution in homogeneous low food conditions. Repeated registrations of single animals and individual registrations of co-occurring animals are compared. The distribution of the single animals is based on 60 registrations of single animals in the ring-chamber. The position was observed for the first time two hours after the animals was inoculated in the chamber, then every fifth minute for one hour; this protocol was repeated with five animals. The distribution of co-occurring animals was registered at one occasion two hours after inoculation in the ring-chamber. The statistics are based on three replicates with 100, 97 and 101 animals. All the 11 sections were given the same low food concentration of 0.015 mg C l^{-1}. If the animals are random distributed, it is expected that the parametric mean is equal to the parametric variance ($1 = \sigma^2/\mu$). $F_{(0.05,10,\infty)} = 1.83$, $1/F_{(0.05,10,\infty)} = 0.546$ (Sokal & Rohlf, 1981, p. 187).

Exp. type	Mean	Var.	$F_s = s^2/-$	p
Single animals	5.455	5.073	0.930	NS
Co-occurring	9.091	2.891	0.318	<0.05
Co-occurring	8.818	6.364	0.722	NS
Co-occurring	9.182	9.564	1.042	NS

resulting in a random distribution. The obtained distributions indicate independence between the observations, although the experiment was based on a time series with pseudoreplicates of only five different animals. When a group of approximately 100 animals was exposed to similar conditions, there was a weak tendency to a more regular distribution than random. This might indicate an attempt to reduce competition as much as possible by avoiding each other. Placed in a food gradient, single daphnids showed a clear ability to locate the highest food density, and they rarely appeared in the sections with less than maximum food. When about 100 animals were placed together in a gradient, the appearance in the different sections became different. When the maximum food concentration was not too high (0.5 mg C l^{-1}) they dispersed in a distribution which was very close to the ideal free distribution. When, however, the highest food concentration was about 2 mg C l^{-1}, this highest concentration was probably above the optimal concentration, and animals not necessarily prefered the highest density, but found the adjacent sections equally preferable. This resulted in a distribution deviating from an ideal free distribution. In addition, in the experiments with 2 mg C l^{-1}, the actual steepness of the food gradient may have been more gentle than it was in the input water. More diffusion of algae to the adjacent sections takes place in a steep gradient, and this might have affected the result. Unfortunately, the outflow food densities were not measured in these experiments.

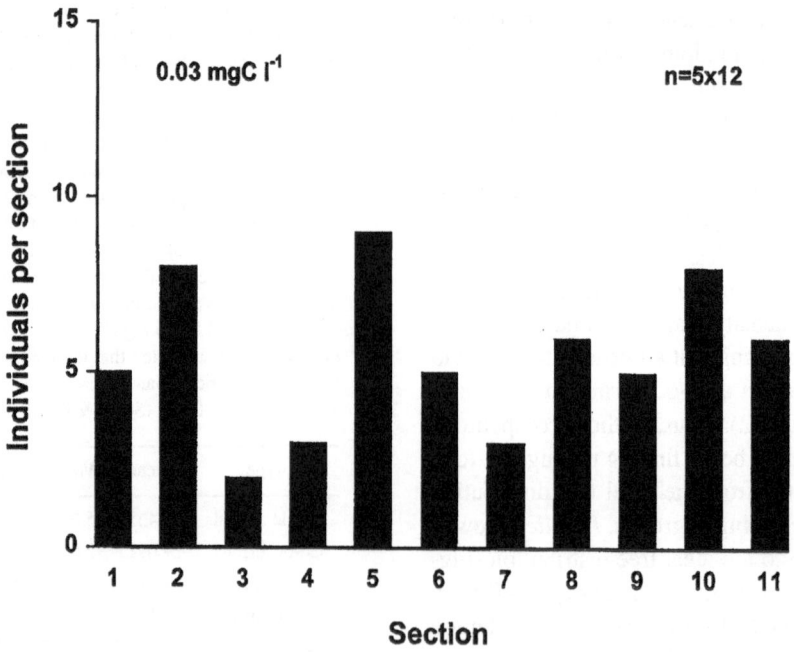

Figure 2. The frequency distribution of single *Daphnia pulex* in 11 sections of the ring chamber at homogeneous low food. In five replicates, single animals were registered every fifth minute for one hour, giving 60 observations. The observations started two hours after the introduction of the animals into the chamber.

Table 2. G-test (Sokal & Rohlf, 1981, p. 705) for deviation from expected 'ideal free distribution' and homogeneous distribution. The distribution of single animals is based on 60 registrations of single animals alone in the ring-chamber. The position of the animals was observed after two hours in the food gradient, every fifth minute for one hour and repeated with a total of five animals. The distributions of co-occuring animals are based on five replicate experiments, each with one registration. Due to multiple testing, the significance level in the single tests has to be corrected. By using a Bonferroni correction for table-wide errors, a difference is judged significant in the present analysis when the *p* value is lower than 0.008 in the individual tests.

Exp. type	Max food mg C l^{-1}	Dilution factor	No of ind. in exp.	Mean per section	Ideal free dist. G-value 10d.f.	Sign	Homogen dist. G-value 10 d.f.	Sign
Repeated single measurements	0.5	2	60	5.46	35.18	<0.001	158.49	<0.001
Co-occurring animals	0.5	2	98	8.91	9.64	NS	98.70	<0.001
Co-occurring animals	0.5	2	88	8.00	4.68	NS	92.88	<0.001
Co-occurring animals	0.5	2	92	8.36	11.08	NS	83.48	<0.001
Co-occurring animals	2.0	4	109	9.91	185.82	<0.001	56.63	<0.001
Co-occurring animals	2.0	4	106	9.64	160.00	<0.001	69.40	<0.001

How do the daphnids sense their environment when they choose their position in a food gradient and among their neighbours? The daphnids are definitively able to find the patches with high food concentrations in the water. Young (1974) indicated that vision is impor-tant in food search in daphnids, and that the degree of scattering of the light may be used as a cue for food density determination. The amount of food in the phar-ynx might also provide the animals with a measure of food concentration (Larsson & Kleiven, 1996) and Van

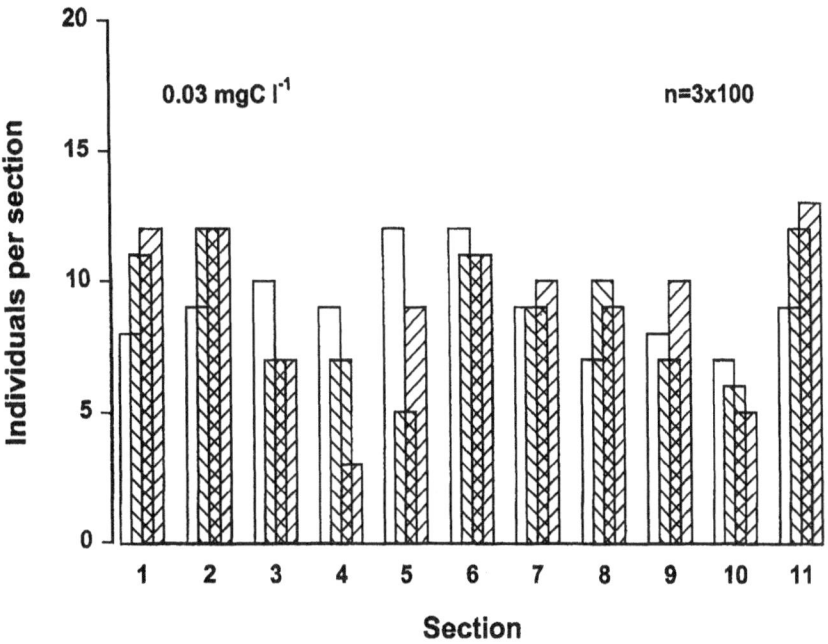

Figure 3. Three independent frequency distributions of 100, 97 and 101 *Daphnia*, respectively. The distribution was registered two hours after introduction of the animals into the chamber.

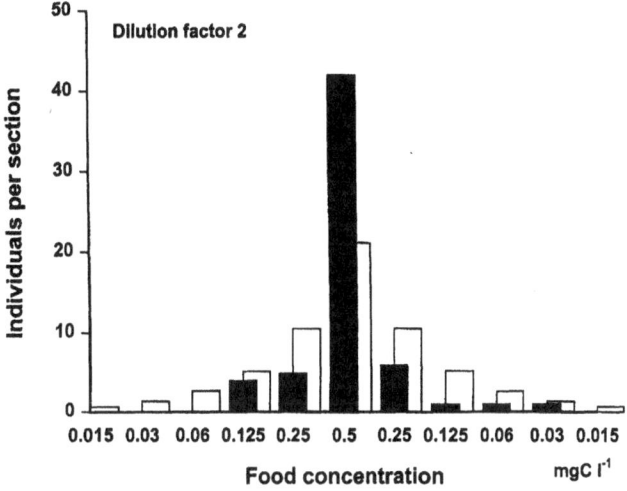

Figure 4. The frequency distribution of single *Daphnia pulex* in 11 sections of the ring chamber in a food density gradient 0.5–0.015 mg C 1^{-1}. Single animals were registered every fifth minute for one hour. Five individuals were observed giving altogether 60 observations. The observations started two hours after introduction of the animals into the chamber. Filled bars represent the observations, open bars show the expected ideal free distribution if 60 animals had been together.

Gool (1996) showed that daphnids can 'smell' algae. Clear evidence for any of these mechanisms and of their relative importance is, however, still lacking.

The ideal free distribution assumes that the animals are not only able to estimate food concentrations, but also that they can estimate the impact of their competitors on the food resources. So far, there are only indications on which mechanisms the daphnids may be using. The simplest explanation (assumed by Larsson & Kleiven, 1996) is that the animals are able to assess their food intake rate. A reduction in food intake may then be a stimulus to move to a place where the food intake rate is better. Daphnids reduce their filtering rate when given higher food densities, demonstrating that

150

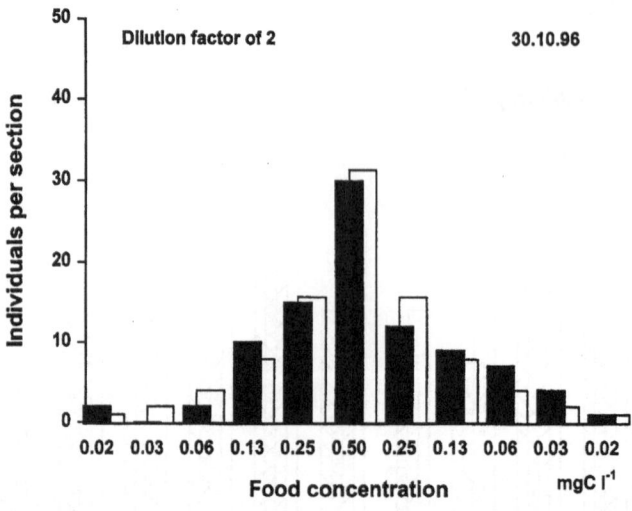

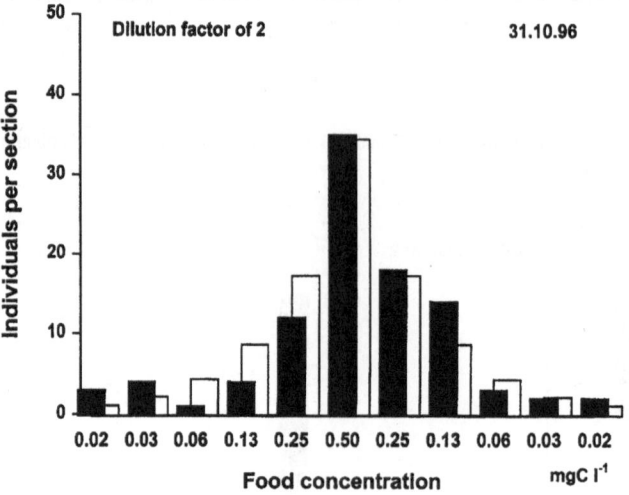

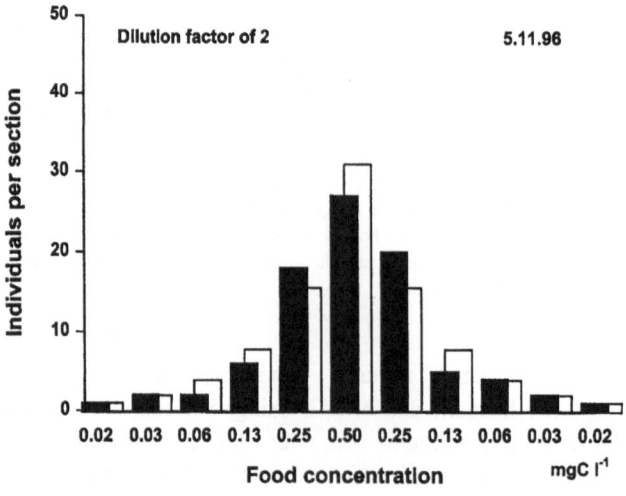

Figure 5. Three independent frequency distributions of three groups of *Daphnia pulex* (98, 88 and 92 individuals, respectively) in the ring-chamber in a food density gradient 0.5–0.015 mg C l^{-1} (dilution factor 2). The distribution was registered two hours after the establishment of the food concentration. Filled bars represent the observations, open bars show the expected ideal free distribution.

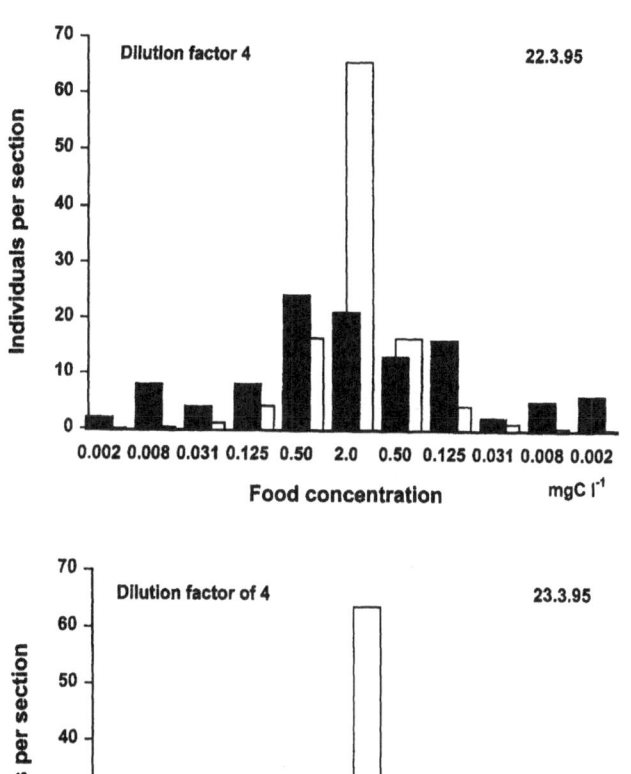

Figure 6. Frequency distributions of two groups of *Daphnia pulex* (109 and 106 individuals, respectively) in the ring-chamber in a food density gradient 2.0–0.002 mg C l^{-1} (dilution factor 4). The distribution was registered two hours after the establishment of the food concentration. Filled bars represents the observations, open bars show the expected ideal free distribution.

they are reacting to algal densities. Average swimming speed also decreases with increasing algal densities. It is, however, not necessarily the food intake rate that is recognized, it could also be the absolute concentration of algae in the water. The food density may be recognized by vision, as assumed by Young (1974). The increased scattering of light with increasing algal density may indeed be used as a measure of food concentration. When the daphnids are exposed to a homogeneous, high food concentration, they show a tendency to aggregate (Kleiven et al., 1996). This indicates that the daphnids are aware of the presence of conspecifics and that they are changing their distribution pattern according to the absolute food concentration even when there is no concentration gradient.

The experiments in the present study were done in a flow-through system with nearly constant food concentration. Theoretically, depletion of the food in the ring-chamber should be so small as to not be recognizable to daphnids. Thus it should be better for all to stay in the highest food concentration, as the single animals did. Instead, the co-occurring daphnids spread out into an ideal free distribution. This strategy indicates that daphnids are considering both food density and the number of competitors when they choose their position in the chamber. The evidence is, however, weak and several new experiments should be done before a clear statement can be made about the mechanisms the daphnids are using. It will then be of particular importance to measure both the in- and out-flow density of algae. Whatever the mechanisms, the daphnids seem able to judge their feeding environment with high precision. This is not surprising when considering the life-history characteristics of daphnids. *Daphnia* has certainly been selected for an efficient use of rich food resources occurring over a very short period. The blooms of algae in temperate lakes are short, normally about one or two months. Within that time, the daphnids can show a very high increase in population density due to their parthenogenetic reproduction. Under such conditions, there will be a strong selection for genotypes that are superior in gathering food, and genotypes that behave according to the expectations in the ideal free distribution principle will on average achieve the highest fitness.

Acknowledgments

This study was started with support form the Max Planck Institute in Plön, Germany. I would like to thank Willi Schröder and Horst Hansen for building the apparatus for this study, Maren Volquardsen and Heinke Clausen for assistance in the experiments, Knut Helge Jensen, Luc De Meester and two anonymous referees for valuable comments on the manuscript, and Jessica Marks for help with the English.

References

Cuddington, K. M. & E. McCauley, 1994. Food-dependent aggregation and mobility of water fleas *Ceriodaphnia dubia* and *Daphnia pulex*. Can. J. Zool. 72: 1217–1226.

Fretwell, S. D. & H. J. Lucas, 1970. On territorial behavior and other factors influencing habitat distribution in birds. Acta Biotheor. 19: 16–36.

Godin, J. G. J. & M. H. A. Keenleyside, 1984. Foraging on patchily distributed prey by a cichlid fish (Teleostei, Cichlidae): a test of the ideal free distribution theory. Anim. Behav. 32: 120–131.

Harper, D. G. C., 1982. Competitive foraging in mallards: 'ideal free' ducks. Anim. Behav. 30: 575–584.

Jakobsen, P. J. & G. H. Johnsen, 1987. Behavioural response of the waterflea *Daphnia pulex* to a gradient in food concentration. Anim. Behav. 35: 48–52.

Kleiven, O. T. , P. Larsson & A. Hobæk, 1996. Direct distributional response in *Daphnia pulex* to a predator kairomone. J. Plankton Res. 18: 1341–1348.

Larsson, P. & K. T. Kleiven, 1996. Food search in *Daphnia*. In Lenz, P. H., D. K. Hartline, J. E. Purcell & D. L. Macmillan (eds), Zooplankton: Sensory Ecology and Physiology, Gordon and Breach Publishers, Amsterdam: 375–387.

Milinski, M., 1979. An evolutionary stable stable feeding strategy in sticklebacks. Z. Tierpsychol. 51: 36–40.

Neary, J., K. Cash & E. McCauley, 1994. Behavioural aggregation of *Daphnia pulex* in response to food gradients. Funct. Ecol. 8: 377–383.

Smith, K. C. & E. R. Macagno, 1990. UV photoreceptors in the common eye of *Daphnia magna* (Crustacea, Branchiopoda). A fourth spectral class in single ommatidia. J. Comp. Physiol. A 166: 597–606.

Sokal, R. R. & F. J. Rohlf, 1981. Biometry. (2 ed.), W. H. Freeman and Company, San Francisco: 859 pp.

Young., S., 1974. Directional differences in the colour sensitivity of *Daphnia magna*. J. exp. Biol. 61: 261–267.

Hydrobiologia **360**: 153–159, 1997.
A. Brancelj, L. De Meester & P. Spaak (eds), Cladocera: The Biology of Model Organisms.
©1997 *Kluwer Academic Publishers.*

Mating behaviour in *Moina brachiata* (Jurine, 1820) (Crustacea, Anomopoda)

László Forró
Department of Zoology, Hungarian Natural History Museum, Baross u. 13, H-1088 Budapest, Hungary

Key words: Moina brachiata, Crustacea, Anomopoda, mating behaviour, mate choice, copulation

Abstract

Information on mating behaviour in Anomopoda is available for very few species only, though mate location and recognition certainly play an important role in maintaining reproductive isolation between species. Ephippial females and males of *Moina brachiata* were observed in a drop of water under a microscope for 10–20 minutes. Different combinations of specimens were used, but copulation was only observed when two males and two ephippial females were placed together. Males were very active, and often tried to grasp a female, whereas females usually made attempts to escape during the entire period of mating. Three phases were recognized: capture, positioning and copulation. The male captured the female from the dorsal side, then moved to the ventral side and took a position with its length axis being perpendicular to that of the female, forming a sort of cross. Thereafter the pair started rotating around the length axis of the female, while the male pushed the postabdomen into the female's brood pouch. Copulation lasted from 16 to 25 seconds. When different kinds of females were used, males showed preference for ephippial females with an empty ephippium and enlarged ovaries. Our results indicate that not only visual and tactile cues may be important in identifying species identity and receptivity of the female, but also chemical signals.

Introduction

Studies on the behaviour of various species of Anomopoda have mostly been carried out at the population level, and very few details of mating behaviour of these species are known (Dodson & Frey, 1991), though some early papers on 'Cladocera' already contain some information. O. F. Müller (1785) recognized the presence of males, described the development of the ephippium, and observed copulation (cited after Scharfenberg, 1911). Jurine (1820) described the copulation of *Daphnia*, and Weismann (1880) provided descriptions of mating in several taxa, including *Moina*. Scharfenberg (1911) investigated the egg development and life cycle of *Daphnia magna* and provided data on mating in this species. Goulden (1966) observed sexual behaviour of co-occurring *Moina* species. In his monograph on Moinidae, Goulden (1968) summarized the knowledge on secondary sex characters and gave a short summary of observations

on copulation. Shan (1969) and Smirnov (1971) dealt with mating of some chydorid anomopods.

Moina brachiata is the commonest of the five *Moina* species occurring in Hungary. It inhabits various types of astatic waters, including sodic ones. Mostly *M. brachiata* is the only species of *Moina* in these water bodies, but in a number of small, temporary pools *M. macrocopa* was found to co-occur with it. Ephippial females and males of both species can often be encountered together in samples. These findings inspired me to start a study of these populations, and of the mating behaviour in both species to find out whether mating behaviour acts as a mechanism in maintaining reproductive isolation. The present paper provides information on the mating behaviour of *M. brachiata*: mating and copulation are described and compared with existing descriptions of Spinicaudata and Anomopoda.

Materials and methods

The specimens used were collected from natural habitats or hatched in the laboratory from dried mud. The animals were placed in a drop of water (diameter about 1 cm) and observed during 10–20 minutes under a WILD M420 stereomicroscope. Their behaviour was filmed with a Panasonic WV-BP-100/G camera fitted to the microscope. A Panasonic VHS video recorder was used in combination with a JVC monitor for recording and for analysis of the videotape.

Different combinations of the specimens were used to observe mating behaviour. It was experienced that mating only occurred if two ephippial females and two males are placed together in the drop of water. Thus, this combination was mainly used to observe mating behaviour, and a total of fifteen experiments were run for ten minutes with different combinations of animals to study mate choice of the male. The combinations were: one male and one ephippial female; two males and two ephippial females; two males and two parthenogenetic females; one male, two ephippial females and one parthenogenetic female; two males; four males; three males and one dead ephippial female; two males, one dead ephippial female and one parthenogenetic female. The animals were videotaped for ten minutes immediately after they had been put together in a drop of water. The first, third, fifth and tenth minutes of the videotapes were used in the analyses. In order to obtain more information on mate preferences, the first five minutes of seven videotaped observations (not used in the experiments mentioned above) were analyzed, in which the two ephippial females and two males combination was employed. The females and males involved in the experiments were kept individually in culture for at least one day before the experiment.

Results

Mating behaviour

The animals placed in the drop of water often swam vigorously around, the males being usually more active than the females. Both males and females were frequently swimming along the edge of the drop. This behaviour could also be seen in much larger drops and in small Petri dishes under the microscope.

Based on observations and analyses of the videotape, four types of interactions between the animals could be established: contact, capture, pursuit and fighting. Contact is a very brief interaction: while swimming around, the animals, both males and females, may touch each other briefly, mostly with the second antennae, and then they swim apart again. During capture, a males approaches a females and tries to grasp her with his first antennae. If successful, capture is the initial phase of the mating process (see below). In a few cases it was observed that a male approached another male and tried to catch him. Pursuit also mainly took place as an interaction between male and female, with the male swimming behind the female, in an effort to get close enough for capture. In many cases pursuit also occurred after capture. Pursuit, as defined here, is characterized by the absence of direct contact between the two specimens. Fighting, as it is defined here, was a rare behaviour occurring between two males: facing each other they were beating very quickly with their second antennae. All these interactions (contact, capture, pursuit and fighting) lasted very short in time. Contacts mostly lasted less than a second, the other three behaviours lasted for 1–3 seconds. When a male encountered a female, the possible sequences of behaviour were contact – capture - mating, capture - mating or capture – pursuit. When a male encountered a male, the possible sequences of behaviour were contact – fighting – pursuit or contact – capture – pursuit. Of the four above mentioned behaviours, contacts occurred most frequently, very often not followed by any of the other behaviours. Capture also occurred sometimes as a single unit, while pursuit and fighting were always preceded by either contact or capture.

Three phases can be differentiated in the mating behaviour: capture, positioning, and copulation. During capture, the male tried to grasp the female. Such an attempt was always made from above and behind the female. Even if the animals were very close to each other frontally or laterally, males did not try a capture. The male used its first antennae to grasp the female around the groove between the head and body carapace, behind her second antennae (Figure 1). Immediately after capture, the male started moving to the ventral side of the female and clinged to the ventral margin of her carapace with the hooks of his first legs. Once secured this way, the male loosened his grip with the first antennae and asssumed in a position where his length axis was perpendicular to that of the female, so that they formed a sort of cross (Figure 2). Thereafter, the pair startd rotating: the male was swimming around and thus turned the female around her length axis. During this circumvolution phase, the male was pushing

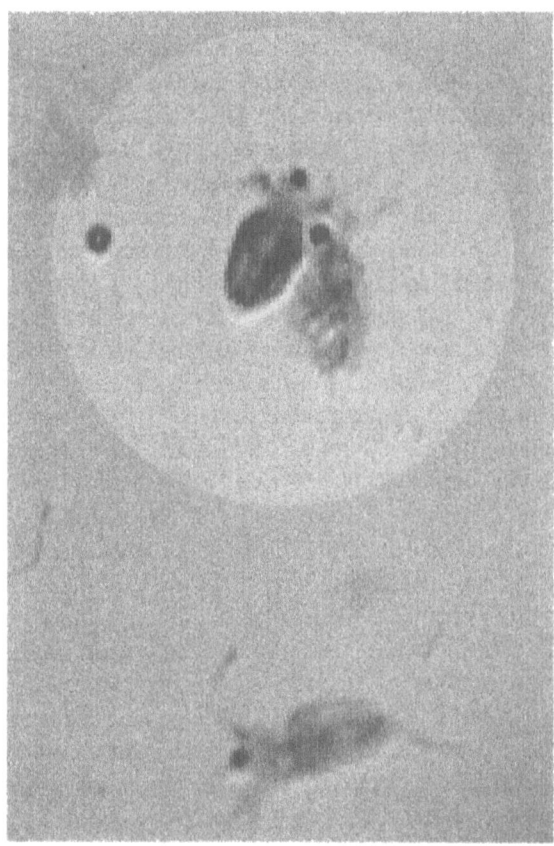

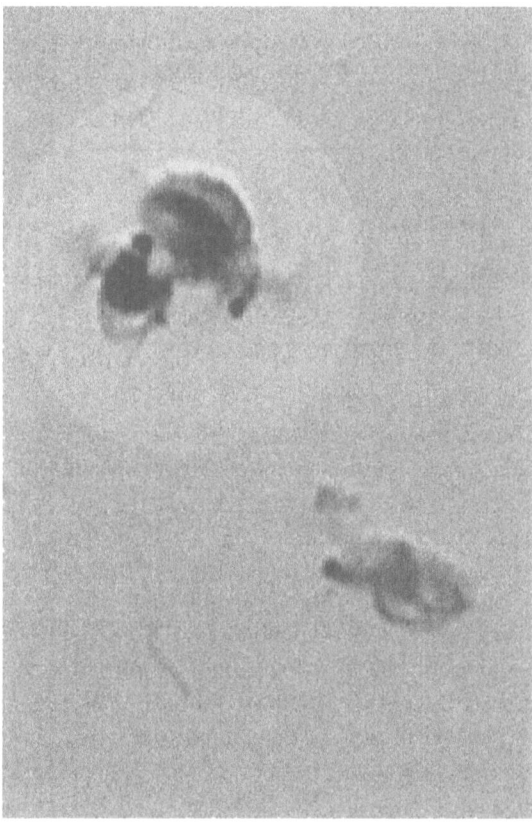

Figure 1. Capture, the first phase of the mating process in *Moina brachiata*: the male is grasping the female.

Figure 2. Circumvolution, the final phase of the mating process in *Moina brachiata*: copulation takes place.

several times his postabdomen between the valves of the female and inserted it into the female's brood pouch to ejaculate the sperm. After the circumvolution, the male and female immediately splitted apart again.

Several mating attempts by a male were observed, but we could observe only five copulations. During the entire mating process, the female was trying to escape by swimming away and by moving vigorously with her postabdomen. Very often the female was able to escape while the male attempted to take his position on the ventral side. In the few cases that the female could get rid of the male while they were rotating, the male continued the circumvolution alone and pushed out his postabdomen.

The duration of the mating process varied between 16 and 25 seconds (mean 19.6). Capture and positioning took only 3–8 seconds, while the circumvolution phase lasted 12–18 seconds. A successful mating probably requires a certain period for circumvolution, while the capture and positioning phases may be very short,

their length very likely depending on the skill of the male.

Mate preferences

Males and different females (parthenogenetic and ephippial ones) were observed for periods of ten minutes in eight different combinations as listed above. Generally the animals were active in the first minutes and they were much less active in the tenth minute. During the observation periods, almost exclusively contacts were seen, other behaviours being very rare. Therefore, the number of contacts in the different combinations will be evaluated. When one male and one ephippial female were placed together in a drop of water, very low activity was observed, and the animals mostly stayed at the edge of the drop and displayed only contacts (1.5 contacts/minute). In the case of two males and two ephippial females, the activity was higher (3.25 contacts/minute between males and 7.5 contacts/minute between males and ephippial

Table 1. Number of contacts and captures in trials with two males and two ephippial females. Seven observations were made, each during a period of five minutes. Numbers refer to mean values for a given period of five minutes; minimum and maximum values are given between brackets.

	Contacts	Captures
Male & male	5.1	0
	(1–12)	
Male & ephippial female	12.1	1
	(5–22)	(0–3)

females). When two males were put together with two parthenogenetic females, the mean number of contacts per minute between males was 2.5, whereas between males and parthenogenetic females 4 contacts/minute were observed. In the experiment with different kinds of females, the male showed greater interest in the ephippial female (3.25 contacts/minute) than in the parthenogenetic one (0.5 contact/minute). When only males were put together the mean number of contacts was 2 (two males) and 0.75 per minute (four males). In the experiment with three males and one dead ephippial female, the males contacted each other less frequently (0.5 contacts/minute) then the dead ephippial female (3.75 contacts/minute). When two males were put together with one parthenogenetic female and one dead ephippial female, the number of contacts between males was 0.5 per minute, between males and the dead ephippial female 1 per minute, and between males and the parthenogenetic female 0.25 per minute.

The number of contacts and captures was counted through the first five minutes in seven trials with the combination of two males and two ephippial females. The mean number of male-ephippial female contacts was 12.1 and the number of male-male contacts was 5.1 (Table 1). Since the probability of a male-ephippial female encounter in this setup is twice as high as that of a male-male encounter, such a difference was expected. However, none of the male-male contacts initiated capture, while seven captures were observed after male-ephippial female encounters.

Discussion

Moina brachiata is a cyclical parthenogen. The first generation hatching from ephippia contains females only, and this is the only generation which invariably and uniformly consists of parthenogenetic females (Grosvenor & Smith, 1913). Any of the succeeding generations may be mixed, containing both parthenogenetic and ephippial females, as well as males. Mature ephippial females have a well developed ephippium and, depending on the species, one or two large eggs in the ovaries wich will be deposited in the ephippium after copulation (Weismann, 1880; Goulden, 1968).

The present observations of the mating behaviour provided very similar results to those of Weismann (1880) and Goulden (1968). There are some differences, however. Weismann (1880) and Goulden (1968) reported that the ventral sides of the male and female are apposed during capture, whereas I found that the couple forms a sort of cross. Also, the circumvolution during copulation was not mentioned in the previous studies. Most likely, the specimens previously observed did not display these characteristics. This is particularly striking in the case of Weismann (1880) since he studied *M. rectirostris*, a synonym of *M. brachiata*. Goulden (1966, 1968) studied copulation in different, American species of *Moina*. During my study, I also made an observation of the copulation of *M. micrura* specimens, and observed that the male of this species too tries to secure himself in a transversal position and make circumvolution (Forró, unpubl.), so this behaviour most likely cannot be considered species specific for *M. brachiata*. A possible explanation for the discrepancy between my observations and previous studies may be that in the present study, the animals were observed in such a small amount of water that they were not able to display their natural behaviour. This is unlikely, however, because I observed one mating with the same characteristics under the microscope in a Petri dish, i.e. in a larger amount of water.

The cross-position and circumvolution do not seem to be restricted to *Moina* species, or not to the Anomopoda. The cross-position was seen by Brauer (1872) in *Leptestheria dahalacensis* and, the cross-position as well as the circumvolution were described in *Cyzicus tetracerus* by Gravier & Mathias (1930) and Mathias (1937). Although Martin et al. (1986) illustrated a non-cross behaviour during mating in a species of *Lynceus*, this drawing was based on a photograph taken during the early stages of mating. Laevicaudatans do in fact form a cross during mating (J. Martin, personal communication). In his description of *Daphnia* mating, Jurine (1820) wrote that 'I noticed other animals that were embraced almost transversely, in such a way as to form a sort of cross'. According to Weismann (1880), succesful fertilization can occur in *Daphnia* only if the male and female are face to face and parallel. However,

Scharfenberg (1911), among the thousands of couples he observed, did not see a single one for a long period in parallel position. Instead, he observed the most males stayed lateral to the female and with their body axis oriented parallel to that of the female. Sometimes there were males on both sides of the female, an observation also reported by Müller (1785). It should be noted that this position is different from that of *Moina*: as can be seen on Figure 2. in Scharfenberg (1911), the length axes of the large female and of the two, much smaller males are parallel, whereas it are the body planes that are at right angles. Although the appearance of the cross-position in *Daphnia* may thus be questioned, it certainly occurs and is followed by circumvolution in at least two species of Spinicaudata.

Table 2 provides a comparison of the duration of mating in various crustacean groups and species. There are considerable differences between species of the different groups. Of the anomopods studied so far, *Moina brachiata* has the shortest mating process. Unfortunately, neither Weismann (1880) nor Goulden (1966, 1968) provided data on the duration of mating. Importantly, within the mating sequence, the precopulatory phase (capture and positioning) is very short, varying between 3–8 seconds. This indicates that the time for locating and identifying a receptive female after an encounter is very short.

The observations of the present study indicate that males are the actors in mating: they initiate the process by grasping the female, while the females invariably try to swim away and escape. Weismann (1880) reported that the female was first trying to escape and then 'submitted to her fate'. In other anomopod species, it was reported that the female was agitated (Jurine, 1820) or stimulated by the activity of the male (Shan, 1969). Some studies observed precopulatory courtship (e.g. Glatzel & Schminke, 1996; Cohen & Morin, 1990) in other microcrustaceans, but this could not be observed during the short mating process of *M. brachiata*.

Our preliminary examinations on mate preferences of the males yielded some indicative data. Our observations revealed that the males positively selected for ephippial females and recognized them as potential mating partners. Both Weismann (1880) and Goulden (1966) noted that *Moina* males almost invariably grasped ephippial females. Goulden (1966, 1968) hypothesized that males might use visual cues in locating ephippial females, and that they recognize ephippial females by the presence of an ephippium. With the medial seta of the antennule, the male might subsequently identify the female as conspecific since the structure and surface sculpture of the ephippium is species specific in *Moina*. Goulden (1968) thus hypothesized that the elaborate ephippium does not merely serve the protection of the egg but is also developed to maintain reproductive isolation among species. The ephippium and the large egg in the ovary of ephippial females have a dark colour and are clearly visible. However, *Moina* species, particularly *M. brachiata*, usually inhabit turbid waters, and visual cues may therefore be of limited value for *M. brachiata* in locating mates. Chemical signals involved in mating behaviour of copepods and rotifers have been described (Snell & Morris, 1993; Snell & Carmona, 1994), and the use of sexual pheromones in Anomopoda is very likely. The possible role of the aesthetascs of the male antennule in chemoreception and mate location was already hypothesized by Scourfield (1905). However, Crease & Hebert (1983) did not find evidence for the presence of sex pheromones in *Daphnia magna*. In their review on chemical communication in zooplankon, Larsson & Dodson (1993) came to the conclusion that there must be minimum one chemical signal involved in the location and attraction of mates in *Daphnia pulex*. Carmona & Snell (1995) demonstrated the presence of glycoproteins on the ovary of *Daphnia obtusa* and *Ceriodaphnia dubia*, and hypothesized that these species specific molecules may be involved in contact recognition of mates, as they are in rotifers. We still lack a coherent view of the role of chemical communication in sexual behaviour of Anomopoda. From the available data on *Moina* mating, it seems likely that chemical signals (diffusible pheromones) are involved in mate location. The identification of the mate might be aided by the surface structure of the ephippium (mechanoreception), and it is also possible that the ovary produces glycoproteins for contact chemoreception in *Moina*. However, further behavioural studies are necessary to provide insight into how contact chemoreception is performed. As a matter of fact, such examinations are also desirable in *Daphnia* and other anomopod genera.

After a successful copulation, when sperm has been ejaculated into the brood pouch, the egg is released from the ovary into the brood chamber and fertilized. Weismann (1880) observed that the spermatozoa of *M. brachiata* desintegrate quickly, while they remain intact for a longer period in *M. macrocopa*. This difference may be related to the fact that the latter species has a two-egg ephippium. Because the ovaries are not functioning simultaneously, some time is necessary until both eggs are released. A given ephip-

Table 2. Mating duration in various crustacean taxa.

Taxon	Duration of mating	Source
Anostraca		
Branchipus schaefferi	6–15 sec	Mathias, 1937
Eubranchipus grubei	1 min	Mathias, 1937
Notostraca		
Lepidurus apus	30 sec	Mathias, 1937
Spinicaudata		
Cyzicus tetracerus	15–20 min	Gravier & Mathias, 1930, Mathias, 1937
Laevicaudata		
Lynceus gracilicornis	5 min	Martin et al., 1986
Anomopoda		
Daphnia spp	max. 8–10 min	Jurine, 1820
Daphnia magna	up to 1 day	Scharfenberg, 1911
Moina brachiata	16–25 sec	this study
Pleuroxus denticulatus	>20 min	Shan, 1969
Copepoda, Harpacticoida		
Parastenocaris phyllura	> 14 minutes	Glatzel & Schminke, 1996
Ostracoda	<1 min to over 30 min	Cohen & Morin, 1990

pial female can produce several ephippia and then switch to parthenogenesis. For instance, Gordo et al. (1994) observed *M. salina* producing a maximum of two ephippia. Hence, an ephippial female of *Moina* may copulate several times, most probably with different males, resulting in the production of genetically diverse offspring. It is conceivable that the continuous attempt to escape by females during mating is a sort of mate selection behaviour.

Acknowledgments

Several colleagues in the Department of Zoology provided technical assistance, and A. Urban made many of the observations. Dr Koen Martens (Brussels) and dr. Alain Thiery (Avignon) provided important information and copies of the relevant literature. Dr Anton Brancelj partially supported my stay in Postojna, and I benefitted from the inspiring discussion at the 4th Cladocera Symposium. Comments from Dr Luc De Meester and two anonymous referees greatly improved the paper. I am very grateful for their kind help. This research was supported by the National Scientific Research Fund (OTKA 017464).

References

Brauer, F., 1872. Beiträge zur Kenntniss der Phyllopoden. Sber. kais. Akad. Wiss., Math.-naturw. Kl., Abt. I, 65: 279–291.

Carmona, M. J. & T. W. Snell, 1995. Glycoproteins in daphnids: Potential signals for mating? Arch. Hydrobiol. 134: 273–279.

Cohen, A. C. & J. G. Morin, 1990. Patterns of reproduction in ostracodes: a review. J. crust. Biol. 10: 184–211.

Crease, T. J. & P. D. N. Hebert, 1983. A test for the production of sexual pheromones by *Daphnia magna* (Crustacea: Cladocera). Freshwat. Biol. 13: 491–496.

Dodson, S. I. & D. G. Frey, 1991. Cladocera and other Branchiopoda. In Thorp, J. H. & A. P. Covich (eds), Ecology and Classification of North American Freshwater Invertebrates. Academic Press, New York: 723–786.

Glatzel, T. & H. K. Schminke, 1996. Mating behaviour of the groundwater copepod *Parastenocaris phyllura* Kiefer, 1938 (Copepoda: Harpacicoida). Contributions to Zoology 66: 103–108.

Gordo, T., L. M. Lubián & J. P. Canavate, 1994. Influence of temperature on growth, reproduction and longevity of *Moina salina* Daday, 1888 (Cladocera, Moinidae). J. Plankton Res. 16: 1513–1523.

Goulden, C. E. 1966. Co-occurrence of moinid Cladocera and possible isolating mechanisms. Verh. int. Ver. Limnol. 16: 1669–1672.

Goulden, C. E., 1968. The systematics and evolution of the Moinidae. Trans. am. phil. Soc. N.S. 58: 1–101.

Gravier, C. & P. Mathias, 1930. Sur la reproduction d'un Crustacé Phyllopode du groupe des Conchostracés (*Cyzicus cycladoides* (Joly)). C.R. acad. Sci. Paris 191: 183–185.

Grosvenor, G. H. & G. Smith, 1913. The life-cycle of *Moina rectirostris*. Quart. J. microsc. Sci. 58: 511–522.

Jurine, L., 1820. Histoire des Monocles, qui se trouvent aux environs de Genève. J. J. Paschoud, Paris, 260 pp.

Larsson, P. & S. Dodson, 1993. Chemical communication in planktonic animals. Arch. Hydrobiol. 129: 129–155.

Martin, J. W., B. E. Felgenhauer & L. G. Abele, 1986. Redescription of the clam shrimp *Lynceus gracilicornis* (Packard) (Branchiopoda, Conchostraca, Lynceidae) from Florida, with notes on its biology. Zool. Scr. 15: 221–232.

Mathias, P., 1937. Biologie des Crustaces Phyllopodes. Hermann & Cie, Paris, 107 pp.

Müller, O. F., 1785. Entomostraca seu Insecta Testacea, quae in aquis Daniae et Norvegiae reperit, descripsit et iconibus illustravit. Lipsiae et Havniae, 135 pp.

Scharfenberg, U., 1911. Studien und Experimente über die Eibildung und den Generationszyklus von *Daphnia magna*. Int. Revue ges. Hydrobiol. Hydrogr. 3, Biologisches Supplement, 42 pp.

Scourfield, D. J., 1905. Die sogenannten Riechstäbchen bei den Cladoceren. Forschungsber. biol. Stat. Plön 12: 340–353.

Shan, R. K., 1969. Life cycle of a chydorid cladoceran, *Pleuroxus denticulatus* Birge. Hydrobiologia 34: 295–302.

Smirnov, N. N., 1971. Chydoridae fauny mira. Fauna SSSR. Nov. Ser. No. 101. Rakoobraznyye. Volume 1. vyp. 2, 531 pp.

Snell, T. W. & P. D. Morris, 1993. Sexual communication in copepods and rotifers. Hydrobiologia 255/256: 109–116.

Snell, T. W. & M. J. Carmona, 1994. Surface glycoproteins in copepods: potential signals for mate recognition. Hydrobiologia 292/293: 255–264.

Weismann, A., 1880. Beiträge zur Naturgeschichte der Daphnoiden. VI. Samen und Begattung der Daphnoiden. Z. wiss. Zool. 33: 55–256.

Hydrobiologia **360**: 161–167, 1997.
A. Brancelj, L. De Meester & P. Spaak (eds), Cladocera: The Biology of Model Organisms.
©1997 *Kluwer Academic Publishers.*

Light-induced swimming of *Daphnia*: can laboratory experiments predict diel vertical migration?

Erik van Gool[1,2]
[1]*Department of Aquatic Ecology, University of Amsterdam*
[2]*Netherlands Institute of Ecology, Centre for Limnology, Rijksstraatweg 6, 3631 AC Nieuwersluis,*
The Netherlands
(e-mail: gool@cl.nioo.knaw.nl)

Key words: Daphnia, phototaxis, swimming behaviour, diel vertical migration, model

Abstract

Vertical displacement velocity of a *Daphnia galeata* × *hyalina* clone was quantified in relation to changes in the relative rate of light change. An increase in the latter variable triggers an enhanced swimming response, and this response is again elicited when a second increase in the rate of relative light increase is applied. Decreases in the rate of light increase affect phototactic swimming in a similar way. The acceleration/deceleration assisted stimulus-response system is an extension of the idea of phototaxis as the underlying behavioural mechanism for vertical migration, and suggests that continuous accelerations in light change also affect vertical displacements observed in the field. A simple diel vertical migration simulation model was used to calculate the vertical displacement of *Daphnia* in relation to the natural light change at sunrise. The calculated vertical displacement fits nicely in the temporal range of the observed averaged downward migration of adult *Daphnia* in Lake Maarsseveen. The calculated migration amplitude, however, is larger than the change in mean population depth observed in nature.

Introduction

The effect of food and chemically mediated signals of predator presence on depth distributions during day and night have often been the focus of studies on diel vertical migration (DVM) of zooplankton. The influence of these factors can relatively easy be measured without knowledge of the underlying behavioural mechanisms. However, if we are to understand DVM, the behavioural mechanism must be described. I consider secondary phototaxis, defined as optically oriented swimming in response to changes in light intensity, as the behavioural mechanism underlying diel vertical migration. Therefore, a first step toward understanding DVM is to describe the involvement of phototaxis as observed in the laboratory for explaining diel migration depth of *Daphnia* under natural conditions.

Since 1989, research in Lake Maarsseveen has been focused on DVM of zooplankton, especially of the hybrid *Daphnia galeata* × *hyalina*. Detailed measure-

ment of the vertical distribution of *Daphnia* showed that the migration pattern in this lake varies seasonally (Ringelberg et al., 1991). In spring, large schools of juvenile Perch (*Perca fluviatilis*) are present in the open water zone, increasing the risk of death to *Daphnia* in the epilimnion. A reduction of this predation risk is often proposed as the ultimate reason for DVM (Zaret & Suffern, 1976; Lampert, 1993).

Besides the field observations at Lake Maarsseveen, laboratory experiments were performed to study the proximate factors of diel vertical migration. Changes in light intensity have been proposed as the main exogenous factor causing vertical displacements of zooplankton at dawn and dusk, in both marine and freshwater environments (reviews by Cushing, 1951; Ringelberg, 1995a). Past studies suggest that the relative rate of change in light intensity and the direction of this change is the stimulus for vertical migration. The stimulus response mechanism proposed by Ringelberg (1964, 1995b) involves relative light changes (RLC) that cause a swimming reaction, leading to the vertical displacements of *Daphnia*. Swimming continues as long as the light changes exceed the RLC-threshold. Based on laboratory estimates of this threshold obtained for *Daphnia galeata* × *hyalina*, supra-threshold changes at dawn last for about two hours (Ringelberg, 1991; Van Gool & Ringelberg, in press), and coincide with the morning descent phase of migration. The relative rate of change in light intensity and the vertical movements of *Daphnia* during a period of migration in Lake Maarsseveen were strongly correlated (Ringelberg & Flik, 1994). Although light-induced swimming of *Daphnia* was found to be enhanced in the presence of a predator kairomone (Ringelberg, 1991; Van Gool & Ringelberg, 1995), displacement velocities in response to RLC in laboratory experiments were still too slow to account for the sometimes extensive migrations observed in the field (Ringelberg, 1991, 1993). Recently, it was found that swimming activity, once started in response to supra-threshold RLC, was enhanced by acceleration in the rate of RLC (Van Gool & Ringelberg, in press). This novel property of the stimulus-response mechanism might cause the sometimes large vertical displacements observed at dawn and dusk.

The objective of the present study was to compare laboratory measurements of phototaxis of *Daphnia* with field measurements of DVM depth in Lake Maarsseveen. First, complementing the data presented in Van Gool & Ringelberg (in press), the effect of a deceleration of the rate of light intensity change on the vertical displacement velocity *Daphnia* was measured. Second, the effect of two successive accelerations in light intensity change rates was studied to determine whether consecutive stimuli cause greater vertical displacements. Finally, swimming in response to the natural light change at sunrise at Lake Maarsseveen was modelled to test the validity of the acceleration/deceleration assisted stimulus-response system for explaining vertical migration amplitude in the field.

Methods

Experiments

The experimental procedures to quantify phototactic behaviour in response to relative light changes (RLC) have been described in detail by Van Gool & Ringelberg (1995, in press). The apparatus consisted of a vertically positioned perspex cylinder (diameter 10 cm, length 118 cm) placed in a perspex jacket filled with water that was kept at a constant temperature ($17 \pm 0.5°C$). The cylinder was illuminated from above by incandescent lamps. Light intensity at the top of the cylinder was $0.082~\mu mol~m^{-2}~s^{-1}$. Between the cylinder and the lamps a frosted perspex sheet was placed to diffuse the light. Relative light changes (RLC) were created by use of a resistance controlled by a computer.

Daphnia galeata × *hyalina* were kept in the experimental cylinder under a L:D cycle of 16:8 hours, with light on at 6:00 h local time. Experiments were carried out between 11:00 and 16:00 hour to reduce any influence from a diel rhythm on the responsiveness.

The daphnids were exposed to the initial light intensity for at least one hour before the start of each experiment. Infra-red (IR) illumination ($\lambda_{(max.)} = 950$ nm) was used for video recordings. This IR radiation had no observable effect on the behaviour of *Daphnia*. The position of an individual daphnid was manually traced on the computer screen, registering the vertical position and time simultaneously by clicks of a cursor. The vertical displacement velocity of individual daphnids was used as the behavioural response variable. Displacement velocity (dv) is defined as the vertical distance over which a daphnid has moved divided by the time needed to move this distance.

Individuals of one clone of the hybrid *Daphnia galeata* × *hyalina* were used. This clone (stock-culture M2, see Reede & Ringelberg, 1995) had been maintained in the laboratory since 1992. It was originally isolated from 15–25 m depth in the open water zone of Lake Maarsseveen during a period of vertical migration. Daphnids were cultured in water from Lake Maarsseveen, which had been circulated over a sand filter for several days to promote the breakdown of organic compounds (such as fish kairomones). This water was also used in the experimental cylinder. One day before exposure to light stimuli, 60 adult *Daphnia* were placed in the experimental cylinder. The algal (*Scenedesmus acuminatus*) concentration in the cylinder was $0.5~mg~C~l^{-1}$, which is well above the incipient limiting level of $0.26~mg~C~l^{-1}$ (Muck & Lampert, 1984). This food concentration was restored one hour before the experiments started.

Stimulus deceleration was studied by exposing daphnids to an initial rate of light increase and then reducing this rate of RLC. Specifically, two minutes after onset of the first light change (RLC1), the rate of light change was decreased to RLC2. Within two minutes after this reduction, the displacement velocity was measured ($dv_{(RLC2/RLC1)}$) for all (five to seven) individual *Daphnia* visible on the computer screen. With the average displacement velocity found in experiments with only the initial RLC rate ($dv_{(RLC1)}$), the decrease in response to the combined stimuli (dv decrease $= dv_{(RLC2/RLC1)}/dv_{(RLC1)}$) was calculated. Various combinations of changes of RLC1 into RLC2 were tested: 0.20–0.10, 0.20–0.15, 0.30–0.10, 0.40–0.08 min^{-1}. Each RLC combination was used three to five times, distributed over at least two different days. On each day, a random sequence of RLC combinations was applied. The 3–5 video recordings of each RLC deceleration combination resulted in the displacement velocity estimate of 20–25 individual daphnids. Linear regression lines with the RLC rate change as independent variable and dv change as the dependent variable were calculated with treatment averages.

In the second series of experiments, the effect of two subsequent RLC accelerations on the swimming response was studied by increasing the rate of RLC twice. Two minutes after onset of the first relative light change (RLC1), the rate of change was increased (RLC2), and 30 seconds later the rate of change was increased again (RLC3). Within two minutes after the switch to the third RLC rate the displacement velocity of all (five to seven) individual *Daphnia* visible on the computer was measured. Various combinations of changes of RLC1 into RLC2 and RLC3 were tested: 0.08–0.15–0.20, 0.08–0.15–0.25, 0.08–0.13–0.30, 0.08–0.13–0.40 min^{-1}. Each RLC-acceleration combination was used 4–5 times, and the displacement velocity of 20–25 individuals was determined. The combinations were assigned in random order at four per day. The relation between the displacement velocity change and the strength of one acceleration in RLC rate as found by Van Gool & Ringelberg (in press, see Figure 2), and the average displacement velocity found in experiments with only one RLC rate ($dv_{(RLC1)}$, Van Gool & Ringelberg, in press) were used to calculate the first increase in displacement velocity ($dv_{(RLC2/RLC1)}/dv_{(RLC1)}$). Then, the measured displacement velocity was divided by $dv_{(RLC2/RLC1)}$ to determine the displacement velocity change elicited by the second RLC acceleration ($dv_{(RLC3/RLC2)}/dv_{(RLC2/RLC1)}$).

Simulation model

Light changes measured at Lake Maarsseveen during sunrise (May 30, 1990) were used as input for a preliminary DVM simulation model. Light was measured 0.30 m below the water surface and light intensity was

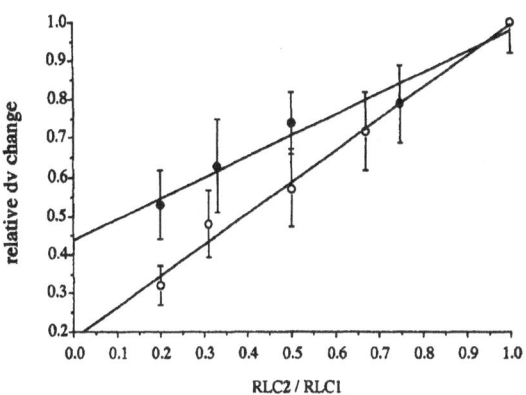

Figure 1. The relation between the ratio of successive relative changes in light intensity (RLC2/RLC1, in which RLC2<RLC1) and the accompanying relative change in displacement velocity ($dv_{(RLC1/RLC2)}/dv_{(RLC1)}$). Each point represents the average dv of 20–25 individuals (error bars: 95% confidence limits of the mean). Regression lines were calculated with these averages. The closed dots represent the found decreases in dv in response to a RLC rate decrease ($y = 0.437 + 0.541x$; $R^2 = 0.96$; $P<0.003$), the open dots represent the changes in dv calculated with dv measured at single RLC rates ($y = 0.183 + 0.812x$; $R^2 = 0.98$; $P<0.001$).

automatically written to a spreadsheet every minute starting 3 h before sunrise. At low light intensities, the intensity was measured with a very sensitive custom-built sensor and a Kipp recorder, and expressed as mV. When light intensity was sufficiently high, light was also measured with a LI-COR quantum sensor (SA) and expressed as μmol m^{-2} s^{-1}. The mV-values were later transformed with a calibration line into μmol m^{-2} s^{-1}. From these light intensities the relative light change rate (RLC $= \frac{1}{I} \times \frac{dI}{dt}$ with RLC the relative light change rate min^{-1}, I the absolute light intensity in μmol m^{-2} s^{-1}, and t the time in min), and the acceleration in these relative changes per minute were calculated ($A = \frac{RLC_n}{RLC_{n-1}}$, with A the acceleration without dimension, RLC_n the rate at time n and RLC_{n-1} the rate at time n-1).

Diel vertical migrations in Lake Maarsseveen are confined to only a short period of time in early summer when juvenile fish are present in the open water zone. The depth change in response to the change in light at sunrise was simulated for such a migration period. For all light change rates exceeding the phototactic response threshold ($R = 0.04$ min^{-1}, Van Gool & Ringelberg, in press), a phototactic swimming response was elicited. Displacement velocity was calculated for the first time at this threshold by substituting the observed RLC-value in an empirically based equation relating the displacement velocity to the prevailing RLC ($dv = 2.71 + 30.74 \times$ RLC, Van Gool & Ringelberg, in press). Further displacement velocities were determined per minute by inserting the RLC changes into the functions found in the present study, thus taking acceleration/deceleration effects into consideration.

Results

Experiments

The regression between the displacement velocity change ($dv_{(RLC2/RLC1)}/dv_{(RLC1)}$) and the reduction in relative light change rate (RLC2/RLC1) shows a positive slope ($R^2 = 0.962$, $n = 4$, $P < 0.003$). Displacement velocity change in response to two successive relative changes in light intensity with decreasing rates (RLC2<RLC1) exceeded the value expected for the case where the reaction in response to the RLC combinations are independent (Figure 1); that is, when the dv measured after the RLC combination ($dv_{(RLC2/RLC1)}$) would have been the same as was measured by exposure to the second rate of RLC ($dv_{(RLC2)}$) only. The latter relation was found by calculating the ratio between $dv_{(RLC2)}$ and $dv_{(RLC1)}$, derived from experiments with only one rate of RLC per test run. The regression line of the displacement velocity change observed with a reduction in RLC rate differed significantly in slope from the regression line of the calculated displacement velocity change without such a reduction (Fs = 11.55, df = 1,6; $P < 0.025$).

In experiments with two successive accelerations (RLC rate increases), the resulting displacement velocity ($dv_{(RLC3/RLC2)}$) was much faster than the velocity that goes with the third RLC rate if given alone ($dv_{(RLC3)}$). In fact, the second RLC acceleration showed the same change in displacement velocity as was observed for the first increase in RLC rate (Figure 2). To show this, first the relation between the RLC and dv, with one RLC rate per test run, was used to calculate the initial displacement velocity ($dv_{(RLC1)}$). Second, the relation for the displacement velocity increase caused by a single RLC rate increase ($dv_{(RLC2/RLC1)}/dv_{(RLC1)} = -0.395 + 1.408 \times$ (RLC2/RLC1), Van Gool & Ringelberg, in press) was used to calculate the displacement velocity before the second RLC rate increase was applied ($dv_{(RLC2/RLC1)}$). Finally, the displacement velocity increase caused by the second RLC acceleration was calculated ($dv_{(RLC3/RLC2)}/dv_{(RLC2/RLC1)}$).

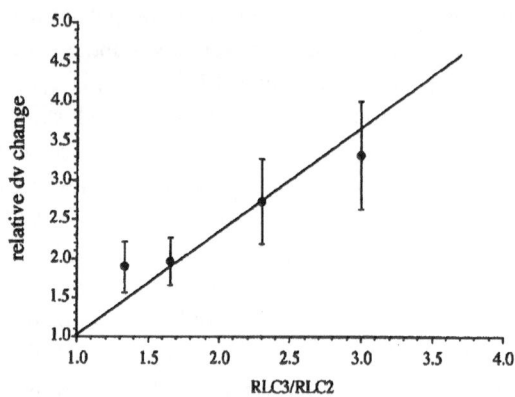

Figure 2. The relation between the ratio of the succeeding relative changes in light intensity of the second RLC rate increase (RLC3/RLC2, with RLC3>RLC2>RLC1) and the accompanying change in displacement velocity ($dv_{(RLC2/RLC3)}/dv_{(RLC2/RLC1)}$). Each point represents the average dv of 20–25 individuals (error bars: 95% confidence limits of the mean). The drawn line is the function relating the increase in RLC to the increase in dv for one acceleration in RLC rate ($y = -0.395 + 1.408x$, Van Gool & Ringelberg, in press)

This latter dv-ratio was plotted as a function of the second RLC increase (RLC3/RLC2; Figure 2). The displacement velocity increases observed in the present experiments are very close to the regression line that was found for a single increase in RLC rate (Van Gool & Ringelberg, in press). The results show that, at least for a second successive increase in RLC rate, the displacement velocity enhancing effect of RLC accelerations is constant.

Simulation model

Because the light intensity measurements contained a lot of noise, as for instance caused by light reflections at the rippling water surface, the moving average over 5 relative light changes (5 min) was used in the model. The light intensity change at sunrise increased gradually up to its maximum about 50 min before sunrise (Figure 3), and then decreased again. Thus, equal values of the RLC are perceived by a daphnid before and after this maximum, but first in an accelerating sequence and then in a decelerating sequence. The simulation resulted in similar displacement velocities on either side of this maximum.

The calculated vertical displacement fits nicely in the temporal range of the observed averaged downward migration of adult *Daphnia* in Lake Maarsseveen (Figure 4). The migration amplitude calculated by the

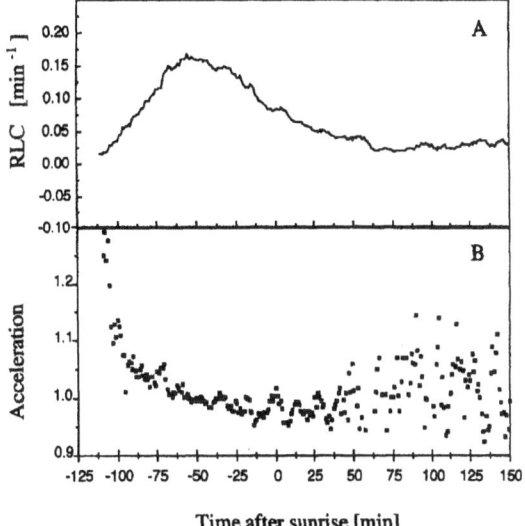

Figure 3. The change in light intensity during sunrise of May 30, 1990, at Lake Maarsseveen, The Netherlands, expressed as relative light change (A; drawn line) and the acceleration in the RLC (B; closed dots) per minute. For calculation methods see text.

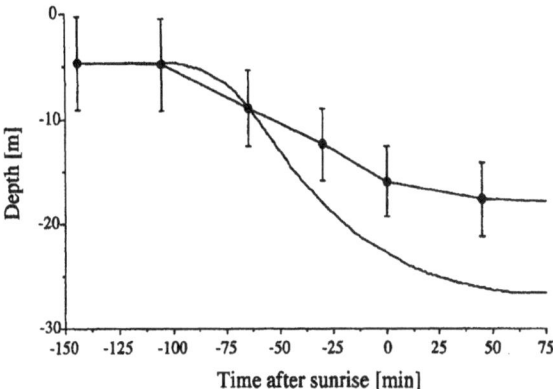

Figure 4. The changes in depth at sunrise (Sunrise is at time = 0). The drawn line without symbols represents the depth change as calculated by the simulation model (see text for explanation). The drawn line with closed dots represents the change in mean population depth of the hybrid *Daphnia galeata* × *hyalina* at different times during dawn at Lake Maarsseveen (June 11, 1992). Error bars: 1 S.D.

simulation model is, however, larger than the change in mean population depth observed in nature.

Discussion

The vertical displacement velocity (dv) as a result of a response to a continuous light intensity change is strongly increased if an acceleration in the light change occurs. I found that the relation between the increase in dv and the acceleration, as expressed by the ratio between the two relative light change rates, remained the same when two successive accelerations were applied. For the simulation model, it was assumed that the enhancing effect would also remain identical for more than two succeeding accelerations. With this assumption in mind, the vertical displacement of an individual as caused by a natural light change was calculated and related to diel vertical migration in Lake Maarsseveen. In the mechanistic model presented by Ringelberg (1995b), swimming velocities as a function of individual relative light changes were too slow to reach the observed day depth when the natural light change was used for the simulation. A calibration factor was needed to simulate natural migration amplitudes. With the presented new acceleration/deceleration assisted stimulus response system, such calibration was unnecessary. Actually observed downward migration of the mean population depth in Lake Maarsseveen (Figure 4) is less than the calculated amplitude. However, in nature, the range within which migration occurs is much larger than the observed change in *mean* population depth. In addition, the model assumed that any change in RLC, no matter how small, resulted in a corresponding change in displacement velocity. Analogous to the threshold for individual RLC rates (Rheobase, see Ringelberg, 1964), it is likely that there is a minimum change needed to elicit an acceleration or deceleration of the displacement velocity.

Altough the effect of environmental variables on phototactic reactions have yet to be incorporated into the model presented here, the preliminary·simulation of DVM caused by natural light change fits within the range of *Daphnia* migration in Lake Maarsseveen. It is important to stress, however, that the change in mean population depth during sunrise is a population parameter, and that not all individuals have to react in a similar way. De Meester & Dumont (1988; De Meester 1991, 1993b) have shown that primarily phototactic behaviour of *Daphnia magna* is clone-dependent. Vertical habitat partitioning has been shown for both pond dwelling (Weider, 1984; De Meester, 1993b) and for lake-dwelling *Daphnia* populations (Muller & Seitz, 1993). The *Daphnia* population in Lake Maarsseveen consists of different clones of the hybrid *Daphnia galeata* × *hyalina* (Spaak & Hoekstra, 1993; Spaak, unpubl. data) that differ in life-history characteristics (Reede & Ringelberg, 1995), and a different migration behaviour may be expected. In addition, changes in phototactic behaviour in response to food (De Meester & Dumont 1988; De Meester 1989) and predator kairomones (De Meester, 1993a) may also be clone-specific. In the clone I studied, the influence of fish kairomone on parameters of the phototactic behaviour was negligible (Van Gool & Ringelberg, in press). Probably, the applied changes in rate of the RLC were strong enough to always elicit a maximum response, even in the absence of fish kairomones. Strong constant RLC also cause phototactic swimming responses in the absence of fish kairomones (Ringelberg, 1964, 1993; Van Gool & Ringelberg, in press).

Ringelberg & Flik (1994) found significant correlations between the RLC at dawn and dusk, and downward and upward migration in Lake Maarsseveen, suggesting that phototaxis was the underlying behavioural mechanism. They carried out a linear regression analysis using RLC as independent variable and dv as the dependent variable, and found slopes of 35.51 and 74.81 for sunrise (June 11, 1992) and sunset (June 22, 1992), respectively. My observations show that, when the effect of RLC acceleration is taken into account, regressions are also significantly linear (slope 120.25). The difference in slope between my result and those presented by Ringelberg & Flik (1994), may result from the fact that they considered average (population) behaviour.

The results of the present DVM simulation model, in which the experimentally found effects of accelerations and decelerations in RLC were included, indicate that yet another step toward a further understanding of the behavioural mechanism of diel vertical migration is made. Extensions that take genotypic and phenotypic variability of phototaxis into account should in the future be used to demonstrate the causes of observed variability in migration amplitude in the field. Moreover, models as the one presented here can be used to calculate costs (caused by a decrease in temperature and food quantity and quality with depth) and benefits (related to the reduction of predation risk by visually hunting predators with depth) of migrations in particular field situations.

Acknowledgments

I thank B. Flik for providing the data on the vertical distribution of *Daphnia* in Lake Maarsseveen (June 11, 1992) and the light measurement at dawn (May 30, 1990), and L. De Meester, R. Gulati, R. Laanbroek, J. Vijverberg and an anonymous reviewer for critical reading and comments on an earlier version the manuscript. I am especially grateful to J. Ringelberg and M. Sabelis for fruitful discussions and significant improvements of the manuscript. The present study was supported by the Life Science Foundation (SLW), which resorts under the Netherlands Organization for Scientific Research (NWO).

References

Cushing, D. H., 1951. The vertical migration of planktonic crustacea. Biol. Rev. 26: 158–192.

De Meester, L., 1989. Phototaxis in *Daphnia*: Interaction of hunger and genotype. Limnol. Oceanogr. 34: 1322–1325.

De Meester, L., 1991. An analysis of the phototactic behaviour of *Daphnia magna* clones and their sexual descendants. Hydrobiologia 225: 217–227.

De Meester, L., 1993a. Genotype, fish-mediated chemicals, and phototactic behaviour in *Daphnia magna*. Ecology 74: 1467–1474.

De Meester, L., 1993b. The vertical distribution of *Daphnia magna* genotypes selected for different phototactic behaviour: Outdoor experiments. Arch. Hydrobiol. Beih. Ergebn. Limnol. 39: 137–155.

De Meester, L. & H. J. Dumont, 1988. The genetics of phototaxis in *Daphnia magna*: Existence of three phenotypes for vertical migration among parthenogenetic females. Hydrobiologia 162: 47–55.

Lampert, W., 1993. Ultimate causes of diel vertical migration of zooplankton: New evidence for the predator avoidance hypothesis. Arch. Hydrobiol. Beih. Ergebn. Limnol. 39: 79–88.

Muck, P. & W. Lampert, 1984. An experimental study on the importance of food conditions for the relative abundance of calanoid copepods and cladocerans. Arch. Hydrobiol. 66: 157–179.

Müller, J. & A. Seitz, 1993. Habitat partitioning and differential vertical migration of some *Daphnia* genotypes in a lake. Arch. Hydrobiol. Beih. Ergeb. Limnol. 39: 167–174.

Reede, T. & J. Ringelberg, 1995. The influence of a fish exudate on two clones of the hybrid *Daphnia galeata* × *hyalina*. Hydrobiologia 307: 207–212.

Ringelberg, J., 1964. The positively phototactic reaction of *Daphnia magna* Straus: A contribution to the understanding of diurnal vertical migration. Neth. J. Sea Res. 2: 319–406.

Ringelberg, J., 1991. Enhancement of the phototactic reaction in *Daphnia hyalina* by a chemical mediated by juvenile perch (*Perca fluviatilis*) J. Plankton Res. 13: 17–25.

Ringelberg, J., 1993. Phototaxis as a behavioural component of diel vertical migration in a pelagic *Daphnia*. Arch. Hydrobiol. Beih. Ergeb. Limnol. 39: 45–55.

Ringelberg, J., 1995a. Changes in light intensity and diel vertical migration: A comparison of marine and freshwater environments. J. Mar. biol. Ass. U.K. 75: 15–25.

Ringelberg, J., 1995b. An account of a preliminary mechanistic model of swimming behaviour in *Daphnia*: Its use in understanding diel vertical migration. Hydrobiologia 307: 161–165.

Ringelberg, J. & B. J. G. Flik, 1994. Increased phototaxis in the field leads to enhanced diel vertical migration. Limnol. Oceanogr. 39: 1855–1864.

Ringelberg, J., B. J. G. Flik, D. Lindenaar & K. Royackers, 1991. Diel vertical migration of *Daphnia hyalina* (sensu latiori) in Lake

Maarsseveen: Part 1. Aspects of seasonal and daily timing. Arch. Hydrobiol. 121: 129–145.

Spaak, P. & J. R. Hoekstra, 1993. Clonal structure of the *Daphnia* population in Lake Maarsseveen: Its implication for diel vertical migration. Arch. Hydrobiol. Beih. Ergeb. Limnol. 39: 157–165.

Van Gool, E. & J. Ringelberg, 1995. Swimmning of *Daphnia galeata × hyalina* in response to changing light intensities: Influence of food availability and predator kairomone. Mar. Fresh. Behav. Physiol. 26: 259–265.

Van Gool, E. & J. Ringelberg (in press). The effect of accelerations in light increase on the phototactic downward swimming of *Daphnia* and the relevance to diel vertical migration. J. Plankton Res.

Weider, L., 1984. Spatial heterogeneity of *Daphnia* genotypes: vertical migration and habitat partitioning. Limnol. Oceanogr. 29: 225–235.

Zaret, T. M. & J. S. Suffern, 1976. Vertical migration in zooplankton as a predator avoidance mechanism. Limnol. Oceanogr. 21: 804–813.

Hydrobiologia **360**: 169–175, 1997.
A. Brancelj, L. De Meester & P. Spaak (eds), Cladocera: The Biology of Model Organisms.
©1997 *Kluwer Academic Publishers.*

The change in phototactic behaviour of a *Daphnia magna* clone in the presence of fish kairomones: the effect of exposure time

L. De Meester[1] & C. Cousyn[2]

[1]*Laboratory of Ecology and Aquaculture, Katholieke Universiteit Leuven, Naamsestraat 59, 3000 Leuven, Belgium*
[2]*Laboratory of Animal Ecology, University of Ghent, K. L. Ledeganckstraat 35, 9000 Gent, Belgium*

Key words: phototaxis, diel vertical migration, induced defence, phenotypic plasticity, time lag, predator avoidance

Abstract

Using a clone that responds to the presence of fish kairomones by a pronounced change in phototactic behaviour, we determined how fast a change to more negatively phototactic behaviour occurs in *Daphnia magna* adults that are exposed to a high concentration of fish kairomones. Kairomone exposed animals showed an approximately linear decrease in the value of the phototactic index with time. Though the response was almost immediate, it took two hours before the difference between fish-induced and control animals was significant. Extrapolation of the observed response indicates that a maximal change in phototactic behaviour, equivalent to animals that have been cultured in the presence of fish kairomones since birth, occurs after about 13 hours exposure. We conclude that the predator-induced change in diel vertical migration of zooplankton is fast, and is fully developed in less than a day. The response time to fish kairomones of *Daphnia* is shorter for phototactic behaviour than for life history traits, which may have important consequences with respect to the evolution of trait-dependence in induced defence responses.

Introduction

Zooplankton, and *Daphnia* in particular, has been shown to exhibit responses to predator-specific chemicals (kairomones) in a variety of traits, including behavioural, morphological and life history traits (Larsson & Dodson, 1993). Numerous studies have focused on these predator-induced defence responses, reporting on benefits and costs of the induced responses (e.g. Riessen, 1984; Havel & Dodson, 1987; Riessen & Sprules, 1990; Loose & Dawidowicz, 1994), genotypic differences in responsiveness (e.g. De Meester, 1993; Weider & Pijanowska, 1993; Reede & Ringelberg, 1995), trait-specificity of the responses (Lüning, 1994; De Meester & Pijanowska, 1997) and characteristics of the chemicals involved (Parejko & Dodson, 1990; Tollrian & Von Elert, 1994). Since natural populations of *Daphnia* harbour genetic variability for both the mean values of specific traits (in the absence or presence of predator kairomones) and predator-induced

phenotypic plasticity responses (see Parejko & Dodson, 1991; De Meester, 1996), a given population can evolve in several ways in response to a predation pressure that is variable in time. The animals (clones) can exhibit appropriate induced plasticity responses without changing the mean value of the trait when not induced. Alternatively, the population can evolve towards a changed mean value of the trait both in the absence and presence of the predator, without a change in the phenotypic plasticity responses (i.e. the shift in mean value when induced) of the individuals (clones). Finally, the animals can exhibit a change in both the trait value and the phenotypic plasticity response. At first sight, one might expect that a change in phenotypic plasticity would be the most effective response, given the superiority of induced defences over constitutive defences when the defence has a cost (Lively, 1986; Adler & Harvell, 1990; Harvell, 1990). However, whether a change in mean value or a change in phenotypic plasticity will be most adaptive will critically depend on the time lag between exposure to predator kairomones and an effective response (Padilla & Adolph, 1996). If the response time is short, a change in the magnitude of the phenotypic plasticity response will be the appropriate way to cope with a change in predation pressure over time. If the response time is long, a change in mean values of traits involved in defence mechanisms may be more appropriate. In the present study, we report on the response time of phototactic behaviour of a *Daphnia magna* clone to the presence of fish chemicals. Phototactic behaviour and diel vertical migration, being behavioural traits, may be expected to show short response times. Ringelberg & Van Gool (1995) recently reported data showing a very fast change in secondary phototaxis (i.e. phototactic behaviour in response to changes in light intensity; Ringelberg, 1964) in a *Daphnia galeata* × *hyalina* clone. Using a different experimental set-up that quantifies primary phototaxis (i.e. phototactic behaviour in the absence of changes in light intensity), we here monitor the change in phototactic behaviour with time of a *D. magna* clone that is known to respond to the presence of fish kairomones with a strong shift in phototactic behaviour.

Material and methods

We used *Daphnia magna* clone $P_1 32,85$ that has been shown in earlier work (De Meester, 1993) to respond to the presence of fish kairomones with a pronounced change in phototactic behaviour. Animals of clone $P_1 32,85$ that are cultured in the presence of fish kairomones are negatively phototactic when tested in a standard experimental set-up such as described in De Meester (1991), whereas animals cultured in the absence of fish kairomones are extremely positively phototactic. The results of De Meester (1993) show that this shift in phototactic behaviour occurs within 24 hours.

Test animals were cultured in 1 l jars in a walk-in culture room kept at 20 °C (range 18–22°C) and provided with a long-day photoperiod (14 h light, 10 h dark). Density was kept low (15–20 adults l^{-1}). Medium consisted of aged tap-water, and one quarter of the medium was refreshed daily. Food level was daily restored to 2.5×10^5 *Scenedesmus acutus* cells ml^{-1}; it thus remained above the incipient limiting level at all times. The unicellular algae were obtained from batch cultures.

The experimental set-up and protocol are modified from previous studies (see description in De Meester, 1991), the main difference being that the experiments in the present study lasted for ten hours. Each experiment was done with different animals, obtained from a different culture or generation, to reduce common environment effects (Lynch & Ennis, 1983). To reduce the impact of an endogenous rhythm on the results (see Ringelberg & Servaas, 1971), all experiments were started between 9 and 10 A.M. The experimental column is a modified version of the one used in previous work, now adapted to enable a flow-through of medium. It consists of an outer and inner column both made of glass. The outer cylinder is about 15 cm high, and has an inlet near the top and an outlet near the bottom. A continuous flow of medium through the column is generated by a peristaltic pump that is connected to the inlet; the outlet is connected to a tube that functions as an overflow, positioned such that the water level remains 2 cm above the top of the inner column. The inner column is 10 cm high and has a fixed bottom and a removable cap, both with a 100 mm plankton gauze. Importantly, the experimental columns are assembled in such a way that the water flow must pass through the inner column, which functions as the observation chamber. Just before an experiment, the combined column is filled with medium to the top of the observation chamber, the experimental animals (10 adult females) are transferred to the chamber, the upper cap is fixed to the inner column, and some medium is gently added to make sure that no air bubbles are entrapped below the plankton gauze. Then, the peristaltic pump is switched on, providing fresh medium at a rate of 18 ml min^{-1} (at this rate, the medium in the experimental column is completely replaced in less than 20 minutes). The observation chamber is externally divided into four compartments of equal height (2.5 cm). Light is provided by a Müster MKL 150 W light source, the glass fibre ending being placed 2 cm above the surface of the water (approximately 4 cm above the top of the observation chamber). Light intensity at the water surface is approximately 500 W m^{-2}. There were two treatments: the behaviour of animals exposed to fish kairomones was compared to that of animals that were not exposed to fish kairomones during the experiment. All animals were cultured in the absence of fish kairomones. Experimental animals were adapted to fresh medium with 1–1.5×10^5 *Scenedesmus* cells ml^{-1} for one hour prior to the experiment. Once transferred to the experimental column, the animals were given a dark adaptation of 20 minutes, after which the light source was switched on. During the observation period of ten minutes, the vertical distribution of the experimental animals was recorded at one-minute intervals. The phototactic behaviour was quantified through a phototactic index, which is modified from De Meester (1991), and is given by the number of animals in the upper 2.5 cm of the inner column minus the number of animals in the lower 2.5 cm of this column, divided by the total number of animals (10). This index ranges from − 1 for extremely negatively phototactic behaviour to + 1 for extremely positively phototactic behaviour. The phototactic index is calculated as the average of the values obtained from the second period of five minutes of each observation period. After the observation period of ten minutes, the light is switched off, and another dark adaptation of 20 minutes is given. In this way, a dark period of 20 minutes and an observation period of ten minutes is alternated during the whole experiment, resulting in one value for phototactic behaviour every 30 minutes. The dark adaptation periods were given to enable us to provide the animals with a similar stimulus every observation period; we only used the value of the phototactic index of the second five minutes of the observation period to enable the animals to exhibit a stable vertical distribution (see De Meester, 1991). Animals in the fish kairomone treatment were exposed to fish kairomones after three hours experimental period (i.e. after six observation periods). Six fish kairomone and five control experiments were carried out, in random order. In all experiments, the medium used in the flow-through system contained $1-1.5 \times 10^5$ *Scenedesmus* cells ml^{-1}. The fish-conditioned medium was prepared by filtering water from a 60 l aquarium with 2 *Leuciscus idus* over 0.45 mm plankton gauze. This medium was prepared just before the start of the experiment. The medium fed to the flow-through system was refreshed every three hours to minimize effects of bacterial degradation of the fish kairomone. The medium used for culturing and preparing fish-conditioned and control medium in experiments consisted of aged tap-water.

In the ten-hour experiments, we expected no difference in phototactic behaviour among fish kairomone and control experiments for the first six observations, as there was no difference in treatment between these two series of experiments during the first three hours. For the subsequent observations, we expected an increasing difference in phototactic behaviour between fish kairomone and control treatments. We randomly paired fish kairomone and control experiments, and calculated the difference between the value

of the phototactic index in the fish kairomone and the control experiment for each time step. As the observed response to the presence of fish kairomones was linear (see Figure 1), we then carried out a linear regression analysis of the difference in value of the phototactic index against time, separately for each pair of experiments, and separately for the time before and after the exposure to fish kairomones (first three hours; remaining seven hours). We then tested whether the mean slope of the replicate experiments differed from zero by a *t*-test. It is expected that the mean slope does not differ from zero during the first three hours, whereas a slope differing significantly from zero is expected for the remaining seven hours of the experiment. In this way, we explicitly test for a significant shift in phototactic behaviour due to a seven-hour exposure to fish kairomones, taking a treatment × time interaction into account.

The results of the ten-hour experiments were compared with those of shorter experiments (involving only one observation period) with animals that were cultured in the presence or absence of fish kairomones their whole life. For the fish kairomone treatment, mothers were transferred to fish-conditioned medium (corresponding to 1 fish in 120 l). Their offspring were used as experimental animals, and were continuously cultured in fish-conditioned medium. Twenty-five percent of the medium was daily refreshed with fish-conditioned tap-water (1 fish in 30 l). These experiments lasted only 15 minutes (five minutes dark adaptation followed by one observation period of ten minutes; see De Meester, 1991), and no flow-through system was established. Six fish kairomone treatment experiments and five control experiments were carried out, each time with different animals.

Results

The results of the ten-hour experiments are presented in Figure 1. As long as no fish kairomones are applied, the animals in the control and fish kairomone treatments behave similarly. As soon as the animals are exposed to fish kairomones, we observe a linear decrease in the value of the phototactic index with time. In the control experiments, the phototactic behaviour remained stable at 0.5–0.7 for the whole experimental period. After ten hours experimental time (seven hours exposure to fish kairomones), the fish-conditioned animals showed an average value of the phototactic behaviour of −0.19. The mean regression slope of the differ-

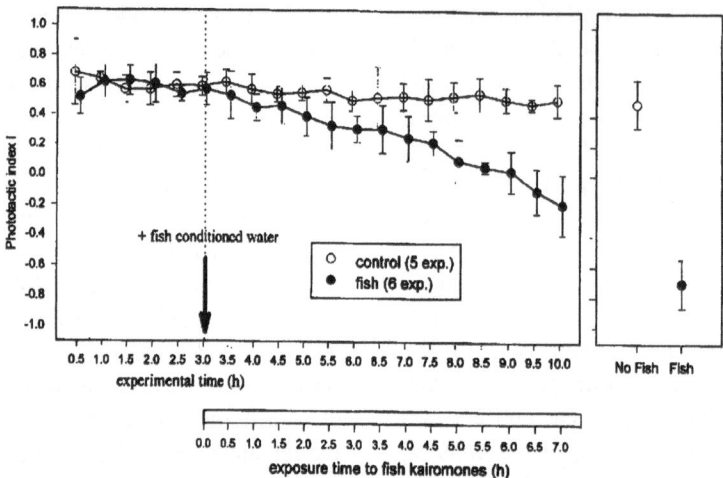

Figure 1. Left panel: The change in the average (± 2 SE) value of the phototactic index I with time in experiments in which the animals are exposed to fish kairomones (full symbols; six experiments) compared to experiments in which the animals are not exposed to fish kairomones (empty symbols; five experiments). In the fish kairomone treatment, the animals were exposed to fish kairomones three hours after the start of the experiment. Experimental time as well as exposure time are indicated. *Right panel*: The average (± 2 SE) value of the phototactic index I of animals that have been cultured in the absence and presence of fish kairomones.

ences between the values of the phototactic index of the fish treatment and control experiments against time was not significantly different from zero for the first three hours (*t*-test, df=5, $p=0.4587$), whereas it was highly significantly different from zero for the seven hours following exposure to fish kairomones (*t*-test, df=5, $p<0.0001$). All linear regressions of the value of the phototactic index against time for the seven hours after exposure to fish kairomones in the fish kairomone experiments revealed a slope significantly different from zero (also after correcting for table-wide errors by sequential Bonferroni, $p<0.05$), with the linear regression explaining 52 to 86% of the variation in the data. After two hours exposure to fish kairomones (five hours experimental time), the phototactic behaviour of the animals in the two treatments was significantly different ($F=5.437$, df=1,9, $p=0.045$). For the control experiments, none of the regressions of the values of the phototactic index against time revealed a slope significantly different from zero ($p>0.05$ in all cases). In the absence of fish kairomones, the animals clearly remained positively phototactic, and the value of the phototactic index ($0.47<I<0.69$) indicates an equally positively phototactic behaviour as that observed in short experiments ($I=0.49$; Figure 1).

Given the linearity of the observed change in phototactic behaviour of fish-exposed animals with time, the time at which a complete induction is obtained can be estimated. The time to full induction is defined as the time needed to reach a value of the phototactic index

similar to that of animals that have been cultured in the presence of fish kairomones during their whole life. In the case of clone $P_1 32,85$, fully induced animals have an average value for the index of phototactic behaviour of -0.70 (2 SE $= 0.18$; Figure 1). Extrapolation of our results yields a time to full induction of approximately 13 hours.

Discussion

A few studies have explicitly dealt with time effects of the induction of more negatively phototactic behaviour in *Daphnia*. De Meester (1993), Loose (1993) and Ringelberg & van Gool (1995) all have provided data that induced animals that are no longer exposed to fish kairomones only gradually drift back to 'normal' behaviour. De Meester (1993) showed that none of 12 different *D. magna* clones drifted back to behaviour similar to that of animals cultured in the absence of fish kairomone within 24 hours after the fish kairomone was removed. It took three to seven days in most clones, and one week after the removal of the stimulus, the behaviour of some clones was still significantly different from that of animals cultured in the absence of fish kairomones. Loose (1993) showed that the diel vertical migration behaviour of a fish-induced *D. galeata × hyalina* clonal population in indoor mesocosms (plankton towers) shifted from a nocturnal migration to a stable epilimnetic distribution in 48 hours. Studying the response to changes in light intensity, Ringelberg & van Gool (1995) observed that fish-induced *D. galeata* × *hyalina* remain sensitized for 5–6 days. Thus, even though these studies focused on different aspects of diel vertical migration behaviour (primary phototaxis, day- and nighttime vertical distribution, secondary phototaxis), they all report essentially similar results. Ringelberg & van Gool (1995) point to the importance of this 'memory' effect, in that it keeps the animals that migrated into the hypolimnion (where there are no fish kairomones) sensitized such that they ascend into the epilimnion at sunset. Furthermore, the time lag also buffers for short term changes in the concentration of fish kairomones associated with the aggregated distribution and shoaling behaviour of fish.

In contrast to the long time lag associated with the change in behaviour when the fish kairomones are removed, daphnids can show a quick change in phototactic behaviour when exposed to the presence of fish kairomones. De Meester (1993) found no difference in behaviour between animals exposed to fish kairomones for 24 h compared to animals exposed to fish kairomones for three to seven days. Indeed, the results of the present experiments suggest that animals show a fully induced behaviour after about 13 h exposure to fish kairomones. Moreover, our results indicate that the response to fish kairomones is initiated almost immediately (in less than 30 minutes) upon exposure, causing a linear decline in the value of the phototactic index I. It took, however, two hours before this change in sensitization to light cues resulted in a significantly altered vertical distribution in the experimental column. Ringelberg & van Gool (1995) report an almost immediate sensitization of a *D. galeata* × *hyalina* clone to the presence of fish kairomones: the responsiveness of the animals to a standard change in light intensity increased dramatically within 15 min upon exposure to fish kairomones. Forward & Rittschof (1993) observed a similarly short response time (<5 min) in their experiments with *Artemia* naupliar larvae.

The value of the phototactic index as determined in our experiments reflects the behaviour of a population (ten adult females). The gradual decrease in the value of the phototactic index for animals that are exposed to fish chemicals can therefore be due to a gradual decrease in the average depth of all individuals (all becoming gradually less positively phototactic) or to a gradual increase in the number of animals that respond to the fish kairomones and become negatively phototactic. To get some insight into which of these two

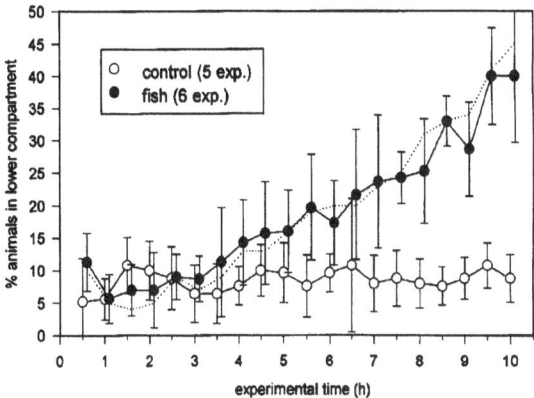

Figure 2. The change in the average (± 2 SE) percentage of animals observed in the lower compartment during the course of experiments in which the animals are exposed to fish kairomones (full symbols; six experiments) and of experiments in which the animals are not exposed to fish kairomones (empty symbols; five experiments). The percentage animals in the lower compartment was determined for the second five minutes of each observation period. In the fish kairomone treatment, the animals were exposed to fish kairomones three hours after the start of the experiment (at 3 h experimental time). The dotted line indicates the expected percentage of animals in the lower compartment for the experiments with exposed animals, given the value of the phototactic index (see Figure 1) and the assumption that the change in value of the phototactic index is entirely due to individuals switching from positively to negatively phototactic behaviour (i.e. without an increase in the number of animals in the middle compartments; the number of animals in the middle compartments was kept constant at the average level observed for animals that were not exposed to fish kairomones).

mechanisms was responsible for the shift in behaviour observed by us, we plotted the change in the percentage of animals that was found in the lower compartment (lower 2.5 cm of the column) during the course of the experiment. It is clear from Figure 2 that the change in the value of the phototactic index can almost completely be explained by the change in the number of animals in the lower compartment rather than by an increase of the number of animals in the middle compartments. This suggests that the change in the value of the phototactic index is largely mediated by a gradual increase in the number of animals that become sensitized by the presence of fish kairomones and turn negatively phototactic.

The time needed for sensitization may be a key factor that differentiates behavioural from morphological and life-history responses to the presence of predator kairomones. Though less well studied, preliminary observations indicate that sensitization of daphnids with respect to 'alertness' (measured, for instance, by the capacity to avoid a pipette handled by an inexperienced human predator) also occurs within a few hours after exposure to fish kairomones (De Meester & Pijanowska, 1997). On the other hand, morphological and life-history responses imply structural modifications or changes in energy allocation patterns, which require more time, typically in the order of days. This has important evolutionary consequences. De Meester (1996), working with *D. magna* clones isolated from three habitats that differed in fish predation pressure, found evidence for local adaptation with respect to the phenotypic plasticity responses in the presence of fish kairomones. The populations studied indeed showed local adaptation related to the induced phenotypic changes in phototactic behaviour rather than to the average values for the trait in the absence of fish kairomones. Using a set of clones isolated from four different habitats that differed in fish predation pressure (including two of the populations studied by De Meester, 1996), Boersma et al. (unpublished data) observed that local adaptation to predation pressure involves mean values of life-history traits (e.g. size at maturity) rather than induced plasticity responses (e.g. the change in size at maturity in the presence of predator kairomones). Though clones isolated from habitats with fish had in general a smaller size at maturity and produced smaller eggs than clones isolated from fishless habitats, clones from all populations responded to the presence of fish kairomones by a similar reduction in the size-related traits. It is conceivable that local adaptation involves mainly plasticity responses only in those cases where the induced shift occurs rapidly. In cases where a longer time lag is involved, the cost of an inducible defence may be too high when there are sudden and pronounced changes in predation pressure in the habitat (see also Padilla & Adolph, 1996). In these populations, it may pay off to be in part constitutively defended: though phenotypic plasticity responses are similar, clones from populations with strong predation pressure are on average more defended than clones from populations with low predation pressure.

Acknowledgments

LDM is a postdoctoral researcher with the National Fund for Scientific Research (Belgium). Part of this study was financially supported by grant VLIM/H/9410 of the Flemish Government. We thank S. Dodson, D. Ebert and E. van Gool for constructive comments on an earlier version of the manuscript, and D. Ebert for very valuable suggestions on the statistical analysis.

References

Adler, F. R. & C. D. Harvell, 1990. Inducible defenses, phenotypic variability and biotic environments. Trends Ecol. Evol. 5: 407–410.

De Meester, L., 1991. An analysis of the phototactic behaviour of *Daphnia magna* clones and their sexual descendants. Hydrobiologia 225: 217–227.

De Meester, L., 1993. Genotype, fish-mediated chemicals and phototaxis in *Daphnia*. Ecology 74: 1467–1474.

De Meester, L., 1996. Evolutionary potential and local genetic differentiation in a phenotypically plastic trait of a cyclical parthenogen, *Daphnia magna*. Evolution 50: 1293–1298.

De Meester, L. & J. Pijanowska, 1997. On the trait-specificity of the response of *Daphnia* genotypes to the chemical presence of a predator. In Lenz, P. H., D. K. Hartline, J. E. Purcell & D. L. Macmillan (eds), Zooplankton: Sensory Ecology and Physiology. Gordon and Breach Publishers, Amsterdam: 407–417.

Forward, R. B. Jr. & D. Rittschof, 1993. Activation of photoresponses of brine shrimp nauplii involved in diel vertical migration by chemical cues from fish. J. Plankton Res. 15: 693–701.

Harvell, C. D., 1990. The ecology and evolution of inducible defenses. Quart. Rev. Biol. 65: 323–340.

Havel, J. E. & S. I. Dodson, 1987. Reproductive costs of *Chaoborus*-induced polymorphism in *Daphnia pulex*. Hydrobiologia 150: 273–281.

Larsson, P. & S. Dodson, 1993. Chemical communication in planktonic animals. Arch. Hydrobiol. 129: 129–155.

Lively, C. M., 1986. Competition, comparative life histories, and maintenance of shell dimorphism in a barnacle. Ecology 67: 858–864.

Loose, C. J., 1993. *Daphnia* diel vertical migration behavior: response to vertebrate predator abundance. Arch. Hydrobiol., Beih. Ergebn. Limnol. 39: 29–36.

Loose, C. J. & P. Dawidowicz, 1994. Trade-offs in diel vertical migration by zooplankton: the costs of predator avoidance. Ecology 75: 2255–2263.

Lüning, J., 1994. Anti-predator defenses in *Daphnia* – are life-history changes always linked to induced neck spines? Oikos 69: 427–436.

Lynch, M. & R. Ennis, 1983. Resource availability, maternal effects and longevity. Exp. Gerontol. 18: 147–165.

Padilla, D. K. & S. C. Adolph, 1996. Plastic inducible morphologies are not always adaptive: the importance of time delays in a stochastic environment. Evol. Ecol. 10: 105–117.

Parejko, K. & S. I. Dodson, 1990. Progress towards characterization of a predator/prey kairomone: *Daphnia pulex* and *Chaoborus americanus*. Hydrobiologia 198: 51–59.

Parejko, K. & S. I. Dodson, 1991. The evolutionary ecology of an antipredator reaction norm: *Daphnia pulex* and *Chaoborus americanus*. Evolution 45: 1665–1674.

Reede, T. & J. Ringelberg, 1995. The influence of a fish exudate on two clones of the hybrid *Daphnia galeata* × *hyalina*. Hydrobiologia 307: 207–212.

Riessen, H. P., 1984. The other side of cyclomorphosis: why *Daphnia* lose their helmets. Limnol. Oceanogr. 29: 1123–1127.

Riessen, H. P. & W. G. Sprules, 1990. Demographic costs of antipredator defenses in *Daphnia pulex*. Ecology 71: 1536–1546.

Ringelberg, J., 1964. The positively phototactic reaction of *Daphnia magna* Straus – a contribution to the understanding of diurnal vertical migration. Neth. J. Sea Res. 2: 319–406.

Ringelberg, J. & H. Servaas, 1971. A circadian rhythm in *Daphnia magna*. Oecologia 6: 289–292.

Ringelberg, J. & E. van Gool, 1995. Migrating *Daphnia* have a memory for fish kairomones. Mar. Freshwat. Behav. Physiol. 26: 249–257.

Tollrian, R. & E. von Elert, 1994. Enrichment and purification of *Chaoborus* kairomone from water: further steps toward its chemical characterization. Limnol. Oceanogr. 39: 788–796.

Weider, L. J. & J. Pijanowska, 1993. Plasticity of *Daphnia* life histories in response to chemical cues from predators. Oikos 67: 385–392.

Hydrobiologia **360**: 177–185, 1997.
A. Brancelj, L. De Meester & P. Spaak (eds), Cladocera: The Biology of Model Organisms.
©1997 *Kluwer Academic Publishers.*

Differential behaviour and shifts in genotype composition during the beginning of a seasonal period of diel vertical migration

Piet Spaak & Joop Ringelberg[1]
EAWAG/ETH, Department of Limnology, Ueberlandstrasse 133, 8600 Dübendorf, Switzerland
[1] *Netherlands Institute of Ecology, Centre for Limnology and University of Amsterdam; Rijksstraatweg 6,*
3631 AC Nieuwersluis, The Netherlands (e-mail: spaak@eawag.ch)

Key words: Daphnia, 0+ perch, genotypes, selective predation, allozyme analysis

Abstract

During the first few weeks of a recurring seasonal period of diel vertical migration in Lake Maarsseveen (The Netherlands), part of the hybrid *Daphnia galeata* × *hyalina* population migrated, while another part remained in the epilimnion. In the epilimnion, 0+ perch prey upon daphnids during daytime. Gradually, the number of adult *Daphnia* in the epilimnion decrease until the epilimnion is nearly devoid of daphnids. The population as a whole may decrease, as in 1991, or may increase as in 1992. Genotype composition, as determined by allozyme analysis, changed substantially within a fortnight in 1992, and one genotype became dominant. Our data are in agreement with the hypothesis that predation on different genotypes (clones) occurs during the beginning of a seasonal period of diel vertical migration, though our data do not allow to exclude alternative explanations.

Introduction

Since the seminal paper by Zaret & Suffern (1976), diel vertical migration (DVM) has been generally considered to be a predator avoidance strategy. Although no definitive statement was made, the suggestion that "only those individuals that exhibited the appropriate behavioural response of vertical migration were able to survive to produce descendants" (p. 811) set the stage for considering genotype selection as a driving force. An equally important paper by Stich & Lampert (1981) demonstrated the existence of species-specific selection. The seasonal increase in migration amplitude in *Daphnia hyalina* was suggested to be generated by "selection pressures eliminating non-migrating animals from the population" (p. 398). By way of precaution, differential responses to changing (non-specified) environmental conditions were added. A third important paper by Gliwicz (1986) seemed to confirm that DVM in zooplankton was a genetically fixed behavioural trait. In an enthusiastic commentary by Huntingford & Metcalfe (1986, p. 682), diel vertical migration was called "a genetically pre-programmed

behavioural change". Since the publication of these papers, phenotypic induction has been demonstrated many times (Ringelberg, 1991b; Loose, 1993; Loose et al., 1993), and some researchers have explicitly recognised that a possible interaction between a genetic component and phenotypic plasticity in DVM behaviour needs to be studied (e.g. in marine habitats: Ohman, 1990; in freshwater systems: Neill, 1992; De Meester, 1993; 1996).

The rapid change from non-migrating to migrating *Daphnia* individuals in Lake Maarsseveen (The Netherlands), which occurs usually within one week, excludes genotype selection as an important factor. The co-occurrence of large shoals of 0+ perch (*Perca fluviatilis*) in the pelagic zone and an experimentally demonstrated enhanced responsiveness to light intensity changes in the presence of an exudate (kairomone) of these perch, made the conclusion of phenotypic induction inevitable (Ringelberg, 1991a; Ringelberg et al., 1991). In several papers, the active role of predator exudates in altering behaviour has been described (De Meester, 1993; Forward, 1993; Loose, 1993; Loose et al., 1993). Now, it is generally accepted that at least some diel vertical migrations are reversible changes in behaviour of individuals and not irreversible changes in genotype. The role of causation by predators is moved to the evolutionary past and selective predation is held to be the ultimate cause of DVM. Ultimate causation is an historical event that can be neither proved nor disproved. However, if a strong trade-off is present, as in DVM, selective predation must be working continually to maintain the necessary traits. Depending on the values of factors that determine progeny, such as food availability and predation rate, the costs of a particular migration might exceed the benefits. The abundance of the migrating genotype(s) will then decrease and other genotypes will take over (Stich & Lampert, 1981; Gliwicz, 1986). For parthenogens with high reproductive rates, this might occur within a short period of time.

Daphnia populations consist of clones representing different genotypes (Lynch & Gabriel, 1983; Hebert, 1987; Lynch, 1987; Spaak, 1996). Some *D. magna* populations consist of genotypes with different (primary) phototactic behaviour (De Meester & Dumont, 1988; De Meester, 1990; 1996) while a genotype-dependent vertical distribution has also been shown for *D. galeata* × *hyalina* hybrids (De Meester et al., 1995). An earlier study on genetic variation in Lake Maarsseveen has indicated an appreciable genetic polymorphism (Spaak & Hoekstra, 1993). Clonal composition may change intra- and interannually, as intraspecific relative fitness may vary. When large numbers of 0+ perch prey upon *Daphnia* (Flik et al., 1997), such severe predation pressure may exert genotype-dependent selection, provided that genotypes differ in characters that determine predation risk. Several field studies (Weider, 1984; Müller & Seitz, 1993; King & Miracle, 1995) have shown indeed that different clones may exhibit different migration patterns. If strong selective predation occurs, the pattern of migration in a population could change rapidly. One could expect that the pattern of change, as well as the accompanying change in population size, varies from year to year, depending on the factors influencing relative fitness and, of course, the genotypic composition of the population. Although several studies addressed the problem of genotype-specific DVM (see above) very little work has been done on genetic shifts during the short period of DVM induction. In the present paper we want to focus on this aspect.

In Lake Maarsseveen, the hybrid *D. galeata* × *hyalina* has become the dominant taxon. Vertical migration occurs during a short period of six to seven weeks, starting in most years at the end of May (Ringelberg et al., 1991). This migration was observed for the first time in 1988. In 1989, a yearly recurring study was started with a weekly sampling program during the crucial migrating period. Day-depth proved to be different from year to year. In the first two years, a large portion of the population stayed within the predation zone, which extended from the water surface to 7.5–8 m. 0+ perch are restricted to this water layer, which makes up the epilimnion (Flik et al., 1997), and we refer to this strata as the predation zone. In 1991, DVM of *D. galeata* × *hyalina* was delayed, indirectly due to a cold spring (Flik & Ringelberg, 1993), but the animals exhibited a much deeper migration. In 1992, a migration amplitude to 1991 was found at the height of the migration period (Ringelberg & Flik, 1994).

In the present paper, we present data on population size and vertical distribution of *D. galeata* × *hyalina* from 1991 and 1992 and on genotype composition of the population in 1992, to gain insight into whether and to what extent a change in genotype frequency contributes to the rapidly changing DVM pattern observed during the first two to three weeks of the seasonal migration period.

Methods

For a physical and chemical description of Lake Maarsseveen, see Swain et al. (1987). Vertical distribution and population size of the *D. galeata* × *hyalina* population were estimated from samples taken at noon and midnight along the length axis of the lake by towing simultaneously, seven (1991) or nine (1992) torpedo nets positioned at equal vertical intervals (2.5 m.). The uppermost net was at a depth of about 1.25 m and the lowest one at 16.25 m in 1991 and at 21.25 m in 1992. Two runs were conducted per sampling bout. Samples were preserved in sugar-formalin and juveniles, adults with eggs and adults without eggs of the hybrid *D. galeata* × *hyalina* were enumerated. Numbers of ind. m^{-2} were calculated for sections of 2.50 m. ($2.5\ m^3$). For the present study, numbers of adults from the three uppermost samples were taken together to represent individuals especially prone to predation by 0+ perch. The adults from all depths below 7.5 m were pooled, and represented individuals that were not exposed to predation by perch during the daytime. Sample sizes ranged from 286 individuals (noon day 184, 1991) to 6431 individuals (night day 161 1992) in both combined samples. The average sample size was 2035 individuals in both samples.

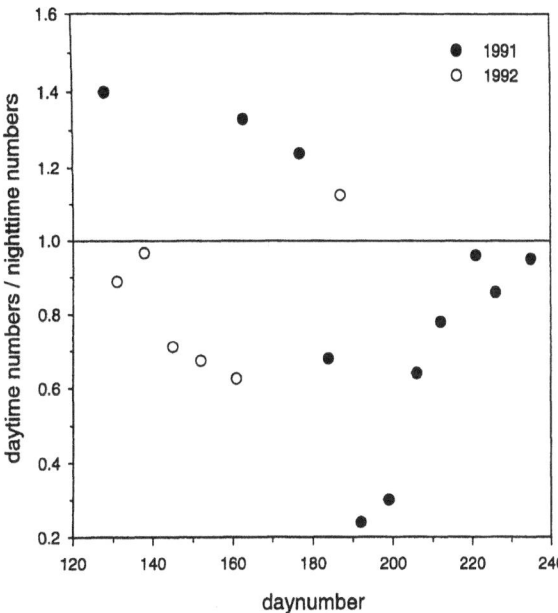

Figure 1. Ratio between total number of adults in the water column around noon and around midnight during a period of DVM in 1991 (closed symbols) and 1992 (open symbols). Day 120 is 30 April. Onset of DVM is on day 184 in 1991 and on day 162 in 1992.

The estimation of population size during a period of diel vertical migration has some problems. As is apparent from Figure 1, the total number of adults, caught at noon during a period of DVM, is often much smaller than the number caught during the night. This is because an increasing number of animals disappear below the depth at which the lowest torpedo net was operating in 1991. In 1992, when two torpedo's were added, more animals were caught during the day than in 1991, resulting in a smaller variation of the day/night ratio of abundance estimates (Figure 1). Although a large part of central Lake Maarsseveen has a depth of 25 m or more, shallower parts along the sampling trajectory prevented the torpedo nets from operating at depths greater than 21.25 m. In fact, the trajectory shown in Figure 1 nicely illustrates the period of DVM and of the change in migration amplitude during this period. Obviously, numbers from night-time sampling represent more closely total population size than those from noon samples. The day-time numbers in the lower portion of the water column (>21.5 m) were corrected by adding the difference between total night-time and daytime numbers. An estimate of the variance in total numbers of adults in the water column was derived from the two daytime and the two night-time sampling series obtained on three pre-migration and three post-migration dates in 1991. Variance is expressed as coefficients of variation (s/$\bar{x}$%) and as standard errors of the mean relative to the mean (s.e./$\bar{x}$%). On average, the resulting values are 21% and 10%, respectively. In 1992, eight samples, taken between 02.00 h and 05.30 h, were also used to estimate the variance. The variance was 20% and the relative standard error was 7%.

We determined clonal composition of the *Daphnia* population in a vertical profile of the lake, by sampling at noon and midnight on 25 May and 10 June 1992. Animals were taken to the laboratory, adult females were frozen in liquid nitrogen and stored for subsequent electrophoretic analysis. Because of the low numbers in some torpedo net samples, we combined samples into five depth categories (0–2.5; 2.5–7.5; 7.5–12.5; 12.5–17.5; 17.5–22.5). Individual adult females were assayed using cellulose acetate electrophoresis following standard methods (Hebert & Beaton, 1989). Generally, 40 animals per depth per sample were assayed for three enzyme loci: phosphoglucomutase (*Pgm*, EC 5.4.2.2), phosphoglucose isomerase (*Pgi*, EC 5.3.1.9) and glutamate oxaloacetate transaminase (*Got*, EC 2.6.1.1). *Got* is a diagnostic marker for distinguishing between *D. galeata* and *D. hyalina* (Wolf & Mort, 1986). *D. galeata* is fixed for the F (fast) allele, while *D. hyalina* is fixed for the S (slow) allele. Interspecific hybrids can be identified as heterozygotes (SF) at *Got*. During the study period, only hybrid *D. galeata* × *hyalina* were found. Variation at the Pgi and Pgm loci made it possible to identify two-locus genotypes. $R \times C$ tests for independence (Sokal & Rohlf, 1995) were used to test if distributions of genotypes were different.

Results

Vertical distributions of *D. galeata* × *hyalina* at noon and midnight for three crucial dates in 1991 are presented in Figure 2. 1991 was exceptional because of a delayed start of the migration period (Flik & Ringelberg, 1993). On 12 June (day number 163), DVM had not yet started since a difference between night-time and daytime relative vertical distributions was not discernible (Figure 2A). Vertical distributions during day and night were completely different when the period of migration was in full progress, as on 3 July (day number 184, Figure 2C). Few individuals were then found in the upper water layers during the day. Adult animals started to arrive in these upper layers about one hour after sunset and had left one hour before sun-

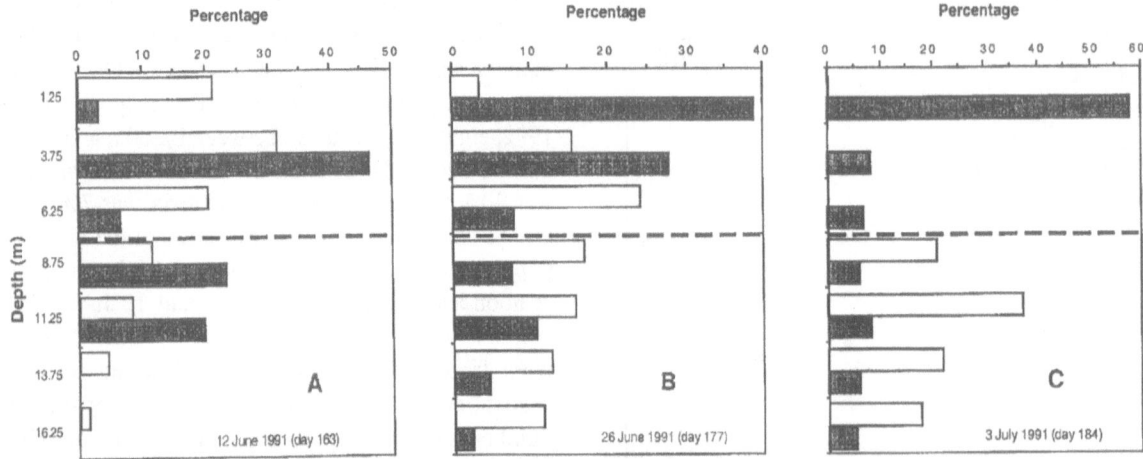

Figure 2. The relative vertical daytime (open bars) and night-time (black bars) distributions of *D. galeata* × *hyalina* in Lake Maarsseveen on three crucial dates in 1991.*

rise (J. Ringelberg, *pers. obs.*). June 26 (day number 177, Figure 2B) is of particular interest because it represents a transient state: part of the population (21%) performed a diel vertical migration, whereas 53% did not leave the epilimnion during the day. Adult numbers had decreased by 63% compared to the night-time estimate obtained two weeks earlier, on the 12th of June (Table 1). However, 26% of the adults were well below the metalimnion (7–9 m) and out of reach of predation during the day. One week later (3 July, day number 184), adult numbers had decreased again and few animals were left in the epilimnion during the day. Migration was at its maximum (Figure 2) and during the subsequent week, the number of adults did not change (Table 1; 11 July, day number 192). Part of the adult population (21–26%) remained in the lower part of the water column.

In 1992, the start of the migration period was at the beginning of June. On the first day of the allozyme analysis (25 May, day number 146), migration had not yet started, but DVM was in full swing during the second sampling period 16 days later (10 June, day number 162; Figure 3). Of the six *Pgi-Pgm* genotypes that were distinguished, two were dominant (MF-SF and MM-MF). We will focus our attention on these two genotypes. Before the migration period started, the genotypic depth distributions were not significantly different between noon and midnight ($R \times C$-test, $G = 9.13$; df = 9; $P = 0.43$; Figure 4). On 10 June, no analysis could be made for the uppermost 2.5 m during daytime, because no *Daphnia* were found. The MF-SF genotype had by far become the most dominant

genotype. The overall abundance of genotype MF-SF changed from 33% to 79% in two weeks. During the night, the MF-SF genotype had a significantly deeper distribution than the MM-MF genotype ($R \times C$-test, G=17.46; df = 4; P<0.002); the distributions were not different during the day.

In 1992, the *Daphnia* population density decreased for three weeks preceding day number 146 and adult numbers were very low on this date, just before the start of the migration period. Population size, however, was certainly underestimated on this date because a heavy *Dinobryon* bloom clogged the sieving surfaces of the torpedo nets. One week later (daynumber 153, 1 June) much higher numbers were found (Table 1). Despite the problematic estimate from the previous week, we conclude that the population had probably increased in size. This increase continued for an additional two weeks. On day number 153 (1 June), part of the population was migrating. Nine days later (day number 162, 10 June) nearly all adults had disappeared from the upper water column and 84% of the adults performed a diel vertical migration.

Discussion

In this study, we analysed the vertical distribution and genotype compositions in *D. galeata* × *hyalina* during the first weeks of a seasonal period of DVM. Monitoring of the population during the start of a seasonal period of migration is crucial for detecting the impact of predation by 0+ perch on the genetic structure of the

Table 1. Adult numbers in upper (U 0 to 7.5 m) and lower (L 7.5 to 16.25 (1991) or 21.25 (1992) m.) part of the water column and the total number (T) of adults in 1991 and 1992. When the number of adults in daytime samples was considerable smaller than that of night samples, it is presumed that this is because part of the animals reside lower than the deepest sampling depth during the day. The corrected number for the lower part of the water column and the corrected total are given between brackets. The percentage of adults that is supposed to be stationary in the upper and lower part, as well as the percentage of migrating animals are mentioned in the third column. * Numbers on day 146, 1992 are underestimated, see text

Day number	Number of adults				Type of behaviour	
	Daytime		Nighttime			
163, 1991	U:	50001	U:	24564		
	L:	20634	L:	28305		
	T:	70635	T:	52869		
177, 1991	U:	10483	U:	14536	53%	upper
	L:	13768	L:	4979	26%	lower
	T:	24251	T:	19515	21%	DVM
184, 1991	U:	105	U:	7093	1%	upper
	L:	6380 (9540)	L:	2552	26%	lower
	T:	[9645]	T:	9645	73%	DVM
192, 1991	U:	17	U:	7792	0.2%	upper
	L:	2464 (9837)	L:	2062	21%	lower
	T:	[9854]	T:	9854	78.8%	DVM
139, 1992	U:	38393	U:	39390		
	L:	9180	L:	9632		
	T:	47573	T:	49022		
146, 1992	U:	5479	U:	932		
	L:	2154	L:	4000		
	T:	7633*	T:	8932*		
153, 1992	U:	4566	U:	25786	6%	upper
	L:	48336 (68222)	L:	47002	65%	lower
	T:	[72788]	T:	72788	29%	DVM
162, 1992	U:	366	U:	118424	1%	upper
	L:	118349 (139846)	L:	21788	15%	lower
	T:	[140212]	T:	140212	84%	DVM

Daphnia population in Lake Maarsseveen. In order to detect selective predation in the field, a prey population must consist of different genotypes, that exhibit heritable behavioural differences. In Lake Maarsseveen, different genotypes were present in the *Daphnia* population in 1991 (Spaak & Hoekstra, 1993) and 1992 (this study). Furthermore, life table studies on Maarsseveen clones (Reede, 1995) have shown that genetic variation in life-history traits is available in this population.

Before the annual period of DVM started, daytime and night-time vertical distributions of the adults were the same in both years (Figure 2A and Figure 3A). This also holds for the vertical distribution of the six genotypes that could be differentiated, by allozyme

analysis in 1992 (Figure 4). As soon as 0+ perch attain a length of about 15 mm (Flik et al., 1997), or the total biomass of these fish surpasses a certain value (Ringelberg et al., 1997) (both variables are coupled), the number of *Daphnia* individuals remaining in the upper part of the water column during the day, starts to decrease (Table 1). This decrease has a two-fold origin. Firstly, *Daphnia* are eaten by 0+ perch that forage at high light intensities present in the upper part of the water column in daytime. Secondly, the *Daphnia* start to migrate. The latter is deduced from the fact that density shifts occur at sunrise and sunset. Since some adult *Daphnia* migrate and some do not migrate on a particular day, behavioural variability exists. The

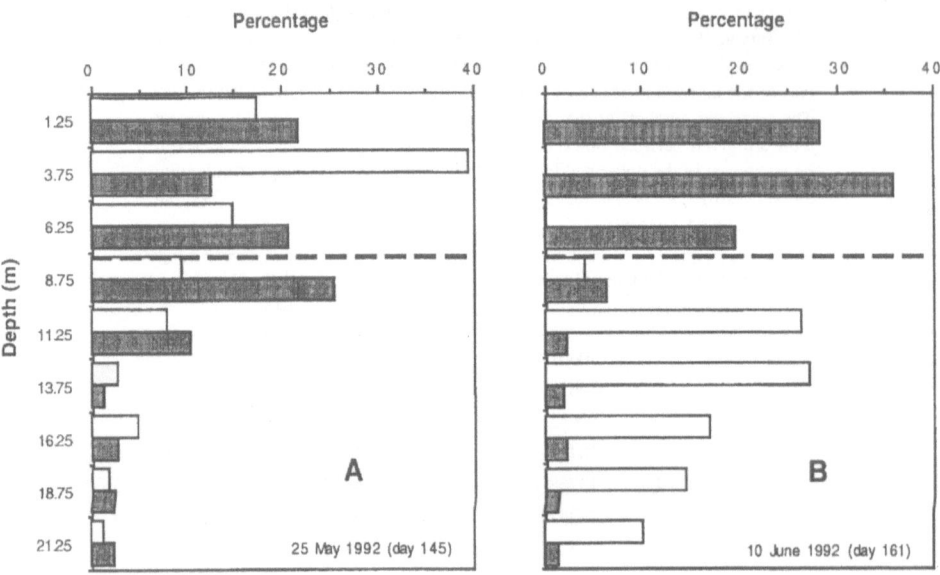

Figure 3. The relative vertical daytime (open bars) and night-time (black bars) distributions of *D. galeata* × *hyalina* in Lake Maarsseveen on two dates in 1992. On both dates allozyme analyses were performed.

crucial point is, however, whether the migrating and non-migrating groups consist of different genotypes. Recently, De Meester et al. (1995) showed experimentally that genotypic variation exists for daytime vertical distribution of *D. galeata* × *hyalina* hybrids. In field studies, clonal variation for DVM behaviour in *Daphnia* has been shown (Weider, 1984; Müller & Seitz, 1993; King & Miracle, 1995). In 1992, during the period the epilimnion became devoid of adult *Daphnia*, the genotypic composition changed substantially: relative numbers of the MF-SF genotype increased while those of the MM-MF genotype decreased.

One might argue that this genotypic shift reflects differential behaviour by different genotypes. The decrease in the abundance of the MM-MF genotype would be explained by differential predation by perch (if this genotype migrates less) while the MF-SF genotype, which supposedly migrated to a greater extent, would have been less vulnerable to predation. If this interpretation of the data is correct, then a different probability of predation mortality caused a shift in the abundance of prey genotypes. However, alternative explanations are possible. For example, vertical migration (to the same extent by both genotypes) is accompanied by changes in environmental conditions such as food availability, temperature, light regime, etc. The new set of conditions may have promoted the relative competitive ability of the MF-SF geno-

type. The existence of interclonal differences in life histories (Reede & Ringelberg, 1995) within the *D. galeata* × *hyalina* population of Lake Maarsseveen may support this hypothesis. Under this scenario too, the presence of predators is the driving force behind the shift in clonal composition. The impact of fish is, however, indirect (a shift in clonal superiority associated with a shift in depth selection behaviour) rather than direct (genotype-dependent fish predation). The time needed for differential competition to become effective is, however, longer than for genotype-dependent predation. Whether the two weeks between our allozyme analyses suffice to bring about the large shift in genotype composition, is thus questionable. Also, these shifts as observed in the field can never be conclusive evidence for genotype-specific selective predation. Proof might be obtained by comparing the genotype composition of *Daphnia* in the stomachs of fish with that in the lake, similar to what Stich & Lampert (1981) did for the case of differential migration of two species. In their study, Stich & Lampert (1981) found more non-migrating *D. galeata* than migrating *D. hyalina* in fish stomachs compared to the relative abundance of these two species in Lake Constance.

In 1991, the density of the *Daphnia* population decreased during the first weeks of the migration period, while in 1992, the population increased (Table 1). This illustrates a point raised in the introduction:

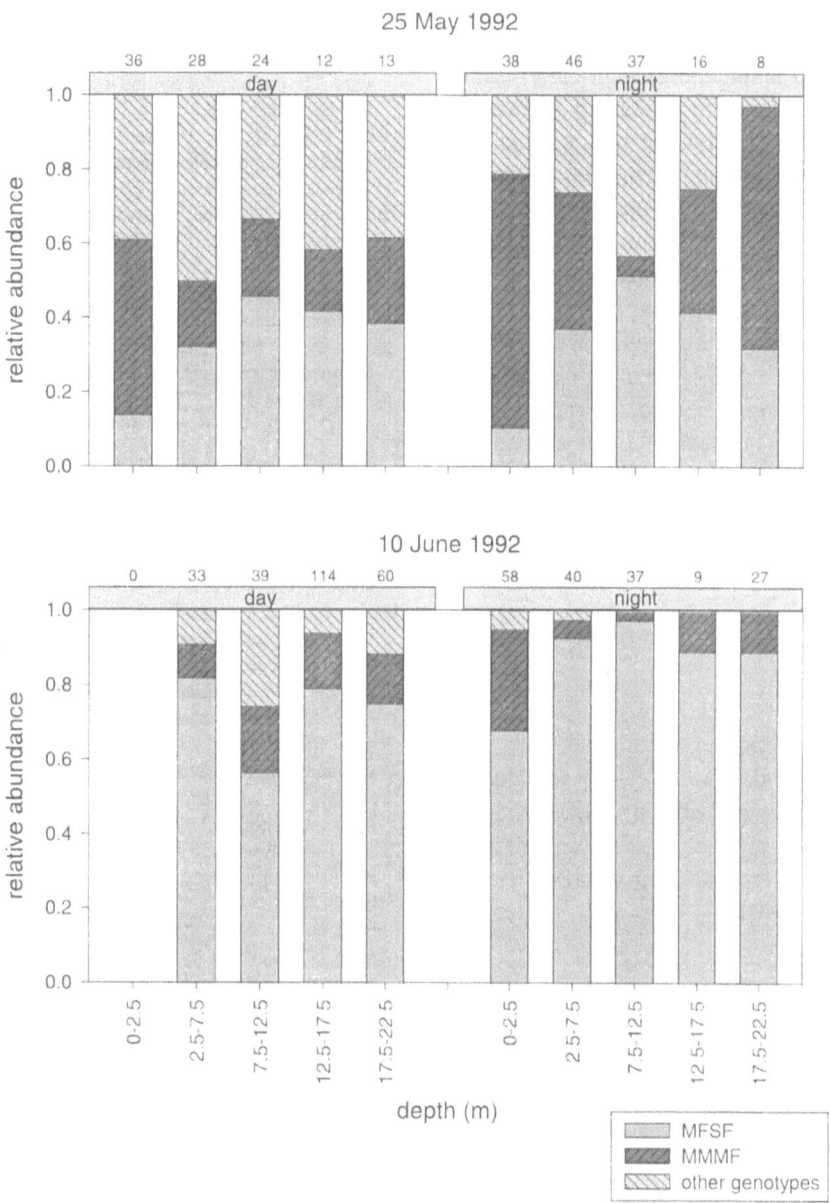

Figure 4. Relative abundance of the two most abundant *Pgi-Pgm* genotypes just before and after the beginning of the seasonal period of DVM in 1992 by *D. galeata* × *hyalina* in Lake Maarsseveen. All other genotypes (MMSF, MFMF, MMFF, MMMM) are lumped together. Numbers above the bars indicate sample sizes.

the change in population sizc, accompanying DVM, may differ from year-to-year, depending on progeny-determining factors and on the genotypic composition of the population. Diel vertical migration decreases the probability of being eaten, but the animals are still prone to predation when they arrive in the predation zone after sunset, or when they remain too

long in thc cpilimnion at dawn. Ycar-to-ycar diffcrences in phototactic behaviour exist (Ringelberg & Flik, 1994), and an increase in the amplitude of DVM was observed from 1989 to 1991 (Ringelberg & Flik, 1994). These differences may be genotype-specific and must be examined in future research.

184

Although genotype-specific predation was not conclusively shown to be operating in our study, our data are suggestive, and it may be (indirectly) responsible for the increased amplitude of migration during the first weeks of the migration period. An alternative hypothesis to explain the rapid shift in migration amplitude at the start of the migration period was also been proposed, i.e. that of an increased phenotypic induction caused by increasing fish kairomone concentrations (Ringelberg et al., 1997). During the first weeks of the migration period, the biomass of the 0+ perch gradually increases due to individual growth. Loose (1993) observed that the day-depth of an experimental *D. galeata* × *hyalina* population increased with increasing kairomone concentration as determined by the number of fish used to condition the medium. In Lake Maarsseveen, it is of course conceivable that both genotypic selection and phenotypic induction operate simultaneously.

Acknowledgements

The critical but constructive remarks of Luc De Meester and Larry Weider on an earlier version of this manuscript are highly appreciated. The Beyerinck-Popping Fund of the Royal Dutch Academy of Arts and Sciences financed the allozyme electrophoresis.

References

De Meester, L., 1990. Evidence for intra-population genetic variability for phototactic behaviour in *Daphnia magna* Straus, 1820. Biol. Jb Dodonaea 58: 84–93.

De Meester, L., 1993. Genotype, fish-mediated chemicals, and phototactic behavior in *Daphnia magna*. Ecology 74: 1467–1474.

De Meester, L., 1996. Evolutionary potential and local genetic differentiation in a phenotypically plastic trait of a cyclical parthenogen, *Daphnia magna*. Evolution 50: 1293–1298.

De Meester, L. & H. J. Dumont, 1988. The genetics of phototaxis in *Daphnia magna*: Existence of three phenotypes for vertical migration among parthenogenetic females. Hydrobiologia 162: 47–55.

De Meester, L., L. J. Weider & R. Tollrian, 1995. Alternative anti-predator defences and genetic polymorphism in a pelagic predator-prey system. Nature 378: 483–485.

Flik, B. J. G., D. Aanen & J. Ringelberg, 1997. The extent of predation by juvenile perch during diel vertical migration of *Daphnia*. Arch. Hydrobiol. Beih. Ergebn. Limnol. 49: 51–58.

Flik, B. J. G. & J. Ringelberg, 1993. Influence of food availability on the initiation of diel vertical migration (DVM) in Lake Maarsseveen. Arch. Hydrobiol. Beih. Ergebn. Limnol. 39: 57–65.

Forward, R. B., 1993. Photoresponses during diel vertical migration of brine shrimp larvae: effect of predator exposure. Arch. Hydrobiol. Beih. Ergebn. Limnol. 39: 37–44.

Gliwicz, M. Z., 1986. Predation and the evolution of vertical migration in zooplankton. Nature 320: 746–748.

Hebert, P. D. N., 1987. Genotypic characteristics of cyclic parthenogens and their obligately asexual derivates. In S. Stearns (ed.), The evolution of sex and its consequences, Birkhäuser Verlag, Basel: 175–195.

Hebert, P. D. N. & M. J. Beaton, 1989. Methodologies for allozyme analysis using cellulose acetate electrophoresis. Helena Laboratories Beaumont, Texas, 32 pp.

Huntingford, F. A. & N. B. Metcalfe, 1986. The evolution of anti-predatory behaviour in zooplankton. Nature 320: 682.

King, C. E. & M. R. Miracle, 1995. Diel vertical migration by *Daphnia longispina* in a Spanish lake: Genetic sources of distributional variation. Limnol. Oceanogr. 40: 226–231.

Loose, C. J., 1993. *Daphnia* diel vertical migration behavior: Response to vertebrate predator abundance. Arch. Hydrobiol. Beih. Ergebn. Limnol. 39: 29–36.

Loose, C. J., E. Von Elert & P. Dawidowicz, 1993. Chemically-induced diel vertical migration in *Daphnia* – a new bioassay for kairomones exuded by fish. Arch. Hydrobiol. 126: 329–337.

Lynch, M., 1987. The consequences of fluctuating selection for isozyme polymorphisms in *Daphnia*. Genetics 115: 657–669.

Lynch, M. & W. Gabriel, 1983. Phenotypic evolution and parthenogenesis. Am. Nat. 122: 745–764.

Müller, J. & A. Seitz, 1993. Habitat partitioning and differential vertical migration of some *Daphnia* genotypes in a lake. Arch. Hydrobiol. Beih. Ergebn. Limnol. 39: 167–174.

Neill, W. E., 1992. Population variation in the ontogeny of predator-induced vertical migration of copepods. Nature 356: 54–57.

Ohman, M. D., 1990. The demographic benefits of diel vertical migration by zooplankton. Ecol. Monogr. 60: 257–281.

Reede, T., 1995. Life history shifts in response to different levels of fish kairomones in *Daphnia*. J. Plankton Res. 17: 1661–1667.

Reede, T. & J. Ringelberg, 1995. The influence of a fish exudate on two clones of the hybrid *Daphnia galeata* × *hyalina*. Hydrobiologia 307: 207–212.

Ringelberg, J., 1991a. Enhancement of the phototactic reaction in *Daphnia-hyalina* by a chemical mediated by juvenile perch (*Perca-fluviatilis*). J. Plankton Res. 13: 17–25.

Ringelberg, J., 1991b. A mechanism of predator-mediated induction of diel vertical migration in *Daphnia hyalina*. J. Plankton Res. 13: 83–89.

Ringelberg, J., B. G. J. Flik, D. Lindenaar & K. Royackers, 1991. Diel vertical migration of *Daphnia hyalina* (sensu latiori) in Lake Maarsseveen: Part: 1. Aspects of seasonal and daily timing. Arch. Hydrobiol. 121: 129–145.

Ringelberg, J. & B. J. G. Flik, 1994. Increased phototaxis in the field leads to enhanced diel vertical migration. Limnol. Oceanogr. 39: 1855–1864.

Ringelberg, J., B. J. G. Flik, D. Aanen & E. Van Gool, 1997. Amplitude of diel vertical migration (DVM) is a function of fish biomass, a hypothesis. Arch. Hydrobiol. Beih. Ergebn. Limnol. 49: 71–78.

Sokal, R. R. & F. J. Rohlf, 1995. Biometry. W.H. Freeman & Co, San Francisco USA, 887 pp.

Spaak, P., 1996. Temporal changes in the genetic structure of the *Daphnia* species complex in Tjeukemeer, with evidence for backcrossing. Heredity 76: 539–548.

Spaak, P. & J. R. Hoekstra, 1993. Clonal structure of the *Daphnia* population in Lake Maarsseveen: its implications for diel vertical migration. Arch. Hydrobiol. Beih. Ergebn. Limnol. 39: 157–165.

Stich, H. B. & W. Lampert, 1981. Predator evasion as an explanation of diurnal vertical migration by zooplankton. Nature 293: 396–398.

Swain, W., R. Lingeman & F. Heinis, 1987. A characterization and description of the Maarsseveen lake system. Hydrobiol. Bull. 21: 5–16.

Weider, L. J., 1984. Spatial heterogeneity of *Daphnia* genotypes: Vertical migration and habitat partitioning. Limnol. Oceanogr. 29: 225–235.

Wolf, H. G. & M. A. Mort, 1986. Interspecific hybridization underlies phenotypic variability in *Daphnia* populations. Oecologia 68: 507–511.

Zaret, T. M. & J. S. Suffern, 1976. Vertical migration in zooplankton as a predator avoidance mechanism. Limnol. Oceanogr. 21: 804–813.

Hydrobiologia **360**: 187–196, 1997.
A. Brancelj, L. De Meester & P. Spaak (eds), Cladocera: The Biology of Model Organisms.
©1997 Kluwer Academic Publishers.

Size distribution of *Daphnia longispina* in the vertical profile

M. D. Boronat & M. R. Miracle

Department of Microbiology and Ecology, University of Valencia, 46100 Burjassot (Valencia), Spain

Key words: Daphnia longispina, toothed morph, size distribution, vertical migration, predation

Abstract

D. longispina of the meromictic lake El Tobar is a round-headed form. It never has a helmet, but in summer a small proportion of immature individuals (0.9–1.2 mm females and males) have one or two neck teeth. The size structure of this *Daphnia* population, as well as the vertical distribution and migration of different size-classes, were studied in September and November of 1991 and April of 1992. The large variation in mean size and size at first reproduction, as well as the occurrence of different patterns of vertical migration are interpreted as responses to different predator situations. At the end of April, when *Daphnia* mortality by visually hunting predators is dominating, a typical nocturnal migration is adopted and size distribution is biased to smaller size classes. In November, when mortality is mainly attributed to the nocturnally migrating *Chaoborus*, *Daphnia* shows a reversed migration pattern. In September, when the population of *Daphnia* is responding to both visual (fish) and non-visual predators *(Chaoborus)*, it adopts a pattern of twilight migration. The presence of neck teeth in vulnerable size classes in September might be an additional adaptive response to *Chaoborus* predation. In September, the size structure of the *Daphnia* population is shifted to larger classes and the vertical distribution of size classes shows a pronounced segregation between juveniles and adults. Juveniles are found closer to the surface, while adults dwell predominantly in the rich, deep waters near the oxicline. This suggests that an additional advantage of the ascent of the adult *Daphnia* exploiting those deep resources is the release of young in more oxygenated and warmer waters. The *Daphnia* population of lake El Tobar is known to be clonally diverse, and the changing frequency of genotypes could play an important part in the observed seasonal differences in behaviour and size.

Introduction

In studies of planktonic cladocerans a succession of related species, hybrids, or 'forms' within a species has often been described. Such shifts in relative abundance of related morphs are to be expected in organisms whose life span is shorter than the periods of environmental change, and particularly when the species reproduce by ameiotic parthenogenesis. Moreover, cladocerans show differential depth distributions and vertical migrations, and the coexistence of genotypes with different vertical distributions and vertical migration patterns has been made evident in different ways (Weider, 1984; De Meester, 1993,1994a; Spaak & Hoekstra, 1993).

Daphnia longispina is a major species in the plankton of Lake El Tobar, a karstic lake in the Iberic moun-

tain ridge (Spain). Using electrophoretic analysis of isoenzymes, it has been shown that both allele and clone frequencies of this *Daphnia* population undergo substantial seasonal variation (King et al., 1995) and that a segregation of genotypes in the vertical profile occurred during stratification (King & Miracle, 1995). It was also made evident that these genotypes showed different patterns of vertical migration. The observation that the *Daphnia* population is genetically heterogeneous raises the question on how this genetic polymorphism is maintained. Is there any important morphometric variation associated with the genotypes that are segregated seasonally or in the vertical profile? Since body size seems to be a major adaptive factor in zooplankton populations (Hall et al., 1976), is there variation in size structure of the *Daphnia* population with season or depth? The meromictic condi-

188

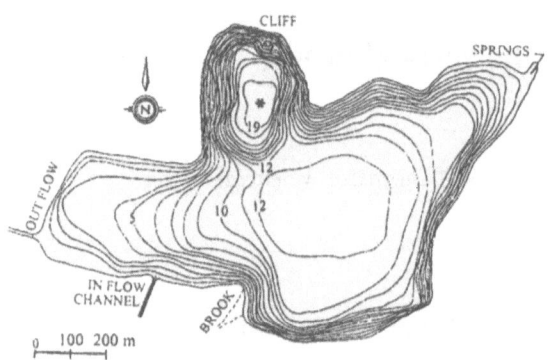

Figure 1. Bathymetric map of lake El Tobar. The sampling point located in the centre of the small meromictic basin is indicated by *.

tion of Lake El Tobar provides the constant presence of an anoxic refuge for *Chaoborus* larvae, thus permitting the coexistence of both fish and *Chaoborus*, which are both potential predators of *Daphnia*. Diel vertical migrations are known to be behavioural adaptive responses to different types of predation (Ohman, 1990). Are there seasonal shifts in vertical migration patterns of *Daphnia* that could be associated with visual and/or non-visual predation pressure? The aim of the present study is to try to answer these questions. We studied the size structure of the *D. longispina* of Lake El Tobar in different seasons, and tried to interpret the differences in distribution and migration of the different size classes along the vertical profile. The variation in size distribution, albeit shaped by food concentration, is expected to be highly modulated by size-dependent mortality, which becomes heavier at opposite ends of the body length range depending on which type of predators dominates the system.

Study site and methods

El Tobar is a solution lake in the karstic Cuenca mountains (Spain, UTM 30TWK806888) at an altitude of 1250 m. The morphometry, hydrology and chemistry of this lake is described in Vicente et al. (1993). It consists of two main basins with a total surface area of 67 ha. Samples were taken in the smaller basin, which is quite isolated from the rest of the lake (Figure 1). It is a circular sink hole of 10 ha, deeper than the main basin, with a maximum depth of 19.5 m and with steep walls. Because it protrudes laterally, it is usually out of the main water flow along the elongated main basin and it is also more sheltered from wind action,

since it lies at the base of a steep cliff. This small limnocrene originated from the general limestone collapse which made the main basin, but also from the additional disolution of a salt-rock stratum underlying the limestone in that area. For this reason, the small basin is subject to crenogenic meromixis. The vertical structure of the meromictic basin of lake El Tobar is thus marked by a stable salt water layer from a depth of 12 m to the bottom. This monimolimnion is anoxic, has a quite constant temperature around 13.5°C and differs in many other characteristics from the mixolimnion (see Vicente et al., 1993). The mixolimnion of this lake is oligotrophic. Information on the composition of the phytoplankton and zooplankton communities of this lake, based on samples taken on the same dates as those of the present study, can be found in Miracle et al. (1993).

The El Tobar *Daphnia* has been identified as *D. longispina* by its morphology (Pejler, 1973; Christie, 1983; Negrea, 1983; Margaritora, 1985; Flössner & Kraus, 1986; Glagolev, 1986; Hrbácek, 1987) and because of its resemblance to other Spanish populations referred to as *D. longispina* (Margalef, 1953; Miracle, 1978; Alonso, 1996). The El Tobar form is colourless, has a concave frons with a prolonged pointed rostrum and a long and flat 'antennule mound'. The highest point of the head is near the eye chamber. It is a round-headed form that never has a helmet, but it has a more or less conspicuous low keel which stretches along the dorsal side of the head and carapace. The abdominal processes are quite long, the first one being almost twice as long as the second.

All samples were taken in the centre of the meromictic basin as described in King et al. (1995). The position of the collecting boat was tightly fixed using ropes attached to three points on the shore. Vertical profiles of temperature, conductivity, oxygen and light penetration were measured *in situ* with the appropriate sensors (WTW and Li-Cor instruments). Vertical profiles of zooplankton (in general, at 1 m intervals) were taken both during the day (around midday) and night (around midnight) on November 19 of 1991 and April 22 of 1992. On September 21–23 of 1991, four profiles were made, two at midday on different days, one at midnight and another just before sunset. Zooplankton was filtered *in situ* from water samples taken with a transparent double Van Dorn bottle (5.4 l capacity and 35 μm mesh) in September, when densities were high, and a Patalas trap (25 l capacity and 100 μm mesh) in November and April. In April, samples were also taken with the Van Dorn bottle at midday. T-tests

Table 1. Some features of lake El Tobar environmental conditions and *Daphnia* populations at different times of the year. SD = Standard deviation

	SEP-91	NOV-91	APR-92
Secchi (m)	8	3.2	2
Irradiance			
$10\ \mu E\ m^{-2}\ s^{-1}$ (depth, m)[1]	11.7	9	6
$1\ \mu E\ m^{-2}\ s^{-1}$ (depth, m)[1]	12.8	12	9.3
Conductivity (mS cm^{-1})			
0–9 m	0.6	0.5	0.5
10 m	1.0	0.5	0.5
11 m	2.1	0.8	3.7
Mean density of *Daphnia* (ind l^{-1})[2]	10 (0–11)	1.4 (0–11)	3 (0–6)
Mean size at first reproduction (mm)	1.48	1.35	1.35
% ovigerous ♀♀ / adult ♀♀	12	37	49
Mean size ± 1 SD			
Total population	1.29 ± 0.35	0.96 ± 0.33	0.93 ± 0.29
Ind. < 1.2 mm	1.05 ± 0.17	0.86 ± 0.18	0.86 ± 0.14
Ind. > 1.2 mm	1.66 ± 0.20	1.65 ± 0.22	1.55 ± 0.14
Mean brood size ± 1 SD	2.20 ± 0.40	3.60 ± 1.10	4.20 ± 1.60

[1] Measured on a cloudless day.

[2] Depth range (m) over which animals occur given in parentheses.

performed on the means of *Daphnia* densities obtained with the Van Dorn and Patalas trap yielded no significant differences, nor did t-tests for paired comparisons over the vertical profile (i.e. depth samples taken with Van Dorn and Patalas, arranged as paired observations).

Formalin preserved samples were counted under an inverted microscope. Body size, as indicated in Figure 2 and Table 1, was measured to the nearest 24.5 μm in all *Daphnia* individuals. Mean size at first reproduction is determined following Lampert (1988), as the size class in which more than 50% of the maximum percentage of ovigerous females is reached.

Results

The morphometric ratios shown in Figure 2 all lie within the range given for *D. longispina* (e.g. in Margaritora, 1985). Although we did not find evidence for cyclomorphosis in the relative length of the head, we did observe the occurrence of a morph with one or two neck teeth, which in other species of *Daphnia* has been related to a defense mechanism against *Chaoborus* predation (Havel & Dodson, 1984; Brancelj et al., 1996). The toothed specimens were found in very low proportions in September, and were restricted to specific size classes. In females, 2.5% of the individuals of the 1.1–

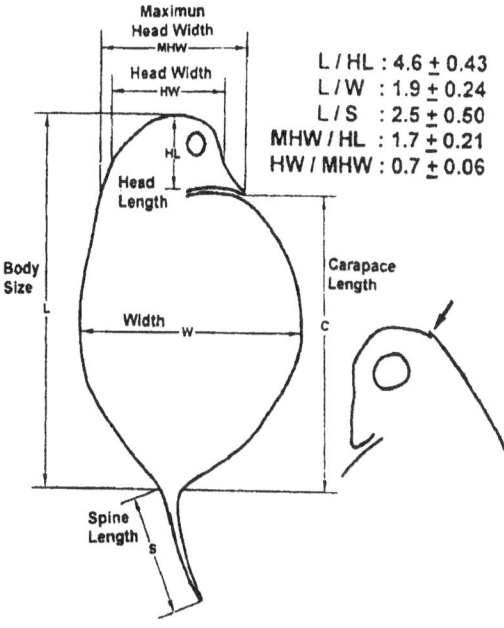

Figure 2. Schematic view of a *Daphnia longispina* adult with mean values of main biometric ratios (means of average values for adults on 11 sampling dates and their standard deviations, total number of measured individuals: 1000). The head of a toothed morph of *D. longispina* is also shown; the 'tooth' is only present in juveniles.

1.2 mm size class were toothed. Only very few toothed individuals were found in other size classes, and none

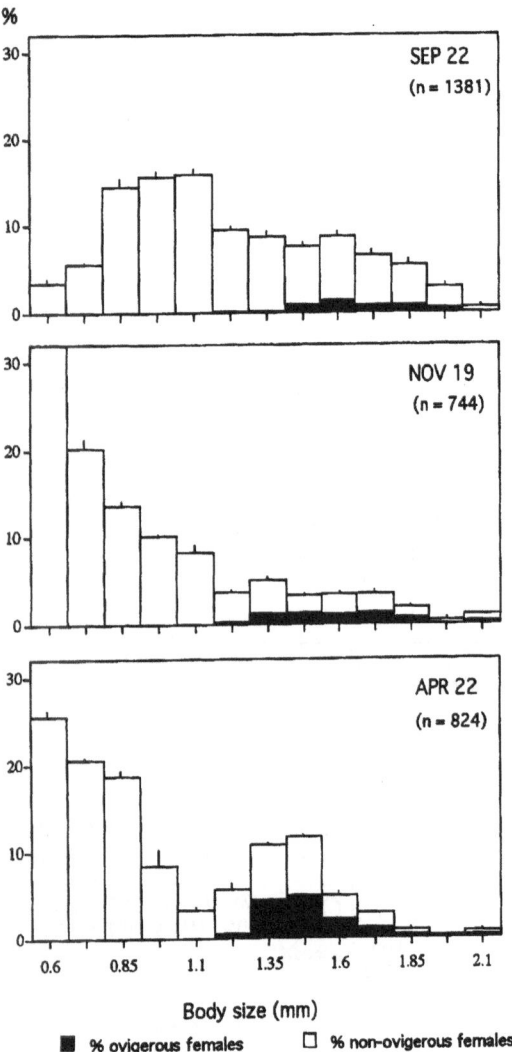

%

SEP 22
(n = 1381)

NOV 19
(n = 744)

APR 22
(n = 824)

0.6   0.85   1.1   1.35   1.6   1.85   2.1

Body size (mm)

■ % ovigerous females   □ % non-ovigerous females

Figure 3. Frequency distributions of *D. longispina* body size for each sampling date. *n* = number of measured individuals; size-classes are indicated by the smallest size. Small vertical lines indicate 1 standard deviation expressing variation among the different vertical profiles (four in September and two in November and April).

Table 2. Results of several one-way ANOVAs for each collection in the vertical profile. ANOVA table testing for the effect of sampling depth on body size in samples taken at different depths. Kendall's τ correlation between mean size and depth is also indicated

| Date, hour | Size | | | |
	F	D.F.	p	τ
21-Sep-91, 17 h.	12.03*	(8, 435)	$0.2 \cdot 10^{-11}$	0.81*
22-Sep-91, 12 h.	14.65*	(8, 322)	$0.2 \cdot 10^{-11}$	0.75*
22-Sep-91, 24 h.	22.93*	(9, 347)	$0.1 \cdot 10^{-11}$	0.85*
23-Sep-91, 12 h.	26.35*	(8, 393)	$0.2 \cdot 10^{-11}$	1.00*
19-Nov-91, 12 h.	3.22*	(11, 361)	$0.3 \cdot 10^{-4}$	–0.06
19-Nov-91, 24 h.	5.49*	(11, 357)	$0.4 \cdot 10^{-8}$	0.60*
22-Apr-92, 12 h.	6.96*	(10, 396)	$0.4 \cdot 10^{-10}$	0.45*
22-Apr-92, 24 h.	1.02	(10, 361)	0.4	–0.09

* Significant at the 5% significance level after sequential Bonferroni correction.

of the smallest *Daphnia* individuals (0.6 mm) showed a dorsal angle or very tiny spines.

Size structure

The size structure of the population showed important differences among samples taken at different seasons (Figure 3; Table 1). At the end of summer (September), when the population density is high, the proportion of adults is very high but egg production is very low. The mean size at first reproduction (1.5 mm) as well as the mean size of the population are large. The proportion of the smallest size classes (0.6–0.7 mm) was very low. The more productive and less competitive conditions during the mixing periods gave rise to another size structure, with a high fecundity, a high proportion of juveniles and a smaller size at first reproduction. After the autumn overturn (November), the population was predominantly composed of young individuals, with a high proportion of small size classes, and the mean fecundity was higher than in September. Dominance by small individuals was also observed during early spring, but the relative abundance of ovigerous females and the mean brood size was much larger than in the September and November samples.

The vertical distribution of four different size classes are plotted in Figure 4. In all cases large females occurred in deeper waters than juveniles. However, during the day, all size classes clearly avoided the surface on all collection dates. In September, when the lake was stratified, the *Daphnia* population showed a bimodal distribution with peaks at 5 and 10 m of depth. In November and April the *Daphnia* were main-

outside the range of 0.9–1.3 mm. Among males, 5% of the individuals were toothed and they too had body sizes ranging from 1.0 to 1.2 mm. Males were not very frequent (at most 2% of the population) and their size, in our samples, ranged from 0.6 to 1.35 mm. The smallest size classes (0.6–0.7 mm) of both males and females showed sometimes another head shape with a scarcely noticeable dorsal angle, similar to the description given by Brancelj et al., 1996. This morph was, however, extremely rare (<0.1%). In the collections taken in November and April, no toothed specimens were found, but at the end of May, a small proportion

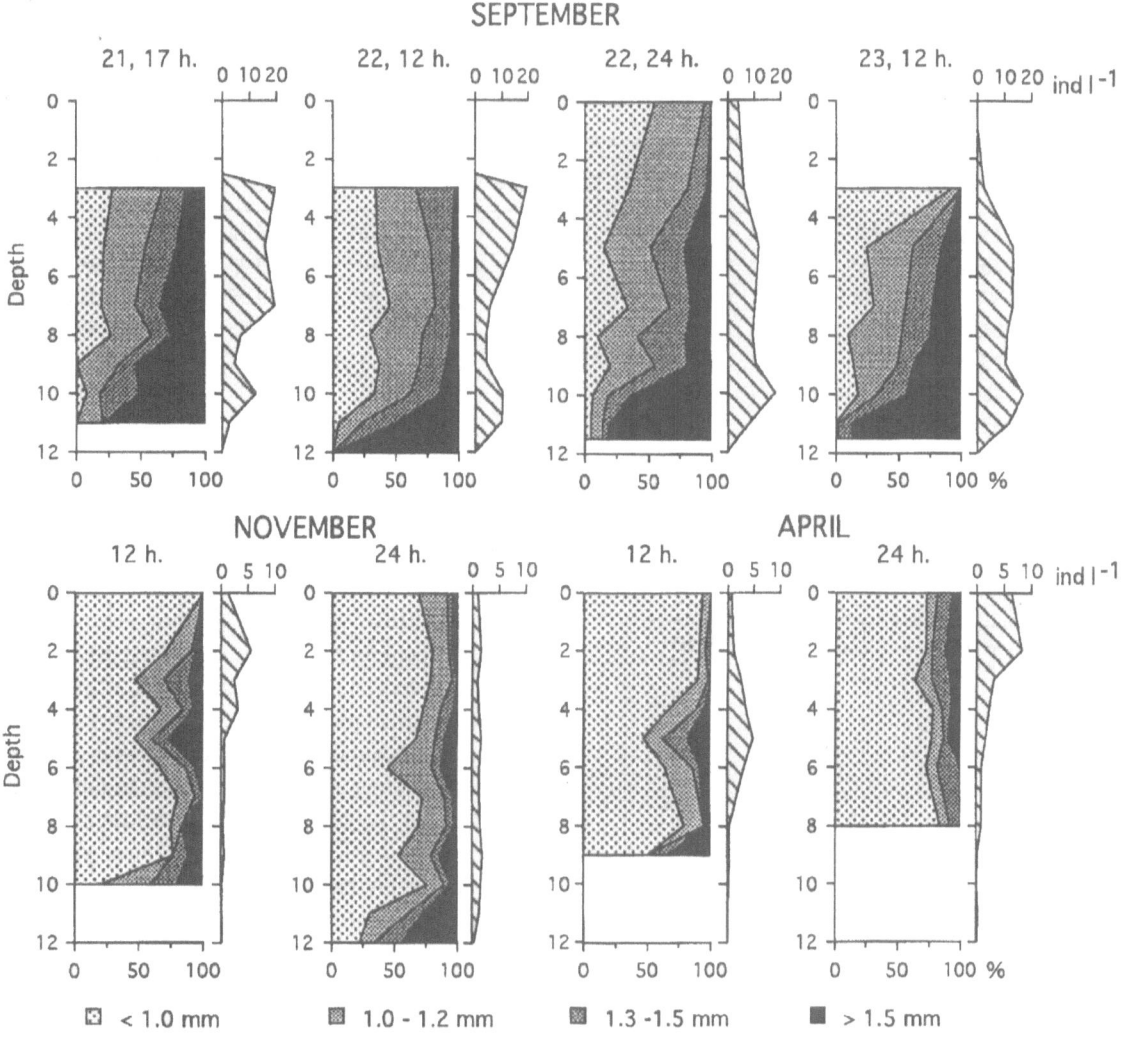

Figure 4. Vertical distribution of four size groups (two for juveniles and two for adults) as relative abundance (%) and of the whole *Daphnia longispina* population as ind. l^{-1}. The percentage distribution is not plotted when the total number of individuals was < 0.2 ind. l^{-1}.

ly found in the upper waters. The vertical distribution of size classes showed a strong size gradient in which the larger individuals were found deepest, especially in September. ANOVA analyses (Table 2) revealed a significant effect of depth on the mean size in all collections except at midnight in April. Kendall's t correlations with the same data (Table 2) showed that mean size significantly increased with depth in September, in all vertical surveys. Weather conditions on the two September dates were very different: September 22 was very cloudy after a stormy break during the night and September 23 was very sunny and calm. The correlation between depth and body size was higher during the second day. In April at noon and in November

at night the correlation was also significant, although much smaller than in September.

Figure 5 shows that the gravity centres of the smaller size classes profiles are in general higher in the water column than those of the bigger ones in September samplings. In November and April, the average depth of the different size classes does not differ much, except in November at night. Standard deviations for mean depths (Table 3) are quite similar for all size classes and, as expected, usually higher at night than at noon, with the exception of the midday collection in the cloudy September 22. We have performed a contingency analysis comparing noon and midnight vertical distributions separately for each size class to answer the

Table 3. Results from χ^2 contingency tests (degrees of freedom in parentheses) testing for a difference in vertical distribution between samples at different times of day, for each size class and for the September 21, 22 and 23, November 19 and April 22 vertical migration surveys. For midday we show the results with the profile taken on a sunny day. The standard deviation (SD) for the depth distributions of individuals of each size class are also indicated. E = Evening, N = Night, D = Day

	September								November			April		
	χ^2 (8)			SD					χ^2 (11)	SD		χ^2 (10)	SD	
μm	21-E/22-N	21-E/23-D	22-N/23-D	21-E	22-D	22-N	23-D			D	N		D	N
615–737	11.29	6.03	10.18	2.28	2.09	2.19	2.19		66.76*	2.65	3.12	49.86*	1.97	2.05
738–860	40.22*	45.10*	12.52	1.95	2.59	3.15	1.92		27.60	2.72	3.35	96.94*	1.89	2.08
861–983	39.45*	46.49*	16.87	1.88	2.50	3.25	2.16		38.66*	2.38	2.98	81.06*	2.19	2.33
984–1106	24.11*	36.14*	19.44	2.00	2.19	2.93	1.99		50.95*	2.24	2.88	43.16*	1.89	1.55
1107–1229	15.58	19.62	−1.98	2.30	2.95	2.62	1.98		30.23*	2.05	3.44	16.56	0.95	2.33
1230–1352	18.44	22.51	13.94	2.59	2.52	2.57	1.70		15.56	2.82	2.98	42.81*	1.72	2.96
1353–1475	−38.01	27.87*	13.69	2.39	2.73	2.23	2.54		27.15	1.39	2.83	31.73*	0.86	3.14
1476–1598	16.75	11.15	−0.32	2.64	3.16	2.09	2.67		15.01	0.71	4.23	34.34*	1.42	1.42
1599–2152	50.85*	57.84*	44.12*	2.21	2.53	1.78	1.68		29.13*	2.37	2.67	33.62*	1.28	1.19

* Significant at the 5% significance level after sequential Bonferroni correction.

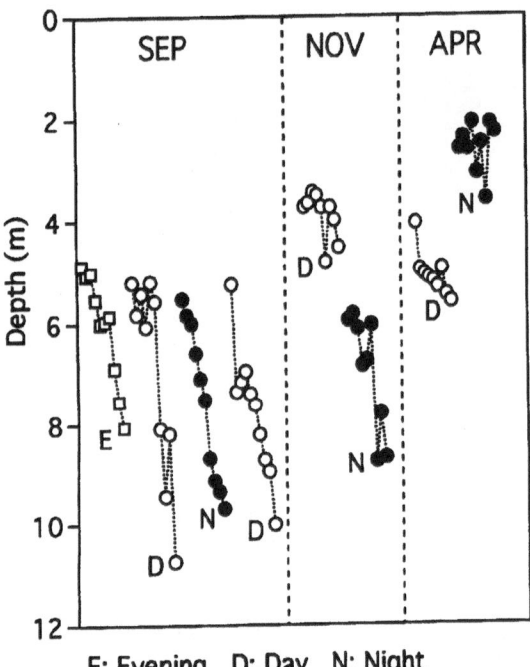

Figure 5. Mean vertical positions of the different size classes of *D. longispina* ordered for each diel collection from the smallest to the largest size-class. Size classes of 122 μm intervals, beginning at 0.6 mm, with the last one grouping all individuals > 1.7 mm for September and > 1.6 mm for the other months. The standard deviations of the ploted means are indicated in Table 3.

question whether or not our data demonstrate vertical migration (Table 3). In September, the evening distribution was significantly different from that of noon and midnight for many size classes, whereas midday and midnight distributions did not show significant differences except for the largest size class. The smallest size class never showed significant differences between day and night samples. In November many size classes showed significant differences between day and night profiles, including the smallest class; the only exceptions were in the size classes with a low number of individuals. In April all size classes showed significant differences, with the only exception being a size class of low density (Figure 3).

The pattern of vertical migration is clearly dissimilar on the different dates as shown by the mean vertical positions of the different size-classes plotted in Figure 6 and the total densities in Figure 4. On April 22 a typical nocturnal migration pattern was shown by all size classes, with the midday population located between 3–6 m of depth, and with the animals migrating to 0–2 m at midnight. The peak of the population at night was at 2 m, the depth showing the maximal temperature gradient. During the day the highest densities were found at 5 and 6 m. On the other dates, the range of the depths occupied by the population was larger. The pattern of migration corresponds to twilight migration in September and reversed migration (nocturnal descent) in November (Figure 5, Table 3). Figure 6 shows that the differences in migration patterns are even more clear when only ovigerous females are considered. These patterns may of course be the result of a mixed assemblage of individuals with different migratory behaviours. The September pattern is the most difficult to interpret since the population showed a bimodal distribution (Figure 4). Added variation due

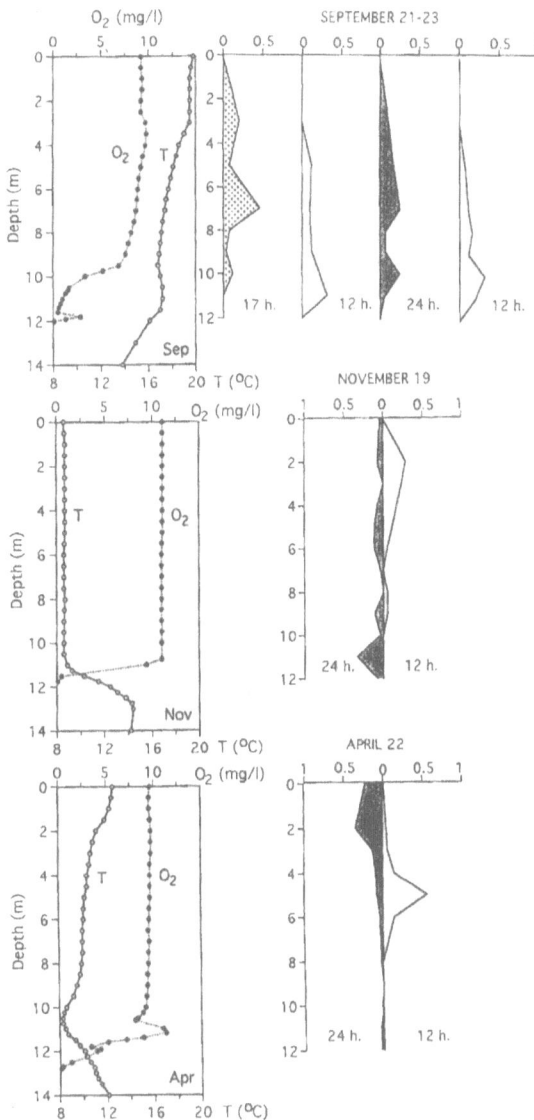

Figure 6. Depth distribution of ovigerous females (relative abundance, %) on the different sampling dates and times. For each date, the vertical distribution of temperature (T) and oxygen (O₂) is also shown in the left side.

in November and April, when light penetration was low.

Discussion

Our results show that in the *D. longispina* population of El Tobar, there is a considerable among-date and among-deph variation in size structure, fecundity and migration behavior, but that morphology does not vary much. The same *Daphnia* population was analyzed genetically by using electrophoretic analysis of isoenzymes and it was shown that 'clonal' frequencies (composite electromorphs of alleles at the AO, PGM and PGI loci) were very variable. Although only three loci were studied, allele frequencies at non selective loci may remain associated with alleles at loci that are more related to clonal fitness, because recombination between clones is infrequent in this parthenogetic population. As a result those allozyme loci can be used as markers. Although we are unable to differentiate morphologically between animals that belong to different clonal groups, the data presented here are from field collections taken simultaneously with those analysed in King et al. (1995) and King & Miracle (1995). In September there were two dominant and equally abundant 'clones' (we will use clone to designate the above mentioned composite electromoph) that together made up about 55% of the population. These clones were segregated vertically, with one clone occurring in the upper waters (density peak at about 5 m) and the other in deeper waters (density peak at about 10 m) showing different diel migration patterns. After autumn overturn, in November, the deep clone almost disappeared and the superficial clone became dominant, constituting 53% of the population. In April 1992, clonal diversity had increased considerably and the population was dominated by different clones than those found in September and November 1991. The higher genetic distance between the April population and the September and November populations is in agreement with differences in population characteristics (Table 1) and in migration patterns. In an experimental study, De Meester (1995) showed that in *D. magna*, genotypes with different phototactic behaviour also differed in life history traits.

From Figures 3 and 6 and Table 1, it is evident that changes in median size and size at maturity do not conform with inverse relations of body size with temperature. Other factors such as size-selective predation must therefore be involved. It has been experimentally

to weather conditions must also be considered: on the cloudy and stormy September 22 most size classes had mean depths around 5–6 m, whereas on September 23 only the smallest individuals remained in these layers and the other size classes migrated to deeper water (Figures 4 and 5). Likewise, when the depth distribution during daytime on the different collection dates are compared, we can observe that in September, when light penetration was high (see Table 1), the mean depth of the population (see Figure 5) was much deeper than demonstrated that invertebrate and fish predation pressure (Cerny & Bytel, 1991; Machácek, 1991; Mumm, 1997) are important factors controlling the size structure of *Daphnia* populations. This is not only due to the removal of selected size classes, but also because the presence of predators induces a phenotypic shift in body size. This has also been reported from field studies comparing lakes with different predator pressures (Gliwicz & Boavida, 1996). Unfortunately, we have only indirect information on fish and invertebrate predation on *Daphnia* in lake El Tobar. *Chaoborus* larvae are abundant in the lake in September and November, and they were caught in net samples as well as in Van Dorn and Patalas samples. It is also known that the lake contains a considerable population of young-of-the-year fish (from several cyprinid species and trout) at certain times of the year (Elvira, B., pers. com.).

The toothed summer form observed in lake El Tobar indicates that the *D. longispina* may be responding to *Chaoborus* predation, since neck teeth have been experimentally induced in other species (*D. pulex*, Havel, 1985; *D. hyalina* Brancelj et al., 1996). A form very similar to that of El Tobar was reported and drawn by Smyly (1955), who named it a 'minnehaha' form of *D. longispina*, since this was the name given to the toothed form of *D. pulex* (*D. pulex minnehaha*, Herrik). However, the neck teeth of *D. pulex* (see Havel, 1985) and *D. rosea* (Negrea, 1983) are somewhat differently located, more in the neck than in the dorsal side of the head as is the case in *D. longispina* (Figure 2) and *D. hyalina* (Brancelj et al., 1996). In the past, spined morphs were not known to be induced by the presence of *Chaoborus* and were described as different varieties (Gurney, 1923; Johnson, 1952), i.e. *D. hyalina var jurassica* and D. *lacustris var vicinia*, the latter probably being a synonym of *D. longispina*. Smyly (1955) commented that the toothed form of *D. longispina* was only found in young individuals and during summer. Christie (1983) obtained clones from toothed as well as round headed forms, and mentioned that they were 'similar to *D. longispina*'. He concluded that they should be genetically different because neonates of toothed forms maintained distinct head shapes and differed in other characteristics (e. g. body size) compared to round headed forms. Our results showed that in September, when toothed forms were found, most neonates were large and the overall size distribution showed a clear trend toward a larger body size compared with the other sampling dates (Figure 3). The relatively high percentage of adults and low fecundity in September (Table 1) can be explained by a reduction

of food availability in the crowded late summer waters. At this time, large eggs and large neonates are expected, since fitness increases with size at birth under food stress (Tessier & Consolatti, 1989). However, experimental results from these and other authors (Tillmann & Lampert, 1984) have shown that for some species of *Daphnia*, size at maturity may become smaller under low food conditions. The observed larger size at maturity in September is therefore likely to be an induced response to *Chaoborus* predation. The data in Lake El Tobar agree with the conclusions of models for optimal resource allocation to growth and reproduction in *Daphnia* (Gabriel & Taylor, 1991) predicting that invertebrate predation selection leads to a delayed age at first reproduction, and that this delay is increased when food is limiting.

Although the larger body size in September and the presence of toothed individuals may reflect a response to *Chaoborus* predation, the animals may at the same time be responding to visual predation of young-of-the-year fish. At this time there is a consistent and strong tendency for an increase in body size with increasing depth (Figure 4). A similar size structure of *Daphnia* populations along the vertical profile has been observed in other lakes (e.g. Wright et al., 1980; Salonen & Lehtovaara, 1992; Brancelj & Blejec, 1994) and is attributed to fish predation.

In lake El Tobar, thermal stratification is very strong in summer and light penetrates deep into the water column. Food resources for *Daphnia* are also stratified, with food concentrations being highest at the thermocline and especially near the oxicline. At the oxicline at the end of summer, there is a slight conductivity increase and a small temperature inversion, thus temperature remains quite high (Figure 6). In September; *Daphnia* showed a peak near the oxicline, consisting mainly of adults (Figure 4). This may suggest that this water layer does not meet juvenile requirements most probably because of the low oxygen concentrations (Figure 6). Vertical migration could be additionally selected in this case to have the offspring born in more oxygenated waters. Keen (1981) reported that the vertical migration of females bearing eggs in late stages of development into warmer waters at night causes a marked periodicity in the release of neonates in these warmer surface waters. In El Tobar also, neonate release might be stimulated when migrating to slightly warmer, but considerably more oxygenated water. The pattern of vertical migration in September is difficult to interpret because it shows a bimodal distribution involving a vertical segregation of at least two main genotypes, one migrating mainly in the epi-metalimnion and the other migrating from the oxicline to the metalimnion (King & Miracle, 1995). Our results suggest an asynchronous migration, but with a fraction of the populations in the different strata exibiting twilight migration (Figures 5 and 6, Table 3). In late summer, when nights become longer, twilight migration seems to be the optimum behaviour to minimize overlap with both *Chaoborus* and visual predators (Salonen & Lehtovaara, 1992).

After autumn overturn, gradients of temperature and oxygen were deeper and sharper, and the lake was isothermal and saturated with oxygen in the upper 12 m. The November *Daphnia* population is genetically less diverse and much reduced in density compared to the situation in September. The diel vertical distribution suggests a reverse migration that could be atributed to non-visual predation (Ohman, 1990). *Chaoborus* larvae were still found in the plankton samples, whereas light penetration had substantially decreased (Table 1), and predation pressure by visual predators is expected to have been reduced because of high mortality rates among young-of-the-year-fish and changes on their feeding habits. Although toothed morphs were not found, this could be due to the fact that the induction may be temperature dependent or because the spine-inducible genotypes have been replaced. *D. pulex* requires temperatures above 10 °C to form spined morphs (Havel, 1985) and in El Tobar November temperatures were below this value in the mixolimnion (i.e. 8.6 °C, Figure 6). If predation pressure by visual predators is reduced for *Daphnia*, but not for *Chaoborus*, *Daphnia* could avoid *Chaoborus* by reverse migration, as *Chaoborus* maintains its nocturnal migration. Morphological or size-related characters to *Chaoborus* predation are then not so important since the behavioural predator avoidance by reverse migration is not constrained.

At the end of April, an incipient thermocline is already established around 2 m and transparency is low. Food availability is concentrated in the upper layers, and so is the *Daphnia* population. We did not observed predacious *Chaoborus* instars in the plankton samples. At this time a typical nocturnal vertical migration is displayed by all size classes. The amplitude of this migration is small (Figures 4, 5 and 6), which may be the result of a compromise involving the avoidance of fish fry while trying to maximize the intake of resources. The downward ordination of the daytime mean depth according to size (Figure 5) agrees with this interpretation.

The pattern of migration is more pronounced in ovigerous females than in non-ovigerous females (Figure 6). This has also been reported by Brancelj & Blejec (1994) who observed greater migrating amplitudes in ovigerous females when compared to immature and to females without ova.

Although we have interpreted our observations as indicating a twilight migration, inverse migration or nocturnal migration depending on sampling date, in all cases part of the population may exhibit a different behaviour. This is especially evident in September. Also on the other dates, however; some individuals apparently migrated in a different pattern from the bulk of the population. The persistence of this variability may be important in allowing this *D. longispina* to cope with seasonal changes. The hypothesis put forward by De Meester (1994a) that *Daphnia* genotypes with different vertical behaviour and/or migration behaviour may coexist in natural populations, agrees with our results on the *Daphnia* population of Lake El Tobar, known to be clonally diverse.

Acknowledgements

We are very grateful to Luc De Meester and an anonymous reviewer for critically reading the manuscript and making valuable comments and language corrections. This study was supported by DGICYT grant NT89-1124.

References

Alonso, M., 1996. Crustacea, Branchiopoda. In Fauna Ibérica, vol. 7. Ramos, M. A. (ed.), Museo Nacional de Ciencias Naturales. CSIC. Madrid. 486 pp.

Brancelj, A. & A. Blejec, 1994. Diurnal vertical migration of *Daphnia hyalina* Leydig, 1860 (Crustacea: Cladocera) in lake Bled (Slovenia) in relation to temperature and predation. Hydrobiologia 284: 125–136.

Brancelj, A., T. Celhar & M. Sisko, 1996. Four different head shapes in *Daphnia hyalina* (Leydig) induced by the presence of larvae of *Chaoborus flavicans* (Meigen). Hydrobiologia 339: 37–45.

Cerny, M. & J. Bytel, 1991. Density and size distribution of *Daphnia* populations at different fish predation levels. Hydrobiologia 225: 199–208.

Christie, P., 1983. A taxonomic reappraisal of the *Daphnia hyalina* complex (Crustacea, Cladocera): An experimental and ecological approach. J. Zool. 199: 75–100.

De Meester, L., 1993. The vertical distribution of *Daphnia magna* genotypes selected for different phototactic behaviour: Outdoor experiments. Arch. Hydrobiol. Beih. Ergebn. Limnol. 39: 137–155.

De Meester, L., 1994a. Habitat partitioning in *Daphnia*: Coexistence of *Daphnia magna* clones differing in phototactic behaviour. In Beaumont, A. R. (ed.), Genetics and Evolution of Aquatic Organisms. Chapman & Hall. London.: 323–335.

De Meester, L., 1994b. Life histories and habitat selection in *Daphnia:* Divergent lifehistories in Daphnia magna clones differing in phototavtic behaviour. Oecologia 97: 333–341.

De Meester, L., 1995. Life history characteristics of *Daphnia magna* clones differing in phototactic behaviour. Hydrobiologia 307: 167–175.

DeMott, W. R., 1995. The influence of prey hardness on *Daphnia*'s selectivity for large prey. Hydrobiologia 307: 127–138.

Flösser, D. & K. Kraus, 1986. On the taxonomy of the *Daphnia hyalina-galeata* complex (Crustacea; Cladocera). Hydrobiologia 137: 97–115.

Gabriel, W. & E. Taylor. 1991. Optimal resource allocation in cladocerans. Verh. int. Limnol. 24: 2784–2787.

Glagolev, S. M., 1986. Species composition of *Daphnia* in Lake Glubokoe with notes on the taxonomy and geographical distribution of some species. Hydrobiologia 141: 55–82.

Gliwicz, Z. M. & J. M. Boavida, 1996. Clutch size and body size at first reproduction in *Daphnia pulicaria* at different levels of food and predation. J. Plankton Res. 18: 863–880.

Gurney, R., 1923. The Crustacean plankton of the English Lake District. J. Linn. Soc. Zool. 35: 411–447.

Hall, D. J., S. T. Threlkeld, C. W. Burns & P. H. Crowley, 1976. The size-efficiency hypothesis and the size structure of zooplankton communities. Ann. Rev. Ecol. Syst. 7: 177–208.

Havel, J. E., 1985. Cyclomorphosis of *Daphnia pulex* spined morphs. Limnol. Oceanogr. 30: 853–861.

Havel, J. E. & S. I. Dodson, 1984. *Chaoborus* predation on typical and spined morphs of *Daphnia pulex*; Behavioral observations. Limnol. Oceanogr. 29: 487–494.

Hrbácek, J., 1987. Systematics and biogeography of *Daphnia* species in the northern temperate regions. In Peters, R. H. & R. de Bernardi (eds), *Daphnia*. Mem. Ist. ital. Idrobiol. 45: 37–76.

Johnson, D. S., 1952. The British species of the genus *Daphnia* (Crustacea, Cladocera). Proc. zool. Soc. Lond. 122: 435–462.

Keen, R., 1981. Vertical migration, hatching rates, and distribution of eggs stages in freshwater zooplankton. J. Therm. Biol. 6: 349–351.

King, C. E. & M. R. Miracle, 1995. Diel vertical migration by *Daphnia longispina* in a Spanish lake: Genetic sources of distributional variation. Limnol. Oceanogr. 40: 226–231.

King, C. E., M. R. Miracle & E. Vicente, 1995. Large Hardy-Weinberg equilibrium deviations in the *Daphnia longispina* of Lake El Tobar. Hydrobiologia 307: 15–23.

Lampert, W., 1988. The relative importance of food limitation and predation in the seasonal cycle of two *Daphnia* species. Verh. int. Ver. Limnol. 23: 713–718.

Machácek, J., 1991. Indirect effect of planktivorous fish on the growth and reproduction of *Daphnia galeata*. Hydrobiologia 225: 193–197.

Margalef, R., 1953. Los crustáceos de las aguas continentales ibéricas. Inst. Forestal de Investigaciones y Experiencias, Madrid, 243 pp.

Margaritora, F., 1985. Cladocera. In Fauna de Italia. Edizioni Calderini, Bologna, 399 pp.

Miracle, M. R., 1978. Composición específica de las comunidades zooplanctónicas de 153 lagos de los Pirineos y su interés biogeográfico. Oecol. Aquat. 3: 167–191.

Miracle, M. R., J. Armengol-Díaz & M. J. Dasí, 1993. Extreme meromixis determines strong differential planktonic vertical distributions. Verh. int. Ver. Limnol. 25: 705–710.

Mumm, H., 1997. Effects of competitors and *Chaoborus* predation on the cladocerans of a eutrophic lake: an enclosure study. Hydrobiologia 360: 253–264.

Negrea, S., 1983. Sur les populations de *Daphnia galeata* sars, 1864 et *Daphnia rosea* sars, 1862 (Cladocera, Daphniidae) de Roumanie. Hydrobiologia 18: 77–92.

Ohman, D. M., 1990. The demographic benefits of diel vertical migration by zooplakton. Ecol. Monogr. 60: 257–281.

Pejler, B., 1973. On the taxonomy of limnoplanktic *Daphnia* species in Northern Sweden. Zoon. 1: 23–27.

Salonen, K. & A. Lehtovaara, 1992. Migrations of haemoglobin-rich *Daphnia longispina* in a small, steeply stratified, humid lake with an anoxic hypolimnion. Hydrobiologia 229: 271–288.

Smyly, W. J. P., 1955. A 'minnehaha' form of *Daphnia longispina* O. F.Müller. J. Quekett Micr. Cl., ser. 4, 4: 217.

Spaak, P. & J. R. Hoekstra, 1993. Clonal structure of the *Daphnia* population in lake Maarsseveen: Its implications for diel vertical migration. Arch. Hydrobiol. Beih. Ergebn. Limnol. 39: 157–165.

Tessier, A. J. & N. L. Consolatti, 1989. Variation in offspring size in *Daphnia* and consequences for individual fitness. Oikos 56: 269–276.

Tillmann, U. & W. Lampert, 1984. Competitive ability of differently sized *Daphnia* species: An experimental test. Freshwat. Ecol. 2: 311–323.

Vicente, E., A. Camacho & M. A. Rodrigo, 1993. Morphometry and physicochemistry of the crenogenic meromictic lake El Tobar (Spain). Verh. int. Ver. Limnol. 25: 698–704.

Weider, L. J., 1984. Spatial heterogeneity of *Daphnia* genotypes: Vertical migration and habitat partitioning. Limnol. Oceanogr. 29: 225–235.

Wrigth, D., W. J. O'Brien & G. L. Vinyard, 1980. Adaptative value of vertical migration: A simulation model argument for the predation hypothesis. In W. C. Kerfoot (ed.), Evolution and Ecology of Zooplankton Communities. The University Press of New England, Hanover (N.H.): 111–121.

Hydrobiologia **360**: 197–203, 1997.
A. Brancelj, L. De Meester & P. Spaak (eds), Cladocera: The Biology of Model Organisms.
©1997 *Kluwer Academic Publishers.*

The trophic role of the marine cladoceran *Penilia avirostris* in the Gulf of Trieste

Lovrenc Lipej, Patricija Mozetič, Valentina Turk & Alenka Malej
Marine Biological Station, Piran, 6330 Slovenia

Key words: Cladocera, grazing, gut fluorescence, food webs, Adriatic Sea

Abstract

The herbivory of the marine cladoceran *Penilia avirostris* was studied in the Gulf of Trieste (Northern Adriatic) from June 1993 to December 1994 using the gut fluorescence method. *P. avirostris* occurred from June to December, but reached its greatest abundance in the summer months. A significant correlation between the gut pigment content and chlorophyll *a* concentration in the surface layer was established. Observations with an epifluorescence microscope revealed that the guts were filled with fluorescing nanoplankton and picoplankton (cyanobacteria). Quantitative estimates indicated that *P. avirostris* grazed less than 5% of the available chlorophyll *a* in more than half of all measurements, but removed most of the available chlorophyll *a* in the surface layer during some periods in September. It can therefore be concluded that *P. avirostris*, together with planktonic protists, plays an important role within the microbial loop.

Introduction

The marine cladoceran *Penilia avirostris* is a dominant member of the coastal Adriatic zooplankton assemblage during summer (Fonda-Umani et al., 1992). In the Adriatic, it was first noted in 1914 by Leder (Fonda-Umani, 1980). During periods of water column stratification, the concentrations of nutrients and chlorophyll above the pycnocline layer are rather low, whereas pico- and nanoplanktonic autotrophs are abundant throughout the water column (Turk, 1992; Malej et al., 1995). As far as the mesozooplanktonic organisms are concerned, only those which are capable of filtering the pico- and nanoplanktonic autotrophs can survive during the stratification period. *P. avirostris* is, in addition to the Appendicularia, one of the rare mesozooplanktonic species that find such conditions favourable.

The life cycle of *P. avirostris* is characterised by an alternation between amphigony and parthenogenesis. Its parthenogenetic breeding manner enables it to undergo a population explosion during the summer months (Della Croce & Venugopal, 1972; Kim et al., 1989), but with the gradual destruction of the thermocline its numbers decrease. At that time, a sexual generation consisting of males and females with resting eggs emerges, so that a new generation can develop in the following year.

In order to assess the trophic role of *P. avirostris* in the Gulf of Trieste, we measured its grazing impact on the chlorophyll *a* standing stock with the aid of the gut fluorescence method. This technique is currently widely used, mostly due to its simplicity and the possibility of studying grazing *in situ* (Kiørboe et al., 1985; Kiørboe & Tiselius, 1987; Peterson et al., 1990). Although *P. avirostris* is a dominant species of the summer zooplankton assemblage in many coastal areas, investigations of its feeding ecology are rather rare (cf. Uye & Onbé, 1993) and mostly done under *in vitro* conditions (Pavlova, 1959; Gore, 1980; Paffenhöffer & Orcutt, 1986; Turner et al., 1988).

Methods

The present study was conducted in the southern part of the Gulf of Trieste (Figure 1). Samples for the estimation of the abundance of *Penilia* were collected every 3 to 4 weeks from May 1989 through December 1994 at station F. Samples for both grazing measurements and abundance assessment were collected approximately weekly from June 1993 to December 1994 at station MBP. The live zooplankton samples were obtained by vertical net tows at the nearshore station MBP. Immediately after sampling, the contents of the cod end were poured into a 30-l container and the samples were transported to the laboratory within 15 minutes. Additional experiments were carried out in 1994 and 1995, when collected samples were stored in liquid nitrogen immediately after capture. Comparison of the two methods of handling showed no statistical differences between them (1994, $n=3$; 1995, $n=12$; one way ANOVA, $P=0.31$). From each living sample, a subsample was collected, which was preserved in 4% neutralised formaldehyde. The samples were always collected in the morning between 9.00 and 10.00. Simultaneously, we collected seawater samples to determine the chlorophyll a concentration and the density of cyanobacteria in the surface layer (at 1 m depth), since *P. avirostris* is considered to be a surface dweller (Specchi, 1969; Wong et al., 1992; Kim & Onbé, 1995). Chlorophyll a (chl a) concentrations were determined fluorometrically (Holm-Hansen et al., 1965). Subsamples (25 ml) were filtered onto 0.22 μm Millipore filters, extracted in 90% acetone, and the fluorescence of the extracts was measured with a Turner fluorometer (Model 112). Autofluorescing cyanobacteria were counted under an epifluorescence microscope (Takahashi et al., 1985).

Conserved samples were checked for the presence of *P. avirostris*. The organisms were counted under a stereomicroscope and converted into individuals per cubic meter of seawater. The body length of at least 50 *P. avirostris* was measured from the tip of the head to the base of the caudal setae as suggested by Della Croce & Bettanin (1965) using an ocular micrometer.

The gut fluorescence method was used to assess grazing (Mackas & Bohrer, 1976), following suggestions given by Morales & Harris (1990) and Morales et al. (1990). Twenty to thirty freshly caught *P. avirostris* were randomly removed from the container as quickly as possible (1–5 min) using a stereomicroscope in dim light. They were placed in tubes with 90% acetone and stored in a dark refrigerator overnight. During periods when there was a dominance of diatoms (*Rhizosolenia, Pseudonitzschia),* cladocerans were rinsed briefly with filtered seawater in order to eliminate diatoms adhering to the carapace. Fluorescence was measured before and after acidification using a Turner 112 fluorometer. The chlorophyll a and phaeopigment

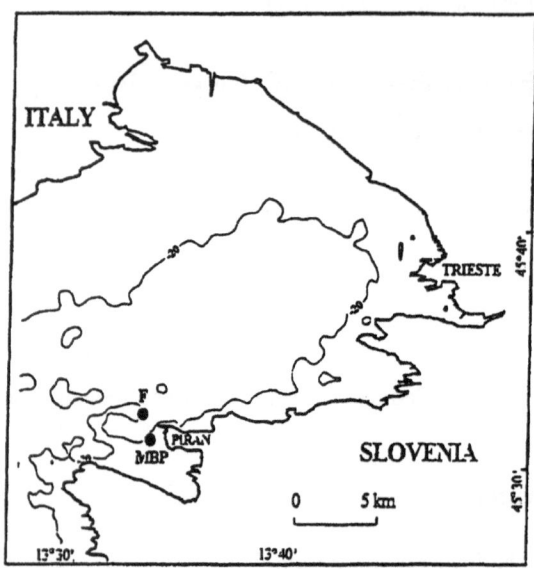

Figure 1. The Gulf of Trieste (Northern Adriatic) with sampling stations.

concentrations were calculated using the equation of Dagg & Wyman (1983). Following the methodological recommendations of Morales & Harris (1990), no corrections for background fluorescence, pigment destruction or differences in molecular weight were applied to the data. Most gut content analyses were carried out in 3–6 replicates, occasionally in 10 replicates. Gut pigment content was expressed as ng Chl $a \cdot$ animal^{-1}.

The gut evacuation rate for *P. avirostris* was calculated from the measurements done on 14 July 1994 (using the equation from Wong et al. (1992)). The seawater temperature at the time of measurements was 25.6 °C. During the initial 50-minute period, 15 to 30 specimens were taken at short time intervals (<10 minutes) from the container of freshly caught cladocerans.

The gut pigment contents were converted to daily ingestion (I) rates (ng chl a ind.$^{-1}$ d^{-1}) by the equation (Dagg & Wyman, 1983)

$$I = k \times G,$$

where k is the gut evacuation rate (min^{-1}) and G is the gut pigment content (ng chl a ind.$^{-1}$). The total daily grazing impact estimate (GI) was calculated from the equation

$$\text{GI}(\%) = ((I \times n) \times 100)/C$$

where n is the abundance of the selected herbivore in the water column (ind. m^{-3}) and C is the concentration of chlorophyll a in the surface layer (μg l^{-1}).

Additionally, we used epifluorescence microscopy to supplement fluorometric data analysis with observations of the guts of live animals.

Results

Chlorophyll a concentration at the surface varied from 0.21 to 5.0 μg l^{-1} during the period from June 1993 to December 1994. During periods of a mixed water column (October–April), chlorophyll a concentrations were generally higher than 0.5 μg l^{-1}, while during periods of a stratified water column (May–September) the chlorophyll a concentrations were generally below 0.5 μg l^{-1}. During periods of water column stratification, the phytoplankton assemblage was dominated by pico- (<2 μm) and nanoplankton (2–10 μm) fractions. Cyanobacteria prevailed in the picoplankton fraction. Their abundance in the surface layer varied from 1.1×10^7 to 1.69×10^8 cells l^{-1}, with highest values occurring in September and November.

Population density

In 1989–1994, *Penilia avirostris* occurred in the Gulf of Trieste from June to December, but its greatest abundance was reached during the period of water column stratification (May–September). Densities varied from 2 to 25 500 ind. m^{-3} at station MBP and up to 11 800 ind. m^{-3} at station F (Figure 2). Although differences between years were distinct, two abundance peaks were recorded during the summer months in each year. The population consisted mostly of parthenogenetic individuals during the summer, while large-sized amphigonic females with resting eggs appeared only from September until December. In December, *P. avirostris* completely disappeared from the plankton community. The size frequency distribution is shown in Figure 3. In July, individuals of *P. avirostris* are more or less equally distributed in all size classes, but by the end of the year larger animals were more numerous.

Population grazing rates

Penilia avirostris grazing as expressed by gut pigment content is shown in Figure 4. Gut pigment content varied from 0.04 to 3.91 ng chl a ind.$^{-1}$ in 1993 and from 0.06 to 3.88 ng chl a ind.$^{-1}$ in 1994. The sea-

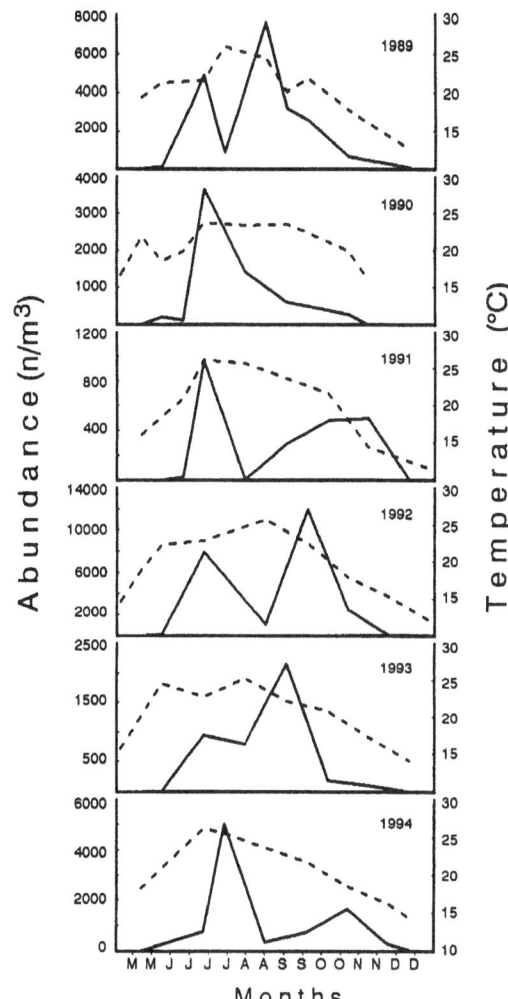

Figure 2. Penilia avirostris abundance (solid line) and surface layer temperature (dotted line) from May 1989 to December 1994 (station F).

sonal dynamics of grazing were similar in both years. In July and August the gut pigment content of *P. avirostris* was low, but gut pigment concentrations tended to increase during the autumn months. The highest values were measured in October 1993 and November 1994. Figure 5 shows the correlation between chlorophyll a concentrations in the surface layer and gut pigment contents of *P. avirostris* (Pearson regression coefficient, $R^2 = 0.82$, $P < 0.001$). Gut pigment contents were low during periods of water column stratification, which were characterized by low chlorophyll a concentrations, and increased with the rise of chlorophyll a biomass (Figure 5). The calculation of daily ingestion rates indicated that during the study period *P. avirostris* ingested from 0.98 to 95.72 ng chl a d^{-1}.

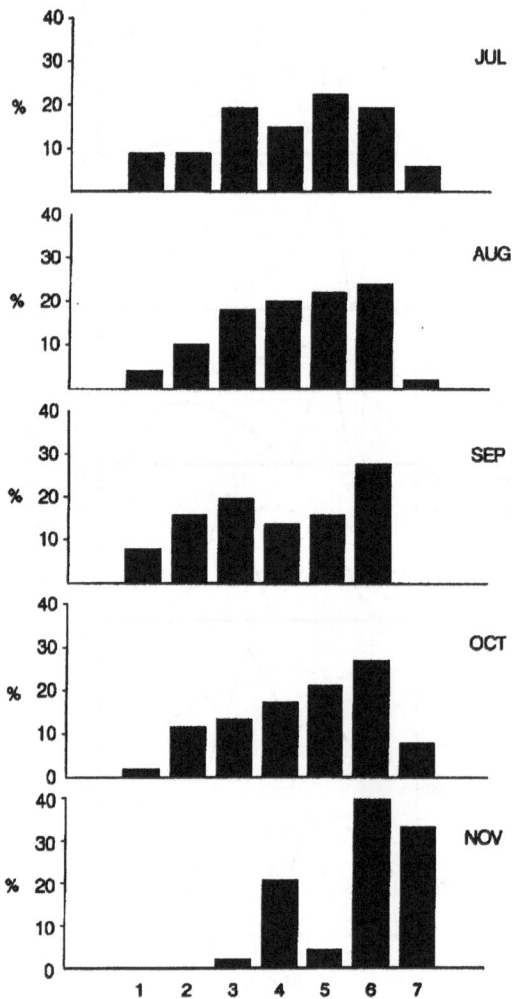

Figure 3. Size frequency distribution (%) of *Penilia avirostris* from July to November 1993. Using a χ^2 distribution, the individuals were assigned to seven size classes: 1: <0.48 mm, 2: 0.49–0.54 mm, 3: 0.55–0.6 mm, 4: 0.61–0.66 mm, 5: 0.67–0.72 mm, 6: 0.73–0.8 mm in 7: >0.8 mm.

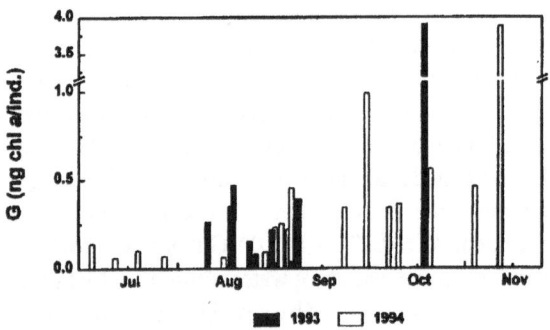

Figure 4. Grazing in terms of gut pigment content (G) in 1993 (black bars) and 1994 (white bars) in the southern part of the Gulf of Trieste (station MBP).

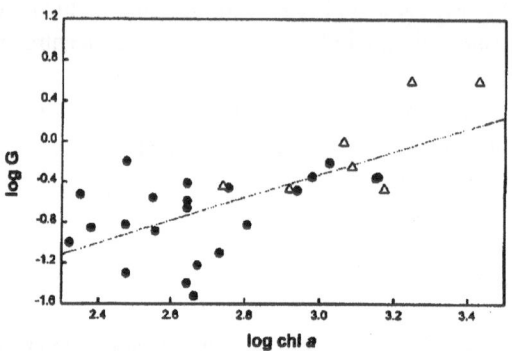

Figure 5. Correlation plot between *Penilia avirostris* gut pigment content and chlorophyll *a* concentration in the surface layer from 1993 to 1994 ($R^2 = 0.82$, $P < 0.001$). White triangles indicate samples obtained during periods of a vertically mixed water column and black circles denote the samples obtained during periods of water column stratification.

In more than half of all measurements, *P. avirostris* grazed less than 5% of the total chlorophyll *a*. But in the period from 5 to 12 September 1994, they grazed up almost the entire quantity of total chlorophyll per day in the surface layer.

Discussion

Penilia avirostris occurred in the Gulf of Trieste in the warmer period of the year, at a temperature ranging from 11 to 27 °C. The maximal recorded abundances of 25 500 ind. m^{-3} are in good agreement with the highest abundances reported by other authors in the Adriatic Sea (Malej, 1979; Corni et al., 1989), in the Indian ocean (Della Croce & Venugopal, 1972) and in the South China Sea (Wong et al., 1992). The seasonal changes in size frequency distribution observed in our study are in good agreement with the observations of Della Croce & Bettanin (1965) and Corni & Cattani (1979), who reported an abundance increment of amphigonic specimens in October and November.

During periods when the water column was well mixed, large, chain-forming diatoms dominated the phytoplankton assemblage with *Pseudonitzschia delicatissima*, *Rhizosolenia fragilissima*, *Chaetoceros* spp., *Thalassionema nitzschioides* being the most prominent species (Malej et al., 1995). Despite the high diatom biomass, their size was obviously not suitable for *P. avirostris*, and thus the grazing impact on the phytoplankton was low. *P. avirostris* is known to feed normally on small particles in the size fraction

<15 μm (Gore, 1980; Paffenhöfer & Orcutt, 1986; Turner et al., 1988). This size-related feeding strategy was also confirmed in our study by observing live *P. avirostris* specimens under the epifluorescent microscope. In addition, these observations confirmed the herbivorous nature of these cladocerans. Fluorescing cyanobacteria and autotrophic nanoplanktonic flagellates were always observed in the digestive tract. On the other hand, we observed neither diatoms nor large dinoflagellates.

Although the gut of *P. avirostris* was always filled with yellow fluorescing cyanobacteria in summer, we did not find a significant correlation between the gut pigment content and the abundance of cyanobacteria in the surface layer. During the period of water column stratification, cyanobacteria may comprise up to 80% of the carbon biomass (Turk, 1992; Delgado et al., 1992). The cyanobacteria play an important role in the diet of tintinnids (Bernard & Rassoulzadegan, 1993) and are considered to be a carbon-rich food source (Marra et al., 1988). Although we did not calculate whether the metabolic requirements of *P. avirostris* are met with the available food during the stratification period, the cladoceran can probably obtain enough energy from other food sources such as heterotrophic flagellates, as previously recorded by Turner et al. (1988), or bacteria (Roff et al., 1995).

Penilia avirostris has been reported to feed efficiently on prey such as heterotrophic microflagellates, small diatoms and autotrophic flagellates (Turner et al., 1988). Small diatoms such as *Skeletonema costatum*, which is the dominant diatom during spring in the Gulf of Trieste, were found in the guts of *P. avirostris* when analyzed with SEM by Kim et al. (1989). They also found a lot of unidentifiable small particles, probably flagellate remains. In our study, we found no diatoms in the guts of *P. avirostris* observed under the epifluorescence microscope.

The gut pigment content of planktonic grazers is generally correlated with the chlorophyll concentration in the ambient water. Our study also confirmed this relationship (Figure 5). The fluorometric analysis data for the period from August 1993 to December 1994 revealed a significant positive correlation between the gut pigment content of *P. avirostris* and the available quantities of chlorophyll *a* in the surface layer, suggesting that ingestion rates increased with increasing chlorophyll concentration in the surface layer.

The asssessment of the grazing impact requires the calculation of gut evacuation rate from *in situ* measurements. Our estimated gut evacuation rate

(0.017 min^{-1}) was significantly lower than that measured in the field by Wong et al. (1992), but it was in good agreement with the value they obtained in the laboratory (0.019 min^{-1}). Our grazing impact estimates may then be underestimated, especially during the period of a mixed water column. Nevertheless, this fact does not change considerably the grazing impact of *Penilia avirostris*, which would still be low during the period of the mixed water column. In the study of Wong et al. (1992), the grazing impact of *P. avirostris* was almost negligible (only 1%) due to the high chlorophyll *a* concentration in Tolo Harbour (mean values 35.8 μg l^{-1} in June and 68.3 μg l^{-1} in November). In our study covering the period from August 1993 to December 1994, *P. avirostris* grazed less than 5% of the total chlorophyll in more than half of all cases. However, during some periods (from 5 to 12 September 1994) they daily grazed up the entire quantity of available chlorophyll in the surface layer. It should be pointed out that chlorophyll levels in the surface layer at that time were low (0.2–0.4 μg l^{-1}) while *P. avirostris* abundances were extremely high (from 4000–25 500 ind. m^{-3}). *P. avirostris* comprised 81% of the zooplankton assemblage during the period of its highest grazing impact. This is consistent with the results of a laboratory study by Paffenhöfer & Orcutt (1986), who demonstrated that *P. avirostris* can reproduce at relatively low particulate concentrations. Moreover, lower gut pigment contents during periods of maximal abundances and low levels of chlorophyll biomass have been previously recorded by Rodriguez et al. (1991) for the freshwater cladoceran *Ceriodaphnia* sp.

The gut pigment contents found in our study were similar to the values obtained from Tolo Harbour, Hong Kong by Wong et al. (1992) which is the only *in situ* research on *P. avirostris*.

The grazing impact during the mixed water column period was generally below 1%. The abundance of *P. avirostris* was also low during that period, when larger diatoms and dinoflagellates dominated the phytoplankton assemblage. The total chlorophyll concentration was unsuitable for grazing impact evaluation, since *P. avirostris* does not generally graze upon particles greater than 15 μm (Gore, 1980; Paffenhöfer & Orcutt, 1986; Turner et al., 1988). We presume that in the period of the mixed water column, *P. avirostris* is more vulnerable to predation, since it lacks an escape ability due to its limited swimming activity (Paffenhöfer & Orcutt, 1986).

Significantly greater grazing impacts of *P. avirostris* were recorded during the period of a stratified water column. At that time, most of the chlorophyll *a* is concentrated in the nanoplankton and picoplankton fractions (Turk, 1992). Characteristic of such conditions is the importance of the microbial loop, within which the major grazing role is played by microplanktonic organisms. On the other hand, the role of herbivorous copepods, the key predators of the classic food chain, is mostly negligible. Generally, copepods are unable to feed efficiently on particles <10 μm (Peterson & Bellantoni, 1987; Peterson et al., 1990). Our data on the grazing impacts of the dominant herbivorous copepods in the Gulf of Trieste, which were measured in the same period (1993–1995), showed that the dominant herbivorous copepods *Acartia clausi* and *Temora* spp. grazed 12% of the available chlorophyll *a* at most, and that grazing impacts were generally substantially lower. The highest grazing impacts were recorded during the autumn blooming of diatoms (Lipej, 1996). During the period of water column stratification, the grazing impacts of both taxa were extremely small (<3%). Our results therefore indicate that copepods in general play a less important trophic role as far as the grazing of phytoplankton is concerned than *Penilia*. It can therefore be concluded that *P. avirostris* plays, together with the planktonic protists, an important role within the microbial loop characteristic for the summer period in the Gulf of Trieste.

Acknowledgments

This study was a part of the Project PALOMA (Production and Accumulation of the Labile Organic Matter in the Adriatic) and was financially supported by the PECO program (Contract N° CIPD-CT94-0106) and the Ministry of Science and Technology of the Republic of Slovenia. We thank Dr Michael Stachowitsch, Prof Thomas Mutz and two anonymous reviewers for their helpful review of the manuscript.

References

Bernard, C. & F. Rassoulzadegan, 1993. The role of picoplankton (Cyanobacteria and plastidic picoflagellates) in the diet of tintinnids. J. Plankton Res. 15: 361–373.

Corni, M. G. & O. Cattani, 1979. Aspetti biologici ed ecologici di *Penilia avirostris* Dana (Cladocera:Sididae) nel plancton di Fano. Nova Thalassia 3: 89–112.

Corni, M. G., I. Ferrari & A. Foresi, 1989. Seasonal succession and heterogonic cycles of Cladocerans at two stations off Fano (Adriatic Sea). Nova Thalassia 10: 119–125.

Dagg, M. J. & K. D. Wyman, 1983. Natural ingestion rates of the copepods *Neocalanus plumcrus* and *N. cristatus* calculated from the gut contents. Mar. Ecol. Progr. Ser. 13: 37–46.

Delgado, M., M. Latasa & M. Estrada, 1992. Variability in the size-fractionated distribution of the phytoplankton across the Catalan front of the north-west Mediterranean. J. Plankon Res. 14: 753–771.

Della Croce, N. & S. Bettanin, 1965. Osservazioni sul ciclo biologico di *Penilia avirostris* Dana nel Golfo di Napoli. Boll. Musei Ist. Biol. Univ. Genova 33: 49–68.

Della Croce, N. & P. Venugopal, 1972. Distribution of marine cladocerans in the Indian Ocean. Mar. Biol. 15: 132–138.

Fonda-Umani, S, 1980. I Cladoceri dell'Adriatico: Un 'review critico'. Nova Thalassia 4: 107–133.

Fonda-Umani, S., P. Franco, E. Ghirardelli & A. Malej, 1992. Outline of oceanography and the plankton of the Adriatic Sea. In Colombo, G., I. Ferrari, V. U. Ceccherelli & R. Rossi (eds), Marine Eutrophication and Population Dynamics. Olsen & Olsen, Fredensborg: 347–365.

Gore, M. A, 1980. Feeding experiments on *Penilia avirostris* Dana (Cladocera:Crustacea). J. exp. mar. Biol. Ecol. 44: 253–260.

Holm-Hansen, O., C. J. Lorenzen, R. W. Holmes & J. D. H. Strickland, 1965. Fluorometric determination of chlorophyll. J. Cons. perm. int. Expl. Mer 30: 3–15.

Kim, S. W. & T. Onbé, 1995. Distribution and zoogeography of the marine cladoceran *Penilia avirostris* in the northwestern Pacific. Bull. Plankton Soc. Japan 42: 19–28.

Kim, S. W., T. Onbé & Y. H. Yoon, 1989. Feeding habits of marine cladocerans in the Inland Sea of Japan. Mar. Biol. 100: 313–318.

Kiørboe, T., F. Mohlenberg & H. U. Rijsgard, 1985. *In situ* feeding rates of planktonic copepods: A comparison of four methods. J. exp. mar. Biol. Ecol. 88: 67–81.

Kiørboe, T. & P. T. Tiselius, 1987. Gut clearance and pigment destruction in a herbivorous copepod, *Acartia tonsa*, and the determination of *in situ* grazing rates. J. Plankton Res. 9: 525–534.

Lipej, L., 1996. Trophic role of planktonic herbivores in the stratified and vertically mixed coastal sea. PhD thesis. University of Ljubljana. xi + 133 pp. (In Slovenian).

Mackas, D. & R. Bohrer, 1976. Fluorescence analysis of zooplankton gut contents and investigations of diel feeding patterns. J. exp. mar. Biol. Ecol. 25: 77–85.

Malej, A, 1979. The zooplankton of the coastal waters in the NE Gulf of Trieste. Nova Thalassia 3: 213–231.

Malej, A., P. Mozetič, V. Malačič, S. Terzič & M. Ahel, 1995. Phytoplankton responses to freshwater inputs in a small semi-enclosed gulf (Gulf of Trieste, Adriatic Sea). Mar. Ecol. Prog. Ser. 120: 111–121.

Marra, J. L., W. Haas & K. R. Heinemann, 1988. Time course of C assimilation and microbial food web. J. exp. mar. Biol. Ecol. 115: 263–280.

Morales, C. E. & R. P. Harris, 1990. A review of the gut fluorescence method for estimating ingestion rates of planktonic herbivores. ICES (International Council for the Exploration of the Sea), C.M. 1990/L:26, Biological Oceanographic Comitee.

Morales, C. E., B. Bautista & R. P. Harris, 1990. Estimates of ingestion in copepod assemblages: gut fluorescence in relation to body size. In Barnes, M. & R. N. Gibson (eds), Trophic Relationships in the Marine Environment. Aberdeen University Press, Aberdeen: 565–577.

Paffenhöfer, G.-A. & J. D. Orcutt, Jr, 1986. Feeding, growth and food conversion of the marine cladoceran *Penilia avirostris*. J. Plankton Res. 8: 741–754.

Pavlova, E. V., 1959. On grazing by *Penilia avirostris* Dana. Tr. Sevastopol. Biol. Stn. 11: 63–71. (In Russian).

Peterson, W. T. & D. C. Bellantoni, 1987. Relationships between water column stratification, phytoplankton cell size and copepod fecundity in Long Island and off central Chile. S. Afr. J. mar. Sci. 5: 411–421.

Peterson, W. T., S. Painting & L. Hutchings, 1990. Diel variation in gut pigment content, diel migration and estimates of grazing impact of copepods in the southern Benguela region in October 1987. J. Plankton Res. 12: 259–281.

Rodriguez, V., F. Echevarria & B. Bautista, 1991. In situ diel variation in gut pigment contents of *Ceriodaphnia* sp. in stratification and destratification periods. J. Plankton Res. 13: 187–196.

Roff, J. C., J. T. Turner, M. K. Weber & R. R. Hopcroft, 1995. Bacterivory by tropical copepod nauplii: extent and possible significance. Aquat. microb. Ecol. 9: 165–175.

Specchi, M., 1969. Influenza della temperatura sulla microdistribuzione superficiale del plancton nel Golfo di Trieste. Pubbl. Staz. Zool. Napoli 37: 338–348.

Takahashi, M., K. Kikuchi & Y. Hara, 1985. Importance of picocyanobacteria biomass (unicellular blue-green algae) in the phytoplankton population of the coastal waters of Japan. Mar. Biol. 89: 63–69.

Turk, V., 1992. The microbial food web: Time scales and nutrient dynamics in the Gulf of Trieste. PhD Thesis. University of Umeå. Sweden.

Turner, J. T., P. A. Tester & R. L. Ferguson, 1988. The marine cladoceran *Penilia avirostris* and the 'microbial loop' of pelagic food webs. Limnol. Oceanogr. 33: 245–255.

Uye, S. & T. Onbé, 1993. Diel variations in gut pigments of marine cladocerans in the Inland Sea of Japan. Bull. Plankton Soc. Japan 40: 67–69.

Wong, C. K., A. L. Chan & K. W. Tang, 1992. Natural ingestion rates and grazing impact of the marine cladoceran *Penilia avirostris* Dana in Tolo Harbour, Hong Kong. J. Plankton Res. 14: 1757–1765.

Hydrobiologia **360**: 205–210, 1997.
A. Brancelj, L. De Meester & P. Spaak (eds), Cladocera: The Biology of Model Organisms.
©1997 *Kluwer Academic Publishers.*

Ecological similarities and differences among littoral species of *Ceriodaphnia*

Radka Pichlová
*Faculty of Biological Sciences, University of South Bohemia, Branišovská 31, 370 05 České Budějovice,
Czech Republic*

Key words: Ceriodaphnia, littoral, coexistence, population dynamics, body length, clutch size

Abstract

The occurrence of six littoral species of *Ceriodaphnia* (*C. affinis, C. laticaudata, C. megops, C. reticulata, C. rotunda* and *C. setosa*) living at the same time and locality is described. Basic population characteristics were studied in all species coexisting in the field, except for *C. setosa*, which was too scarce. Abundances, proportions of gamogenetic individuals, mean body lengths and clutch sizes were compared among species. Dissolved oxygen, temperature, chlorophyll-*a* and total seston volume were measured and related to population changes. *C. laticaudata* and *C. rotunda* were closely similar to each other in many features and showed significant differences to other species. *C. reticulata* and *C. megops* seem to be similar in some features, while *C. affinis* was ecologically different.

Introduction

In contrast with limnetic species, field observations and laboratory experiments on littoral species of cladocerans are still relatively scarce, even for species of common occurrence. As a result, life history strategies of littoral species are little known. Species of the genus *Ceriodaphnia* living in the littoral macrophyte zone of shallow water bodies belong among these lesser known cladocerans. Except for faunistic studies, in which these species are just on the checklist of zooplankton samples, only a small number of papers dealing with the study of their biology is available. In addition, the species were usually studied separately, not in coexistence with other *Ceriodaphnia* species (Shuba & Costa, 1972; O'Brien, 1974; Gophen, 1976; Cowgill et al., 1985). However, the occurrence of three or more species of the genus living together in the same time was described several times for *Ceriodaphnia* (Hensick, 1955; Burgis, 1967; Lim, 1976; Chengalath, 1982). Since I have recorded six species living all together in one locality, I have investigated differences and similarities among these species with respect to some of the population characteristics, in an effort to shed light on the mechanism of coexistence.

Study area

This study was carried out in a littoral macrophyte zone of a small shallow fishpond, Starý Pálenec, (area approx. 1 ha, max. depth 1.5 m) near the southbohemian town of Blatná, Czech Republic. The vegetation, composed of *Glyceria* sp., *Typha latifolia, Sparganium* sp., covered about 50% of the study area which was approximately 10 × 3 m in size. The water depth in the study area varied between 10 and 50 cm, and there was a thick layer of organic detritus on the bottom. The locality was mostly shaded by shore trees during the day. The pond was empty during winter and early spring, filled one week before the start of my observations and stocked with yearlings of common carp. The water level did not change considerably throughout the season.

Except for *Ceriodaphnia* species, the dense community of cladocerans also consisted of *Daphnia longispina* O. F. Müller, *Scapholeberis mucronata* (O. F. Müller), *Megafenestra aurita* (S. Fischer), *Simocephalus vetulus* (O. F. Müller), *Acroperus harpae* Baird and other Chydoridae. Prevalent phytoplankton species were Bacillariophyceae, Chlorophyceae and *Cryptomonas* sp.

Material and methods

Samples were taken at two-week intervals between June and August 1992, in the morning (9 h–11 h). Since it was not possible to use the classic limnological techniques that are common in pelagial investigations for a study of a littoral habitat, some modifications were made. Zooplankton samples were collected with a plastic beaker (volume approximately 700 ml) and put into a 10 l calibrated bucket, the mouth of which was protected with a sieve (mesh size 3 mm) to keep out macrophytes, leaves and water insects. Samples were taken while walking through the study area, taking care of sampling the whole water column (about one third close to the surface, one third of the samples were taken in the middle of the water column and one third close to the bottom). The total collected volume on each sampling occasion was 25 l. Samples were sieved alive immediately over mesh sizes of 100 μm, 350 μm and 1000 μm, to separate individuals of different size (it was found in preliminary studies that the size class of 350–1000 μm encompasses all adult *Ceriodaphnia*). The sample of the size class of 350–1000 μm was rapidly preserved by immersing the sieve in 40% formalin, to avoid losses of eggs and embryos (Hořická, 1983). The other fractions were preserved with 4% formalin.

Abundance data of the different species were collected by counting subsamples in a Sedgwick-Rafter chamber under a microscope. For each *Ceriodaphnia* species, juveniles, parthenogenetic and ephippial females and males were recorded separately. Within the size range of 350–1000 μm, total length (from the crown of the head to the tip of the spine at the posterior-dorsal angle of the carapace) of adult parthenogenetic females was measured and the number of eggs or embryos in their brood pouches was counted and presented as the clutch size. About 90–100 females of each sample and species, if present, were processed in this way. Samples with less than ten adult females were not processed at all (that is why there are some missing data in Figure 5 and Figure 6).

A representative top-to-bottom water sample was taken from the study region on each occasion, using a plastic tube (with a diameter of 4 cm and a length of 50 cm) for assessment of chlorophyll-*a* content, total seston volume and phytoplankton community composition. Dissolved oxygen was measured using Bruhns's modification of Winkler's titration method in samples taken with a special plastic pump to avoid contact with the air. The temperature of the water was mea-

sured with a mercury thermometer approximately 5 cm below the surface.

Results and discussion

Ceriodaphnia *species in the study area*

Six species of the genus *Ceriodaphnia* were found in the study area: *C. affinis* Lilljeborg, *C. laticaudata* P. E. Müller, *C. megops* G. O. Sars, *C. reticulata* (Jurine), *C. rotunda* G. O. Sars, and *C. setosa* Matile. In Europe, all species found by me, except *C. affinis*, occur typically in the littoral zone of water bodies like Starý Pálenec (Šrámek-Hušek, 1962; Hudec, 1993); *C. affinis* lives both in the littoral and in the pelagial of shallow waters (Straškraba, 1963; Hudec, 1993).

The changes in the population densities during the season are shown in Figure 3. The populations of all six species started to grow at the same time (the pond was filled one week before starting observations). *C. rotunda* and *C. laticaudata* occurred in low numbers, up to 8 individuals per liter, reached their maximum in the beginning of July and then disappeared. The very rare species, *C. setosa*, was recorded only on four sampling dates; there were only a few individuals per sample, so it was not included in further analysis. The other species occurred in much higher numbers (up to 200 individual per liter), reached peak densities later in the season and disappeared at the end of August.

Gamogenetic individuals appeared from 21.6. to 1.8 in all species, except of, in which gamogenesis was not recorded at all (Figure 4). Populations of *C. affinis* are supposed to migrate between the littoral and pelagic zones of fishponds during the year (Straškraba, 1963). Unfortunately, I have no data from the pelagic zone and therefore can not state whether gamogenesis in this species did or did not occur in the pelagic zone. The ratio of males to females, as well as the ratio of ephippial females to adult parthenogenetic females, were high in *C. laticaudata*, *C. rotunda* and *C. megops* (Figure 4) during the whole time high population densities. This is probably genetically programmed in view of the fact that these species are found usually in very shallow habitats which frequently dry up. Ephippial females in *C. reticulata* occurred only before the population began to decline.

Environmental correlates

It was found that all studied species show a similar pattern throughout the season in relation to environmental characteristics, so they will be discussed together here. The similar pattern is apparent in the abundances (Figure 3), mean body lengths of females carrying brood (Figure 5), and their mean clutch sizes (Figure 6). All these population parameters dropped or stagnated when dissolved oxygen concentration decreased sharply at the beginning of July (Figure 1). Gamogenesis started at the end of June, when the oxygen levels began to decline (Figure 4). The positive influence of temperature in the range of 15–25 °C on growth and reproduction of *Ceriodaphnia* species was found by Burgis (1967), Nováková (1976), Cowgill et al. (1985), Anderson & Benke (1994). However, the observed decrease and stagnation at the beginning of July does not seem to be the result of a decrease in temperature of only two degrees; more probably it was connected with the rapid decrease of dissolved oxygen (this depletion of dissolved oxygen may be caused by an enhancement of destructive processes in the water column and in the pond sediment due to the high temperatures in June, while considerable growth of autotrophs did not occur until later (Figure 2)). Especially remarkable is the conspicuous decrease of mean clutch size in all the species, while mean body length did not change substantially (except in *C. laticaudata*). Provided that the responses to environmental characteristics are similar in related cladocerans, as are the genera *Ceriodaphnia* and *Daphnia*, my observations correspond very well with the results of Hanazato & Dodson (1995) in *Daphnia pulex*. Low oxygen levels in their experiments reduced clutch size, while body length did not show any relationship with oxygen content.

The amount of dissolved oxygen might have an influence on the abundances of *C. laticaudata*, *C. rotunda* and *C. setosa*. At the end of July, when the oxygen concentration reached over about 7 mg per liter, their populations disappeared suddenly. These species apparently prefer biotopes with a relatively low concentration of oxygen (Burgis, 1967; Kamiński, 1979). These biotopes typically have a high amount of detritus, so the species can be found there because of their food preference for decaying vegetation and detritus. Similarly to Burgis (1967), I observed that keeping these species for a longer time in the laboratory was very difficult without the presence of detritus and decaying plants. Burgis (1967) observed also a

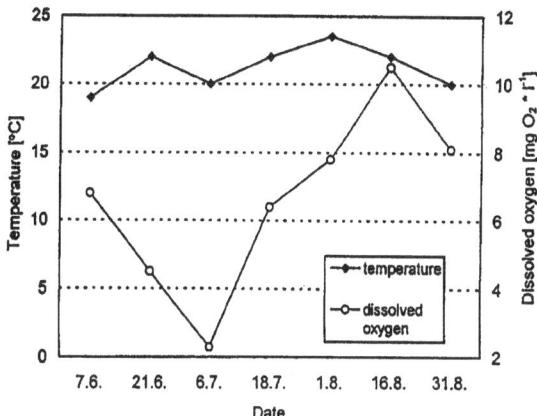

Figure 1. Seasonal change in the temperature and in the dissolved oxygen concentration in Starý Pálenec.

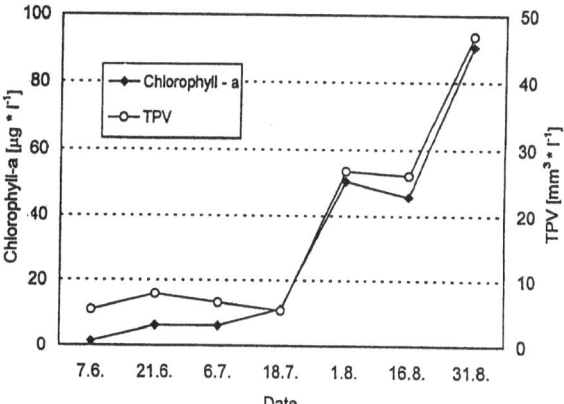

Figure 2. Seasonal change in the chlorophyll-*a* content and in the total volume of seston in Starý Pálenec.

conspicuous decrease in populations of *C. laticaudata* when oxygen content increased.

Chlorophyll-*a* content and total seston volume (Figure 2) show the same trend, which means that algae and not detritus made up the largest part of the volume of seston particles in the water column. The increase of dissolved oxygen from mid July coincides with an increase in chlorophyll-*a* content. The changes of chlorophyll-*a* do not seem to have a great influence on the *Ceriodaphnia* abundance, which is consistent with the results of Rodriguez et al. (1991) and Boersma & Vijverberg (1996). This is probably caused by the diet of the studied *Ceriodaphnia* species. They are both herbivorous and detritivorous; bacteria and detritus make up an important part of their diet (Gophen et al., 1974; Forsyth & James, 1984; Schoenberg, 1989). However, there is an apparent positive influ-

Table 1. Body length and clutch size in parthenogenetic females of the *Ceriodaphnia* species that coexist in Starý Pálenec (*n* = number of females; *min, max* = ranges of values; *CV* = coefficient of variance; *r* = correlation coefficient between length and clutch size; *ro* = overall correlation coeficient for each species.

Species	Date	*n*	Length (mm) Min	Max	CV	Clutch size Min	Max	CV	*r*	*r₀*
C. affinis	21.6	99	0.605	0.952	0.10	2	11	0.42	0.76	
	6.7	93	0.694	1.020	0.08	1	6	0.41	0.50	
	18.7	96	0.661	1.060	0.09	2	6	0.33	0.52	0.34
	1.8	97	0.582	1.010	0.14	2	12	0.46	0.80	
	16.8	100	0.504	0.806	0.10	2	7	0.32	0.52	
C. laticaudata	21.6	10	0.571	0.784	0.10	2	6	0.44	0.65	
	6.7	71	0.526	0.840	0.09	1	5	0.31	0.57	0.43
	18.7	31	0.594	0.907	0.11	2	4	0.22	0.77	
C. megops	21.6	97	0.694	1.010	0.09	1	7	0.46	0.74	
	6.7	83	0.683	0.974	0.07	1	6	0.40	−0.46	
	18.7	92	0.661	1.030	0.07	1	6	0.30	0.43	0.28
	1.8	87	0.616	1.100	0.10	1	8	0.37	0.68	
	16.8	99	0.538	0.974	0.11	2	8	0.41	0.56	
C. reticulata	21.6	86	0.605	0.952	0.10	2	8	0.38	0.63	
	6.7	94	0.571	0.952	0.13	1	6	0.32	0.33	
	18.7	99	0.605	0.997	0.13	2	4	0.29	0.31	0.19
	1.8	94	0.538	1.01	0.13	2	12	0.44	0.62	
	16.8	83	0.470	0.784	0.10	2	8	0.40	0.63	
C. rotunda	21.6	50	0.560	0.818	0.09	2	6	0.34	0.41	
	6.7	39	0.582	0.773	0.06	1	6	0.43	−0.02	0.15
	18.7	14	0.582	0.683	0.04	2	3	0.13	0.63	

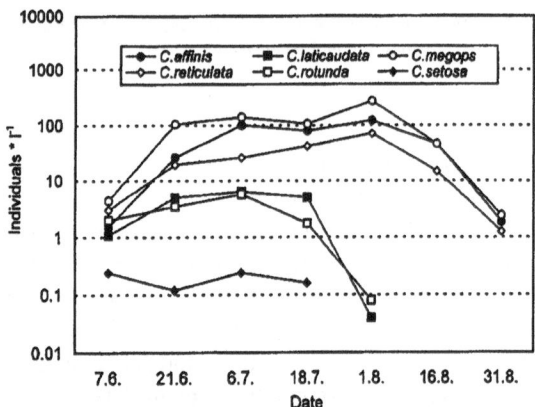

Figure 3. Seasonal changes in the abundance of the *Ceriodaphnia* species living in Starý Pálenec.

ence of chlorophyll-*a* content on mean clutch size, which agrees with the observations of Burgis (1967).

It was not possible to explain the rapid decrease of abundances at the end of September in *C. affinis,*

C. megops and *C. reticulata* by any of the measured environmental characteristics – temperature, dissolved oxygen and chlorophyll-*a* were still high enough not to cause extinction of the populations.

Differences among species

In spite of many similarities in response to environmental variables, there were consistent differences among *Ceriodaphnia* species. Mean length of females carrying broods (Figure 5) was highest in *C. megops,* and lowest in *C. laticaudata* and *C. rotunda.* The highest mean clutch size was observed in *C. affinis,* the lowest in *C. laticaudata* and *C. rotunda* (Figure 6). Although the S. E. plotted in Figure 5 and Figure 6 do not overlap very much (note that *n* was mostly close to 100), the variations in both body length and clutch size were rather high (Table 1). The variability (CV and S. E.) seems to be higher in clutch size but this can be just a consequence of a different measurement scale. A nonparametric Kruskal-Wallis test was

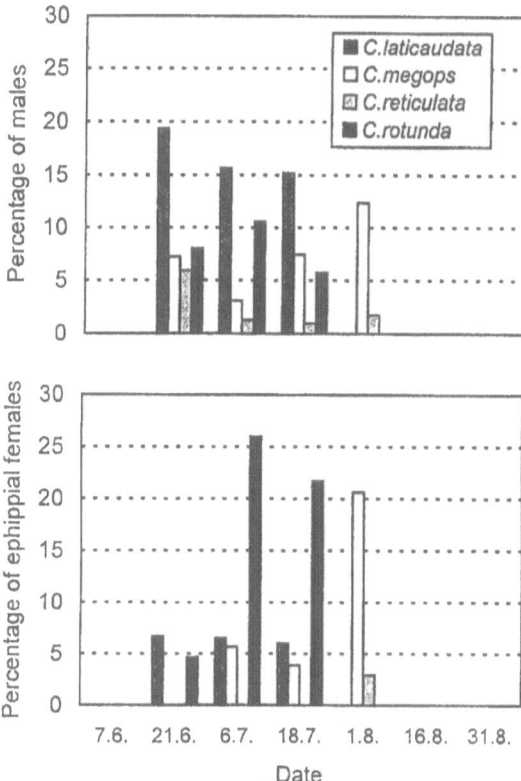

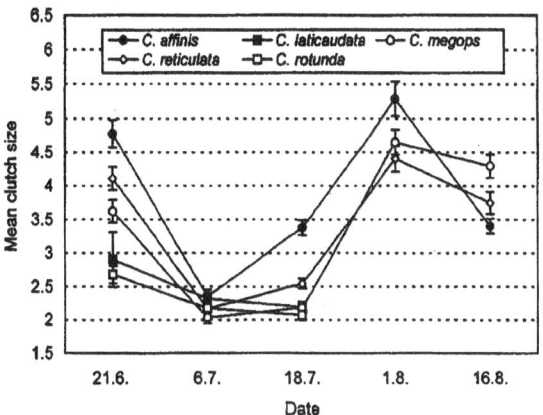

Figure 6. Seasonal changes in the mean clutch size (± S.E.) of the parthenogenetic females of the *Ceriodaphnia* species that coexist in Starý Pálenec.

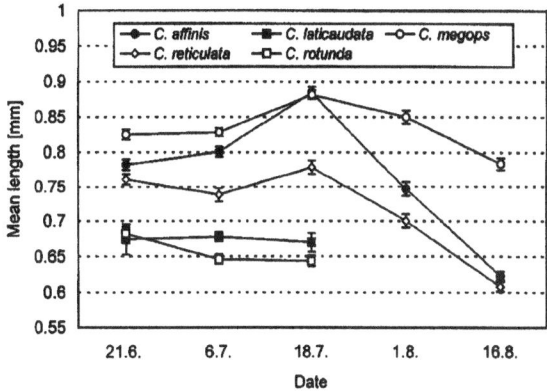

Figure 4. Seasonal change in the occurrence of male and ephippial female in the *Ceriodaphnia* species that coexist in Starý Pálenec.

Figure 5. Seasonal changes in the mean body length of females with brood (± S.E.) of the *Ceriodaphnia* species that coexist in Starý Pálenec.

used to test the null hypothesis that differences in body length and clutch size among species in all females with brood of the respective species are not significant. After rejecting H_0 in both cases (at $\alpha = 0.05$), a Dunn nonparametric multiple comparison for unequal sam-

ple sizes with ties (Zar, 1984) was run to determine between which species significant differences occur. I found that *C. laticaudata* and *C. rotunda* did not differ significantly either in body length or in clutch size, whereas *C. megops* and *C. reticulata* did not differ in clutch sizes. All other among species differences were significant. The differences in length agree with Burgis (1967) for *C. reticulata* and *C. megops*, but she found differences in clutch sizes too and close similarities in the length of *C. reticulata* and *C. laticaudata*.

Pearson correlation coefficients were calculated between length and clutch size for each population of each species for each date (Table 1). This method was preferred to linear regression since the functional dependence between these variables does not seem to be strong in some cases, as they can be influenced by different factors independently of each other (Hanazato & Dodson, 1995). The *r* coefficient shifted considerably in most species during the course of the season due to uneven variations in clutch size and body length (Figures 5 and 6) The changes in *r* were parallel to changes in clutch size for *C. affinis*, *C. megops* and *C. reticulata* (Figure 6). The significance of the differences among *r* for each species throughout the season and the significance of the differences among *r* for each species at any one date was determined using the method of comparing more than two correlation coefficients as outlined by Zar (1984). A nonsignificant difference (at $\alpha = 0.05$) was found in *C. laticaudata*, which means that the relation between body length and clutch size seems to be stable in *C. laticaudata*. In addition, at the end of the season (date 16.8.) the relationship between body length and clutch size was very similar in *C. affinis*, *C. reticulata* and *C. megops*. The correlation analysis was also calculated for all individuals of a given, to evaluate the overall correlation between body length and clutch size (Table 1). The highest overall relation was found in *C. laticaudata*; in *C. rotunda* and *C. reticulata* body length and clutch size related very weakly.

Conclusion

The studied species showed different features in their population characteristics when coexisting under the same environmental conditions, suggestive of species-specific responses. *C. laticaudata* seem to be very similar to *C. rotunda*, whereas *C. megops* and *C. reticulata* show some similarities, and *C. affinis* has an apparently different biology. It is conceivable that the coexistence of the studied species in one locality is possible due to the presence of different microhabitats, which were not recognised by the sampling method used. Although my study is preliminary, it confirmed some results of Burgis (1967) and indicated some interesting properties in the biology of littoral *Ceriodaphnia* species.

Acknowledgments

This study was carried out at the Faculty of Science, Charles University in Prague and forms a part of the author's master thesis. I thank Vladimír Kořínek for guidance throughout this work and for remarks on early drafts of the manuscript. I am especially grateful to Dorothy B. Berner for her valuable and substantial comments on the manuscript and for encouragement in completing the paper. Keith Edwards is acknowledged for improving the English, and two anonymous reviewers for their comments.

References

Anderson, D. H. & A. C. Benke, 1994. Growth and reproduction of the cladoceran *Ceriodaphnia dubia* from a forested floodplain swamp. Limnol. Oceanogr. 39: 1517–1527.

Boersma, M. & J. Vijverberg, 1996. Food effects on life history traits and seasonal dynamics of *Ceriodaphnia pulchella*. Freshwat. Biol. 35: 25–34.

Burgis, M. J., 1967. A quantitative study of reproduction in some species of *Ceriodaphnia* (Crustacea: Cladocera). J. anim. Ecol. 36: 61–75.

Chengalath, R., 1982. A faunistic and ecological survey of the littoral Cladocera of Canada. Can. J. Zool. 60: 2668–2682.

Cowgill, U. M., K. I. Keating & I. T. Takahashi, 1985. Fecundity and longevity of *Ceriodaphnia dubia/affinis* in relation to diet at different temperatures. J. Crust. Biol. 5: 420–429.

Forsyth, D. J. & M. R. James, 1984. Zooplankton grazing on lake bacterioplankton and phytoplankton. J. Plankton Res. 6: 803–810.

Gophen, M., 1976. Temperature dependence of food intake, ammonia excretion and respiration in *Ceriodaphnia reticulata* (Jurine) (Lake Kinneret, Israel). Freshwat. Biol. 6: 451–455.

Gophen, M., B. Z. Cavari & T. Berman, 1974. Zooplankton feeding on differentially labelled algae and bacteria. Nature 247: 393–340.

Hanazato, T., S. I. Dodson, 1995. Synergistic effects of low oxygen concentration, predator kairomone, and a pesticide on the cladoceran *Daphnia pulex*. Limnol. Oceanogr. 40: 700–709.

Hensiek, W., 1955. Das Vorkommen der Cladoceren im Dümmer im Sommer 1952. Arch. Hydrobiol. 50: 160–187.

Hořická, Z., 1983. Reaction of the population of *Daphnia pulicaria* and *Daphnia galeata* (Cladocera) on the changes of seston concentration within spring period. Msc. Thesis, Faculty of Science, Charles University, Prague: 119 pp (in Czech).

Hudec, I., 1993. Notes to the distribution of the genus *Ceriodaphnia* (Crustacea: Anomopoda, Daphniidae) in Slovakia. 1st part: *C. reticulata*, *C. affinis*, *C. megops*. Biológia Bratislava 48: 485–491.

Kamiński, K., 1979. *Ceriodaphnia setosa* Matile, 1890 (Cladocera) – rarely observed species in Poland. Przegl. Zool. 23: 41–44 (in Polish).

Lim, R. P., 1976. Community description, population dynamics and production of Cladocera with special reference to the littoral region of Pinehurst lake, Ontario. PhD. Thesis, University of Waterloo: 350 pp.

Nováková, J., 1976. Development and growth of three species of the genus *Ceriodaphnia*. Rigor. Thesis. Faculty of Science, Charles University, Prague, 46 pp (in Czech).

O'Brien, J. W., 1974. Filtering rate of *Ceriodaphnia reticulata* in pond waters of varying phytoplankton concentrations. Am. Midl. Nat. 91: 209–512.

Rodriguez, V., F. Echevarria & B. Bautista, 1991. In situ diel variation in gut pigments of *Ceriodaphnia* sp. in stratification and destratification periods. J. Plankton Res. 13: 187–196.

Schoenberg, S. A., 1989. Effects of algal concentration, bacterial size and water chemistry on the ingestion of natural bacteria by cladocerans. J. Plankton Res. 11: 1273–1295.

Shuba, T. & R. R. Costa, 1972. Development and growth of *Ceriodaphnia reticulata* embryos. Trans. am. Microsc. Soc. 91: 429–435.

Šrámek-Hušek, R., 1962. Fauna ČSSR. Lupenonožci-Branchiopoda, Sv. 16, ČSAV, Prague, 470 pp (in Czech).

Straškraba, M., 1963. Share of the littoral region in the productivity of two fishponds in Southern Bohemia. Rozpravy ČSAV, MPV 73, 13, 64 pp.

Zar, J. H., 1984. Biostatistical analysis. 2nd edn. Prentice-Hall, Englewood Cliffs, N.J. 718 pp.

Hydrobiologia **360**: 211–221, 1997.
A. Brancelj, L. De Meester & P. Spaak (eds), Cladocera: The Biology of Model Organisms.
©1997 Kluwer Academic Publishers.

The effect of different fish communities on the cladoceran plankton assemblages of the Kis-Balaton Reservoir, Hungary

János Korponai[1], Kálmán Mátyás[1], Gábor Paulovits[2], István Tátrai[2] & Nóra Kovács[3]
[1] West-Transdanubian District Water Authority, Dept. Kis-Balaton, H-8360 Keszthely, Csík F. sétány 1, Hungary
[2] Balaton Limnological Research Institute, Hungarian Academy of Sciences, H-8237 Tihany, POB 35, Hungary
[3] University of Veszprém, School of Environmental Engineering, H-8201 Veszprém, POB 158, Hungary

Key words: Kis-Balaton Reservoir, zooplankton, cladocerans, Daphnia sp.

Abstract

In 1995 the authors studied the effect of different fish communities on the structure of the cladoceran plankton in a shallow hypertrophic lake. After a fish kill of 1991, different fish communities developed in the Kazetta and the outer area of the Kis-Balaton reservoir. In the outer area of the reservoir, the densities of plankton feeding fish species were considerably higher than in the Kazetta. These differences induced changes in the structure and dynamics of the cladoceran plankton. The biomass of small-bodied cladocerans (mainly *Bosmina longispina*) was higher and the biomass of the large-bodied cladocerans (*D. hyalina, D. magna*) was lower in the western and eastern part of Kis-Balaton reservoir than in the Kazetta. A peak in cladoceran biomass in the Kis-Balaton reservoir was observed during the summer, close or during a bloom of filamentous cyanobacteria, whereas in the Kazetta a peak was observed during the spring, before the bloom of cyanobacteria. The adult females of *D. hyalina* were larger and produced more eggs in the Kazetta than in the outer area of the reservoir.

Introduction

In general, the water quality of lakes and reservoirs can be controlled by either reducing the external nutrient load or by manipulating trophic relationships. In order to influence eutrophication processes, the simultaneous application of these two strategies may be the most effective (Benndorf, 1988). Biotic factors and especially the structure of fish communities, have at least an equally important influence on the structure of zooplankton assemblages as the abiotic environment (Hrbáček et al., 1961). According to Carpenter et al. (1985), the predator organisms can control the population dynamics and abundance of the lower trophic levels. It has been observed in certain cases that a decrease in biomass of planktivorous fish the biomass of zooplankton increases (Hrbáček et al., 1961; Carpenter et al., 1987; Luecke et al., 1992). Within the zooplankton community, the relative abundance of cladocerans, as well as the average size of daphnids increases with a decrease in predation pressure

by fish (Giussani & Galanti, 1992). With an increase in biomass of planktivorous fish the biomass of the crustacean plankton decreases, and in turn that of the phytoplankton increases (Carpenter et al., 1987; Riemann et al., 1990; Vanni et al., 1990). At the same time the size of adult daphnid females gets smaller, as a response to an increase in the predation pressure by positively size selective fish (Machaček, 1991, 1993; Stibor, 1992; Taylor & Gabriel, 1992).

In addition, blooms of cyanobacteria, which appear regularly in eutrophic lakes, may significantly influence the structure of the zooplankton community (Richman et al., 1984; Benndorf & Hennig, 1989; Fulton III & Jones, 1991; Nauwerck, 1991; Lathrop & Carpenter, 1992). Two main factors might determine whether cyanobacteria are used as prey: the biochemical properties of the different species or strains (toxic effect), and the shape and size of the colonies (de Bernardi & Giussani, 1990). Filamentous cyanobacteria reduce the growth rate of *Daphnia* species. Their effect is, however, somewhat smaller on small species.

Table 1. The main chemical parameters of KBWPS I. The data refer to fortnightly averages from 1995 (*W*: Western part, *E*: Eastern part, *K*: Kazetta, +: $p<0.05$, ++: $p<0.001$).

Chemical component ($mg\,l^{-1}$)	Stations W	E	K	p
COD_{Mn}	13.055	17.175	23.565	++
BOD	4.453	6.355	5.542	n.s.
DO	10.26	10.275	8.669	n.s.
pH	8.405	8.41	8.373	n.s.
conductivity (μS)	582	538.9	584.51	n.s.
TSM	40.4	39.35	33.078	n.s.
NH_4-N	0.056	0.077	0.421	++
NO_3-N	0.506	0.372	0.232	n.s.
PO_3-P	0.013	0.007	0.069	++
chl-*a* ($mg\,m^{-3}$)	92.431	111.577	79.862	n.s.
TN	2.695	3.083	3.912	++
TP	0.23	0.251	0.336	+

When the filament density reaches the critical concentration, zooplankton is not able to control their growth anymore (bottleneck effect; Gliwicz, 1990a).

The Kis-Balaton Water Protection System (KBW-PS) was constructed on the lower section of the River Zala as a main inflow for the reduction of the external nutrient load to Lake Balaton (Figure 1). The KBW-PS consists of two main parts. The first part (KBW-PS I) was constructed in 1985. The KBWPS I is a shallow (area = 18 km², volume = 21 million m³, average depth = 1.14 m) and hypertrophic reservoir. It consists of three part: a western basin (W), a eastern basin (E) and the Kazetta (K). The Kazetta (surface area = 286 ha, mean depth = 1.2 m) operates independently from the other parts of KBWPS I (Pomogyi, 1993). In 1995, the Kazetta significantly differed from the other parts of the reservoir in four chemical parameters: chemical oxygen demand (COD_{Mn}), dissolved inorganic phosphor (PO_3-P), NH_4-N and total nitrogen (TN) (Kruskall-Wallis Test, $P<0.05$) (Table 1).

Zooplankton assemblages in the KBWPS I became homogenous quite soon after its construction (Gulyás & Forró, 1992; Korponai & Bancsi, 1996). Prussian carp (*Carassius auratus gibelio*) became the most important fish species in the reservoir. Although the prussian carp feeds mostly on benthic fauna and detritus (Pintér, 1989), the adults are able to filterfeed on planktonic crustaceans (Briylińska, 1991; Csányi et al., 1996; Tátrai, pers. comm.).

In the Kazetta, drastic fish kills were induced by *Aeromonas punctata* between 1991 and 1993. About 500 kg ha^{-1} (mainly prussian carp and other planktivorous fish) were killed during that time. Following the fish kill, water quality and the structure of planktonic communities of the Kazetta significantly changed with respect to the other parts of the KBWPS I. Large bodied cladocerans (*D. magna*, *D. hyalina*) appeared and became dominant (Mátyás, 1993; Mátyás et al., 1995).

The main aim of the present study was to investigate to what extent zooplankton assemblages had changed in the Kazetta after the fish kill, and to find out the most important biological factors controlling the water quality of the KBWPS I system.

Materials and methods

Water samples were collected fortnightly from three stations representative of the KBWPS I system in 1995 (Figure 1). For zooplankton analysis, 10–20 l of water from the whole water column were filtered through a plankton net of 65 μm mesh and were preserved with Lugol solution. The quantitative analyses were performed using inverted microscope. For the species composition, usually 10% of the sample volume was counted. In cases where less than 100 individuals were counted, the entire sample was assessed. The length of the first 20 individuals encountered in each sample was measured with an ocular micrometer and their biomass was estimated using length – dry weight regression relationships (Downing & Rigler, 1984).

In the case of *D. hyalina*, the body length of the first 20–25 egg-bearing females was measured from the top of the head to the base of the tail spine. The number of eggs per individual was also recorded. *D. magna* was excluded from this investigation, because they lost their eggs during fixation. Differences among the populations from the three sampling sites were tested for by *t*-test.

Integrated water samples for phytoplankton analysis were taken fortnightly and were preserved with Lugol solution and counted using Utermöhl protocol. Biovolumes were computed from the numerical density of each species and its cell volume. Chlorophyll-*a* concentration was determined spectrophotometrically after methanol extraction (Iwamura et al., 1970).

Fish stock was sampled four times between March–October 1995 sampling a surface of 0.05 to 0.25 ha using electric fishing gear.

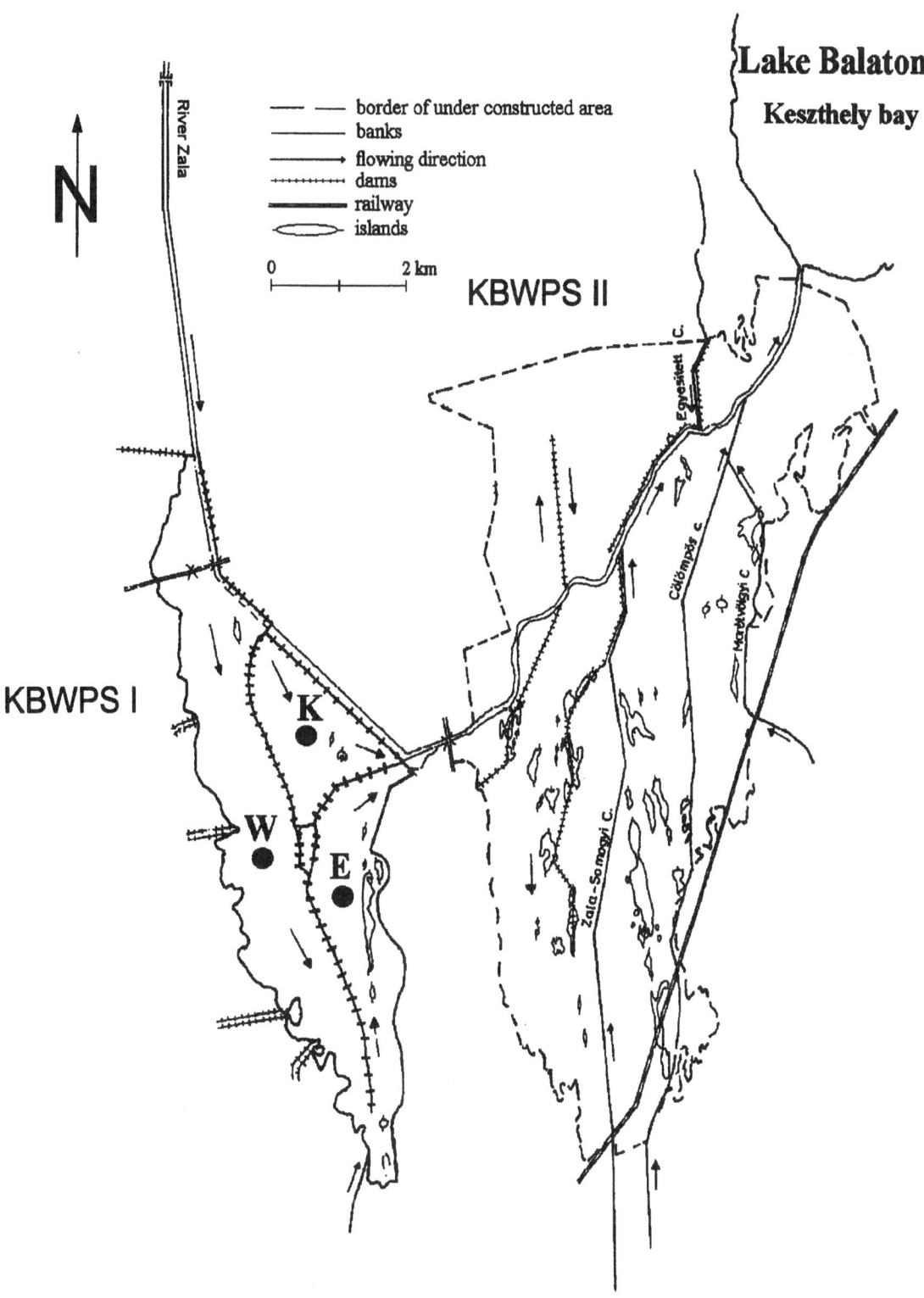

Figure 1. map of study area. The closed circles represent the sampling statins (*W*: Western part, *E*: eastern part, *K*: Kazetta).

Results

Fish

In the Kazetta, 293 individuals of 16 fish species were collected during spring and summer. The estimated density varied between 390 and 712 ind ha^{-1}. The dominant species were roach (*Rutilus rutilus*, 30%), bream (*Abramis brama*, 23%) and prussian carp (*Carassius auratus gibelio*, 18.5%). There were altogether 12 species of plankton, detritus and benthos feeders, and these represented 94% of the estimated total abundance. The four main predators (pike (*Esox lucius*), asp (*Aspius aspius*), pike-pearch (*Stizostedion lucioperca*) and wels (*Silurus glanis*)) represented only 6% of the total fish biomass, with pike-pearch (2.5%) being the most abundant.

In the autumn, the density of planktivorous fish (bleak (*Alburnus alburnus*), pumpkinseed (*Lepomis gibbosus*), perch (*Perca fluviatilis*), false rasbora (*Pseudorasbora parva*)) and their offspring reached 9600 ind ha^{-1}, corresponding to a biomass of 10 kg ha^{-1}. The density of benthos and detritus feeders (bream, roach, white bream (*Blicca bjoerkna*), prussian carp, crucian carp (*Carassius carassius*), tench (*Tinca tinca*), ruffe (*Gymnocephalus cernuus*), and 1+ age classes of carp (*Cyprinus carpio*) amounted to 670 ind ha^{-1}, with a biomass of approx. 127 kg ha^{-1}. The density of predators was 130 ind ha^{-1} and the biomass was estimated to be 48 kg ha^{-1}. The prey: predator biomass ratio was 137 kg ha^{-1}: 48 kg ha^{-1} (approx. 3:1) while the proportion between plankton and detritus feeders was 10 kg ha^{-1}: 127 kg ha^{-1} (approx. 1:13).

In the eastern and western part of KBWPS I, 23 species were collected. The sample consisted of 417 individuals. In terms of abundance, the dominant species were prussian carp (31%) and roach (16%). The total density of the plankton feeders exceeded 10 000 ind ha^{-1}, with an estimated biomass of 95 kg ha^{-1}. The density of benthos and detritus feeders was 236 ind ha^{-1}, with an estimated biomass of 276 kg ha^{-1}. The density of predators was 118 ind ha^{-1}, with an estimated biomass of 47 kg ha^{-1}. The prey: predator biomass ratio was 89: 11 (approx. 9:1), while that of plankton to benthos feeders was 26: 74 (approx. 1:3).

Zooplankton

In total, 13 cladoceran species were found in KBWRS I (Table 2). *Bosmina longirostris*, *Daphnia hyalina*,

Table 2. Occurrence of cladoceran species in KBWPS I, 1995. (*W*: Western part, *E.*: Eastern part, *K*: Kazetta)

	W	E	K
Alona guttata			+
A. quadrangularis	+		+
A. rectangula	+	+	+
Bosmina longirostris	+	+	+
Chydorus sphaericus	+	+	+
Daphnia hyalina	+	+	+
D. magna			+
Ilyocryptus sordidus	+		+
Leptodora kindtii	+	+	
Macrothrix laticornis	+		+
Moina micrura	+	+	+
Pleuroxus trigonellus			+
Simocephalus vetulus	+		

Chydorus sphaericus and *Moina micrura* were the most abundant species.

In station W of KBWRS I, *B. longirostris* reached population densities between 200 ind m^{-3} and 445 10^3 ind m^{-3}, the latter being found in August (Figure 2). Four maxima were observed during the studied period. *D. hyalina* was present throughout the year and its population density ranged from 1000 to 214.4 10^3 ind m^{-3} (in October). *C. sphaericus* was also found during the entire year, although in relatively low numbers (100–2500 ind m^{-3}).

In station E of KBWPS I, *B. longirostris*, was the most abundant cladoceran, with a mean population density of approx. 200 10^3 ind m^{-3} and a maximum in June. *D. hyalina* was much less abundant (0.75– 50 10^3 ind m^{-3}) and its maximum population density was recorded in July. *C. sphaericus* was present only in spring and autumn, in very low numbers (0.10–37 10^3 ind m^{-3}).

In the Kazetta, relatively high numbers of *D. hyalina* were found (0.8–173 10^3 ind m^{-3}), with a peak in April and one at the end of July. The density of *B. longirostris* varied between 100 ind m^{-3} and 558 10^3 ind m^{-3}, with a peak density in June. *D. magna* was also present in relatively high numbers (2–65 10^3 ind m^{-3}), with a peak in June. *C. sphaericus* was more abundant in the Kazetta than in the other stations (0.05 10^3, 59 10^3 ind m^{-3}).

We subdivided cladocerans into the small-bodied species (*B. longirostris*, *C. sphaericus* and *M. micrura*) and the large-bodied species (*D. hyalina* and *D. magna*). The changes in biomass of the two groups

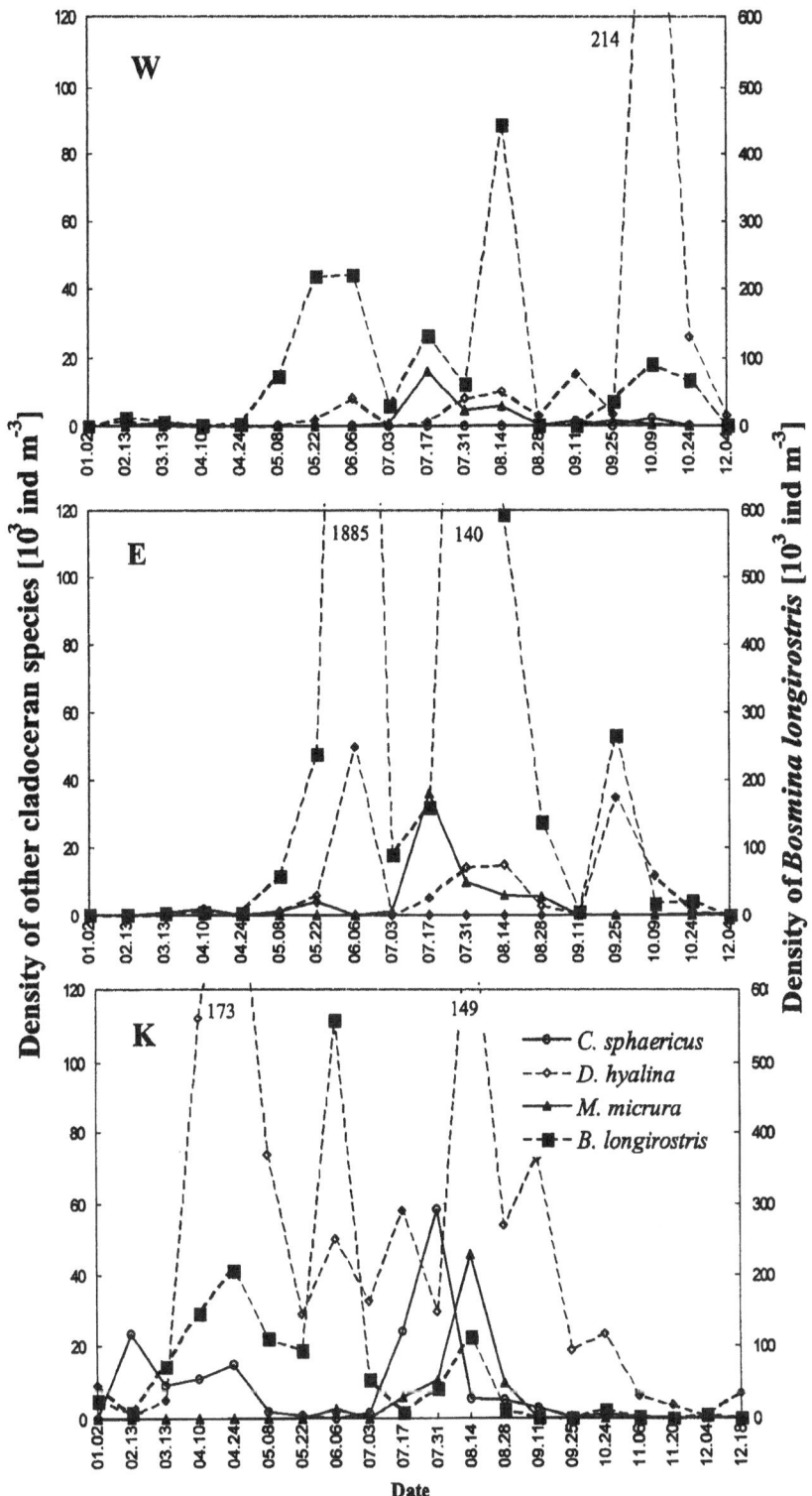

Figure 2. Changes in population density of the most abundant cladocerans in the different parts of Kis-balaton Water Protection System I (KBWPS I) during the course of 1995 (*W*: Western part, *E*: Eastern part, *K*: Kazetta).

216

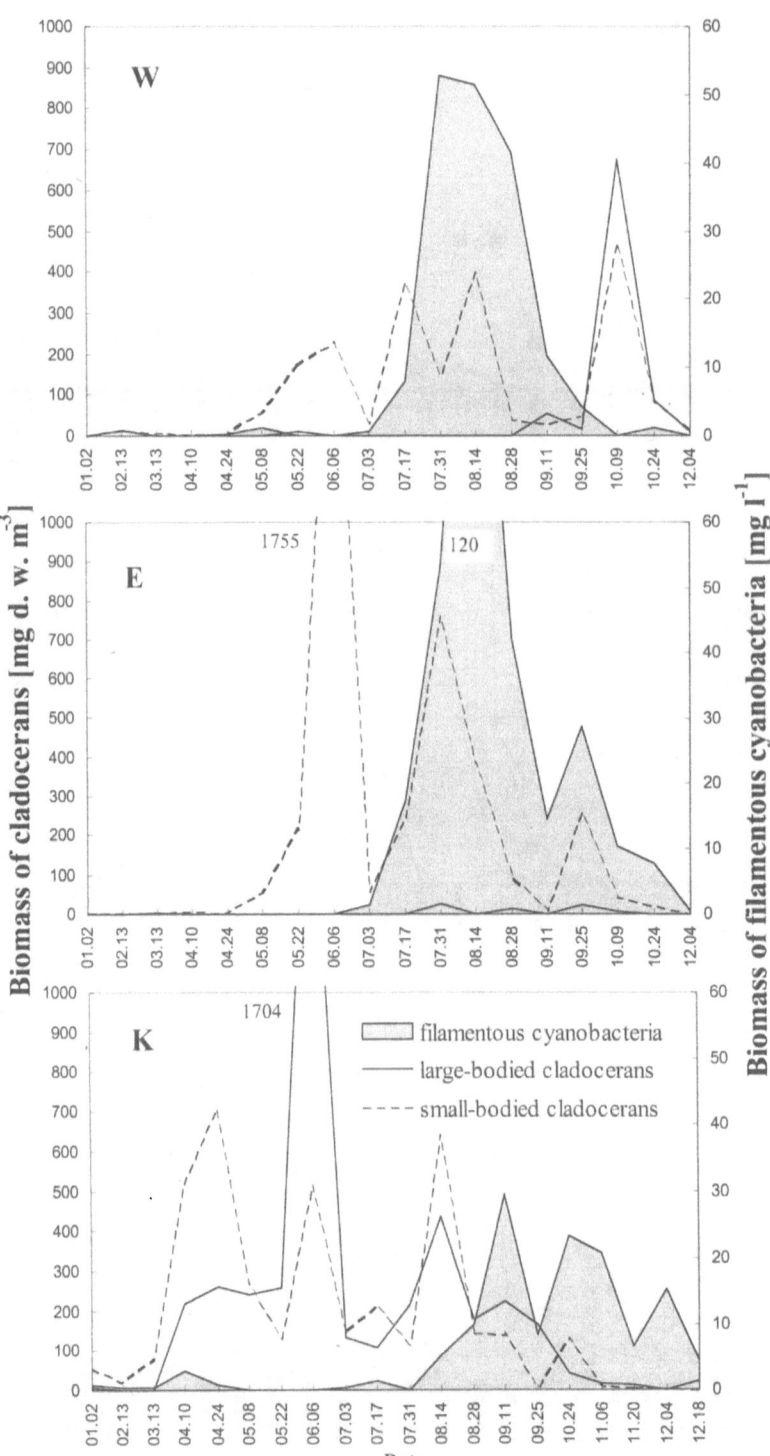

Figure 3. Changes in biomass of large and small bodied cladocerans and of filamentous cyanobacteria in the three parts of KBWPS I during the course of 1995 (*W*: Western part, *E*: Eastern part, *K*: Kazetta).

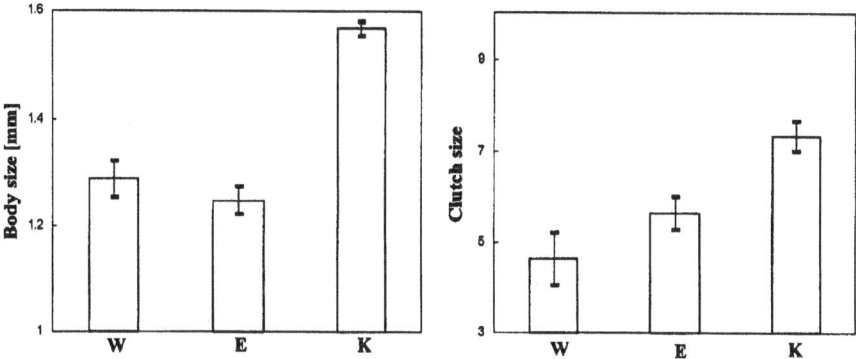

Figure 4. Mean body length and number of eggs (mean ± SD) of *D. hyalina* females in different part of KBWPS I (*W*: Western part; *E*: Eastern part, *K*: Kazetta).

with showed different seasonal pattern in the different stations (Figure 3). In station W, the biomass of small-bodied cladocerans was highest in October, when that of cyanobacteria was low. However, values measured during the bloom exceeded 100 mg d.w. m^{-3}. The biomass maximum of large-bodied cladocerans was recorded in October, following the bloom of cyanobacteria. In station E, the biomass of the small cladocerans peaked before the cyanobacteria bloom, and the biomass of the large cladocerans was low during the bloom. In the Kazetta, the biomass maxima of both groups preceded the bloom of cyanobacteria. The gravid *D. hyalina* females found in the different parts of KBWPS I system differed in length and number of eggs (Figure 4). In the eastern and western part, gravid females started to produce eggs at the same body size (*t*-test, $t = 0.956$, $P = 0.171$) and the average number of eggs/female (clutch size) was also similar (*t*-test, $t = -1.58$, $P = 0.069$). In the Kazetta, adult *D. hyalina* females were significantly larger in size (*t*-test, $t = -4.652$, $P < 0.001$), and carried on average more eggs (*t*-test, $t = -12.456$, $P < 0.001$) when compared to the combined sample of eastern and western sample.

Phytoplankton

The annual average phytoplankton biomass was high in the eastern and western part of KBWPS I compared to that of the Kazetta (Figure 5). At the beginning of the year, phytoplankton was characterised by relatively low biomass and a dominance of unicellular diatoms (*Stephanodiscus hantschii*). In summer, filamentous cyanobacteria became dominant and caused extended surface blooms. In the beginning, these phytoplankton blooms were dominated by *Anabaena flos-*

aquae, then *Oscillatoria agardhii* also appeared in high numbers, followed by *Cylindrospermopsis raciborskii*. In autumn, cyanobacteria became less abundant and unicellular dominated the phytoplankton community again.

In the Kazetta, the average phytoplankton biomass in spring was dominated by *Cryptomonas* sp. and unicellular diatoms whereas in summer it was dominated by *Chlorococcales* and *Cryptomonas* sp.. At the end of summer, cyanobacteria became dominant. *Cryptomonas* sp. again became important at the end of the year.

Discussion

According to the biomanipulation concept (Benndorf, 1988) and the trophic cascade hypotesis (Carpenter et al., 1985), any change in the biomass or the relative abundance of planktivorous versus piscivorous fish leads to changes in the biomass of herbivorous zooplankton. If the biomass of planktivorous fish decreases, the biomass of large cladocerans increases, and the phytoplankton biomass decreases. The final result is may be clear water. The top-down effect of fish, however depends on the trophic condition of the water body: it is clearly visible in less eutrophic lakes, but in highly eutrophic lakes its effects is limited (McQueen et al., 1986). A clear water phase often appears during the spring maximum of zooplankton, when the predation pressure of fish is lòw due to the lack of feeding pressure from 0+ fish (Lampert, 1988).

A clear water phase occurred only in the Kazetta in spring and early summer when the biomass of cladocerans was high enough to control the growth of phytoplankton (Figure 6). In the western and eastern stations

218

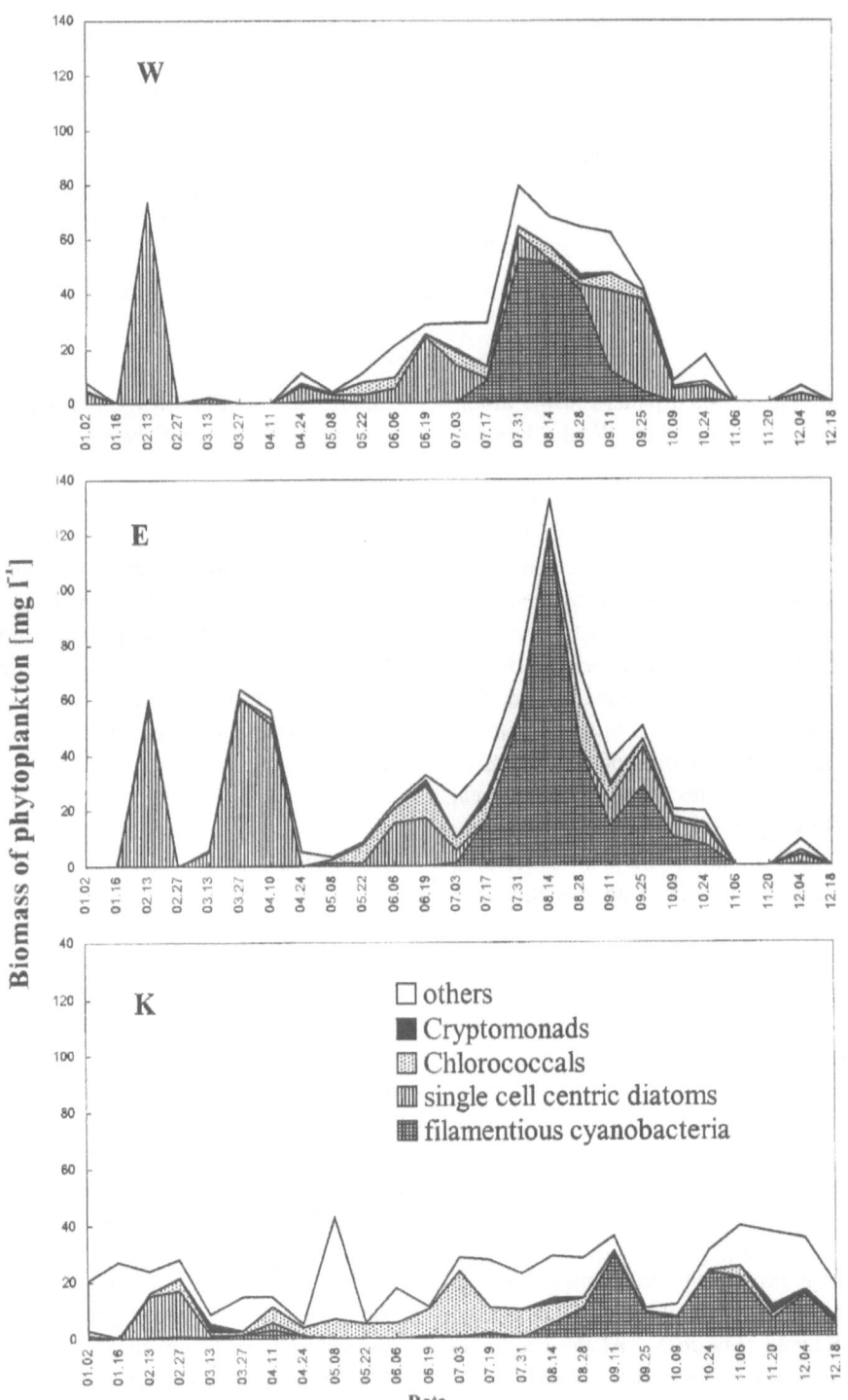

Figure 5. Seasonal changes in phytoplankton biomass and community composition in the three parts of KBWPS I (*W*: Western part, *E*: Eastern part, *K*: Kazetta) during the course of 1995.

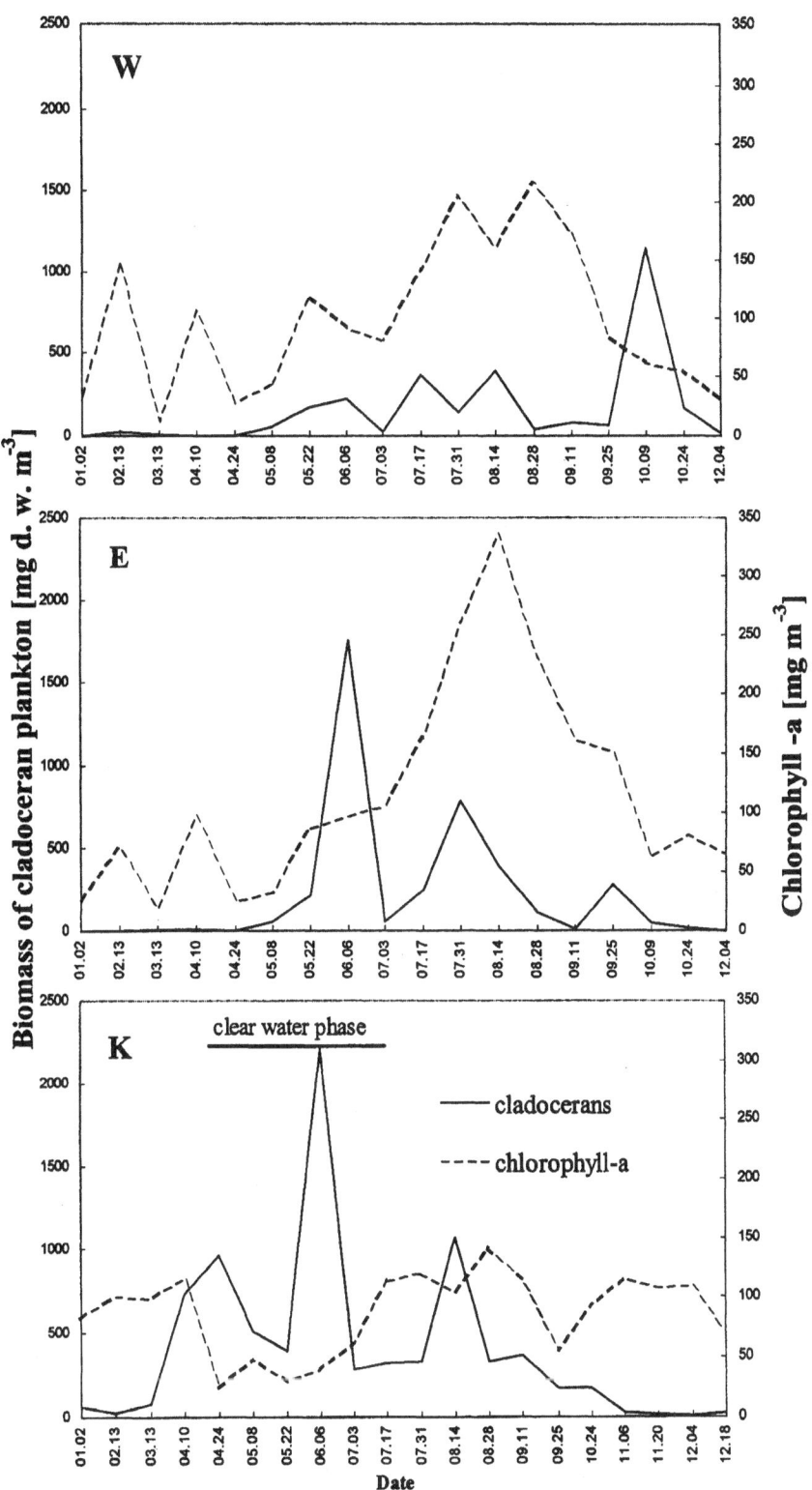

Figure 6. Seasonal changes in chlorophyll-*a* and cladoceran biomass in the three parts of KBWPS I (*W*: Western part, *E*: Eastern part, *K*: kazetta) during the course of 1995.

of KBWPS I, the low chlorophyll-*a* concentration in spring may have been influenced by other environmental factors rather than by zooplankton grazing. In all the three stations of the KBWPS I system, a clear water phase could be observed in spring, when phytoplankton was mainly composed by edible algal species and the phytoplankton community was dominated by unicellular diatoms and *Cryptomonas* sp.. During this period, cladoceran biomass in stations W and E was low whereas in the Kazetta, the biomass of *B. longirostris* and *D. hyalina* was high. Algal biomass was not lower and cladoceran species appeared earlier in the Kazetta than in the other parts of KBWPS I. As a result of reduced fish predation, *Daphnia* species were larger in size and produced more eggs in the Kazetta than in the KBWPS I (see also Machaček, 1991, 1993; Stibor, 1992).

Gliwicz (1990b) showed that at high levels of abundance, *Aphanizomenon* sp. reduces the growth rate of large *Daphnia* species (*D. magna, D. pulicaria*), whereas small species (*D. cucullata, D. hyalina*) tend to grow faster. According to Dawidowicz et al., (1988), the amount of filamentous cyanobacteria necessary to interfere with *Daphnia* growth was 150 mg m^{-3} filamentous blue-greens chlorophyll-*a* (ca 50 mg l^{-1} biomass). In our system, in addition to the effect of fish predation (the top-down control mechanism), interference from cyanobacteria may also have been an important factor. As a large species are more sensitive to this interference than small ones (Gliwicz, *loc. cit.*), this might may the dominance of small cladocerans in the stations of KBWPS I where cyanobacteria were most abundant and the appearance of *D. magna* in the Kazetta where the biomass of cyanobacteria was low.

Acknowledgments

The authors thank their colleagues from the Laboratory of West-Transdanubian District Water Authority for their help with data collection and collaboration as well as to Dr István Varanka for his help with fisheries.

References

Benndorf, J., 1988. Objectives and unsolved problems in ecotechnology and biomanipulation: A preface. Limnologica 19: 5–8.

Benndorf, J. & M. Henning, 1989. *Daphnia* and toxic blooms of *Microcystis aeruginosa* in Bautzen Reservoir (GDR). Int. Revue ges. Hydrobiol. 74: 233–248.

Brylińska, M., 1991. Ryby słodkowodne Polski (Freshwater fish of Poland). Państwowe Wydawnictwo Naukowe. Warsawa, 429 pp (in Polish).

Carpenter, S. R., J. F. Kitchell & J. R. Hodgson, 1985. Cascading trophic interactions and lake productivity. Fish predation and herbivory can regulate lake ecosystem. Bioscience 35: 634–639.

Carpenter, S. R., J. F. Kitchell, J. R. Hodgson, P. A. Cochran, J. J. Elser, M. M. Elser, D. M. Lodge, D. Kretchmer, X. He & C. von Ende, 1987. Regulation of lake primary productivity by food-web structure. Ecology 68: 1863–1876.

Csányi, B., J. Németh & P. Gulyás, 1996. Haltáplálék-szervezetek vizsgálata a KBVR víztereiben. (Study of fish food items in different habitats of KBWPS). In Pomogyi, P. (ed.), Összefoglaló értékelés a Kis-Balaton Védörendszer 1991–1995 közötti kutatási eredményeiröl. Kis-Balaton Ankét, 1996, Keszthely. 423–436. (in Hungarian).

de Bernardi & G. Giussani, 1990. Are blue-green algae a suitable food for zooplankton? An overview. Hydrobiologia 200/201: 29–41.

Downing, J. A. & F. H. Rigler (eds), 1984. A Manual on Methods for the Assessment of Secondary Productivity in Fresh Waters. IBP Handbook 17. 2nd edn. Blackwell. Oxford, 501 pp.

Dawidowicz, P., Z. M. Gliwicz & R. D. Gulati, 1988. Can *Daphnia* prevent a blue-green algal bloom in hypertrophic lakes? A laboratory test. Limnologica 19: 21–26.

Fulton III, R. S. & R. C. Jones, 1991. Growth and reproductive responses of *Daphnia* to cyanobacterial blooms on the Potomac River. Int. Revue ges. Hydrobiol. 76: 5–19.

Gliwicz, Z. M., 1990a. *Daphnia* growth at different concentration of blue-green filaments. Arch. Hydrobiol. 120: 51–65.

Gliwicz, Z. M., 1990b. Why do cladocerans fail to control algal blooms? Hydrobiologia 200/201: 83–97.

Gulyás, P. & L. Forró, 1992. Composition and abundance of microcrustacean fauna in the Upper Reservoir (Hídvégi-tó) of the Kis-Balaton. Miscnea zool. hung. 7: 39–51.

Giussani, G. & G. Galanti, 1992. Experience in eutrophication recovery by biomanipulation. In Guilizzoni, P., G. Tartari & G. Giussani (eds). Limnology in Italy. Mem. Ist. ital. Idrobiol. 50: 397–416.

Hrbáček, J., M. Dvořakova, V. Kořínek & L. Procházkóva, 1961. Demonstration of the effect of the fish stock on the species composition of zooplankton and the intensity of metabolism of the whole plankton association. Verh. int. Ver. Limnol. 14: 192–195.

Iwamura, T., H. Nagai & S. Ichimura, 1970. Improved methods for determining contents of chlorophyll, protein, ribonucleic acid and deoxyribonucleic acid in planktonic population. Int. Revue ges. Hydrobiol. 55: 131–147.

Lampert, W., 1988. The relationship between zooplankton biomass and grazing: A review. Limnologica 19: 11–20.

Lathrop, R. C. & S. R. Carpenter, 1992. Zooplankton and their relationship to phytoplankton. In Kitchell, J. F. (ed.), Food Web Management. A Case Study of Lake Mendota. Springer Verlag, New York: 127–150.

Luecke, C., L. G. Rudstam & Y. Allen, 1992. Interannual patterns of planktivory 1987–89: An analysis of vertebrate and invertebrate planktivores. In Kitchell, J. F. (ed.), Food Web Management. A Case Study of Lake Mendota. Springer Verlag, New York: 275–301.

Korponai, J. & I. Bancsi, 1996. A Kis-Balaton Védörendszer zooplankton szerkezetének változása 1991–1995 (Changes in zooplankton community between 1991–95 in KBWPS). In Pomogyi, P. (ed.), Összefoglaló értékelés a Kis-Balaton Védörendszer 1991–1995 közötti kutatási eredményeiröl. Kis-Balaton Ankét, 1996, Keszthely. 263–274. (in Hungarian).

Macháček, J., 1991. Indirect effect of planktivorous fish on the growth and reproduction of *Daphnia galeata*. Hydrobiologia 225: 193–197.

Macháček, J., 1993. Comparison of the response of *Daphnia galeata* and *Daphnia obtusa* to fish-produced chemical substrate. Limnol. Oceanogr. 38: 1544–1550.

Mátyás, K., 1993. Effect of food web changes on the primary producers in a shallow pond of the Kis-Balaton Water Protection System. 5th International Conference on The Conservation and Management of Lakes 'Strategies for Lake Ecosystem beyond 2000' – 17–21 May, 1993, Stresa, Italy. Proceedings 148–151.

Mátyás, K., J. Korponai & P. Pomogyi, 1995. Effect of fish removal on the water quality of 'Cassette' of Kis-Balaton Water Protection System (KBWPS), Hungary. 6th International Conference on The Conservation and Management of Lakes 'Harmonising Human Life With Lakes' – Oktober 23–27, 1995, Kasumigaura. Proceedings 1: 328–991.

McQueen, D. J., J. R. Post & E. L. Mills, 1986. Trophic relationship in freshwater pelagic ecosystems. Can. J. Fish. aquat. Sci. 43: 1571–1578.

Nauwerck, A., 1991. Zooplankton changes in Mondsee. Verh. int. Ver. Limnol. 24: 974–979.

Pintér, K., 1989. Magyarország halai biológiájuk és hasznosításuk. (Fish fauna of Hungary and its exploitation). Akadémiai Kiadó, Budapest, 202 pp (in Hungarian).

Pomogyi, P., 1993: Nutrient retention by the Kis-Balaton Water Protection System. Hydrobiologia 251: 309–320.

Richman, S., M. D. Bailiff, L. J. Mackey & D. W. Bolgrien, 1984. Zooplankton standing stock, species composition and size distribution along a trophic gradient in Green Bay, Lake Michigan. Verh. int. Ver. Limnol. 22: 475–487.

Riemann, B., K. Christoffersen, H. J. Jensen, J. P. Müller, C. Lindegaard & S. Bosselmann, 1990. Ecological consequences of a manual reduction of roach and bream in a eutrophic, temperate lake. Hydrobiologia 200/201: 241–250.

Stibor, H., 1992. Predator induced life-history shifts in a freshwater cladoceran. Oecologia 92: 162–165.

Taylor, B. E. & W. Gabriel, 1992. To grow or not to grow: optimal resource allocation for *Daphnia*. Am. Nat. 139: 248–266.

Vanni, M. J., C. Luecke, J. F. Kitchell & J. J. Magnuson, 1990. Effect of planktivorous fish mass mortality on the plankton community of Lake Mendota, Wisconsin: implication for biomanipulation. Hydrobiologia 200/201: 329–336.

Hydrobiologia **360**: 223–232, 1997.
A. Brancelj, L. De Meester & P. Spaak (eds), Cladocera: The Biology of Model Organisms.
©1997 *Kluwer Academic Publishers.*

Importance of water-level fluctuation on population dynamics of cladocerans in a hypertrophic reservoir (Lake Arancio, south-west Sicily, Italy)

Luigi Naselli-Flores & Rossella Barone
Dipartimento di Scienze Botaniche, Via Archirafi, 38, 90123 Palermo, Italy
(e-mail: luigi.naselli@mbox.unipa.it)

Key words: Zooplankton, man-made lakes, phytoplankton, fish predation, water-level fluctuations

Abstract

During a period of three years (1990–1991 and 1993), we studied the population dynamics of planktonic cladocerans in a hypertrophic reservoir. Weekly sampling revealed that the five most common species followed a trend which reflects the peculiar hydrological characteristics of the reservoir and their key position in the pelagic food web. In particular, 1991 was characterized by a strong water inflow which probably interfered with the reproductive activities of the dominant fish population (*Rutilus rubilio*) and reduced the concentration of inedible planktonic algae allowing the development of small Chlorococcales. This event was associated with higher population densities of *Daphnia hyalina* compared to the other years of the survey. In spring 1991, an extended clear-water phase was observed and Secchi disk depth increased to 6 m, whereas in the other years it did not surpass 1 m. In addition, the *D. hyalina* population persisted throughout the summer in 1991, whereas it started to decline at the end of June in the other years. This development of *D. hyalina* probably influenced the population dynamics of the other cladoceran species in the reservoir, and in particular reduced the summer growth of *Diaphanosoma lacustris* and delayed the occurrence of *Bosmina longirostris*. Stomach analysis indicated that *D. hyalina* is the preferred food item of juvenile (less than two months old) *R. rubilio*. Overall, the hydrology of the reservoir was observed to interact with the trophic processes in the pelagic environment of the ecosystem in at least two different ways: via bottom-up processes, influencing phytoplankton dynamics, and via top-down processes, regulating the predation efficiency of the planktivores.

Introduction

Pronounced fluctuations in water level over a short time are an important characteristic of many reservoirs. Fluctuations in water level may affect phytoplankton biomass and species composition by influencing both underwater light climate (Barone & Naselli Flores, 1994; Naselli Flores & Barone, 1996) and nutrient dynamics (Kimmel et al., 1990).

With respect to fish populations, reproductive success has been generally found to increase as reservoirs fill, since flooded vegetation offers better sites for spawning (O'Brien, 1990). In some cases, however, it was observed that the increased siltation, associated with the filling of the reservoir, damages the spawned eggs (Hassler, 1970). It has also been reported that

a drawdown may seriously damage nursery areas and dry out the eggs (Kubecka, 1993).

It is widely known that zooplankton occupies a key position in pelagic food webs. As a result, zooplankton biomass and species composition may be influenced by water mass movements which affect food availability (bottom-up effects) and predation pressure (top-down effects). Due to size-dependent predation rates, a high predation pressure from fish may substantially affect the zooplankton assemblage favouring a selective shift from large to small bodied species and individuals (Brooks & Dodson, 1965). Ultimately, the phytoplankton community may be influenced by both grazing and nutrient recycling by zooplankton (Kitchell & Carpenter, 1993). Cladoceran population dynamics often play an important role in lacustrine ecosystems, since cladocerans are the principal source of food for fish fry (Zalewski et al., 1990) and important grazers of phytoplankton (Peters & de Bernardi, 1987).

In the present paper we study the pelagic community of hypertrophic reservoir Lake Arancio throughout 1990–1991 and 1993, and analyze the effects caused by water-level fluctuations on the population dynamics of pelagic cladocerans.

Description of the site studied

Lake Arancio is a hypertrophic water body (Barone & Naselli Flores, 1994) located in south-west Sicily (Italy); it was created in 1951, by damming a torrent to retain water for irrigation and for recreational purposes.

The climate of the region is Mediterranean semi-arid with an average annual rainfall of 550 mm in the years 1970–1991 (Regione Siciliana, 1996). Following a spring and summer period of nearly complete drought, heavy rains, generally concentrated in a few winter days, account for 90% of the annual rainfall. In this period, the outlets of the reservoir are not active and the volume expands quickly. In the spring and summer seasons, the volume contracts due to intense evaporation and out-flow through outlets for irrigation (Barone et al., 1993).

The catchment area of the lake covers a surface of 138 km^2 and the lake surface is 3.5 km^2 at maximum storage capacity, which corresponds to a volume of 32×10^6 m^3 and to a maximum depth of 30 m at the dam. Nevertheless, these values have been rarely attained since the creation of the dam. In the last twenty years, the maximum volume recorded was 24×10^6 m^3 (in May 1979), while the minimum (1.5×10^6 m^3) was reached in September 1982 when the reservoir had nearly dried up. The years in which the present investigation was carried out were characterized by a maximum volume of 6.5×10^6 m^3 (1990), 10.5×10^6 m^3 (1991) and 22×10^6 m^3 (1993). Figure 1A shows that in February 1991, a strong water inflow increased the lake volume by about 2×10^6 m^3 in a few days.

Although the lake is potentially monomictic, it was always circulating during the course of the investigation, and its mean depth never exceeded 7.5 m. These shallow lake features are principally due to the high out-flow rates during summer, so that no stable thermocline could be formed.

Due to the continuous fluctuations in water-level, the reservoir has no submerged vegetation. The only vegetation present are flooded terrestrial herbs in springtime, when the reservoir reaches its maximum volume. The bottom of the littoral zone is mainly composed of stones, gravel and sand, whereas the bottom of the open water zone is covered by mud.

Material and methods

Phytoplankton

Sampling methods used for the study of phytoplankton dynamics in 1990/91 and 1993 and the methods used to measure the main physical and chemical parameters are described by Barone & Naselli Flores (1994; data from 1990–1991) and Naselli Flores & Barone (1996; data from 1993). In the present paper the algal biomass was divided into two size classes (≤ 30 μm and > 30 μm) according to the greatest axial linear dimension of the different species, which was estimated by measuring single cells of at least 30 specimens of each species. Algae > 30 μm are considered inedible due to their size.

Zooplankton

Zooplankton was sampled weekly, from the 25th of May 1990 to the end of December 1991 and from the beginning of January to the end of December 1993. Samples were taken both in the littoral zone and at a fixed station near the middle of the basin, with a 21 cm diameter net of 75 μm and 125 μm mesh, respectively. A comparison with samples taken at other stations revealed that the choosen station is representative of the open water zone (Naselli Flores & Barone, unpubl. data). Three vertical tows were made from the bottom to the surface. The collected specimens were immediately concentrated on a 60 μm mesh gauze and fixed in 95% ethanol.

Population densities of the zooplankton were estimated using a custom-made counting chamber and a Zeiss microscope. From each sample, a subsample representing 10% of the volume was taken and all individuals as well as the number of parthenogenetic and resting eggs were identified and counted. For each species, males and females were counted separately. Wet weight biomass was evaluated from recorded abundances using biovolume estimates (De Bernardi, 1974). The body length of *Daphnia* was measured from the top of the head to the base of the tail spine in 200 individuals from each sample.

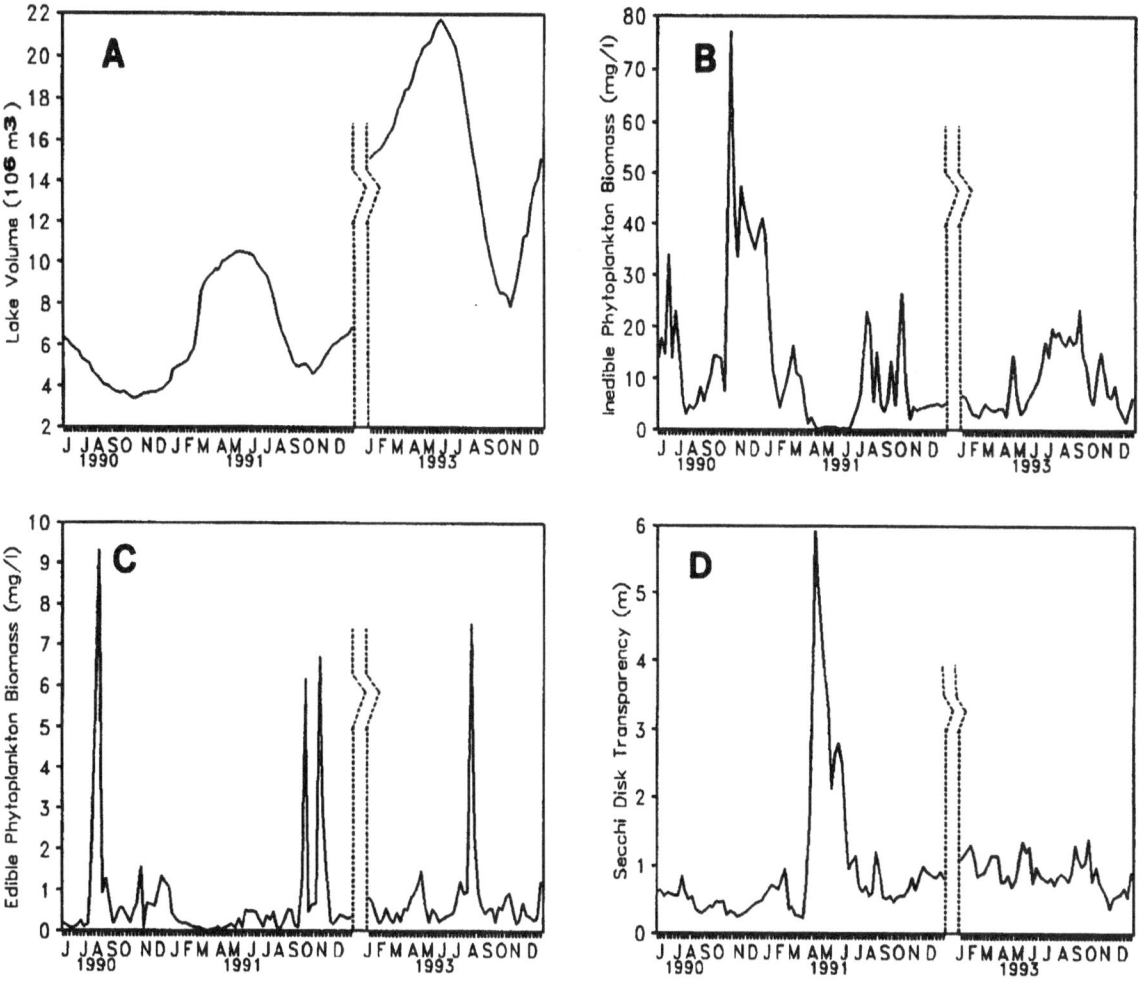

Figure 1. Volume (A), inedible phytoplankton biomass (B), edible phytoplankton biomass (C) and Secchi disk transparency depth (C) trends during the years of survey. The time interval between vertical broken lines was not sampled. Modified from Barone & Naselli Flores (1994) and Naselli Flores & Barone (1996).

Fish

The fish community was sampled weekly, from April to October 1993, in five stations located in the littoral zone, using a 20 m trawl net with 3 mm mesh. The net was towed by hand, forming a semicircle, from the shoreline to a depth of about 1.5 m. Population densities of fish species were estimated using the removal method (Seber & Le Cren, 1967). All specimens caught were immediately frozen on dry ice and returned to the laboratory where the entire gut content of 10% of the specimens with a standard length <50 mm was analyzed. Fish were then preserved in 95% ethanol.

The importance of each dietary item of underyearling fish was estimated by measuring prey items in the gut and calculating individual prey volumes using empirical relationships based on volume calculation of simple solids choosen in accordance to the shape of the prey item. In the case of very young fish (<15 mm SL) it was usually impossible to quantify the whole gut content. To facilitate interpretation, the results of gut content analyses were pooled into monthly samples.

Results

Phytoplankton

The lake generally exhibits high values of wet weight phytoplankton biomass (Figures 1B and C), generally above 10 mg l^{-1}. In particular, the phytoplankton assemblage was dominated by large inedible algae (mainly *Closterium* spp. in 1990/91 and *Mougeotia* sp. in 1993). Inedible phytoplankton biomass values increased up to 77 mg l^{-1} in autumn 1990. After this peak, a continuous decrease in biomass took place until the beginnning of April 1991, when a long clear water phase started with biomass values <0.5 mg l^{-1} (Figure 1D). Secchi disk transparency depth then changed from less than 0.5 m to 5.9 m. Edible phytoplankton (mainly small Chlorococcal algae) generally showed low biomass values (<2 mg l^{-1}), except in the middle of August 1990 and August 1993 when it peaked to values >8 mg l^{-1}. In 1991 a double biomass peak of edible algae was recorded at the end of the summer season.

Zooplankton

Similar to the phytoplankton, the zooplankton also showed high densities. Wet weight biomass values of most samples ranged between 10 and 50 mg l^{-1}, but much higher values were recorded during summer 1990 and spring 1993 (Figures 2A and B). These changes were observed both in the littoral and in the open water zone, which differ only slightly in terms of species composition and biomass. Cyclopoid copepods, and in particular *Tropocyclops prasinus* (Fischer), often dominated the community. Numerically, rotifers were the most abundant zooplankton organisms. *Brachionus* spp. and *Keratella* spp. generally dominated the rotifer community. In general, cladocerans represented less than 30% of the zooplankton biomass, except during spring and summer 1991 when their biomass exceeded 50% of the total zooplankton biomass.

Seven cladoceran species were identified in the zooplankton assemblage. Two of them (*Macrothryx hirsuticornis* Norman & Brady and *Alona rectangula* Sars) are benthic organisms that were recorded sporadically and in low numbers. Five pelagic species were recorded: *Daphnia hyalina* Leydig, *Diaphanosoma lacustris* Korinek, *Bosmina longirostris* (O. F. Müller), *Daphnia (Ctenodaphnia) magna* Straus and *Moina brachiata*, (Jurine). Their densities are reported in Figure 3.

Daphnia hyalina was nearly always present in the lake. Population densities started to increase in the winter season (Figure 3A), reached its highest values (>30 ind. l^{-1}) at the end of spring and then fairly quickly decreased in abundance again. In 1993, *D. hyalina* was not caught from the end of July until mid September and in 1990, the species was rare from August until November. In contrast, population densities in

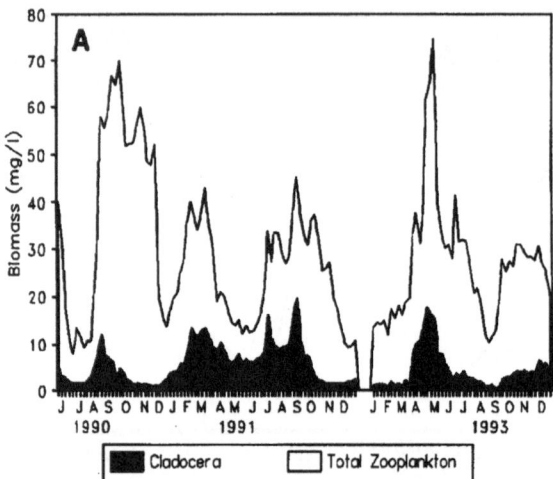

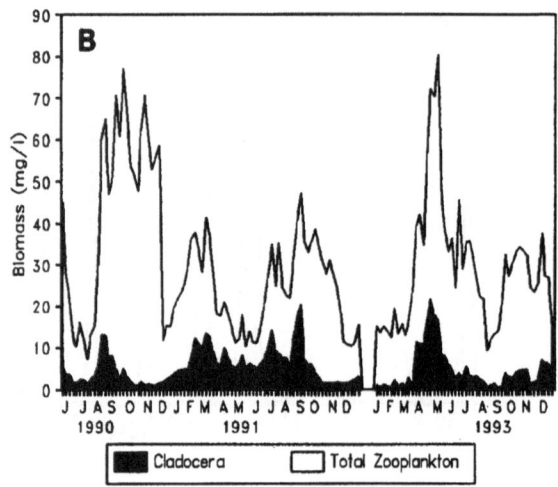

Figure 2. Changes in total microcrustacean zooplankton and cladoceran wet biomass in Lake Arancio during the course of 1990–1993 in the littoral zone (A) and in the open water (B).

1991 were high during the summer, reaching values above 70 ind. l^{-1} at the end of July. In that year, the *D. hyalina* population started to decrease in density from August onwards, and almost completely disappeared by November, when the population growth turned positive again. In 1990 and in 1993, males were predominantly observed in June and July, whereas in 1991, their peak density occurred in August and September.

In 1991, the increase in population density of *D. hyalina* was interrupted at the end of April and the beginning of May. During this period, we observed a pronounced peak density of *D. magna* (Figure 3B), reaching over 30 ind. l^{-1} at the beginning of May. A

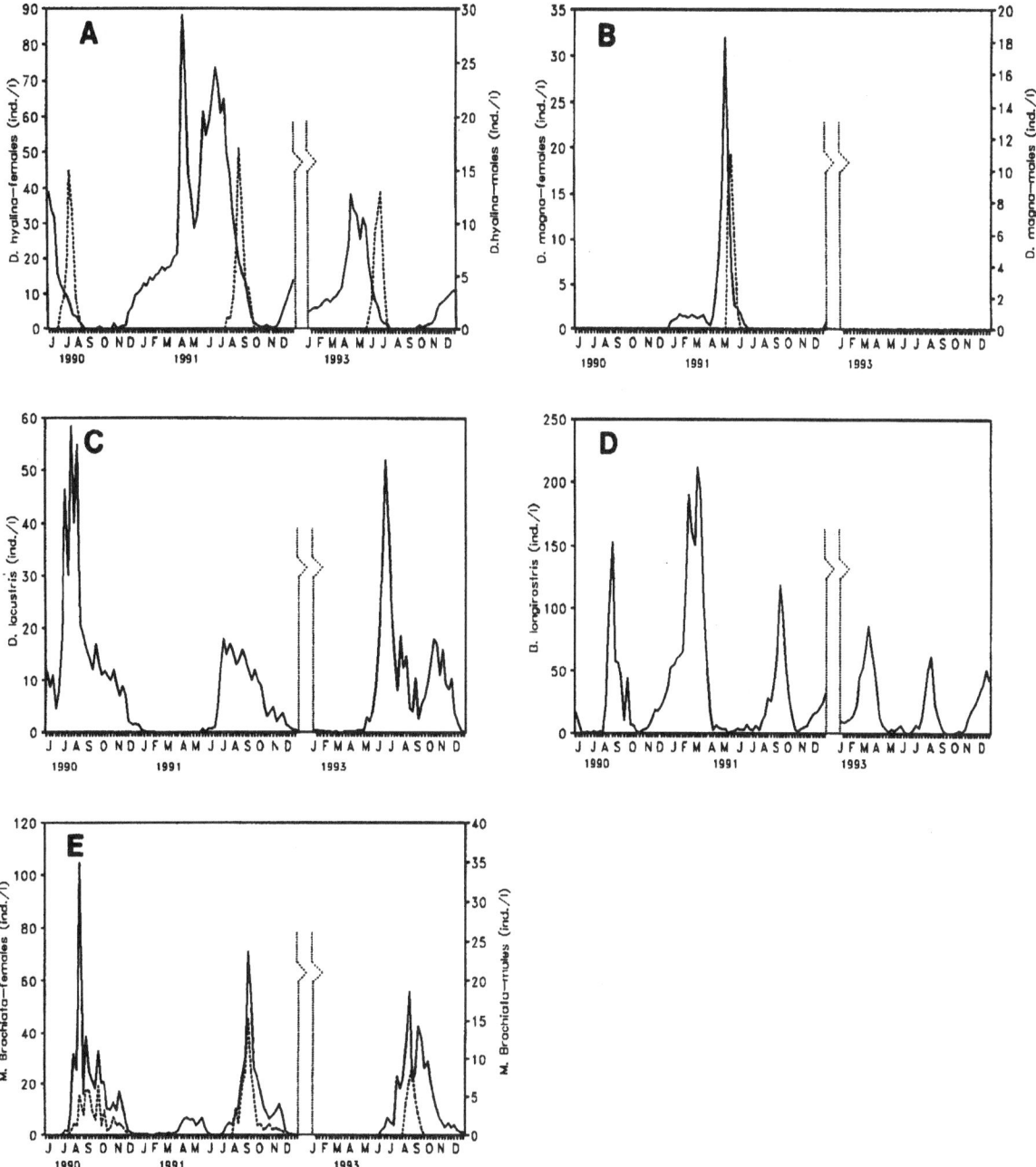

Figure 3. Changes in density of the different pelagic cladocerans in Lake Arancio during the course of 1990–1993. Dotted lines represent males; solid lines represent females.

peak of *D. magna* males was recorded three weeks later than that of parthenogenetic females.

Diaphanosoma lacustris in general reached highest population densities in July–September, after the decline of *D. hyalina* (Figure 3C). Values were above 50 ind. l^{-1} in 1990 and 1993, whereas in 1991 they did not reach 20 ind. l^{-1}. Males were never recorded.

Bosmina longirostris was recorded troughout thc year but showed density peaks (>150 ind. l^{-1}) in early spring and in late summer (Figure 3D). *Moina brachiata* occurred in the summer season with densities up to 100 ind. l^{-1} (Figure 3E). Males generally appeared when densities of females were highest. This species also showed only high densities after the decline of *D. hyalina*.

The average body size of *D. hyalina* changed over the three years. In particular, 1990 and 1993 were characterized by a marked reduction in the size of ovigerous females compared to 1991 (Figure 4A). In 1990 and 1993, very small (average length <750 μm, with standard deviation ranging between 5 and 12% of the mean) ovigerous females were observed with a low number of parthenogenetic eggs (mostly one or two eggs, Figure 4B). In addition, the number of ovigerous parthenogenetic females was generally less than 35% of the total population (Figure 4C). Females carrying resting eggs (Figure 4D) and males (Figure 3A) appeared at the end of June. Ephippial females often made up more than 20% of the total population (Figure 4D). In spring 1991, the population on average was characterized by large specimens (average length >1200 μm, with standard deviation ranging between 9 and 16% of the mean) carrying relatively large parthenogenetic broods (up to 7 eggs per brood). Moreover, females with parthenogenetic eggs often represented 40% or more of the total population, and males and ephippial females only appeared from August onwards, their percentage remaining rather low (<15%).

We observed a positive exponential relationship between average body length (in mm) of *D. hyalina* and average clutch size (number of parthenogenetic eggs):

$$F = e^{2.6L - 1.80},$$

where F is the number of eggs and L the body length of *D. hyalina* ($r = 0.97$; $n = 89$; $p < 0.001$).

Fish

An average fish biomass of 200 kg fresh weight ha^{-1} ($\pm 11\%$) was estimated in the five stations sampled in 1993. There were no strong differences between stations, neither in biomass nor in species composition of the fish community.

The fish fry community biomass ranged between 3 and 20 kg ha^{-1} during the sampling period. The fish fry community was numerically dominated (93–100%) by *Rutilus rubilio* (Bonaparte). The other species present in the samples were *Gambusia holbrooki* Girard and *Carassius auratus* (L.).

The growth of 0+ fishes versus time showed a linear relationship:

$$SL = 0.147\,D + 13.4$$

where SL is the standard length (in mm) and D is time since hatching (in days; $r = 0.96$, $n = 27$, $p < 0.001$). The highest density of *R. rubilio* was recorded in the first half of June (Figure 5) when the juvenile fish reached an average standard length of approximately 25 mm.

The diet of fish caught in April, May and June (Figure 6) consisted primarily of *Daphnia hyalina* (>45%). Other important prey items were phytoplankton and rotifers, especially in April when the fish were less than 15 mm long. The cyclopoid copepod *Tropocyclops prasinus* and statoblasts of the bryozoan *Plumatella fungosa* (Pallas), two abundant taxa in the reservoir, were also present in the diet.

In July, *D. hyalina* was still an important prey item, representing approximately 35% of the gut content. By that time, benthic prey (mainly chironomid larvae) appeared in the diet. The benthic feeding mode is also reflected by the presence of silt particles and detritus of macrophytic origin in the gut. The shift from planktonic to benthic prey items in July was probably due to the low densities of cladocerans and copepods in the water body. In the subsequent months, the fish showed a diet basically consisting of fragments of macrophytes, benthic filamentous algae and chironomid larvae, except for a relatively high consumption (>20%) of *Tropocyclops prasinus* in September. Silt particles and fragments of macrophytes always constituted the highest percentage of the gut contents, and zooplankton virtually disappeared from the diet.

Discussion

Since the classical paper by Brooks & Dodson (1965), many studies have focused on trophic relationships in freshwater lacustrine environments; and the interactions between fish, zooplankton and phytoplankton. The important role of fish fry as a principal consumer of zooplankton, and of large cladocerans such as *Daphnia* in temperate reservoirs has been pointed out by Zalewski et al. (1990). These authors observed that an increase in the average surface of a reservoir during the spawning and post-spawning periods generally improved the diversity of littoral habitats, enabling larger densities of juvenile fish to occur. Although this suggests that predation pressure on the zooplankton would increase, Tatrai et al. (1995) observed that the average number of prey consumed per unit of time by planktivorous fish was reduced when habitat complexity increased.

Due to its remarkable annual fluctuations in water level, Lake Arancio exhibits a shoreline covered by

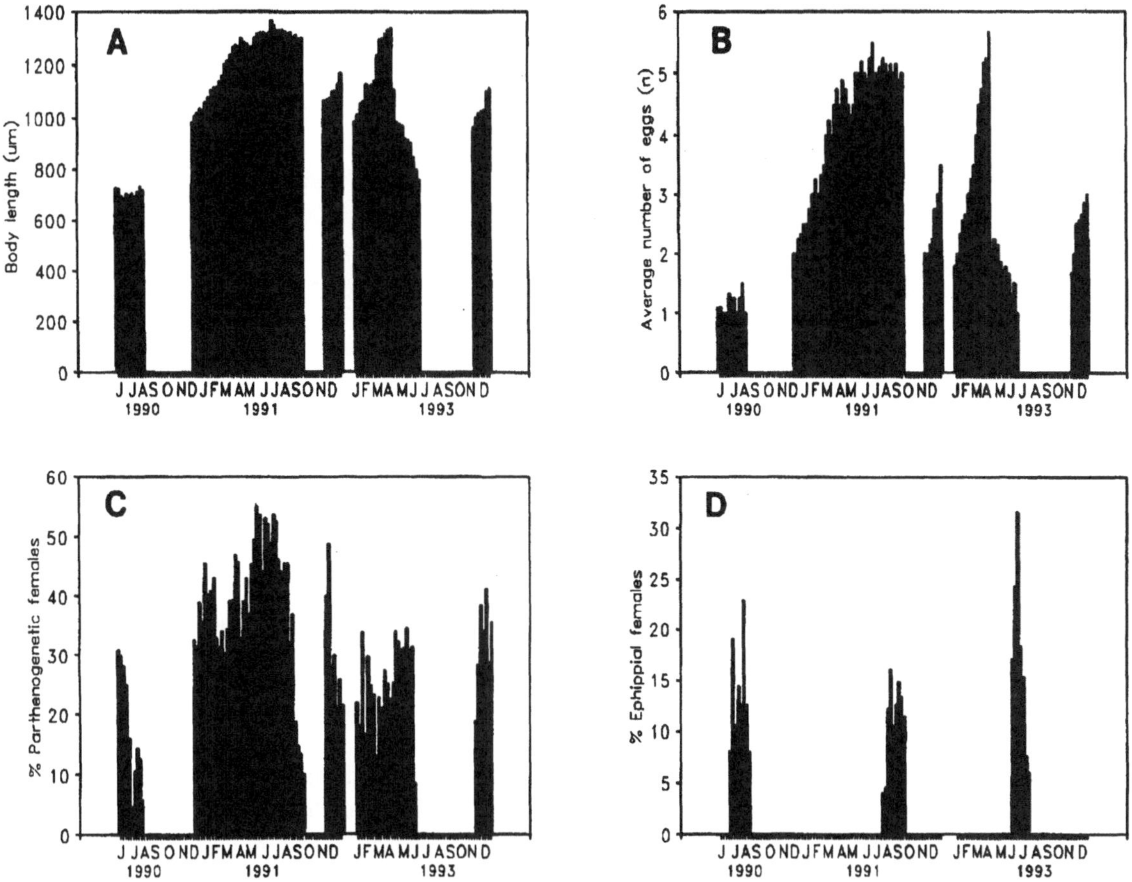

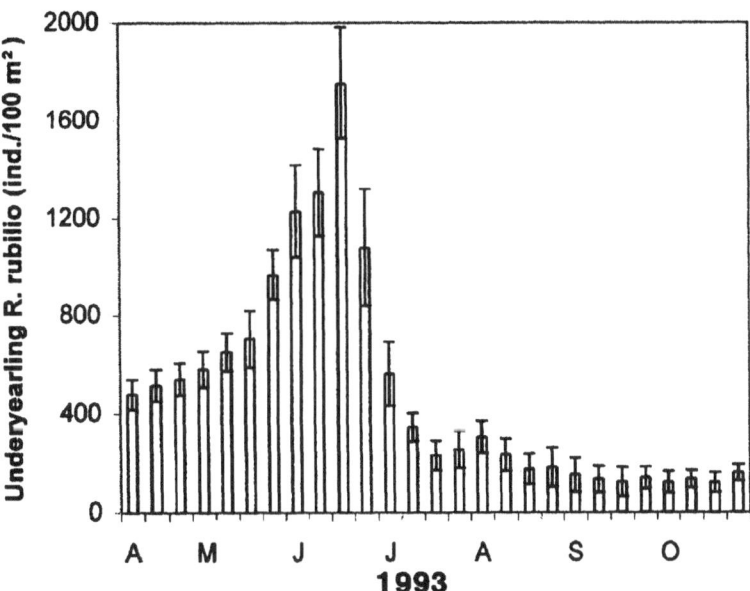

Figure 4. Seasonal variations in body length (A), average number of eggs per female (B), percentage of ovigerous females (C) and percentage of ephippial females (D) in *Daphnia hyalina* in Lake Arancio.

Figure 5. Density of underyearling *Rutilius rubilio* in 1993. Error bars represent one standard deviation.

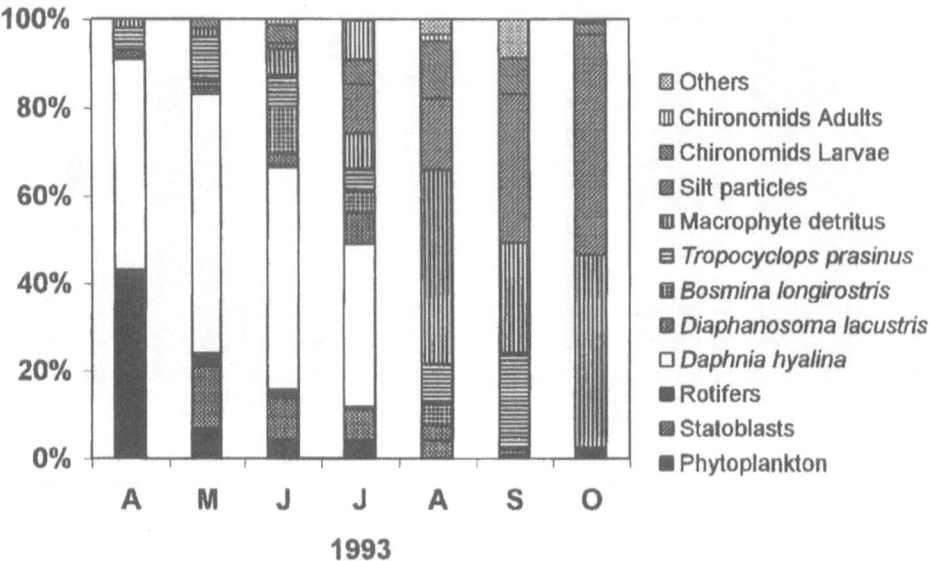

Figure 6. Seasonal changes in the relative composition of gut contents (%, biovolume) of underyearling *Rutilus rubilio* in 1993.

stones and sand, without any macrophytes. Thus, an increase in the average surface of the reservoir does not improve the diversity of the littoral environment. In addition, it was observed that silt deposition on fish eggs as a result of inflow increased egg mortality and thus decreased spawning success (Ploskey, 1981). In particular, a sedimentation rate of 1 mm per day may cause the loss of 97% of the spawned eggs (Hassler, 1970). The strong flooding which affected Lake Arancio at the end of winter 1991 caused a large flow of allochtonous matter from the catchment area into the reservoir. As a result, Secchi disk depth in the reservoir was reduced considerably during the first half of March (<0.25 cm). This reduced transparency was the only remarkable feature of the physical characteristics of the reservoir in 1991, since no differences were observed in the seasonal trends of temperature, conductivity, or in any other physical or chemical variable measured (Naselli Flores & Barone, 1996).

Since *R. rubilio* hatchlings have a body length of approximately 9 mm (Zerunian, 1981) and since the fry caught by us in early April (1993) measured 15 mm, it seems reasonable to hypothesize that, in 1991, egg deposition of *R. rubilio* took place contemporary to or just before the decrease in transparency. This may have caused a strong decrease in the hatching success and early survival of the fish in that year. This is also suggested by the fact that the 2+ age class in the adult

population in 1993 is poorly represented (Naselli Flores, unpubl. data).

Although a high diversity in prey items was present in the guts of underyearling *R. rubilio*, the fish showed a strong preference for *D. hyalina* as long as this prey item was abundant in the reservoir. *D. hyalina* still represented a considerable part of the gut contents in the middle of July, even though the population density of the species in the reservoir was already reduced substantially. Our data therefore strongly suggest that predation by underyearling *R. rubilio* had a strong impact on the population dynamics and size distribution of *D. hyalina*. *Bosmina longirostris* and *Diaphanosoma lacustris* were rarely present in the guts. In contrast, the copepod *Tropocyclops prasinus* was always present, but in small amounts.

As it was observed by Barone & Naselli Flores (1994), the strong inflow in 1991 also caused the dilution of the dominant large planktonic algae (*Closterium* spp.) and allowed the development of small Chlorococcales (mainly *Oocystis* sp. and *Ankyra* spp.). Though these taxa are characterized by high growth rates, biomass values remained low for a considerable period causing a long clear water phase (with Secchi depth values of 6 m).

The decrease of fish fry density together with the increase of food availability seem to be the most plausible causes of the increase in body size and density of *D. hyalina* in 1991, and its occurrence throughout the summer of that year. The lack of predation by fish also allowed the development of a *Daphnia magna* population, which was recorded in 1991 only. A short period of high densities of this species caused a momentary decrease in the density of *D. hyalina*.

Our results also indicate that *Diaphanosoma lacustris* suffered from competition exerted by the larger *Daphnia* in the summer season. In 1991, the population density of *Diaphanosoma* was about one third of that recorded in 1990 and 1993. *Bosmina longirostris* also slightly delayed its density peak in summer 1991, probably due to the longer period of occurrence of *D. hyalina*. Finally, whereas edible phytoplankton showed a density peak just after the population decline of *Daphnia* in August 1990 and 1993, it did not show a density peak in August 1991, but only bloomed in September and November, again just after the population decline of *Daphnia*. Together, these results indicate that water level fluctuations can have pronounced impacts on bottom-up and top-down interactions in artificial reservoirs. Our results are in agreement with the observations of a survey of 21 Sicilian dam reservoirs (Naselli Flores & Barone, 1994). In this survey, a dominance of large cladocerans was recorded in those reservoirs that were subjected to strong fluctuations in water level.

Concluding remarks

As stated by Straškraba et al. (1993), there is a need to understand how reservoir ecosystems couple with hydrological processes. This topic is often ignored in lake studies because of their more stable hydrological regimes, but it becomes very important in environments subject to pronounced and short-term changes in water level.

One of the most important factors governing cladoceran dynamics in Lake Arancio appears to be the strong fluctuations in water level which act on the zooplankton community of the reservoir ecosystem in at least two different ways:

1 Bottom-up forces: the continuous variation in water depth may cause modifications of the mixing depth-euphotic depth ratio and changes in turbulence, both of which affect phytoplankton succession (Reynolds, 1989; Naselli Flores & Barone, 1996). In addition, the rapid variation in flushing rate acts as an intermediate disturbance which may allow the coexistence of small edible *r*-strategists with larger inedible species (see Padisák et al., 1993).

2 Top-down forces, due to the influence of fluctuations in water level on predation pressure exerted by planktivorous fish. In the reservoir studied, a decrease in transparency due to the transport of allochtonous matter from the surroundings of the reservoir may have caused a strong decrease in the reproductive success of fish. As a result, predation pressure exerted by fish fry on the zooplankton was reduced. A high predation pressure may result in a reduction of the effect of competition and grazing, thereby influencing the structure of the ecosystem profoundly (Soranno et al., 1993).

Our results are significant with respect to biomanipulation as applied to reservoir water quality management. For instance, lowering the water level to dry out the spawned eggs results in a decrease in fish fry population density (Kubecka, 1993; Zalewski et al., 1990), and, as a result, the increased density of large cladocerans may improve the water quality of the ecosystem.

Acknowledgments

The authors thank two anonymous referees for constructive comments on a previous version of the manuscript. They also acknowledge Luc De Meester for his useful suggestions and linguistic help with the English text.

References

Barone, R. & L. Naselli Flores, 1994. Phytoplankton dynamics in a shallow, hypertrophic reservoir (Lake Arancio, Sicily). Hydrobiologia 289: 199–214.

Barone, R., S. Calvo, L. Naselli Flores & G. Viviani, 1993. Thermal analysis of a Sicilian dam reservoir. Verh. int. Ver. Limnol. 25: 105–110.

Brooks, J. L. & S. I. Dodson, 1965. Predation, body size and composition of plankton. Science 150: 28–35.

De Bernardi, R., 1974. Popolamento Zooplanctonico. In Tonolli, L. (ed.), Indagini ecologiche sul lago d'Endine. Ist. ital. Idrobiol., Pallanza: 225–259.

Hassler, T. G., 1970. Environmental influences on early development and year-class strength of northern pike in lakes Oahe and Sharpe, South Dakota. Trans. am. Fish. Soc. 99: 369–375.

Kimmel, B. L., O. T. Lind & L. J. Paulson, 1990. Reservoir Primary Production. In Thornton, K. W., B. L. Kimmel & F. E. Payne (eds), Reservoir Limnology: Ecological Perspectives, John Wiley & Sons, New York: 133–193.

Kitchell, J. F. & S. R. Carpenter, 1993. Cascading trophic interactions. In Carpenter, S. R. & J. F. Kitchell (eds), The Trophic

232

Cascade in Lakes. Cambridge University Press, Cambridge: 1–14.

Kubecka, J., 1993. Succesion of fish communities in reservoirs of Central and Eastern Europe. In Straškraba, M., J. G. Tundisi & A. Duncan (eds), Comparative Reservoir Limnology and Water Quality Management. Dev. Hydrobiol. 77. Kluwer Academic Publishers, Dordrecht: 153–168.

Naselli Flores, L. & R. Barone, 1994. Relationships between trophic state and plankton community structure in 21 Sicilian dam reservoirs. Hydrobiologia 275/276: 197–205.

Naselli Flores, L. & R. Barone, 1996. Phytoplankton and underwater light climate in a hypertrophic reservoir (Lake Arancio, Sicily). Giorn. Bot. Ital. 129: 1288–1292.

O'Brien, W. J., 1990. Perspectives on Fish in Reservoir Limnology. In Thornton, K. W., B. L. Kimmel & F. E. Payne (eds), Reservoir Limnology: Ecological Perspectives. John Wiley & Sons, New York: 209–225.

Padisák, J., C. S. Reynolds & U. Sommer (eds), 1993. Intermediate Disturbance Hypothesis in Phytoplankton Ecology. Dev. Hydrobiol. 81. Kluwer Academic Publishers, Dordrecht, 199 pp. Reprinted from Hydrobiologia 249.

Peters, R. H. & R. de Bernardi (eds), 1987. *Daphnia*. Ist. ital. Idrobiol., Pallanza, 502 pp.

Ploskey, G. R., 1981. Factors affecting fish production and fishing quality in new reservoirs, with guidance on timber clearing, basin preparation, and filling. Technical Report E-81-11, Fish and Wildlife Service, National Reservoir Research Program, Vicksburg (MS), 35 pp.

Regione Siciliana, 1996. Annali Idrologici 1991. Assessorato Lavori Pubblici, Servizio Idrografico, Palermo, 202 pp.

Reynolds, C. S., 1989. Physical determinants of phytoplankton succession. In Sommer, U. (ed.), Plankton Ecology: Succession in Plankton Communities. Springer Verlag, Berlin: 9–56.

Seber, G. A. F. & E. D. Le Cren, 1967. Estimating population parameters from catches large relative to the population. J. anim. Ecol. 36: 631–643.

Soranno, P. A., S. R. Carpenter & M. M. Elser, 1993. Zooplankton community dynamics. In Carpenter, S. R. & J. F. Kitchell (eds), The Trophic Cascade in Lakes. Cambridge University Press, Cambridge: 116–152.

Straškraba, M., J. G. Tundisi & A. Duncan, 1993. State-of-the-art of reservoir limnology and water quality management. In Straškraba, M., J. G. Tundisi & A. Duncan (eds), Comparative Reservoir Limnology and Water Quality Management. Dev. Hydrobiol. 77. Kluwer Academic Publishers, Dordrecht: 213–288.

Tatrai, I., G. Giussani, M. Manca & R. de Bernardi, 1995. An experimental study of Lago Maggiore zooplankton consumption by bleak (*Alburnus alburnus alborella*) in different simulated habitats. Mem. Ist. ital. Idrobiol. 53: 75–84.

Zalewski, M., B. Brewinska-Zaras & P. Frankiewicz, 1990. Fry communities as a biomanipulating tool in a temperate lowland reservoir. Arch. Hydrobiol. Beih. Ergebn. Limnol. 33: 763–774.

Zerunian, S., 1981. Il comportamento riproduttivo del triotto *Rutilus rubilio* (Bp.) (Pisces, Cyprinidae). Boll. Mus. Civ. St. Nat. Verona 8: 265–273.

Hydrobiologia **360**: 233–242, 1997.
© 1997 *Kluwer Academic Publishers.*

Long-term dynamics of small-bodied and large-bodied cladocerans during the eutrophication of a shallow reservoir, with special attention for *Chydorus sphaericus* *

J. Vijverberg[1] & M. Boersma[2]

[1]*Netherlands Institute of Ecology, Centre for Limnology, Rijksstraatweg 6, 3631 AC Nieuwersluis, The Netherlands (e-mail: vijverberg@cl.nioo.knaw.nl)*
[2]*Max-Planck-Institut für Limnologie, P.O. Box 165, D-24302 Plön, Germany*

Key words: Chydorus, body size, zooplankton community, eutrophication, detrital food web, Cyanobacteria

Abstract

Eutrophication in Tjeukemeer involved a gradual increase in chlorophyll concentrations from ca. 30 mg m^{-3} in 1968–69 to 125 mg m^{-3} in 1976. From 1976 onwards, chlorophyll concentrations remained at a high level fluctuating between 100–225 mg m^{-3}. Hillbricht-Illkowska (1977) hypothesized that small-bodied species will become increasingly abundant and dominant over large-bodied species with increasing eutrophication. We tested this hypothesis using observations from life history experiments on *Chydorus sphaericus*, combined with data from 25 years of field observations on the population dynamics of cladocerans in Tjeukemeer.

In life history experiments with *C. sphaericus*, the fitness measure r in treatments with natural lake seston and laboratory cultured green algae was significantly higher on lake seston from Tjeukemeer, containing a high proportion of detritus. This suggests that detrital particles are good quality food for *C. sphaericus*. Field observations during the period 1968–1976 showed that all three categories of cladocerans: *C. sphaericus*, 'other' small-bodied cladocerans (predominantly *Bosmina* spp.) and large-bodied cladocerans (predominantly *Daphnia galeata*), increased in biomass with increasing chlorophyll concentration. However, of these three cladoceran categories only *C. sphaericus* showed a distinct and significant increase whereas the other two only showed a marginally significant increase. During the period 1977–1992, both 'other' small-bodied cladocerans and *C. sphaericus* significantly decreased in biomass with increasing chlorophyll concentration, whereas the biomass of the large-bodied cladocerans significantly increased with increasing chlorophyll content. These observations are not in agreement with the hypothesis that small-bodied zooplankton become increasingly abundant with increasing eutrophication. We suggest that the observed trends are partially caused by a food effect, and partially caused by predation pressure. *Daphnia* shows a better response to the increase in detritus and filaments of Cyanobacteria than small-bodied cladocerans, but is more vulnerable to fish predation. Densities of 0+ zooplanktivorous fish show strong annual fluctuations in Tjeukemeer, and because of hydrological conditions, 0+ fish abundance in this lake is probably negatively related to chlorophyll content.

Introduction

In the pelagic zone of eutrophic temperate lakes and reservoirs, small-bodied cladoceran species like *Bosmina* spp. and *Chydorus sphaericus* are generally more abundant than large-bodied species (i.e. *Daphnia* spp.). However, whereas the role of *Daphnia* spp. in the pelagic food web has figured in numerous studies, much less is known about the quantitative role of small-bodied cladocerans in the food web (e.g. DeMott & Kerfoot, 1982; Balseiro et al., 1992).

---

* Publication no. 2336 of The Netherlands Institute of Ecology, Centre for Limnology.

It is widely accepted that the pelagic zone of lakes of low productivity tends to be dominated by large-bodied cladocerans and that the size of the dominant forms decreases with increasing trophy (Brooks, 1969; Nilsson & Pejler, 1973; Pejler, 1975; Rankin et al., 1979; Sprules & Knoechel, 1983; Ewald, 1991). Interestingly, this decrease in body size is often to a large extent the result of a niche shift by small-bodied species (e.g. *Chydorus sphaericus*, *Ceriodaphnia quadrangula*), which are originally restricted to the littoral zone, but invade the pelagic zone when the lake becomes more eutrophic (Pejler, 1975). In eutrophic lakes, detrital particles with attached bacteria are often more important than algae as food for zooplankton. These particles occur in concentrations four to six times higher than those of live algae (Mann, 1988; Meijer et al., 1990; Gons et al., 1992). Even if the detrital particles contain only 17–25% of the energetic content of living algae, they probably realise more than half of the energetic requirements of the consumers on a bulk basis (Kerfoot & Kirk, 1991).

The small-bodied *C. sphaericus* often appears as a common planker in eutrophic waters where extensive Cyanobacteria blooms are prevalent (Gannon, 1972). *C. sphaericus* belongs to the Chydoridae, which are adapted to creeping along submerged surfaces, either macrophytes or bottom substrates, and are poor swimmers (Fryer, 1968), although some species occasionally leave their substrate (Whiteside, 1974). As most Chydoridae remain in close contact with a substrate, they belong to the microbenthos rather than to the zooplankton. *C. sphaericus* is an exception, as it has two alternative ways of life. It can be found in the littoral zones of lakes among macrophyte vegetation and on bottom substrates that are rich in organic material (Goulden, 1971; Keen, 1973; Daggett & Davis, 1974; Whiteside, 1974; Williams, 1982), as well as in the water column in the open water zone of eutrophic lakes and ponds (Cummins et al., 1969; Franken & Franken, 1978; Pedros-Alio & Brock, 1985; Rognerud & Kjellberg, 1990; Vijverberg et al., 1990; Ewald, 1991).

The foodweb dynamics of Tjeukemeer have been studied intensively from 1968 to 1992 (Vijverberg et al., 1993). During the first ten years of the study, the lake progressively became more eutrophic (Moed & Hoogveld, 1982). Floating macrophyte fields, which up to 1970 occupied approximately 10% of the lake's surface area, disappeared in the late summer of 1971 and did not reappear since (pers. obs. J. Vijverberg). At the same time filamentous Cyanobacteria increased.

Total algal biomass, measured as chlorophyll concentration, progressively increased, and the relative abundance of Cyanobacteria (dominated by *Oscillatoria* spp.) increased from ca. 1000 filaments ml^{-1} in 1968–69 to around 5000 filaments ml^{-1} in 1971, and reached a maximum level of around 50 000 filaments ml^{-1} in 1976–1978 (Moed & Hoogveld, 1982). Since 1972 *Oscillatoria* spp. have dominated the phytoplankton community in the lake, both in numbers and in biovolume.

Because most small-bodied cladoceran species inhabiting the pelagic zone of lakes and reservoirs are of littoral origin, they are probably better adapted to detrital food than large-bodied pelagic cladocerans. Furthermore, because of their narrower carapace gape, the feeding process in small-bodied cladocerans is less inhibited by the presence of large filamentous Cyanobacteria than the feeding process in large-bodied cladocerans, because there is a lower risk of filaments entering the food chamber and entangling the thoracic appendages (Gliwicz & Siedlar, 1980). Largely based on these two arguments, Hillbricht-Illkowska (1977) hypothesized that small-bodied zooplankton species will become increasingly abundant and dominant over large-bodied species as lakes become more eutrophic. We tested this hypothesis by addressing three questions: (1) is the intrinsic rate of population increase (r) of *C. sphaericus* on a medium of natural seston from Tjeukemeer with algae and a high detritus content higher than on algal food alone?, (2) does the biomass of small-bodied cladocerans increase at a faster rate than that of large-bodied cladocerans under conditions of progressive eutrophication?, and (3) does *C. sphaericus*, belonging to a taxonomic group of benthic organisms, which is supposed to be preconditioned to detrital food, increase faster in biomass than other small-bodied cladocerans such as *Bosmina* spp.?

Material and methods

Study area

Tjeukemeer, situated in the North of the Netherlands, is a shallow (mean depth = 1.5 m), eutrophic freshwater lake with a surface area of 2150 ha. The lake is part of an interconnected system of waterbodies, which act as a reservoir for the surrounding polders. The system receives water from the nearby IJsselmeer (120 000 ha) during the growing season (April to September) and polderwater, which is rich in humic compounds, during the winter. The littoral zone of the lake is poorly developed, covering only 1% of the lake's surface area.

Laboratory experiments

C. sphaericus was cultured on two different types of food: (1) natural seston from lake Tjeukemeer, and (2) a 1:1 (by Carbon mass) mixture of the green algae *Chlamydomonas globosa* and *Scenedesmus obliquus*, supplied at a total food level of 1 mg C l^{-1}, a concentration which is well above the incipient limiting level of a small-bodied cladoceran (Duncan, 1989). For the medium with natural seston, lake water from Tjeukemeer (1990, April–May) was collected fresh every day and sieved over a 76 μm mesh plankton gauze in order to remove crustacean zooplankton. *C. globosa* and *S. obliquus* were cultured axenically in 2 litre flow-through systems (Boersma & Vijverberg, 1994). The algae were harvested daily from the overflow bottles of these continuous cultures. The culture medium of the algae was removed by centrifuging twice for 20 min at 3000 r.p.m., followed by rinsing with distilled water. The algae were resuspended in 0.45 μm-filtered lake water from Tjeukemeer. The algal density was measured using a haemacytometer, counting a minimum of 500 cells. The carbon content of *C. globosa* was 2.52×10^{-11} g C cell^{-1}, and that of *S. obliquus* was 2.30×10^{-11} g C cell^{-1}, resulting in carbon: dry weight ratios of 0.53 and 0.50, respectively. *S. obliquus* was usually unicellular. Both algal species had a maximum length of around 15 μm. Fresh media were prepared daily.

Experimental animals were kept at 17.5 °C and a light: dark regime of 16:8 h. Parthenogenetic females of *C. sphaericus* were randomly collected from the field about 4 weeks before the start of the experiment and kept on 76 μm mesh filtered lake water. As soon as the field caught females produced their second batch of newborns, the mothers were removed. The first and second generation animals were reared on filtered lake water to maturity. The offspring produced by the animals of the second generation were transferred individually to 100 ml tubes for acclimation to the two food types. The animals were selected in such a way that the clonal composition for the two food treatments was the same. Newborns produced in these cultures were used for the experiments.

For each treatment, thirty neonates were collected within 12 h of birth and placed individually in 100-ml test tubes. The animals were inspected and transferred to clean tubes with fresh medium every other day. We

Table 1. Model parameters for regressions of log$_{10}$-transformed annual mean cladoceran biomass of large-bodied cladocerans (Clad. Large), *C. sphaericus* (*Chydorus*), and 'other' small-bodied cladocerans (Clad. Small) as dependent variables and time (year) as the independent variable. Regressions are calculated for two periods: (1) period 1968–1976, the period of increasing eutrophication, and (2) period 1977–1992, the period of hypertrophic state. See also Figure 4.

Period	Group	Slope	Intercept	r^2	P
1968–1976	Clad. Large	0.04	2.34	0.23	0.19
	Clad. Small	0.05	1.95	0.46	0.045
	Chydorus	0.21	0.50	0.49	0.035
1977–1992	Clad. Large	−0.03	2.98	0.12	0.19
	Clad. Small	0.02	2.24	0.07	0.32
	Chydorus	0.01	1.71	0.01	0.72

recorded the time needed to reach maturity as the first day on which there were eggs present in the brood chamber. Once the animals reached maturity, we measured the number of eggs per brood and the duration of the adult instars, which is equal to the development time of the eggs. The life table experiment was continued until the animals reached the fourth adult instar. The intrinsic rate of population increase (r) was estimated using the Euler equation, and the standard error of r was computed using the jackknife method (Meyer et al., 1986). r was taken as a measure for fitness (Stearns, 1992).

Field work

Zooplankton was sampled from 1968 to 1992 with a 5-litre Friedinger closing sampler at five fixed stations, generally at fortnightly intervals during April–September. Samples were taken at two depths, one just below the surface and the other just above the bottom. The samples were pooled, concentrated by filtering through a 120-μm mesh plankton gauze and preserved in 4% formaldehyde. Densities of animals per species and length-frequency distributions were established based on a one-tenth subsample. Mean annual biomass was calculated over the growing season (April–September) using the length/weight relationship for each species given in Table 1 of Vijverberg et al. (1990).

Chlorophyll-*a* was taken as a measure for total phytoplankton biomass, and was estimated from spectrophotometer readings after extraction of the pigments in 80% ethanol at 75 °C for 5 min (Moed & Hallegraeff, 1978).

Data analysis

For a detailed analyses of the changes in the cladoceran community structure we distinguished three categories of cladocerans: (1) large-bodied species (mainly *D. galeata*), (2) *C. sphaericus*, and (3) 'other' small-bodied cladocerans (mainly *Bosmina* spp.). The changes in population density of these categories were investigated for two trends: (1) changes in cladoceran biomass with time (years), and (2) changes in cladoceran biomass with phytoplankton biomass (chlorophyll content). The field data were analysed by linear regression. Although time series are generally dependent, we believe that this is not the case with the present field data. In our data set every data point (year) represents a mean value for the growing season (April–September) of a specific year, based upon approximately 11 samples collected during that period. Slopes of regression lines were compared and tested for parallelism using a sub-routine of the ANOVA-MANOVA routine of Statistica (Statsoft, 1992). In the analysis we used species or zooplankton size category as the independent variable, zooplankton biomass as the dependent variable, and phytoplankton biomass as the co-variable.

Results

Laboratory experiments

In the life history experiments with *C. sphaericus*, we always observed two eggs per brood. The fitness measure r calculated for *C. sphaericus* on lake seston and algae-medium is presented in Figure 1. Lake seston resulted in a significantly higher r compared with cultured green algae by t-test ($t = 4.20$; df $= 20$; $P < 0.001$).

Field observations

From the start of the study, chlorophyll concentrations tended to increase annually. Starting at ca. 30 mg m^{-3} in 1968–69, chlorophyll-a concentrations reached a maximum annual mean of ca. 225 mg m^{-3} in 1979 (Figure 2). Between 1979 and 1992 the chlorophyll-a concentrations remained at a high level, and fluctuated between 100 and 225 mg m^{-3} (mean ca. 160 mg m^{-3}). The total cladoceran biomass roughly followed the same trend, with an increase in biomass up to 1976,

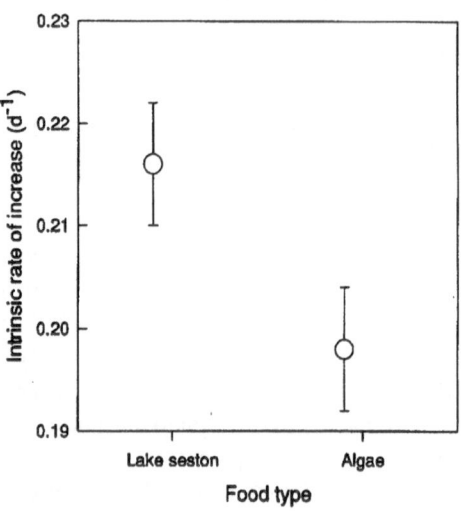

Figure 1. The intrinsic rate of population increase (r, d^{-1}) of *Chydorus sphaericus* in different food treatments. Two food types were used: (1) lake seston from Tjeukemeer (Lake seston), and (2) a 1:1 mixture of *Chlamydomonas globosa* and *Scenedesmus obliquus* at a total concentration of 1.0 mg C l^{-1} (Algae). Error bars indicate 95% confidence limits of the means.

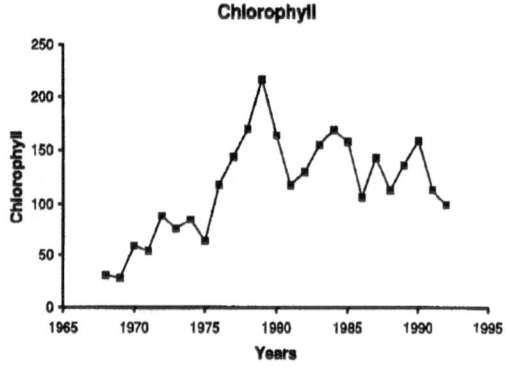

Figure 2. Change of annual mean chlorophyll content (April–September, mg m^{-3}) over time in Tjeukemeer (1968–1992).

followed by fluctuations around a high biomass level of ca. 1200 mg dry wt m^{-3} (Figure 3). The cladoceran biomass fluctuations during 1977–1992 are very large (range: 400–2400 mg dry wt m^{-3}), much larger than the observed fluctuations in chlorophyll concentrations (range: 100–220 mg m^{-3}).

During the whole study period, both small-bodied cladocerans and large-bodied cladocerans were present in substantial numbers. The group of small-bodied cladocerans was dominated by three species, *C. sphaericus*, *Bosmina coregoni*, and *B. longirostris*,

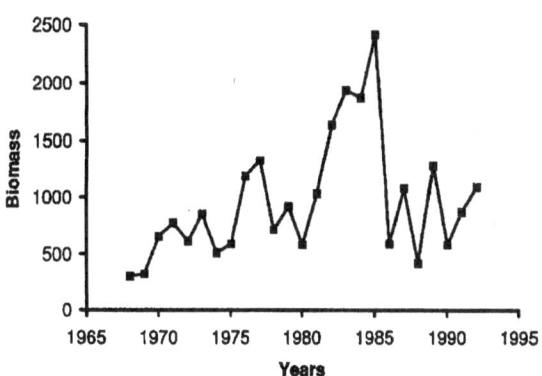

Total Cladoceran Biomass

Figure 3. Change of annual mean cladoceran biomass (April–September, mg dry wt m^{-3}) over time in Tjeukemeer (1968–1992).

Table 2. Model parameters for regressions of log$_{10}$-transformed annual mean cladoceran biomass of large-bodied cladocerans (Clad. Large), *C. sphaericus* (*Chydorus*), and 'other' small-bodied cladocerans (Clad. Small) as dependent variables and log$_{10}$-transformed mean chlorophyll-*a* content as the independent variable. Regressions are calculated for two periods: (1) period 1968–1976, the period of increasing eutrophication, and (2) period 1977–1992, the period of hypertrophic state. See also Figure 5.

Period	Group	Slope	Intercept	r^2	P
1968–1976	Clad. Large	0.71	1.27	0.41	0.07
	Clad. Small	0.59	1.15	0.37	0.08
	Chydorus	2.98	−3.77	0.59	0.02
1977–1992	Clad. Large	2.20	−1.97	0.32	0.03
	Clad. Small	−2.15	7.02	0.36	0.01
	Chydorus	−2.61	7.39	0.40	0.01

whereas the group of large-bodied cladocerans was dominated by one species only, *Daphnia galeata*. The changes in population density of three categories, *C. sphaericus*, 'other' small-bodied and large-bodied cladocerans, were investigated for two trends: (1) changes in cladoceran biomass with time (years), and (2) changes in cladoceran biomass with phytoplankton biomass (chlorophyll content). The results of the regression analyses are shown in Tables 1 and 2, and Figures 4 and 5. During the first period (1968–1976), biomass of both *C. sphaericus* and 'other' small-bodied cladocerans increased with time, but large-bodied cladocerans did not significantly increase (Figure 4a, Table 1). During the period 1977–1992, biomass of all three categories did not significantly change with time (Figure 4b, Table 1). When we compare the changes in biomass of cladocerans in relation to the chlorophyll concentration during the period 1968–1976, all three categories of cladocerans increased in biomass with increasing chlorophyll concentration. However, of these three cladoceran categories only *C. sphaericus* showed a distinct and significant increase whereas the other two only showed a marginally significant increase (Figure 5a, Table 2). Therefore, *C. sphaericus* does clearly increase with a higher rate with increasing chlorophyll concentration than 'other' small-bodied cladocerans and large-bodied cladocerans (pairwise comparison of slopes; $P = 0.03$ and $P = 0.04$, respectively). In the period 1977–1992, both *C. sphaericus* and 'other' small-bodied cladocerans and *C. sphaericus* significantly decreased in biomass with increasing chlorophyll concentration, whereas the biomass of large-bodied cladocerans showed a significant pos-

itive relationship with chlorophyll concentration (Figure 5b, Table 2). Therefore, large-bodied cladocerans differ significantly in their response to increasing algal biomass from *C. sphaericus* and 'other' small-bodied cladocerans ($P < 0.01$ for both comparisons).

Discussion

We observed that *C. sphaericus* adults in the life table experiments always carried two eggs per brood. This agrees with Robertson (1988) who states that the Chydoridae, with the exception of two sub-families, produce a maximum of two eggs per brood, and only rarely less than two. In life history experiments with *C. sphaericus* the fitness measure *r* was sigificantly higher on natural lake seston from Tjeukemeer, containing ca. 80% detritus versus 20% algae (Gons et al., 1992), than on laboratory cultured green algae. This suggests that detrital particles are good quality food for *C. sphaericus*, even better than a mixed medium of two different green algae (*Chlamydomonas* sp. and *Scenedesmus* sp.) which is generally a good quality food for *Daphnia* (Vijverberg, 1989).

We studied the changes in cladoceran abundance in terms of biomass. However, production would have been a better indicator than biomass of the functional role of these organisms in an ecosystem. Production is a measure of flow of energy or organic matter within a community, thus also of the success of a species or functional group (guild) in a given ecosystem (Benke, 1993). We chose biomass because it is

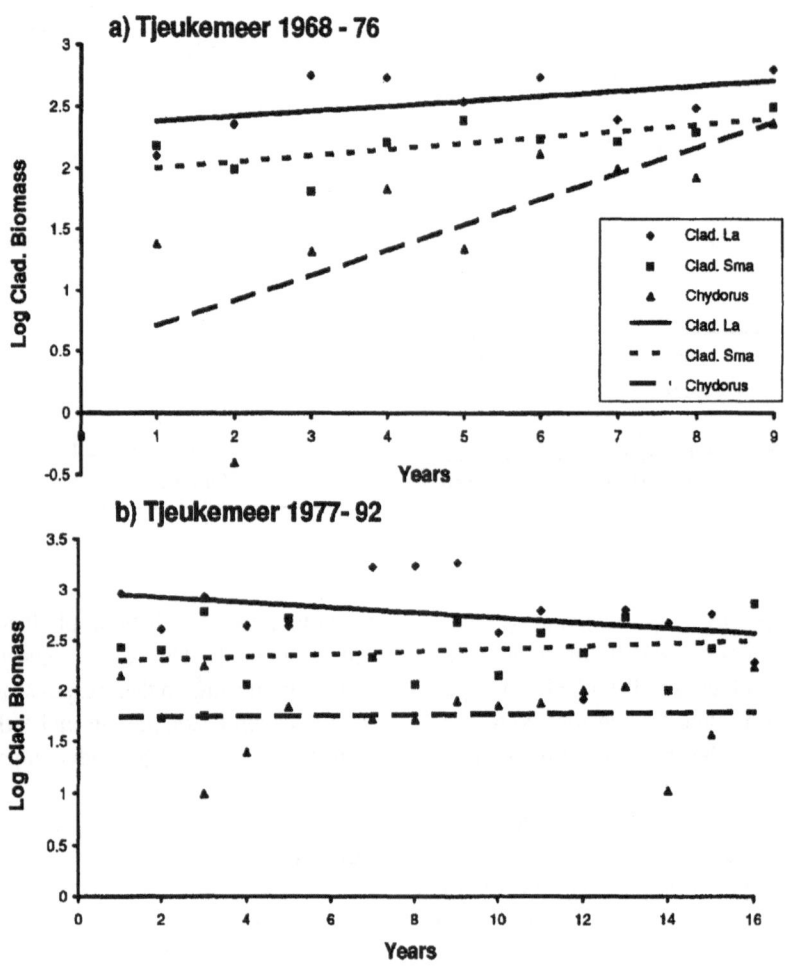

Figure 4. Relation between log_{10}-transformed annual mean cladoceran biomass of large-bodied cladocerans (Clad. La), *C. sphaericus* (Chydorus), and 'other' small-bodied cladocerans (Clad. Sma) as dependent variables and time (year) as the independent variable. Regressions are given for two periods: (a) period 1968–1976, the period of increasing eutrophication, and (b) period 1977–1992, the period of hypertrophic state. For summary statistics see Table 1.

relatively easy to measure, whereas production studies imply the quantification of dynamical parameters such as growth and development, for which we did not have data for the whole study period. However, using species specific P/B ratios we can roughly calculate production. Species specific P/B ratios are based on field observations in Tjeukemeer during the years 1969–1971 and on laboratory observations on growth and development using fresh Tjeukemeer lake seston, carried out during 1970–1973 and in 1982 (Vijverberg 1980; Vijverberg & Richter, 1982; Vijverberg & Koelewijn, 1991; Vijverberg, unpubl.). We observed that the P/B ratios of small-bodied species were about two times higher than those of large-bodied species.

Thus, using biomass as a variable we underestimated the impact of the small-bodied species in the system. Taking the P/B ratio's into account, approximately two thirds of the total cladoceran production was realised by small-bodied species and only one third by large-bodied species (mainly *D. galeata*) during the study period.

We tested the hypothesis of Hillbricht-Illkowska (1977) that smaller species will become increasingly more abundant and more dominant over larger species during eutrophication, because small cladoceran species such as *C. sphaericus* are better adapted to detrital food than larger ones, and because filter feeding efficiency of small-bodied cladocerans is less

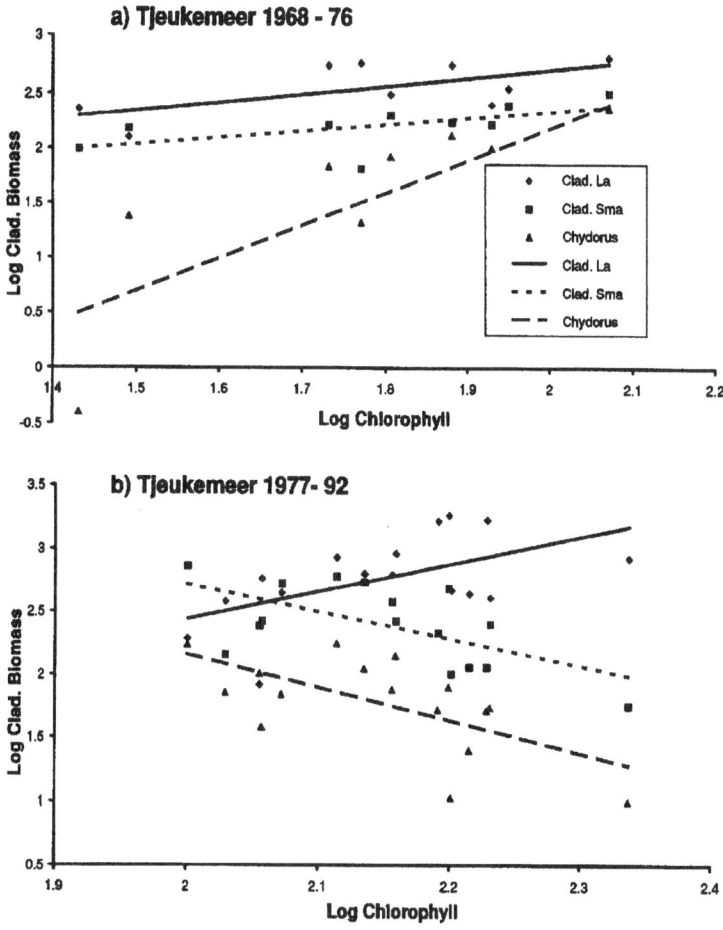

Figure 5. Relation between $\log_{10}$-transformed annual mean cladoceran biomass of large-bodied cladocerans (Clad. La), *C. sphaericus* (Chydorus), and 'other' small-bodied cladocerans (Clad. Sma) as dependent variables and $\log_{10}$-transformed mean annual chlorophyll concentrations the independent variable. Regressions are given for two periods: (a) period 1968–1976, the period of increasing eutrophication, and (b) period 1977–1987, the period of hypertrophic state. For summary statistics see Table 2.

hampered by the presence of large colonial Cyanobacteria than that of large-bodied cladocerans. The results of the first period of our study (1968–1976), the period of increasing phytoplankton biomass, supports the hypothesis of Hillbricht-Illkowska (1977): biomass of all small-bodied cladocerans increased with time, but biomass of large-bodied cladocerans did not. Although all three categories of cladocerans increased in biomass with increasing chlorophyll concentration. Only *C. sphaericus* showed a distinctly significant increase whereas the other two only showed a marginally significant increase with increasing chlorophyll concentration. Therefore, *C. sphaericus* does clearly increase with a higher rate with increasing chlorophyll concentration than 'other' small-bodied and large-bodied

cladocerans. The results of our life history experiments also support the idea that *C. sphaericus* is able to utilize detritus efficiently as a resource for growth and production. Our hypothesis that *C. sphaericus,* given its phylogeny, is better adapted to eutrophic conditions than the other small-bodied species, is also supported by the observations during this period of increasing chlorophyll concentration, as *C. sphaericus* then performs better than the other cladocerans. However, the observations during the second period (1977–1992), the period of hypertrophic state, do not support the hypothesis of Hillbricht-Illkowska (1977) nor our '*Chydorus* preadaptation' hypothesis. During this period, we observed a decrease rather than a further increase in the biomass of small-bodied species with increasing chlorophyll concentrations, with both *C. sphaericus* and 'other' small-bodied cladocerans showing the same relationship. Furthermore, instead of a decrease of large-bodied cladocerans with increasing phytoplankton biomass, we observed a significant increase. Such an increase in *Daphnia* biomass with increasing phytoplankton biomass, which was dominated by filamentous *Oscillatoria* spp., was not expected. In environments with high concentrations of filamentous Cyanobacteria, small-bodied cladocerans are expected to perform better than large-bodied cladocerans (Webster & Peters, 1978; Hawkins & Lampert, 1989). This has been explained by differences in the carapace gape width. The narrow carapace width in small-bodied cladocerans decreases the probability that filaments are taken in and may become entangled in the filtering limbs to obstruct the feeding process. Rejection movements with the terminal claw remove the filaments, but also the particles already collected, thus reducing the efficiency of feeding and resulting in an increased energetic cost (Burns, 1968; Gliwicz & Siedlar, 1980; Porter & McDonough, 1984). However, inhibition of *Daphnia* feeding by filamentous Cyanobacteria is not a general phenomenon. In several cases high densities of filamentous Cyanobacteria did not inhibit consumption of filaments or other algae present in the phytoplankton assemblage (Holm et al., 1983; Knisely & Geller, 1986; Schaffner et al., 1994; Epp, 1996). Furthermore, life history experiments show that *Daphnia* is growing and reproducing well on a diet of pure cultured *Oscillatoria limnetica*, although not as good as on a high quality diet with *Scenedesmus* as food (Repka, 1996). *Daphnia* also grows and reproduces well on natural seston from Tjeukemeer, which contains a high proportion of *Oscillatoria* filaments: Boersma (1995) observed that growth rate and fecundity for *Daphnia* fed natural seston were only a small fraction lower than for *Daphnia* fed a high quality diet (*Scenedesmus*). Under natural conditions, *Daphnia* often persists during blooms of filamentous Cyanobacteria (Mills & Forney, 1988; present study), most likely by a combination of selectively avoiding some filamentous species and successfully consuming others.

Since it is generally accepted that a high abundance of large-bodied cladocerans is dependent upon a low predation pressure by zooplanktivorous fish (e.g. Hrbacek et al., 1961), we checked for a negative correlation between the mean chlorophyll concentration during the growing season of a specific year and the biomass of 0+ fish in September of the same year, using 12

years of observations on 0+ fish of Tjeukemeer over the period 1976-1987 (W. L. T. van Densen, unpublished). Within the group of zooplanktivorous fish, the 0+ fish are the most important group of vertebrate predators in terms of predation pressure on the zooplankton (Vijverberg et al., 1990). The biomass stock estimate of 0+ fish in September is a reliable measure for the mean biomass of 0+ fish during the growing season (Lammens et al., 1985). There appeared to be a negative relationship between the mean chlorophyll content as the independent variable, and 0+ fish biomass as the dependent variable (slope $= -3.5$, $r^2 = 0.27$, $P = 0.083$, $N = 12$). Although this relationship is only marginally significant, it suggests that fish predation pressure was higher when chlorophyll content was lower. Such a relationship between 0+ fish abundance and chlorophyll content is not unexpected since 0+ fish stocks in Tjeukemeer are unstable due to fisheries management measures affecting the stocks of the main piscivore, pikeperch (*Stizostedion lucioperca*), and hydrological conditions. Until the autumn of 1976, the biomass stock of the pikeperch was kept at a moderate level by an intensive and effective commercial fisheries with gill-nets. This fisheries was abruptly stopped during the winter of 1976/77 when the commercial fisheries in the Frisian Lake District sold their fishing rights for 'scale fish' to the sport fisheries (Lammens, 1988). Since legal regulations forbid sport fisheries to use gill-nets and angling is much less effective, this resulted in a ca. four times higher biomass stock of piscivorous pikeperch, which in turn resulted in an increased predation pressure upon the 0+ fish and a reduction of the local recruitment of smelt, the dominant fish species in the 0+ age group (Lammens et al., 1985). Therefore, since 1977 passive migration of fish larvae from the nearby IJsselmeer became much more important for the recruitment of young fish in Tjeukemeer (Densen & Vijverberg, 1982). Generally, during spring and summer, waterflow is from the IJsselmeer into the Frisian Lakes System, whereas the direction of waterflow is reversed during autumn and winter. Much depends, however, on weather conditions, with a higher rain fall resulting in less inflow of IJsselmeer water in the Frisian Lakes. Most fish larvae enter the lake during spring. Since IJsselmeer water contains less algae than Tjeukemeer water it is not surprising that 0+ fish abundance and chorophyll content are negatively correlated. We also hypothesize that under hypertrophic conditions with a relatively high chlorophyll content, a high abundance of filamentous Cyanobacteria, and a large proportion of detrital particles in the seston, large daphnids may be competitive dominant over the small-bodied cladocerans. To fully understand the observed trends in zooplankton biomass, experimental research is necessary to investigate the competitive interactions among large-bodied and small bodied-cladocerans under hypertrophic conditions.

Acknowledgments

Part of this study was supported by the Life Science foundation (SLW), which is subsidized by the Netherlands Organisation for Scientific Research (NWO). We thank Aafje Landman, Peter Mac Gillavry and Theo Frank for their assistance in the laboratory, Guus Postema for help with data analysis, Thijs de Boer and Hans Hoogveld for making the Tjeukemeer chlorophyll-a data available to us, and Wim van Densen (WAU) for stimulating discussions and for providing data on 0+ fish. Erik van Gool, Ramesh Gulati, Riks Laanbroek, Sari Repka, Joop Ringelberg and Arnold Veen are acknowledged for their comments on the manuscript. We also appreciate the constructive criticism of Luc De Meester and two other, anonymous, referees.

References

Balseiro, E. G., B. E. Modenutti & C. P. Queimalinos, 1992. The coexistence of *Bosmina* and *Ceriodaphnia* in a south Andes lake – an analysis of demographic responses. Freshwat. Biol. 28: 93–101.

Benke, A. C., 1993. Edgardo Baldi Memorial Lecture: Concepts and patterns of invertebrate production in running waters. Verh. int. Ver. Limnol. 25: 15–38.

Boersma, M., 1995. Competition as a factor regulating population dynamics of *Daphnia* species in Tjeukemeer. Oecologia 103: 309–318.

Boersma, M. & J. Vijverberg, 1994. Possible toxic effects on *Daphnia* resulting from the green alga *Scenedesmus obliquus*. Hydrobiologia 294: 99–103.

Brooks, J. L., 1969. Eutrophication and changes in the composition of zooplankton. In Rohlich, G. A. (ed.), Eutrophication: Causes, Consequences, Correctives. Proc. of the International Symposium. Natl Acad. Sci. Washington, D.C.: 36–255.

Burns, C. W., 1968. Direct observations of mechanisms regulating feeding behavior of *Daphnia* in lake water. Int. Revue ges. Hydrobiol. 53: 83–100.

Cummins, K. W., R. R. Costa, R. E. Rowe, G. A. Moshiri, R. M. Scanlon & R. K. Zajdel, 1969. Ecological energetics of a natural population of the predaceous zooplankter *Leptodora kindtii* (Focke) (Cladocera). Oikos 20: 189–223.

Daggett, R. F. & C. C. Davis, 1974. A seasonal quantitative study of the littoral Cladocera and Copepoda in a bog pond and an acid marsh in New Foundland. Int. Revue ges. Hydrobiol. 59: 667–683.

DeMott, W. R. & W. C. Kerfoot, 1982. Competition among cladocerans: Nature of the interactions between *Bosmina* and *Daphnia*. Ecology 6: 1949–1966.

Densen, W. L. T. van & J. Vijverberg, 1982. The relationship between 0+ fish density, zooplankton size and the vulnerability of pikeperch, *Stizostedion lucioperca*, to angling in the Frisian lakes. Hydrobiologia 95: 321–336.

Duncan, A., 1989. Food limitation and body size in the life cycles of planktonic rotifers and cladocerans. Hydrobiologia 186/187: 11–28.

Epp, G. T., 1996. Grazing on filamentous cyanobacteria by *Daphnia pulicaria*. Limnol. Oceanogr. 41: 560–567.

Ewald, S., 1991. Long-term changes of crustacean plankton during successful restoration of Lake Schlachtensee (Berlin-West). Verh. int. Ver. Limnol. 24: 866–872.

Franken, W. & M. Franken, 1978. Limnologische Untersuchungen am Grossen Bullensee, einem sauren Heidesee Norddeutschlands. II. Zooplankton. Arch. Hydrobiol. 54: 80–100.

Fryer, G., 1968. Evolution and adaptive radiation in the Chydoridae (Crustacea: Cladocera): A study in comparative functional morphology and ecology. Phil. Trans. r. Soc., London, Series B 254: 221–385.

Gannon, J. E., 1972. Effects of eutrophication and fish predation on recent changes in zooplankton Crustacea species composition in Lake Michigan. Trans. Am. micros. Soc. 91: 82–85.

Gliwicz, Z. M. & E. Siedlar, 1980. Food size limitation and algae interfering with food collection in *Daphnia*. Arch. Hydrobiol. 88: 155–177.

Gons, H. J., T. Burger-Wiersma, J. H. Otten & M. Rijkeboer, 1992. Coupling of phytoplankton and detritus in a shallow, eutrophic lake (Lake Loosdrecht, The Netherlands). Hydrobiologia 233: 51–59.

Goulden, C. E., 1971. Environmental control of the abundance and distribution of the chydorid Cladocera. Limnol. Oceanogr. 16: 320–331.

Hawkins, P. & W. Lampert, 1989. The effect of *Daphnia* body size on filtering rate inhibition, in the presence of a filamentous cyanobacterium. Limnol. Oceanogr. 34: 1084–1088.

Hillbricht-Ilkowska, A., 1977. Trophic relations and energy flow in pelagic plankton. Pol. Ecol. Stud. 3: 3–98.

Holm, N. P., G. Ganf & J. Shapiro, 1983. Feeding and assimilation rates of *Daphnia pulex* fed *Aphanizomenon flos-aquae*. Limnol. Oceanogr. 28: 677–687.

Hrbacek, J., M. Dvorakova, V. Korinek & L. Prochazkova, 1961. Demonstration of the effect of the fish stock on the species composition of zooplankton and the intensity of metabolism of the whole plankton association. Verh. int. Ver. Limnol. 14: 192–195.

Keen, R., 1973. A probabilistic approach to the dynamics of natural populations of the Chydoridae (Cladocera, Crustacea). Ecology 54: 524–534.

Kerfoot, W. C. & K. L. Kirk, 1991. Degree of taste discrimination among suspension-feeding cladocerans and copepods – implications for detritivory and herbivory. Limnol. Oceanogr. 36: 1107–1123.

Knisely, K. & W. Geller, 1986. Selective feeding of four zooplankton species on natural lake phytoplankton. Oecologia 69: 86–94.

Lammens, E. H. R. R., 1988. Trophic interactions in the hypertrophic lake Tjeukemeer: top-down and bottom-up effects in relation to hydrology, predation and bioturbation during the period 1974–1985. Limnologica 19: 81–85.

Lammens, E. H. R. R., H. W. de Nie, J. Vijverberg & W. L. T. van Densen, 1985. Resource partitioning and niche shifts of bream

(*Abramis brama*) and eel (*Anguilla anguilla*) mediated by predation of smelt *(Osmerus eperlanus)* on *Daphnia hyalina*. Can. J. Fish. aquat. Sci. 42: 1342–1351.

Mann, K. H., 1988. Production and use of detritus in various freshwater, estuarine, and coastal marine ecosystems. Limnol. Oceanogr. 33: 910–930.

Meijer, M. L., E. H. R. R. Lammens, A. J. P. Raat, M. P. Grimm & S. H. Hosper, 1990. Impacts of cyprinids on zooplankton and algae in the drainable ponds. Hydrobiologia 191: 275–284.

Meyer, J. S., C. G. Ingersoll, L. L. McDonald & M. S. Boyce, 1986. Estimating uncertainty in population growth rates: Jackknife vs Bootstrap techniques. Ecology 67: 1156–1166.

Mills, E. L. & J. L. Forney, 1988. Trophic dynamics and development of freshwater pelagic food webs. In Carpenter, S. R. (ed.), Complex Interactions in Lake Communities, Springer, Berlin: 11–30.

Moed, J. R. & G. M. Hallegraeff, 1978. Some problems in the estimation of chlorophyll-*a* and phaeopigments from pre- and postacidification spectrophotometric measurements. Int. Revue ges. Hydrobiol. 63: 787–800.

Moed, J. R. & H. L. Hoogveld, 1982. The algal periodicity in Tjeukemeer during 1968–1978. Hydrobiologia 95: 223–234.

Nilsson, N. A. & B. Pejler, 1973. On the relationship between fish fauna and zooplankton composition in North Swedish lakes. Rep. Inst. Freshw. Res. Drottningholm 53: 51–77.

Pedros-Alio, C. & T. D. Brock, 1985. Zooplankton dynamics in Lake Mendota: Short-term *versus* long-term changes. Freshwat. Biol. 15: 89–94.

Pejler B., 1975. On long-term stability of zooplankton composition. Rep. Inst. Freshw. Res. Drottningholm 54: 107–117.

Porter, K. G. & R. McDonough, 1984. The energetic cost of response to blue-green algal filaments by cladocerans. Limnol. Oceanogr. 29: 365–369.

Rankin, D. P., H. J. Ashton & O. D. Kennedy, 1979. Crustacean zooplankton abundance and species composition in six experimentally fertilized British Columbia lakes. Fish. Mar. Sec. Tech. Rep. 897, 27 pp.

Repka, S., 1996. Inter- and intraspecific differences in *Daphnia* life histories in response to two food sources: the green alga *Scenedesmus* and the filamentous cyanobacterium *Oscillatoria*. J. Plankton Res.18: 1213–1223.

Robertson, A. L., 1988. Life history of some species of Chydoridae (Cladocera: Crustacea). Freshwat. Biol. 20: 75–84.

Rognerud, S. & G. Kjellberg, 1990. Long-term dynamics of the zooplankton community in Lake Mjøsa, the largest lake in Norway. Verh. int. Ver. Limnol. 24: 580–585.

Schafner, W. R., N. G. Hairston Jr & R. W. Howarth, 1994. Feeding rates and filament clipping by crustacean zooplankton consuming Cyanobacteria. Verh. int. Ver. Limnol. 25: 2375–2381.

Sprules, W. G. & R. Knoechel, 1983. Lake ecosystem dynamics based on functional representations of trophic components, p. 383–403. In Meyers, D. G. & J. R. Strickler (eds), Trophic Interactions within Aquatic Ecosystems, Westview Press, Boulder, Colorado, 472 pp.

Statsoft, 1992. CSS: Statistica. Volume I, Conventions and Statistics I. Statsoft Inc., Tulsa, Oklahoma, 679 pp.

Stearns, S. C., 1992. The evolution of life histories. Oxford University Press, Oxford.

Vijverberg, J., 1980. Effect of temperature in laboratory studies on development and growth of Cladocera and Copepoda from Tjeukemeer, The Netherlands. Freshwat. Biol. 10: 317–340.

Vijverberg, J., 1989. Culture techniques for studies of growth, development and reproduction of copepods and cladocerans under laboratory and in situ conditions: a review. Freshwat. Biol. 21: 317–373.

Vijverberg, J., M. Boersma, W. L. T. van Densen, W. Hoogenboezem, E. H. H. R. Lammens & W. M. Mooij, 1990. Seasonal variation in the interactions between piscivorous fish, planktivorous fish and zooplankton in a shallow eutrophic lake. Hydrobiologia 207: 279–286.

Vijverberg, J., R. D. Gulati & W. M. Mooij, 1993. Food-web studies in shallow eutrophic lakes by the Netherlands Institute of Ecology: Main results, knowledge gaps and new perspectives. Neth. J. aquat. Ecol. 27: 35–49.

Vijverberg, J. & H. P. Koelewijn, 1991. Size dependent mortality and production of *Diaphanosoma brachyurum* (Lieven) in an eutrophic lake. Verh. int. Ver. Limnol. 24: 2768–2771.

Vijverberg, J. & A. F. Richter, 1982. Population dynamics and production of *Daphnia hyalina* Leydig and *Daphnia cucullata* Sars in Tjeukemeer. Hydrobiologia 95: 235–259.

Webster, K. E. & R. H. Peters, 1978. Some size-dependent inhibitions of larger cladoceran filterfeeders in filamentous suspensions. Limnol. Oceanogr. 23: 1238–1245.

Whiteside, M. C., 1974. Chydorid (Cladocera) ecology: Seasonal patterns and abundance of populations in Elk Lake, Minnesota. Ecology 55: 538–550.

Williams, J. B., 1982. Temporal and spatial patterns of abundance of the Chydoridae (Cladocera) in Lake Itasca, Minnesota. Ecology 63: 345–353.

Hydrobiologia **360**: 243–252, 1997.
A. Brancelj, L. De Meester & P. Spaak (eds), Cladocera: The Biology of Model Organisms.
© 1997 *Kluwer Academic Publishers.*

Assessment of the importance of fish predation *versus* copepod predation on life history traits of *Daphnia hyalina*

Maria-José Caramujo, M. Cristina Crispim & Maria-José Boavida*
Dept. Zoologia, Faculdade de Ciências, Univ. Lisboa
[1] *Dept. Zoologia and Centro de Biologia Ambiental, Faculdade de Ciências, Univ. Lisboa, Campo Grande C2.*
1700 Lisboa. Portugal. (*author for correspondence)

Received September 1997; accepted September 97

Key words: Acanthocyclops robustus, cladocerans, Anomopoda, tail spine elongation

Abstract

Daphnia hyalina is a cladoceran present through the whole year except for late summer in Maranhão, a meso-eutrophic reservoir in central Portugal. Apart from the influence of food, both vertebrate and invertebrate predation pressures seem to have an effect on *D. hyalina* population dynamics. Enclosure experiments were designed to assess the relative importance of both types of predation. After the summer crash, *D. hyalina* reached higher numbers in the fishless enclosures than in the lake despite of high predation pressure upon juveniles by *Acanthocyclops robustus*. Fish predation upon the largest individuals, especially large egg bearing females, was responsible for the lower fertility of the open water population when compared with the enclosure population. In the enclosures an increase in tail spine length was observed. The longer tail spine probably offered protection from copepod predation, allowing at least some of the juveniles to coexist with their potential predator and reach the adult stage, less susceptible to copepod predation.

Introduction

Vertebrate and invertebrate predators have a widely recognized impact on zooplankton communities through size selective predation. The selective removal of larger prey individuals by vertebrate predators (Zaret, 1980) is counteracted by the selective removal of smaller individuals by invertebrate predators (Gliwicz & Pijanowska, 1989). In a long term enclosure study Gliwicz & Lampert (1993) found a total depletion of small prey species and a peculiar age structure in large bodied species in the presence of natural densities of the copepod *Acanthocyclops robustus*. Old individuals were found to be overwhelmingly dominant and neonates nearly absent. The feeding rate of this copepod has been shown to decrease with prey body size (Gliwicz & Umana, 1994). Although prey body size is relevant (Smyly, 1970; Dodson, 1974; Brandl & Fernando, 1975) other morphological and behavioural features unrelated to body size might be of extreme importance (Li & Li, 1979).

In the presence of waterborne cues produced by *Chaoborus*, cladocerans produce progeny with neckteeth or enlarged crests (helmets) (Havel, 1985; Havel & Dodson, 1984). Another common response is the presence of elongated tail spines in juveniles (Lüning, 1992) and the production of larger newborns (Lüning, 1995). Tail spine is important in stabilizing the swimming behaviour to allow straight-line movement. Mort (1986) observed that *D. galeata mendotae* lacking tail spines were perceived at a greater distance by *Chaoborus* than the daphnids with tail spines. Cyclopoid copepods also have the ability to induce tail spine elongation in daphnids (Crispim & Boavida, *in prep.*; Caramujo, *unpubl. data*). *Daphnia* are active swimmers and their jerky hop and sink mode of swimming creates enough disturbance to be detected by cyclopoids such as *Mesocyclops* (Williamson, 1983). *A. robustus* is a cruising search and attack predator like *Mesocyclops*, and probably detects its prey in a similar way. *Daphnia middendorffiana* was found to co-exist with the predaceous calanoid *Heterocope septentrion-*

alis in Arctic ponds (Haney & Buchanan, 1987) and with *Parabroteas sarsi* in Southern Andes (Balseiro & Vega, 1994). In both cases tail spine proved to be an important deterrent against copepod predation. Dodson (1974) suggested that phenotypic modifications against predators would serve as post-capture defense. However, in the case of tail spine elongation this defense seems to decrease detection by the predator.

Earlier laboratory experiments with *A. robustus* and *D. hyalina* by Crispim & Boavida *(in prep.)* showed that the presence of the copepods induces tail spine elongation in *Daphnia*, allowing the prey to coexist with the predator. In order to assess whether the same trend could be observed in the field, an enclosure experiment excluding the effect of fish was undertaken. The morphology of the daphnids was studied to investigate whether the presence of copepods whould induce changes. Furthermore, we studied the population dynamics to investigate the importance of both vertebrate and invertebrate predation on the *Daphnia* population in Maranhão Reservoir.

Material and methods

Study site

Maranhão Reservoir has been subjected to eutrofication during the past decades (Boavida & Marques, *submitted*), with average chlorophyll-*a* values in the upper 10 meters ranging from 22.36 mg l^{-1} in mid September to 0.45 mg l^{-1} in December. During our study, Secchi disk values oscillated around 1 m until December when a value of 2.1 m was reached and then increased smoothly to a maximum of 2.9 m in February.

The main vertebrate predators on *Daphnia* are the fish species *Lepomis gibbosus* and *Cyprinus carpio*. The only invertebrate predator present is the cyclopoid copepod *Acanthocyclops robustus*. The copepod community also includes the preferentially detritivorous cyclopoid *Thermocyclops dybowskii*, which is present from spring to autumn and the herbivorous calanoid *Copidodiaptomus numidicus*, which is present throughout the year. Other cladoceran genera in the reservoir are *Diaphnanosoma* and *Bosmina*.

Enclosures and sampling

Preliminary sampling with van Dorn bottles showed that most of the *Daphnia* population could be found in the upper 10 meters of the reservoir. Vertical migration was not detected. Two polystyrene enclosures (0.4 m in diameter, 10.5 m long) with the bottom covered by a 100 μm mesh size net were suspended in the reservoir. *Copidodiaptomus numidicus* and small zooplankon hatching from diapausing eggs in the sediments or diapausing young cyclopoid copepods could enter the enclosures from the bottom. On August 29 the zooplankton assemblage was collected in the reservoir and transferred to the enclosures at lake density. A week later the population densities were estimated and that day was considered as day one for the experiment – September 5. It was assumed that the species not present in the lake on August 29 were allowed to colonize the enclosures from the bottom but it is also possible that some individuals have entered through the open top placed 20 cm above the water level.

On each sampling date, one vertical haul (10 m long) per enclosure was taken with a Wisconsin type net of 80 μm mesh size. Two replicate hauls of 10 m long were taken from the lake. Sampling dates were August 22, September 5, 13 and 22, October 3, 9 and 16, November 7, December 12, January 10, February 9 and March 9. The animals were anaesthetized with carbonated water and preserved in sugar saturated formaldehyde. Copepodid stage five and adult copepods (C V to C VI) were counted in the whole sample, whereas copepodid stages one to four (C I to C IV) and nauplii of each species were counted in subsamples. Adult total length was measured from the anterior to the posterior edge of the carapace, excluding the furcal setae. Twenty females and males of *Copidodiaptomus numidicus* and 50 females and males of the more variable *Acanthocyclops robustus* were measured under the microscope. If present, 100 *Daphnia* per sample were measured. Body length of *Daphnia* was measured from the anterior edge of the carapace to the base of the tail spine and the tail spine was measured separately. The number of eggs and embryos in the pouch of 50 or all egg bearing females was counted. The smallest adult size class (primipara) was determined as the size class in which at least 5% of the total number of egg bearing females was observed. The minimum amount of 5% ensured that the smallest adult size class would not be set by a single precociously reproducing female. Juveniles (*JUV*), adults with eggs (*AD+*) or adults without eggs (*AD–*) were counted separately under a dissecting microscope, and the proportion of each 0.1 mm size class was estimated from size measurements. The whole sample was counted or, when the density of *Daphnia* was high, two subsamples of 1/5 of the total volume were analysed.

Numerical procedures and statistics

All calculations were based on one sample from each of the replicate enclosures and two replicate samples from the reservoir. Average clutch size (CLS) was based on the number of eggs in the brood pouch of 50 egg bearing females for each sampling date. Fecundity (*F*) was calculated as the average number of eggs per adult female. Egg development time (*D*) intervals were calculated according to the Plankton Ecology Group (PEG) model as presented in Bottrell et al. (1976), using the maximum and minimum temperatures on each sampling date (the vertical distribution of individuals was not recorded).

Birth rate (*b*) intervals were calculated applying Paloheimo's (1974) equation:

$$b = (1/D) \ln(1 + (E(N)),$$

where *D* and 1/*D* are the egg development time and rate respectively, *E* is the number of eggs and *N* the total number of animals. In order to relate the equation to environmental conditions, Polishchuk (1994) rewrote the equation in terms of fecundity (*F*) and proportion of adults (*A*) in the total population:

$$b = (V) \ln(1 + FA),$$

where $F = \{(CLS) * (AD+)\}$/Total adults and *V* is the egg development rate (per day). (AD+) is the number of egg-carrying females.

To quantify how changes in major environmental factors such as food, predation, and temperature determining key population features (*F*, *A* and *V*), translate into changes in *b*, Polishchuk (1994) used an approach based on the total derivative. Each triplet of the moving successive entries in *V*, *F*, *A* and *b* that correspond to two adjacent sampling intervals was approximated by a square polynomial. Then, the derivatives were estimated at the middle sampling point. Con$A = VF/(1 + FA)(dA/dt)$, Con$F = VA/(1 + FA)(dF/dt)$ and Con$V = \ln(1 + FA)(dV/dt)$ were computed at this same middle point. The whole procedure was repeated for the next moving triplet, shifted each time by one sampling point. The ratio of the numerical estimate of the total SumCon = Con$V$ + Con$F$ + Con$V$ to that of d*b*/d*t* was used to judge the efficiency of the numerical procedure. The triplets where the discrepancy between SumCon and d*b*/d*t* was more than 2-fold were excluded from the analysis.

To make sure that changes in birth rate accounted for by Con*A* were actually predator induced, the directions of changes in *A* were compared to the directions

in the population's total death rate, *d*, calculated as the difference between *b* and *r*, the per capita rate of population change observed (Polishchuk, 1995):

$$r = (\ln N_{t=1} - \ln N_{t=0})/(t = 0 - t = 1)$$

where N_t is the total population at each time *t*.

$$d = b - r.$$

Sampling depleted the population in the enclosures by 9% on each sampling date. Therefore, the actual population remaining in the enclosures was corrected before calculations.

Since egg development time is always less than half the sampling interval, changes in *A* and *d* have nearly identical intervals. Because of the qualitative nature of such comparisons, only the direction of changes, that is their signs, needs to be examined (Polishchuk, 1995).

Since we have two replicate samples from the lake instead of two replicate experimental units, results from statistical analysis comparing enclosures and the lake should be interpreted cautiously due to simple pseudoreplication.

ANOVA was applied to the clutch size of *C. numidicus* on January and February with treatment (enclosure/outside) as factor. ANCOVA were performed on the spine length of the juveniles (0.4–0.6 mm body lengths) and primiparae (first adult size class) of *Daphnia* from the reservoir and from the enclosures with body length as covariate and date as factor. Spearman's coefficient of rank correlation was calculated to measure the intensity of association observed between spine length and body length of juveniles and of primiparae, separately for the lake and the enclosures. Spine/body ratios of juveniles were angular transformed (arcsin) and correlated with the densities of *A. robustus* densities at the enclosures and at the lake.

Results

Copidodiaptomus numidicus

This species was present in low numbers during August and September. In the reservoir, the population size increased by the end of September and reached a maximum on October 3 followed by a decrease until November 7 when a new minimum was reached. After December a slow recovery took place and all post-embryonic developmental stages attained high numbers (Figure 1).

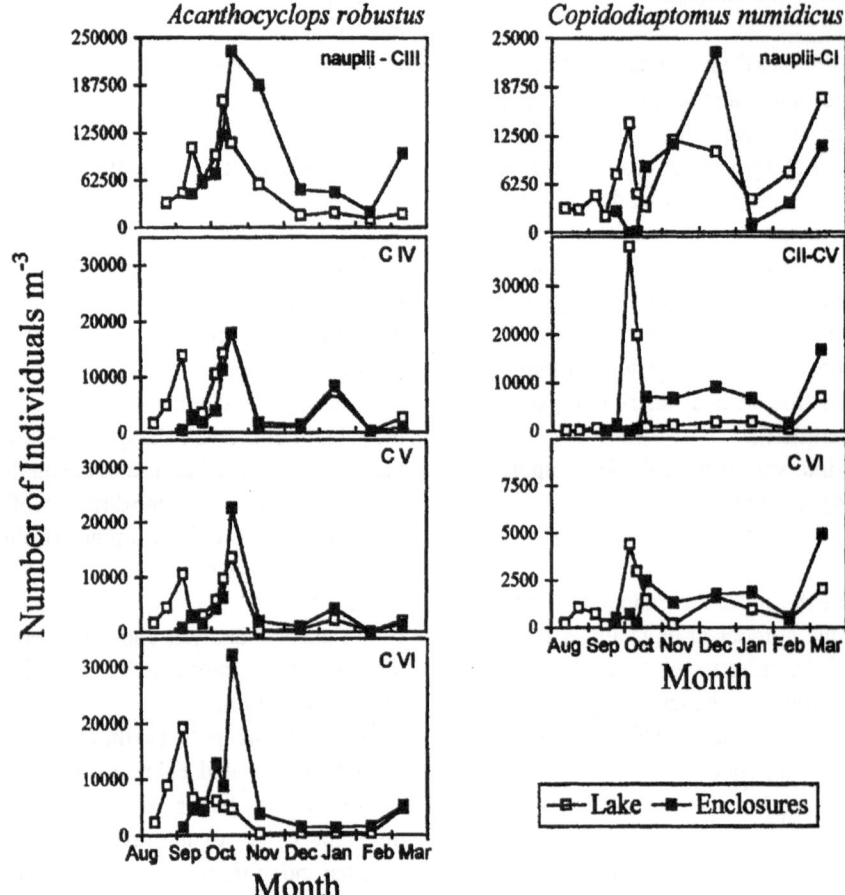

Figure 1. Changes in densities of different instars of the major copepod populations in Maranhão reservoir and in the enclosures. C I to C VI stand for copepodid stage one to six (adults). Means of two replicate enclosures and two replicate samples from the lake are given.

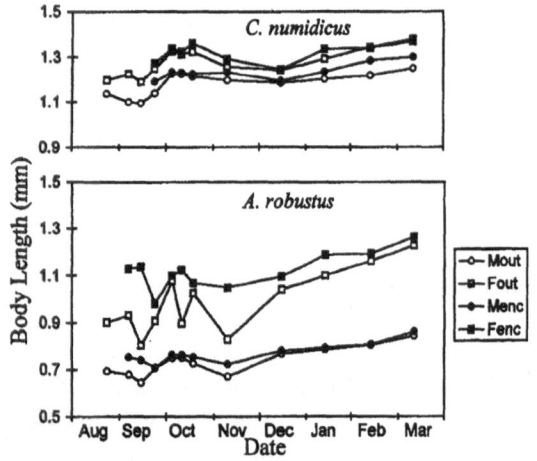

Figure 2. Changes in body length of adult females (F) and males (M) of *A. robustus* and *C. numidicus* in the reservoir (out) and in the enclosures (enc).

In the enclosures, the population was established later (after October 3) and high numbers of later developmental stages were reached on October 16. The population from then on reached higher numbers than in the reservoir. After January, although copepodids IV–V and adults were in lower numbers in the reservoir than in the enclosures, densities of nauplii and copepodid I were lower inside the enclosures than in the reservoir. Apparently, larger developmental stages were favoured in the enclosures. The lower number of juveniles in the enclosures was not a reflection of poor feeding conditions since clutch sizes were significantly higher in the enclosures (7 and 21.25 respectively, $F = 33.04$; df = 21; $p = 0.01$). Adults tended always to be larger inside the enclosures than in the reservoir (Figure 2), but only for males was this difference significant. There was a general increase in size until October 16, followed by a decrease until December and a new increase until the end of the experiment.

Acanthocyclops robustus

In the reservoir a population peak of *A. robustus* was observed on September 5, followed by a second peak on October 3 for early developmental stages (until CIII) and October 16 for later stages. Adult (CVI) females did not register this second peak (Figure 1).

All post-embryonic developmental stages were found in high numbers inside the enclosures. After September 5 a large population was established and densities continued to increase until October 16. After this date, all developmental stages decreased in density despite of the availability of *C. numidicus* nauplii and CI as prey to CV and to CVI of *A. robustus*. These stages prey heavily on younger developmental stages of *C. numidicus*. Actually, in preserved samples, it was rather common to observe parts of the calanoid in the mouth parts of the cyclopoid.

The adults of *A. robustus* were always larger in the enclosures than in the reservoir (Figure 2).

Daphnia hyalina

Population structure

D. hyalina appeared simultaneously in the reservoir and in the enclosures at the end of September. Abundances increased first in the enclosures (October 9) and one week later in the reservoir (Figure 3). The presence of males in the reservoir on October 9 indicated that the small population of *Daphnia* was under some sort of stress. The average clutch size was five eggs but only half the adults had eggs and the proportion of adults in the population was less than 20%. After a brief recovery on October 16, both the populations inside and outside the enclosures decreased in November. Before November, adults in the reservoir never exceeded 1.3 mm in body length, while in the enclosures adults of up to 1.9 mm were observed. The primipara size class was similar in the enclosures and the reservoir. In November, primiparous females were 27% smaller than on October 16 in the reservoir. However, inside the enclosures large adults were constantly present. A reversed shift was observed in March. All these shifts were preceded by the presence of males. The population size in the enclosures was always at least twice as large as outside.

Size structure

Because of the presence of *A. robustus*, we expected that the *Daphnia* population would be depleted

Table 1. Values of the Spearman's coefficient of correlation between the spine/body ratio of juveniles and the copepod densities in the enclosures. Copepods in the stages IV (CIV) to adult (CVI). * $p < 0.05$ and ** $p \leq 0.001$

Copepod	Angular transformed Spine/ Body ratio of juveniles	
Developmental Stages	All Copepod Densities $n = 154$	Copepod Densities < 10 ind. 1^{-1} $n = 135$
CIV – CVI	0.259*	0.356**
CV – CVI	0.313*	0.430**
CVI (adults)	0.252*	0.351**

of juveniles inside the enclosures. However, only the two first size classes were observed in relatively low numbers compared to the other juvenile size classes (Figure 3). Except for February and March, juveniles of the upper size classes were represented in higher numbers in the enclosures than in the reservoir, and the adults were also larger in the enclosures than in the reservoir. The primiparae tail spine length was correlated with body length both in the reservoir and in the enclosures ($r = 0.857$ and $r = 0.839$, respectively, df = 7; $p < 0.05$). The juvenile body length was more stable than that of the primiparae (Figure 4) during the sampling period. Juvenile spine length was not correlated with body length. Outside the enclosures the juveniles from February, March and October 9 had significantly larger spines than the ones in the period from October 16 until January (smaller difference for October 9 *vs.* October 16, $F = 4.70$, df = 2; $p < 0.05$). Inside the enclosures juvenile spine length oscillated strongly from date to date. The angular transformed spine/body ratio of juveniles in the enclosures was positively correlated with the densities of the stages CIV–CVI, CV–CVI and CVI of *A. robustus* (Table 1). This correlation is stronger when we use only densities below 10 000 individuals m^{-3}. The ratio of juvenile spine to body length thus seems to increase with copepod density until a certain value is reached (Figure 5). In the reservoir there was no significant correlation with *A. robustus* density.

Primiparous body length was also correlated with the length of *C. numidicus* females in the enclosures ($r = 0.821$, df = 7; $p < 0.05$) and with the length of both male and female *C. numidicus* in the reservoir ($r = 0.786$, df = 7; $p < 0.05$).

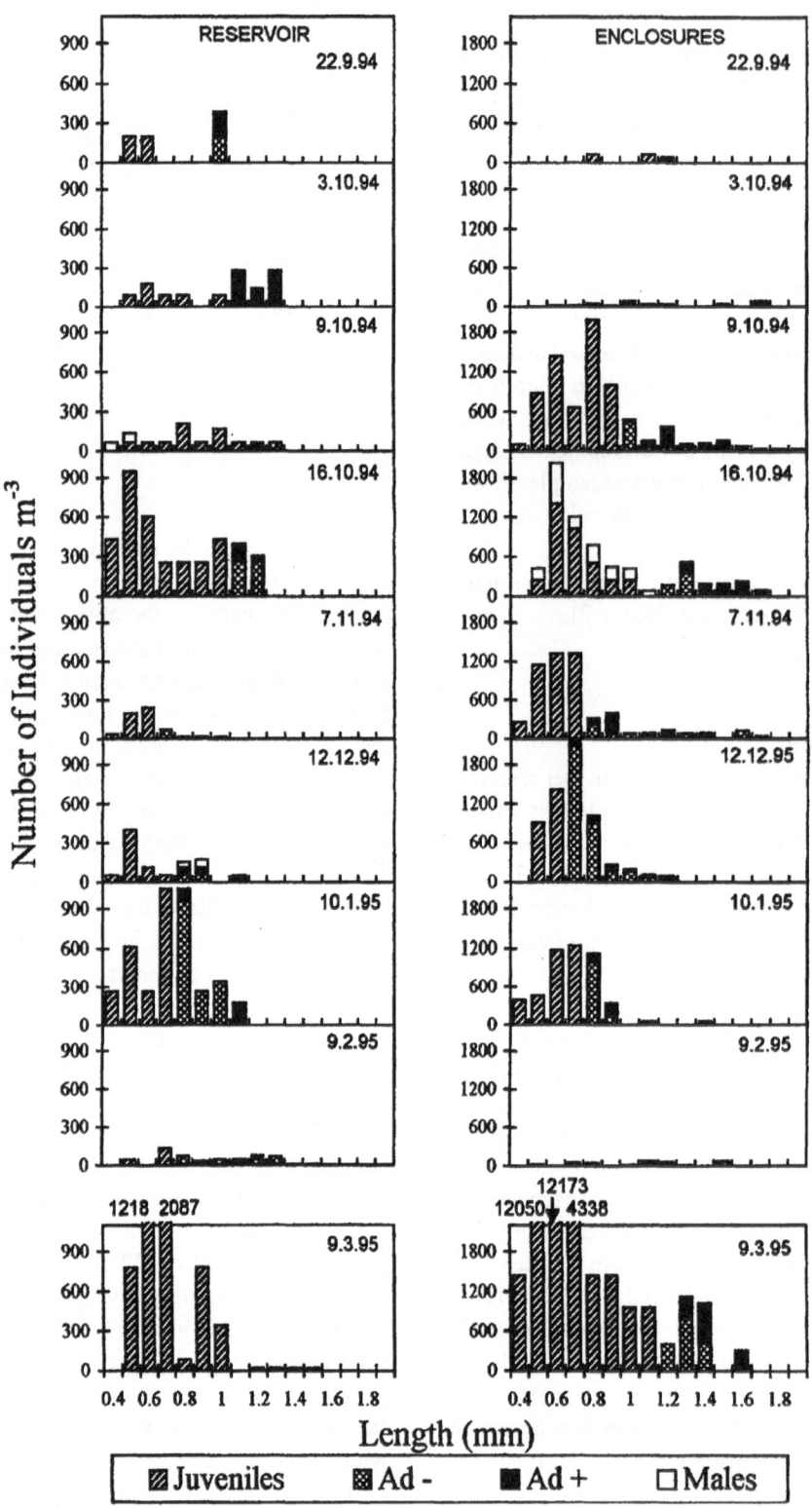

Figure 3. Body size classes (in mm) of *Daphnia* population in the reservoir and enclosures at each sampling date. *Ad–* and *Ad+* stand for adults without eggs and adults with eggs, respectively.

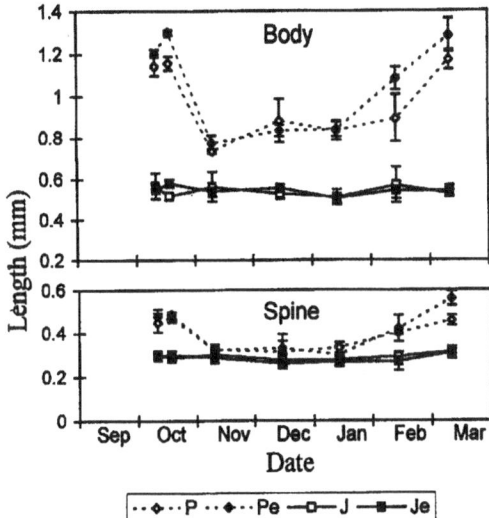

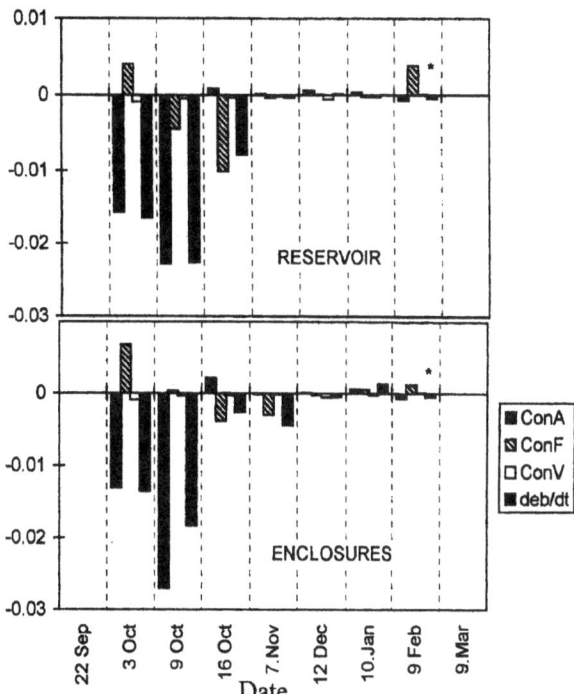

Figure 4. Changes in spine and body lengths of primiparae and juveniles of *Daphnia* in the reservoir (P and J, respectively) and in the enclosures (Pe and Je). Error bars indicate 95% confidence limits for the average. *n* ranged from 20 to 50 individuals.

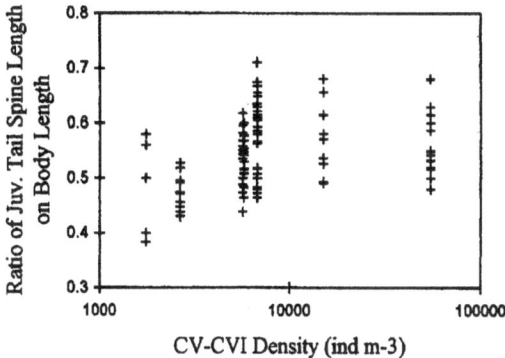

Figure 5. Relation between the ratio of spine length on body length of *Daphnia* juveniles and *A. robustus* population density. Note the logarithmic scale for *A. robustus* density.

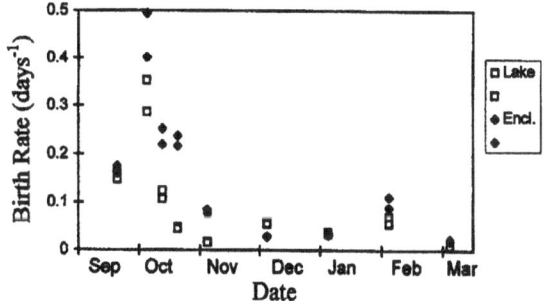

Figure 6. Maxima and minima birth rates of the *Daphnia* populations in the reservoir and in the enclosures. Values were obtained using maximum and minimum egg development times at each sampling date.

Figure 7. Analysis of birth rate of *Daphnia* in terms of contributions of changes in egg development time (ConV), fecundity (Con *F*) and proportion of adults (ConA) to changes in birth rate (d*b*/d*t*). One interval around February (∗) was excluded from the analysis because of a >2-fold discrepancy between SumCon = ConV + ConF + ConA and d*b*/d*t*.

Table 2. Relationships between changes in the proportion of adults (*A*) and the total population death rate (*d*) for all the periods in which analysis of birth rate (see Figure 7) suggests the effect of predation on birth rate dynamics. Changes in *A* are shown in left parentheses and changes in d in right ones. The first sign in each parenthesis marks the direction of change (−, decrease; +, increase) of the corresponding parameter over the sampling interval just before the indicated date; the second sign refers to the interval just after that date

Date	Reservoir	Enclosures
3 October	$(+, -)$ vs. $(+, -)$	$(+, -)$ vs. $(-, +)$
9 October	$(-, +)$ vs. $(-, +)$	$(-, +)$ vs. $(+, -)$
12 December	$(+, +)$ vs. $(+, +)$	
10 January	$(+, +)$ vs. $(+, -)$	$(+, +)$ vs. $(+, -)$

Birth rate

High birth rate values were recorded in the beginning of the experiment, but after November a strong decrease was observed and from then on the values were always low (Figure 6). Egg development rate (*V*) was always a minor contributor to the changes in birth rate (Figure 7). The only exception was observed in the enclosures in December. In the reservoir, changes in birth rate were controlled by the proportion of adults (A) in September, December and January. On October 16 and November, fecundity (F) was the main controlling factor of birth rate. In the enclosures, A had a similar role as outside at the beginning of the experiment. After October 16, changes in birth rate in the enclosures were controlled by F. In the enclosures, changes in A were different from the changes in death rate (Table 2). In the reservoir, on October 3 and 16 and in December the changes in A and death rate had the same sign.

Discussion

Different growth stages of A. robustus, as well as of other cyclopoid copepods, show variable patterns of feeding (e.g. Brandl & Fernando, 1986). Daphnia may therefore interact in two ways with the copepods in Maranhão Reservoir. There can be a prey-predator relationship with CV copepodids and adults of A. robustus and competition with the naupliar and early copepodid stages. Our study did not show direct evidence of the competitive interaction. However, the sharp decrease of the densities of nauplii and early copepodids (CI – CIII) in February which was associated with severe food depletion (Secchi disk value of 2.9 m), suggests a similar effect of food shortage on both early developmental stages of the cyclopoid and Daphnia. Competition must also have been important between Daphnia and the calanoid C. numidicus. Calanoids that are taxonomically close to C. numidicus are known to act as macrofilters, whereas D. hyalina is considered a microfilter (Horn, 1985). Nevertheless, competition with the young developmental stages of the copepod must have been keen. Naupliar instars of C. numidicus were recorded at high densities after the decrease in the Daphnia population in November and the densities decreased in January when the Daphnia population increased. The correlation found between body size of primiparous Daphnia and adults of C. numidicus in the enclosures may also indicate the importance of the effect of food availability on both populations. In the reservoir, a significant correlation may indicate the importance of either food availability or similar trends of predation intensity upon the two taxa.

The changes in birth rate (b) were controlled both by the proportion of adults (A) and by fecundity (F). In the enclosures until October 16, ConA controlled b probably due to the additions of juveniles to the population since ConF was in the opposite direction and death ratio (d) did not follow the trend of A. The

same could have happened on October 3 in the reservoir but on October 9 both ConA and ConF were in the same direction. F is mainly dependent on food resources but sometimes also on the age or size structure of the adult animals, as well as on the predation pressure from vertebrate predators. When fish controls F, both F and A are likely to change simultaneously (Polishchuk, 1994). The evidence for vertebrate predation is further supported by a similar trend of d relative to that of A. Gliwicz (1981) suggests that females with eggs are more vulnerable to vertebrate predation since they are more conspicuous and their ability to escape the predator is reduced by their cargo of eggs. From September until October 16, adult Daphnia never exceeded 1.3 mm in body length in the reservoir whereas 1.9 mm was reached in the enclosures. In October, we also observed a relatively low density of large developmental stages of A. robustus indicative of vertebrate predation. In March, there was again a smaller proportion of adult Daphnia in the reservoir than in the enclosures, suggestive of the importance of vertebrate predation. From October 16 until the end of the experiment, changes in birth rate were mainly controlled by changes in F. The decrease of the primipara size class in November could be the result of food limitation (Taylor & Gabriel, 1992; Boersma, 1995a), although the effect of fish kairomones can not be completely overruled (Vanni 1987).

Hrbácek (1977) stated that, in the field, food limited Daphnia populations maximize numbers instead of individual biomass. However, the decrease in body length is known to be disadvantageous in the case of invertebrate predation. Furthermore, offspring size is dependent on the size of the mother (Boersma, 1995b) which would increase the vulnerability of juveniles to copepod predation. Nevertheless, in the present study no clear evidence was found that A. robustus preyed heavily on neonate daphnids (as observed e.g. by Gliwicz and Lampert, 1993). Irrespective of changes in body length, juvenile tail spine length varied in a way correlated with the densities of A. robustus. Balseiro & Vega (1994) showed that the feeding rate of the calanoid copepod Parabroteas sarsi was negatively correlated with the total body length (body plus spine lengths) of Daphnia middendorffiana. The presence of spined morphs of rotifers has been connected with protection against copepod predation (Stemberger & Gilbert, 1984; Marinone & Zagarese, 1991). In Daphnia, tail spines and helmets have been referred to as important stabilizers of swimming, rendering the daphnids somewhat cryptic to the mechanoreceptors of

Chaoborus and copepods and enabling them to swim faster to better escape their predators (Grant & Bayly, 1981; Mort, 1986; Lüning, 1992). It was indeed observed in previous experiments (Caramujo, unpubl. data) that first and second instar juveniles with long tail spines are preyed upon by adult cyclopoids at a lower rate than juveniles with regular spine length.

A gradient of copepod density apparently resulted in a gradient effect on tail spine elongation. Juvenile tail spines rapidly increased in length at low copepod densities (maximum of 10 000 ind. m^{-3}) until a size of approximately 0.36 mm was reached. At higher copepod densities, this size was only slightly increased despite of large increases in copepod density. There seems to be a maximum for the elongation of the tail spine relative to the body length of the daphnid. It would be interesting to know precisely at which density of the predatory copepods such threshold tail spine length would be reached.

During our experiment, the only important predation impact controlling the *Daphnia* population seems to have originated from vertebrate predators. Evidence for strong invertebrate predation in the reservoir was not found. In the enclosures, elongation of the tail spine of juvenile daphnids seems to have enabled them to coexist with the invertebrate predator. However, further evidence is needed to validate this hypothesis. The use of enclosures to follow the growth of *Daphnia* in the presence and absence of the copepod predator would be a suitable experimental design. Furthermore, the role of fish kairomones on tail spine induction in *Daphnia* by copepods needs clarification.

Acknowledgements

We would like to thank two anonymous reviewers whose comments helped to improve considerably an earlier version of this manuscript, and Dr Piet Spaak for editorial assistance and important suggestions. This study was partially supported by a doctoral grant BD-5358/95 awarded to M. J. Caramujo by J.N.I.C.T., Portugal.

References

Balseiro, E. G. & M. Vega, 1994. Vulnerability of *Daphnia middendorffiana* to *Parabroteas sarsi* predation: the role of the tail spine. J. Plankton Res. 16: 783–793.

Boersma, M., 1995a. Competition in natural populations of *Daphnia*. Oecologia 103: 309–318.

Boersma, M., 1995b. The allocation of resources to reproduction *in Daphnia galeata*: against the odds? Ecology 76: 1251–1261.

Brandl, Z. & C. H. Fernando, 1975. Investigations on the feeding of carnivorous cyclopoids. Verh. int. Ver. Limnol. 19: 2959–2965.

Brandl, Z. & C. H. Fernando, 1986. Feeding and food consumption by *Mesocyclops edax*. Proc. Second Int. Conf. on Copepoda, Natl. Mus. Nat. Sciences Canada, Syllogeus No. 58: 254–258.

Bottrell, H. H., A. Duncan, Z. M. Gliwicz, F. Grygierek, A. Herzig, A. Hillbricht-Ilkowska, A. Kurasawa, P. Larsson & T. Weglenska, 1976. A review of some problems in zooplankton studies. Norw. J. Zool. 24: 419–456.

Dodson, S. I., 1974. Adaptive change in plankton morphology in response to size selective predation: A new hypothesis of cyclomorphosis. Limnol. Oceanogr. 19: 721–729.

Gliwicz, Z. M., 1981. Food and predation in limiting clutch size of cladocerans. Verh. int. Ver. Limnol. 21: 1562–1566.

Gliwicz, Z. M. & W. Lampert, 1993. Body-size related survival of cladocerans in a trophic gradient: An enclosure study. Arch. Hydrobiol. 129: 1–23.

Gliwicz, Z. M. & J. Pijanowska, 1989. The role of predation in zooplankton succession. In U. Sommer (ed.), Plankton Ecology. Springer-Verlag, Berlin: 253–296.

Gliwicz, Z. M. & G. Umana, 1994. Cladoceran body size and vulnerability to copepod predation. Limnol. Oceanogr. 39: 419–424.

Grant, J. W. G. & I. A. E. Bayly, 1981. Predator induced crests in morphs of the *Daphnia carinata* King complex. Limnol. Oceanogr. 15: 721–729.

Haney, J. F. & C. Buchanan, 1987. Distribution and biogeagraphy of *Daphnia* in the arctic. In R. H. Peters & R. DeBernardi (eds), *Daphnia*. Mem. Ist. Ital. Idrobiol. 45: 77–105.

Harbácek, J., 1977. Competition and predation in relation to species composition of freshwater zooplankton, mainly Cladocera. In J. J. Cairns (ed.), Aquatic Microbial Communities. Garland Publishing, Inc., New York: 308–353.

Havel, J. E., 1985. Cyclomorphosis of *Daphnia pulex* spined morphs. Limnol. Oceanogr. 30: 853–861.

Havel, J. E. & S. I. Dodson, 1984. *Chaoborus* predation on typical and spined morphs of *Daphnia pulex*: Behavioral observations. Limnol. Oceanogr. 29: 487–494.

Horn, W., 1985. Investigations into the food selectivity of the planktic crustaceans *Daphnia hyalina*, *Eudiaptomus gracilis* and *Cyclops vicinus*. Int. Revue ges. Hydrobiol. 70: 603–612.

Li, J. L. & H. W. Li, 1979. Species-specific factors affecting predator-prey interactions of the copepod *Acanthocyclops vernalis* with its natural prey. Limnol. Oceanogr. 24: 613–626.

Lüning, J., 1992. Phenotypic plasticity of *Daphnia pulex* in the presence of invertebrate predators: morphological and life history responses. Oecologia 92: 383–390.

Lüning, J., 1995. Life-history responses to *Chaoborus* os spined and unspined *Daphnia pulex*. J. Plankton Res. 17: 71–84.

Marinone, M. A. & H. E. Zagarese, 1991. A field and laboratory study on factors affecting polymorphism in the rotifer *Keratella tropica*. Oecologia 86: 372–377.

Mort, M., 1986. *Chaoborus* predation and the function of phenotypic variation in *Daphnia*. Hydrobiologia 133: 39–44.

Paloheimo, J. E., 1974. Calculation of instantaneous birth rate. Limnol. Oceanogr. 19: 692–694.

Polishchuk, L. V., 1994. Cladoceran birth rate dynamics: does population characteristic analysis reflect environmental control? Verh. int. Ver. Limnol. 25: 2369–2371.

Polishchuk, L. V., 1995. Direct positive effect of invertebrate predators on birth rate in *Daphnia* with a new method of birth rate analysis. Limnol. Oceanogr. 40: 483–489.

Smyly, W. J. P., 1970. Observations on the rate of development, longevity and fecundity of *Acanthocyclops viridis* (Jurine) (Copepoda, Cyclopoida) in relation to type of prey. Crustaceana 18: 21–36.

Stemberger, R. S. & J. J. Gilbert, 1984. Spine development in the rotifer *Keratella cochlearis*: induction by cyclopoid copepods and *Asplanchna*. Freshwat. Biol. 14: 639–647.

Taylor, B. E. & W. Gabriel, 1992. To grow or not to grow: optimal resource allocation for *Daphnia*. Am. Nat. 139: 248–266.

Vanni, M. J., 1987. Effects of food availability and fish predation on a zooplankton community. Ecol. Monogr. 57: 61–88.

Williamson, C. E., 1983. Behavioral interactions between a cyclopoid copepod predator and its prey. J. Plankton Res. 5: 701–711.

Zaret, T., 1980. Predation in freshwater communities. Yale University Press, New Haven, Connecticut, 187 pp.

Hydrobiologia **360**: 253–264, 1997.
A. Brancelj, L. De Meester & P. Spaak (eds), Cladocera: The Biology of Model Organisms.
©1997 *Kluwer Academic Publishers.*

Effects of competitors and *Chaoborus* predation on the cladocerans of a eutrophic lake: an enclosure study

Heike Mumm
*Max Planck Institute for Limnology, P.O. Box 165, D-24302 Plön, Germany**
** Present address: University of Konstanz, D-78457 Konstanz, Germany, Heike.Mumm@uni-konstanz.de*

Key words: cladocera, *Daphnia*, *Chaoborus*, size-selective predation, competition, birth rate, death rate

Abstract

The role of large laboratory grown food competitors of the genus *Daphnia* as well as the predation impact of *Chaoborus* on the cladoceran community of an eutrophic lake was assessed in five *in situ* enclosure experiments. The hypothesis tested was that the outcome of competition and gape-limited predation on cladocerans is size dependent. According to the generally accepted assumptions on competition and invertebrate predation, large-bodied cladoceran taxa were expected to be less affected by competing congeners and by *Chaoborus* than were small-bodied taxa. Effects of the predator upon an assemblage of differently sized cladoceran taxa were much more pronounced than effects of competition. There was a tendency of predation and competition impact to decrease with cladoceran size, but predation pressure was also low for some small cladocerans and high for some large cladocerans. The general trends were further obscured by factors not or indirectly linked to body size.

Introduction

Gape-limited predation by invertebrate predators such as *Chaoborus* is considered a factor favouring large-bodied zooplankton (Zaret, 1980). *Chaoborus* tends to remove small prey items as strike efficiency decreases with increasing prey size. The role of vertebrate predation has been the subject of experimental studies since its importance for the structure of freshwater zooplankton communities was first emphasised (Hrbáček et al., 1962). Size-selective predation by visually foraging fish preferring large prey has often been demonstrated to be the cause for the relative scarcity of large-bodied zooplankton (e.g. Galbraith, 1967; Hall et al., 1970). When fish are absent, the better competitive abilities enable large *Daphnia* to be the dominating grazers (Brooks & Dodson, 1965; Hall et al., 1976), as the food concentration threshold for growth decreases with size (Gliwicz, 1990). The results of whole-lake studies have recently lead to the hypothesis that vertebrate and invertebrate predation act alternately because invertebrate predators are themselves subject to fish predation (Carpenter & Kitchell, 1992). As a consequence, the role of

invertebrate predation and competition is expected to increase simultaneously when fish predation declines. Whether competitive exclusion or invertebrate predation is the main reason for the dominance of large daphnids in fish free environments still requires examination. If competitive ability of cladocerans as well as non-susceptibility to invertebrate predation were strictly increasing with body size, the impact of both factors should increase within a range of cladoceran species of decreasing size, small species being more frequently or strongly affected. While predation should primarily influence death rates, competition is likely to exhibit the most pronounced effects on birth rates because fecundity of cladocerans is dependent on food supply. When food resources decline due to competing organisms, less energy can be allocated to egg production. The aim of this experimental study was to separate the influence of competition from invertebrate predation under fish exclusion on a multi-species cladoceran community of different body sizes. In addition I wanted to test the hypothesis that the impact of predation and competition increase with decreasing cladoceran body size.

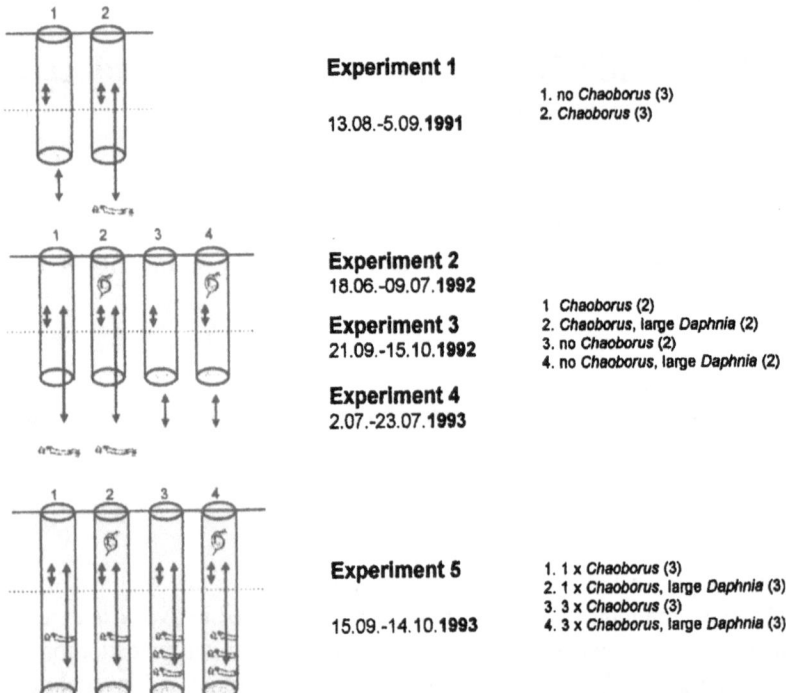

Figure 1. Design of the five enclosure experiments carried out in Plußsee. Number of replicates in parentheses.

Table 1. Mean body length [μm] of cladocerans in Plußsee at the beginning of experiment 5

Taxa	Mean	Standard deviation	n
Large *Daphnia*	956.6	372	149
D. 'longispina'	815.7	226	114
Diaphanosoma	557.3	103	139
Bosmina	452.0	87	99
Ceriodaphnia	393.5	98	103
Chydorus	272.5	56	91

Material and methods

Experiments were carried out in Plußsee, a small eutrophic lake in Northern Germany with a maximum depth of 28 m. Its sheltered location and the surrounding forest promote the development of a stable and shallow metalimnion in summer with an anoxic hypolimnion. The most important invertebrate predators are the larvae of the phantom midge *Chaoborus flavicans*. Planktivorous fish are abundant. Plußsee is inhabited by several differently sized cladoceran taxa. The cladocerans present, in the order of decreas-

ing body size (Table 1), are *Daphnia 'longispina'*, a species complex of *D. galeata*, *D. cucullata* and the hybrids C × G (*D. cucullata × galeata*), G × H (*D. galeata × hyalina*) and C × H (*D. cucullata × hyalina*) (Spaak, 1995), *Diaphanosoma brachyurum*, *Bosmina coregoni* (replaced by *Bosmina longirostris* in early summer), *Ceriodaphnia quadrangula* and *Chydorus sphaericus*. In three subsequent years five enclosure experiments were carried out (Figure 1) during the period of thermal stratification (when the hypolimnion was anoxic) in order to investigate the role of competition and invertebrate predation in different seasons and to account for interannual variation. Polyethylene enclosure bags of 1 m diameter and 10 to 14 m length (depending on the oxygen profile at the beginning of the experiment) were exposed in the lake reaching far through the oxicline, thus cutting out a lake cylinder with its entire plankton (Mumm & Sell, 1995). Zooplankton was trapped inside and fishes were kept outside the bags since the only possible entrance was located in the anoxic layer. In experiments 1 to 4 some of the enclosures were open at the bottom and the others were closed by a 1000 μm mesh. Late *Chaoborus* larvae of the third and fourth instar staying in the anoxic hypolimnion by day could enter the open bags during their upward migration at dusk but could not pass the mesh of the closed bags. Thus *Chaoborus* larvae established themselves in the open enclosures according to their natural density. The experimental design allowed a natural behaviour and encounter rate of predator and prey. In the last experiment only bags closed by a 100 μm net were used and Chaoborus was inoculated at the beginning of the experiment. Three-fold versus natural *Chaoborus* densities was tested since higher *Chaoborus* numbers can be expected at a lower level of fish predation.

In temperate regions *Chaoborus flavicans* overwinter as 4th instar larvae. They pupate, mature and lay eggs mainly from late June through August. The biomass of the late instar larvae declines from the end of June, reaches a minimum in August when early larval stages are present and increases towards September (Parma, 1971). The enclosure experiments covered the periods of strong *Chaoborus* predation by late instars before the summer decline and after the new generation had hatched and grown.

In experiments 2 to 5 competition by large bodied cladocerans was tested by adding two laboratory grown clones of *D. pulex* and *D. pulicaria* originating from pond populations which have been grown in the laboratory for several years (hereafter called large *Daphnia*). The bags were inoculated with these large *Daphnia* not naturally occurring in the lake at 0.15 to 0.5 animals l^{-1} to mimic the invasion of a new species into a lake (Vanni, 1988). Oxygen and temperature profiles in the enclosures were registered and samples were taken twice a week by vertical tows from the end of the bags to the top with a 100 μm mesh size plankton net. In the first two years a simple net with a 0.25 m opening diameter was used but replaced by a plankton net of 0.1 m opening diameter equipped with a sampling cone because of its better performance. Cladocerans were enumerated and their eggs were counted in a Bogorov chamber under a dissecting microscope. If present, at least 100 animals were counted. All abundances were calculated as numbers per net tow (whole water column) because density estimates on a volume basis would be biased by variation in epilimnion thickness during the experiments. Population growth rates were estimated as

$$r = \frac{\ln N_2 - \ln N_1}{t_2 - t_1},$$

with N_1 and N_2 being the absolute population sizes of two subsequent sampling days t_1 and t_2. Birth rates

Table 2. Abundance of *Chaoborus* in the enclosures. Experiments with odd numbers were done in late summer, experiments with even numbers in early summer

Year	Experiment	*Chaoborus*	Number [m^{-2}]	SE
1991	1	no	133	133
		yes	2411	329
1992	2	no	229	51
		yes	537	175
	3	no	0	0
		yes	1076	207
1993	4	no	0	–
		yes	629	–
	5	natural	750	–
		3× natural	2250	–

were estimated by the egg ratio method (Paloheimo, 1974):

$$b = \ln(E + 1)/D,$$

where E is the egg ratio (average number of eggs per animal) and D the egg development time in days. Egg development times were estimated after regressions given in Bottrell et al. (1976). Death rates were then estimated as the difference of birth rates and population growth rates. Birth and death rates were not calculated for large *Daphnia* and a few other cladocerans in some of the experiments due to low absolute numbers.

One and two way repeated measures anovas were applied to estimate the effects of *Chaoborus* in experiment 1 or *Chaoborus* and large *Daphnia* in experiments 2 to 5 on each cladoceran taxon. As time was used as a repeated measure, the test of the between subjects effect did not rely on the assumption of independence of consecutive samples (e.g. Winer, 1971). Since the degrees of freedom for the between subjects tests were not inflated by sampling frequency, for clearness only probabilities of the treatment effects are given in Tables 3–5. *Chaoborus* × large *Daphnia* interaction terms were not significant unless otherwise stated in the text. The within subjects effects (time and interactions with time) are given in Mumm (1996). All abundances were log transformed before statistical analysis in order to homogenise variances (Sokal & Rohlf, 1994). The tests for the effect of *Chaoborus* were based on all sampling dates. To account for the delay until the establishment of large *Daphnia* in the bags, the tests for the competition effect used only the data after stocking (experiments 2, 3 and 4). To correct for possible initial differences due to the delay, the log transformed values of the sampling prior to stocking were subtracted from each following value of a specific enclosure bag (Liber et al., 1992). In experiment 2, the abundances of *Daphnia 'longispina'* differed significantly between treatments on the first sampling day. In this case, a correction was made by subtracting the log transformed values of the first sampling from the following ones.

The migration of the *Chaoborus* larvae into the open bags at sunset could be documented with the help of an echosounder (Mumm & Sell, 1995). *Chaoborus* abundances in the bags were determined by night catches towards the end of experiment 1, 2 and 3 and estimated from lake densities for experiment 4. In the fifth experiment, initial densities were already known from the stocking.

Results

Abundance

Larval densities in the *Chaoborus* treatments (open bags) of experiments 1 to 4 ranged between 537 and 2411 animals m^{-2}. Larvae in the closed bags were 0 to 229 m^{-2} (Table 2). Larvae found in the closed bags were exclusively young instars whose predation impact on cladocerans is negligible. In the last experiment, where natural and three-fold *Chaoborus* treatments were established, larval densities were 712 and 2112 animals m^{-2} respectively. Large *Daphnia* survived in all bags of the four experiments in which a competition effect was tested (Figure 2). In experiments 2 to 4, their populations could grow to higher densities than those inoculated only in the non-*Chaoborus* treatments. In the last experiment, the low-*Chaoborus* treatments had higher densities of large *Daphnia* in the first period but were surpassed by the three-fold *Chaoborus* treatments in the end. *Daphnia 'longispina'* developed similarly in all bags of the five experiments. Despite the slightly higher numbers in the non- or low-*Chaoborus* treatments (except in experiment 2), there was no significant *Chaoborus* effect in any experiment (Table 3). Large *Daphnia* neither had a significant effect on the density of *Daphnia 'longispina'* (Table 3). *Diaphanosoma* abundance was clearly affected by *Chaoborus* (Figure 3). In the early summer experiments 2 and 4, abundance of *Diaphanosoma* in the *Chaoborus* treatments decreased after about two weeks while in late summer experiments differences could be found from the second or third sampling day onwards. Only in the second experiment, the *Chaoborus* effect was not significant

(Table 3). There was no effect of large *Daphnia* in any of the experiments, but in experiment 2 a significant interaction term of *Chaoborus* × large *Daphnia* ($p < 0.001$) was present. *Bosmina* were reduced in number by *Chaoborus* in all but the first experiment. The presence of large *Daphnia* did not result in any change of *Bosmina* abundance. *Ceriodaphnia* only occurred in the late summer experiments 1, 3 and 5. *Chaoborus* had a negative effect on *Ceriodaphnia* in all of these three experiments (Figure 4). A negative effect due to the presence of large *Daphnia* was ascertained for experiment 3. *Chaoborus* exhibited a negative influence on *Chydorus*, the smallest cladoceran, only in the second experiment and when larval density was raised to the three-fold natural level in experiment 5. Although the tendency of reduced abundance in the large *Daphnia* treatments was much clearer than for other cladocerans, the *Daphnia* effect was only significant in the second experiment. In this early summer experiment there was also a significant *Chaoborus* × *Daphnia* interaction effect in *Chydorus* as was the case in *Diaphanosoma*.

Death rates and birth rates

Chaoborus' impact on death rates of *Daphnia 'longispina'* was significant in the 4th experiment (Table 4). *Diaphanosoma* was influenced in three out of five experiments. *Bosmina* and *Ceriodaphnia* death rates were influenced by *Chaoborus* in all calculable cases. In the summer experiment 3, *Bosmina*'s death rate was not higher but reduced in the presence of *Chaoborus*. The smallest cladoceran *Chydorus* had higher mortality rates only in the last two experiments. When three-fold versus natural *Chaoborus* density was tested in the last experiment, striking effects on mortality rates were noted for all cladocerans except for *Daphnia 'longispina'*. In the presence of the large *Daphnia* death rates of *Diaphanosoma* and *Chydorus* increased during the second experiment.

An effect of large *Daphnia* on birth rates was found in two cases (Table 5). In the early summer experiments 2 and 4, birth rates of *Daphnia 'longispina'* and *Bosmina* were lower in the presence of the competitors. Birth rates of *Daphnia 'longispina'* and *Diaphanosoma* increased due to *Chaoborus* in the last experiment.

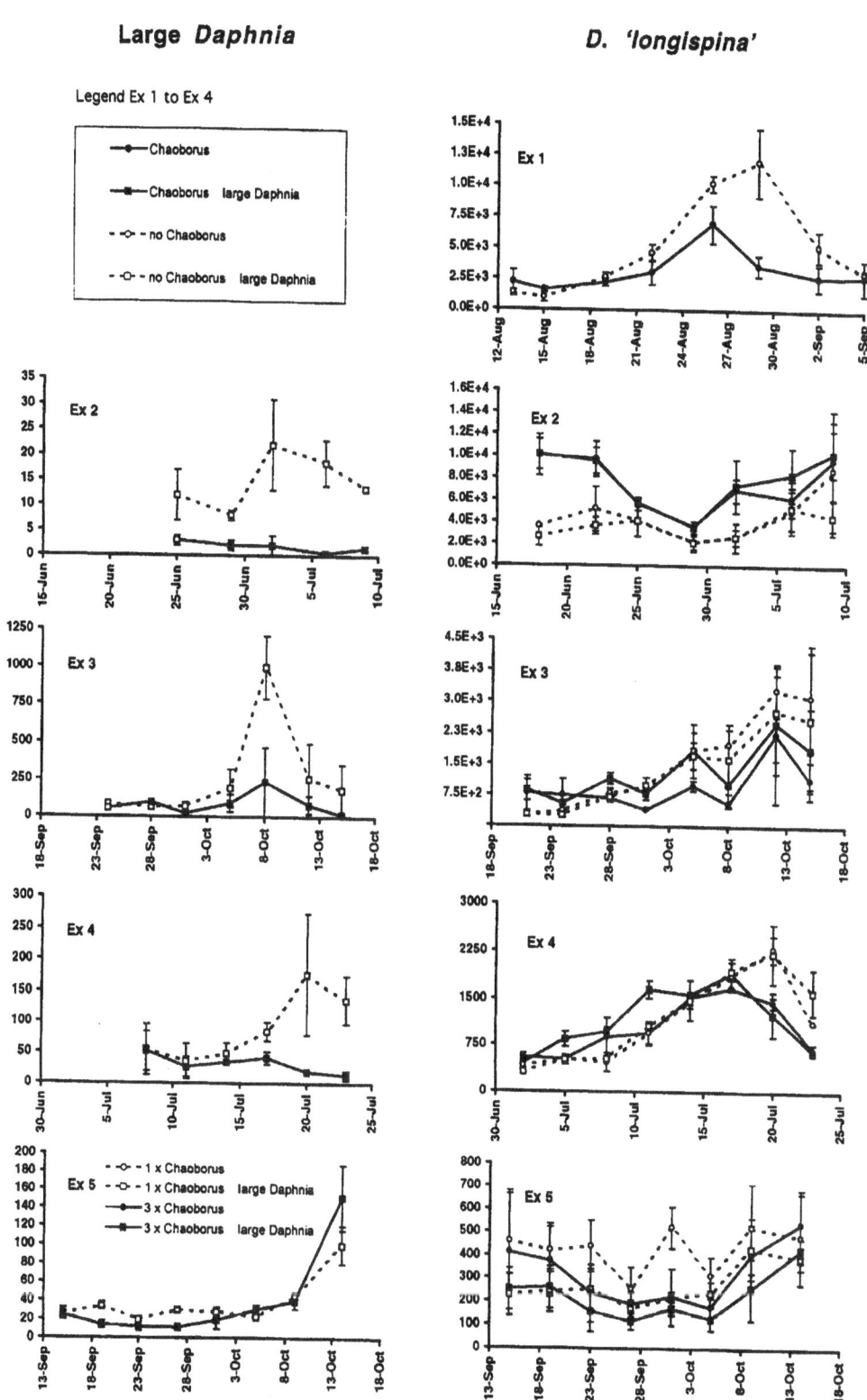

Figure 2. Abundance of large *Daphnia* and *Daphnia 'longispina'* in the five enclosure experiments. Mean numbers per net tow are plotted. Error bars indicate standard errors.

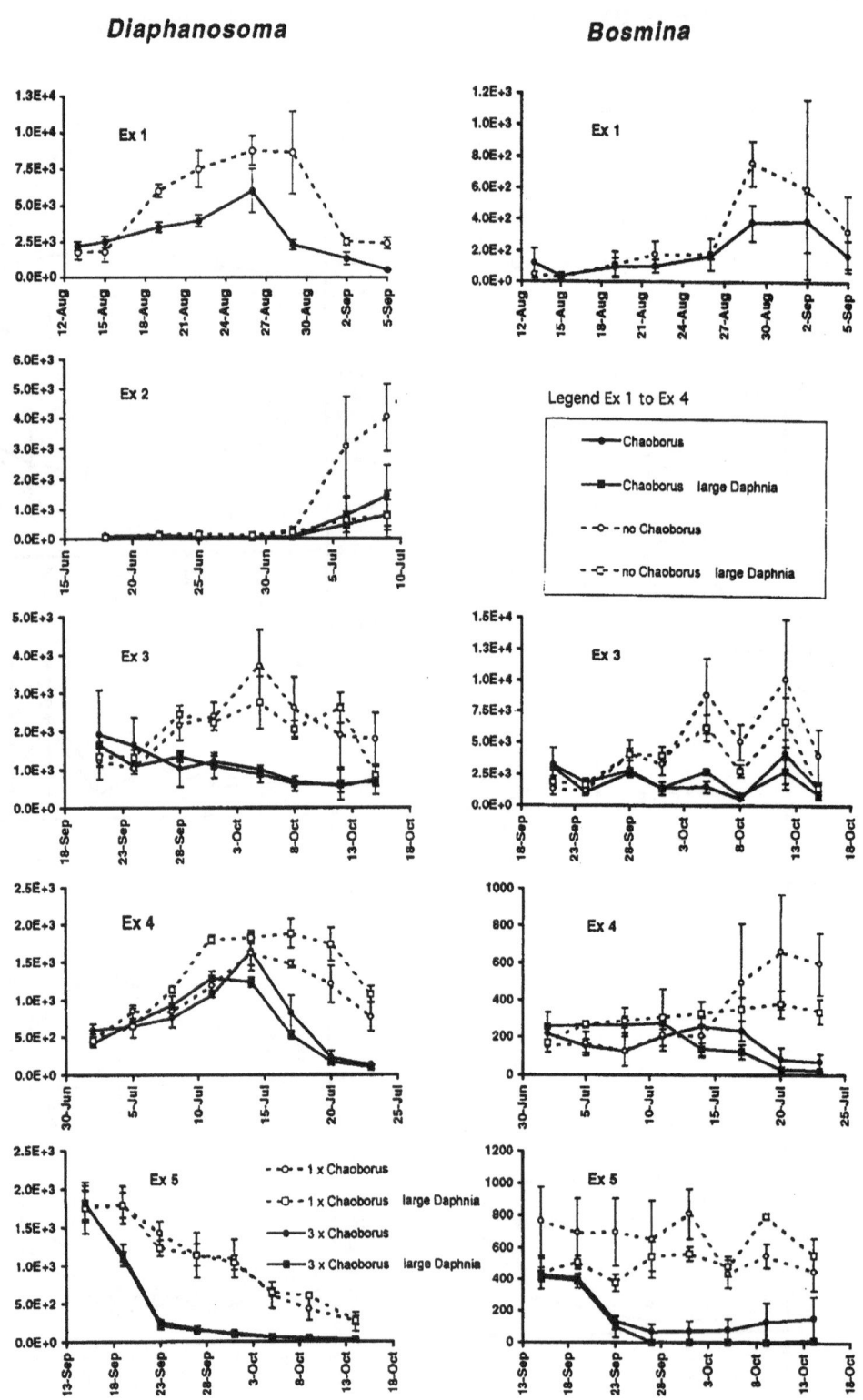

Figure 3. Abundance of *Diaphanosoma* and *Bosmina* during the five enclosure experiments. See Figure 2 for details.

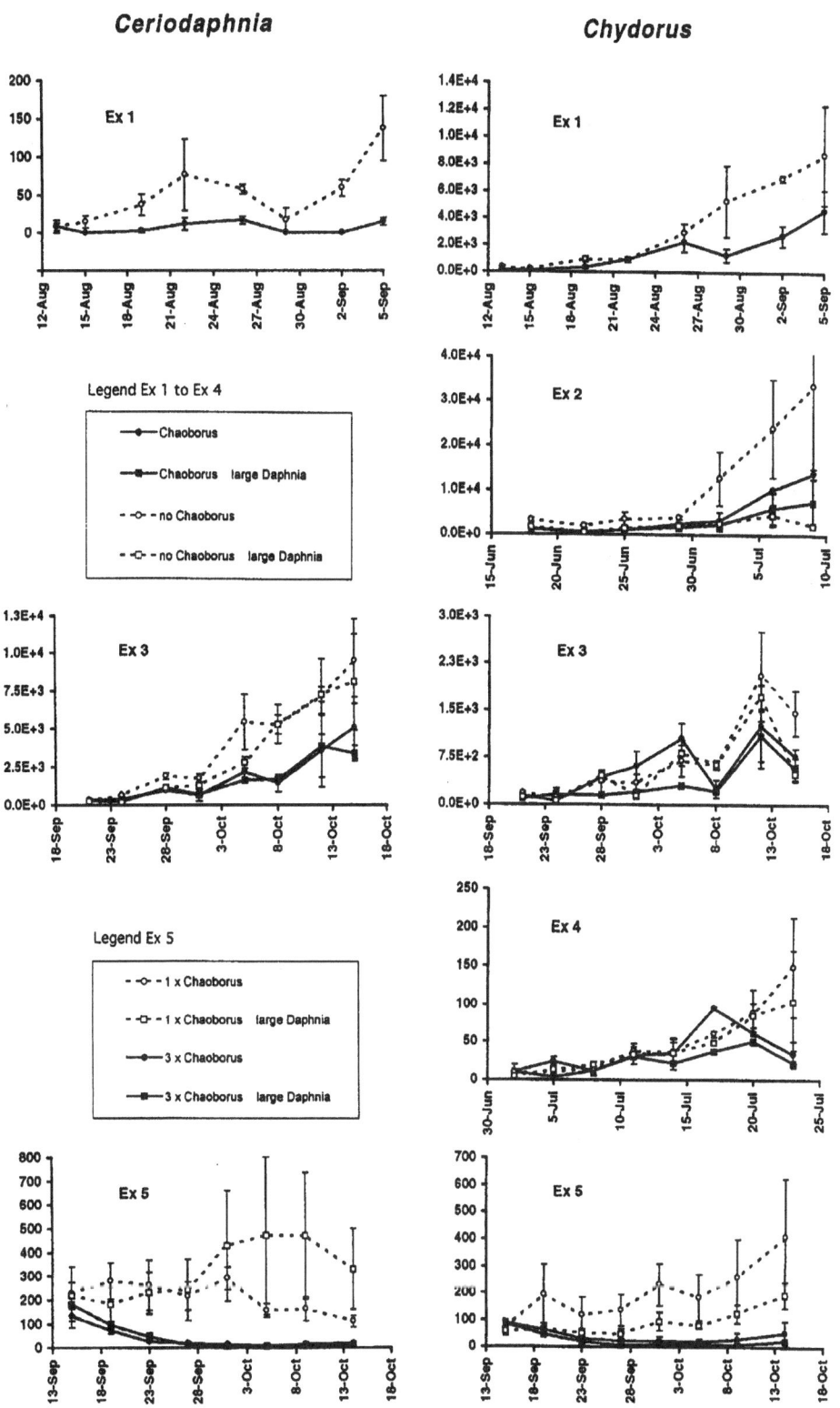

Figure 4. Abundance of *Ceriodaphnia* and *Chydorus* in experiments one to five. See Figure 2 for details.

Table 3. Effects of *Chaoborus* and large *Daphnia* on cladoceran abundances. P: Probabilities of main effects as derived from one-way anova (ex 1) and two-way Anova (ex 4–5) with repeated measures. + denotes a significant positive, – a negative effect of the factor *Chaoborus* or large *Daphnia*. /: species lacking. The cladocerans are listed according to decreasing size

Experiment	Large Daphnia	Daphnia 'longispina'	Diaphanosoma	Bosmina	Ceriodaphnia	Chydorus
Effect of Chaoborus						
1	/	0.076	0.001−**	0.999	0.009−**	0.100
2	0.007−**	0.056	0.128	/	/	0.047−*
3	0.020−*	0.876	0.037−*	0.006−**	0.021−*	0.221
4	0.019−*	0.848	0.001−**	0.050−*	/	0.145
5	0.273	0.254	<0.001−***	<0.001***	<0.001−***	0.002−**
Effect of Large Daphnia						
2		0.698	0.156	/	/	0.005−**
3		0.847	0.907	0.933	0.040−*	0.469
4		0.112	0.198	0.081	/	0.734
5		0.651	0.530	0.163	0.743	0.150

Table 4. Effects of *Chaoborus* and large *Daphnia* on death rates. See Table 3 for further explanation

Experiment	Daphnia 'longispina'	Diaphanosoma	Bosmina	Ceriodaphnia	Chydorus
Effect of Chaoborus					
1	0.730	0.198	/	/	0.636
2	0.382	0.001+**	/	/	0.812
3	0.089	0.227	0.035−*	0.007+**	0.339
4	0.004+**	0.001+**	0.009+**	/	0.026+*
5	0.074	<0.001+***	0.003+**	<0.001+***	0.003**
Effect of Large Daphnia					
2	0.968	0.046+*	/	/	0.002+**
3	0.319	0.379	0.771	0.512	0.679
4	0.977	0.646	0.135	/	0.201
5	0.152	0.338	0.631	0.474	0.437

Discussion

The larval densities in the *Chaoborus* treatments, ranging from 537 to 2411 animals m^{-2}, can be considered low to intermediate when compared to literature data of fourth instar larvae which range from almost zero to 11000 animals m^{-2} in some shallow lakes (e.g. Goldspink & Scott, 1971; Schiemer et al., 1974; Hillbricht-Ilkowska et al., 1975; Wagner-Döbler, 1988).

Both, competition and invertebrate predation influenced the cladoceran assemblage of Plußsee. All cladocerans were affected by *Chaoborus* in at least one experiment. The impact of competition by large *Daphnia* was much less pronounced. As expected, the effect of *Chaoborus* on the abundances of the natural Plußsee cladocerans showed a tendency of size dependence. While the abundance of the largest Plußsee cladoceran, *Daphnia 'longispina'*, was not significantly reduced in any of the five experiments, *Diaphanosoma* was reduced in 4 out of 5 cases and the small cladoceran *Ceriodaphnia* was affected in all experiments where it was present. *Bosmina* and *Chydorus*, however, did not completely fit to this pattern. The lack of a *Chaoborus* effect on *Bosmina* in experiment 2 could be assigned to the very low absolute numbers of *Bosmina* increasing the type 2 error. For *Chydorus,* other factors besides size could have been of importance. *Chydorus* is a taxon not solely occurring in the pelagic zone. It could

Table 5. Effects of *Chaoborus* and large *Daphnia* on birth rates. See Table 3 for further explanation

Experiment	Daphnia 'longispina'	Diaphanosoma	Bosmina	Ceriodaphnia	Chydorus
Effect of *Chaoborus*					
1	0.130	0.472	/	/	0.780
2	0.510	0.814	/	/	0.887
3	0.937	0.854	0.869	0.395	0.426
4	0.086	0.052	0.490	/	0.254
5	0.011+*	0.006+**	0.789	0.566	0.303
Effect of Large *Daphnia*					
2	0.030−*	0.052		/	0.826
3	0.367	0.394	0.606	0.615	0.799
4	0.993	0.861	0.024−*	/	0.856
5	0.746	0.611	0.867	0.252	0.246

have preferentially dwelled on the enclosure walls, thus avoiding predation by *Chaoborus*. In addition, akinesis ('deadman response') has been suggested as an effective mode of tactile predator avoidance for *Chydorus* as well as for *Bosmina* (Kerfoot et al., 1980; Swift, 1992). When attacked by a predator, the animals cease all movements and cannot be redetected in case of an unsuccessful first strike. Finally, both *Chydorus* and *Bosmina* are spherical in shape and difficult to manipulate by *Chaoborus*.

The significant effect of *Chaoborus* on the large *Daphnia* in experiments 2 to 4 was unexpected and contrasted to the absence of any effect on the smaller *Daphnia* 'longispina'. Population growth rates can be influenced by predation and food supply but neither mortality nor food availability was anticipated to cause a bigger effect on the large *Daphnia* than on *Daphnia* 'longispina'. The vulnerabilty of zooplankton to *Chaoborus* can be described by an optimum function resulting from the encounter probability of predator and prey (increases with prey size) and the strike efficiency (decreases with prey size; Pastorok, 1981). Shei et al. (1988) found that 98% of *D. rosea* eaten by *Chaoborus flavicans* were smaller than 1.25 mm. The change from positive to negative prey length selection for *D. longispina* was found to take place at 1 mm body length (Krylov, 1993). Strike efficiency of *Chaoborus* is limited by the size of the capture apparatus i.e. mouth width (Riessen, 1990), which is approximately half the head capsule length (e.g. Mumm & Sell, 1995). In Plußsee, fourth instar head capsule length ranged between 994 and 1303 μm (Voss, 1995). It can therefore be estimated that the prey vulnerability curve of

fourth instar *Chaoborus flavicans* larvae in Plußsee has its maximum well below 1 mm body length of the prey. A considerable proportion of the large *Daphnia* must therefore have been above *Chaoborus*' preferred prey size, whereas a higher proportion of the smaller *Daphnia* 'longispina' fell within the preferred size range. A food effect on large *Daphnia* was not likey because the same *Daphnia* clones had been used in an enclosure study in nearby Schöhsee (Gliwicz & Lampert, 1993), where they survived and even monopolized the resources to the detriment of all smaller cladocerans under a range of oligotrophic to eutrophic conditions even when inedible Cyanobacteria filaments were present.

The genus *Daphnia* is known to be able to react to predator presence with morphological changes (e.g. Havel, 1987) as well as life history shifts (e.g. Dodson & Havel, 1988; Stibor, 1992) which provide a direct protection or an escape in size. The *D. pulex* clone used in this study reacts with neck teeth formation in young instars when exposed to *Chaoborus* kairomones (Tollrian, 1994). Neckteeth were, however, never seen in the present enclosure study and the minimum inducing kairomone concentration given in Tollrian (1994) indicates that natural *Chaoborus* densities in Plußsee remain too low to induce morphological changes in this clone. If comparable larval densities are required to induce morphological changes or changes in life history for both large *Daphnia* clones used in this study, induced changes were not likely to occur in the *Chaoborus* treatments and neonates were thus not protected. When three-fold *Chaoborus* densities were applied, the abundances of large *Daph-*

nia in the high *Chaoborus* treatments were equivalent to the abundances found in the low *Chaoborus* treatments or even slightly higher. This suggests that the population development of large *Daphnia* depended upon responses controlled by *Chaoborus* abundance and exposure time. *Daphnia 'longispina'*, having coexisted with *Chaoborus* in the Plußsee may have been able to react to lower *Chaoborus* densities than the large *Daphnia*, thereby reducing their mortality risk.

The presence of large *Daphnia* in general had less effect upon the cladoceran assemblage than *Chaoborus* predation. However, an inoculation with a higher density of competitors or the continuation of the experiments for a longer period might have resulted in a stronger effect. As such, it cannot be stated beyond any doubt that competition is a minor factor favouring large bodied cladocerans when fish are absent.

The smallest cladocerans *Ceriodaphnia* and *Chydorus* did suffer from competition by large *Daphnia*, which conforms to the size dependence of competition impact (Hall et al., 1976).

Effects on mortality rates were generally in accordance with the effects on abundances. *Chaoborus* affected *Diaphanosoma* more than it affected the larger *Daphnia 'longispina'*. *Ceriodaphnia* was affected in all cases when birth and death rates could be estimated. *Chydorus* was not influenced in the first three experiments and the mortality rates of *Bosmina* even decreased in the presence of predators in experiment 3. The pattern of prey vulnerability to *Chaoborus* suggested by Pastorok (1981) has often been demonstrated. While most authors have only focused on the decline in vulnerability with larger prey sizes, the decline for prey items below the maximum has not received appropriate attention. Not only prey items which are too large to be eaten are safe from predation, but also very small prey which are seldom encountered or unattractive to the predator. This is a probable reason why the size-effect pattern of predation is broken by the small prey types *Chydorus* and *Bosmina*. The fact that *Daphnia 'longispina'* had an increased mortality rate due to *Chaoborus* in one experiment while its abundance was not significantly reduced illustrates that *Daphnia 'longispina''* is able to compensate for *Chaoborus* predation e.g. by life history shifts or increased birth rate as verified in experiment 5.

Competition effects on mortality rates have rarely been reported (Goulden et al., 1982; Vanni, 1987). A direct effect of large *Daphnia* on mortality rates has often been suggested for rotifers (e.g. Burns & Gilbert, 1986) but mechanical interference on organisms caught by the filter current is not a likely explanation for interaction among cladocerans. Mortality effects by competitors could only derive indirectly from a shortage in food. As cladocerans first reduce egg production when food becomes limiting, birth rates should be lowered before mortality rates are affected. This was not the case for *Diaphanosoma* and *Chydorus* in the early summer experiment 2, showing increased mortality rates while birth rates remained unaffected. However changes in egg development time due to changing temperatures may have masked the decrease in birth rates. An increase in mortality rate by competition cannot be considered common but may occur at times when food is strongly limiting, as is often the case in early summer, possibly in combination with additional stress factors (e.g. parasitism).

Effects of *Chaoborus* on birth rates were not found except when natural against three-fold larval densities were applied in experiment 5. Birth rates increased in *Daphnia 'longispina'* and *Diaphanosoma* in the high predation treatments. The cause can be either direct or indirect. If small individuals are selectively removed by the predator, this leads to a high proportion of adults. A low amount of young, non-reproducing animals increases the egg ratio. As birth rate linearly depends on the egg ratio, a removal of small animals will result in an increased birth rate. The indirect effect of *Chaoborus* derives from the removal of competing grazers leading to better food conditions for the remaining animals. Abundances and death rates showed that fewer cladoceran grazers were present in the high than in the low predation treatments of experiment 5. As a result, secchi disk transparency was lower and the biomass of some edible algae was higher in the high than in the low *Chaoborus* treatments. Apparently, the larger Plußsee cladoceran taxa *Daphnia 'longispina'* and *Diaphanosoma* took advantage of the better food conditions, confirming the high efficiency of these grazers compared to the smaller *Bosmina* and *Chydorus*.

It can be concluded that there was a general trend of the effect of *Chaoborus* predation to decrease with prey body size, but the pattern was strongly blurred. The small, round-shaped *Bosmina* and *Chydorus* were less affected than expected, suggesting an effect of body shape and behavioural traits, but probably also reflecting the declining effect of predation towards very small prey size. The large laboratory grown *Daphnia* were more strongly affected than expected. The results show that the prey vulnerability curve to *Chaoborus* predation proposed by Pastorok (1981), with decreasing vulnerability not only for big prey but also for very small prey, should no longer be overlooked because of its consequences for zooplankton assemblages that cover a broad body size range. A declining effect of competition by large *Daphnia* with cladoceran size was only weakly supported, presumably due to the low population densities of the competitors.

Acknowledgements

I thank the staff of the Max Planck Institute for Limnology in Plön, my scientific advisor Prof. Dr W. Lampert and two anonymous reviewers for substantially improving the manuscript. Many thanks to Sue Mitchell for improving the language and to Piet Spaak for editorial support. This study was financed by the BMFT (German Ministry of Research and Technology, now BMBF) grant number 0339434 A.

References

Bottrell, J., A. Duncan, Z. M. Gliwicz, E. Grygierek, A. Herzig, A. Hillbricht-Ilkowska, H. Kurasawa, P. Larsson & T. Weglenska, 1976. A review of some problems in zooplankton production studies. Norw. J. Zool. 24: 419–456. Brooks, J. L. & S. I. Dodson, 1965. Predation, body size, and the composition of plankton. Science 150: 28–35.

Burns, C. W. & J. J. Gilbert, 1986. Direct observations of the mechanism of interference between *Daphnia* and *Keratella cochlearis*. Limnol. Oceanogr. 31: 859–866.

Carpenter, S. R., J. F. Kitchell & J. R. Hodgson, 1985. Cascading trophic interactions and lake productivity. BioScience 35: 634–635.

Dodson, S. I. & J. E. Havel, 1988. Indirect prey effects: some morphological and life history responses of *Daphnia pulex* exposed to *Notonecta undulata*. Limnol. Oceanogr. 33: 1274–1285.

Galbraith, M. G., jr., 1967. Size-selective predation on *Daphnia* by rainbow trout and yellow perch. Trans. am. Fish. Soc. 96: 1–10.

Gliwicz, Z. M. & W. Lampert, 1993. Body-size related survival of cladocerans in a trophic gradient: an enclosure study. Arch. Hydrobiol. 129: 1–23.

Gliwicz, Z. M., 1990. Food thresholds and body size in cladocerans. Nature 343: 638–640.

Goldspink, C. R. & D. B. C. Scott, 1971. Vertical migration of *Chaoborus flavicans* in a Scottish loch. Freshwat. Biol. 1: 411–421.

Goulden, C. E., L. L. Henry & A. J. Tessier, 1982. Body size, energy reserves, and competitive ability in three species of cladocera. Ecology 63: 1780–1789.

Hall, D. J., W. E. Cooper & E. E. Werner, 1970. An experimental approach to the production dynamics and structure of freshwater animal communities. Limnol. Oceanogr. 15: 839–928.

Hall, D. J., S. T. Threlkeld, C. W. Burns & P. H. Crowley, 1976. The size-efficiency hypothesis and the size structure of zooplankton communities. Ann. Rev. Ecol. Syst. 1: 177–208.

Havel, J. E., 1987. Predator-induced defences: a review. In W. C. Kerfoot & A. Sih (eds), Predation: Direct and Indirect Impacts on Aquatic Communities. The University Press of New England, Hanover (N.H.): 263–278.

Hillbricht-Ilkowska, A., Z. Kajak, J. Ejsmont-Karabin, A. Karabin & J. Rybak, 1975. Ecosystem of the Mikolajskie Lake. The utilisation of the consumers production by invertebrate predators in pelagic and profundal zones. Pol. Arch. Hydrobiol. 22: 53–64.

Hrbácek, J., M. Dvorakova, V. Korinek & L. Prochazkova, 1962. Demonstration of the effect of the fish stock on the species composition and the intensity of metabolism of the whole plankton association. Verh. int. Ver. Limnol. 14: 192–195.

Kerfoot, W. C., D. L. Kellogg jr. & J. R. Strickler, 1980. Visual observation of live zooplankters: evasion, escape and chemical defences. In W. C. Kerfoot (ed.), Evolution and Ecology of Zooplankton Communities. The University Press of New England (N.H.): 10–27.

Krylov, P. I., 1993. Density-dependent predation of *Chaoborus flavicans* on *Daphnia longispina* in a small lake: the effect of prey size. Hydrobiologia 239: 131–140.

Liber, K., Kaushik, N. K., Solomon, K. R. & J. H. Carey, 1992. Experimental designs for aquatic mesocosm studies: a comparison of the 'ANOVA' and 'Regression' design for assessing the impact of tetrachlorophenol on zooplankton populations in limnocorrals. Exp. Toxicol. Chem. 11: 61–77.

Mumm, H., 1996. Zooplanktonentwicklung im Plußsee: invertebrate Räuber, die Wirkung der Biomanipulation und Langzeittrends. Dissertation, Christian-Albrechts-Universität, Kiel.

Mumm, H. & A. F. Sell, 1995. Estimating the impact of *Chaoborus* predation on zooplankton: a new design for *in situ* enclosures studies. Arch. Hydrobiol. 134: 195–206.

Paloheimo, J. E., 1974. Calculations of instantaneous birth rate. Limnol. Oceanogr. 19: 692–694.

Parma, S., 1971. *Chaoborus flavicans* (Meigen) (Diptera, Chaoboridae): an autecological study. Dissertation, Rijksuniversiteit Groningen.

Pastorok, R. A., 1981. Prey vulnerability and size selection by *Chaoborus* larvae. Ecology 62: 1311–1324.

Riessen, H. P., 1990. Demographic analysis of *Chaoborus* predation by *Daphnia pulex*. Verh. int. Ver. Limnol. 24: 339–343.

Schiemer, F., E. Dolezal, E. Gnaiger & A. Jantsch, 1974. Beobachtungen über Verteilung, tageszeitliche Wanderungen und Nahrungsaufnahmeraten von *Chaoborus flavicans* (Meigen) im Goggausee. Carinthia 2: 165–196.

Shei, P., T. Iwakuma & F. Koichi, 1988. Feeding of *Chaoborus flavicans* larvae (Diptera: Chaoboridae) on *Ceratium hirundinella* and *Daphnia rosea* in a eutrophic pond. Jap. J. Limnol. 49: 227–236.

Sokal, R. & F. J. Rohlf, 1994. Biometry: the principles and practice of statistics in biological research. W. H. Freeman & Co., San Francisco, 887 pp.

Spaak, P., 1995. Sexual reproduction in *Daphnia*: Interspecific differences in a hybrid species complex. Oecologia 104: 501–507.

Stibor, H., 1992. Predator induced life-history shifts in a freshwater cladoceran. Oecologia 92: 162–165.

Swift, M. C., 1992. Prey capture by the four larval instars of *Chaoborus crystallinus*. Limnol. Oceanogr. 37: 14–24.

Tollrian, R., 1994. Induktion von Verteidigungsstrukturen bei *Daphnia pulex* durch ein von *Chaoborus*-Larven abgegebenes Kairomon. Verlag Shaker. Dissertation, Christian-Albrechts-Universität, Kiel.

Vanni, M. J., 1987. Effects of food availability and fish predation on a zooplankton community. Ecol. Monogr. 57: 61–88.

Vanni, M. J., 1988. Freshwater zooplankton community structure: introduction of large invertebrate predators and large herbivores

264

to a small-species community. Can. J. Fish. aquat. Sci. 45: 1758–1770.

Voss, S., 1995. Zeitliche und räumliche Dynamik der *Chaoborus*-Larven im Plußsee. Master's thesis. University of Hannover.

Wagner-Döbler, I., 1988. Vertical migration of *Chaoborus flavicans* (Diptera, Chaoboridae): the control of day and night depth by environmental parameters. Arch. Hydrobiol. 114: 251–274.

Winer, B. J., 1971. Statistical principles in experimental design. 2nd edn., McGraw-Hill Kogakusha Ltd, 907 pp.

Zaret, T. M., 1980. Predation and freshwater communities. Yale University Press, New Haven, 187 pp.

Hydrobiologia **360**: 265–275, 1997.
A. Brancelj, L. De Meester & P. Spaak (eds), Cladocera: The Biology of Model Organisms.
©1997 Kluwer Academic Publishers.

The relevance of size efficiency to biomanipulation theory: a field test under hypertrophic conditions

Steven Declerck[1], Luc De Meester[2], Nicole Podoor[3] & José M. Conde-Porcuna[4]

[1] Laboratory of Animal Ecology, State University of Ghent, K.L. Ledeganckstraat 35, B-9000 Gent, Belgium
[2] Laboratory of Ecology and Aquaculture, Catholic University of Leuven, Naamsestraat 59, B-3000 Leuven, Belgium
[3] Instituut voor Plantkunde, Catholic University of Leuven, Kardinaal Mercierlaan 92, B-3030 Leuven (Heverlee), Belgium
[4] Departamento de Biologia Animal y Ecologia, Universidad de Granada, Campus Fuentenueva s/n, E-18071 Granada, Spain

Key words: Size Efficiency Hypothesis, trophic cascade, top-down control, biomanipulation

Abstract

The superiority of large zooplankton in suppressing phytoplankton growth has often been inferred from the Size Efficiency Hypothesis (S.E.H.). The S.E.H. has originally been formulated to account for the competitive superiority of large to small zooplankton under food limiting conditions. Extrapolation of its predictions to the suppression of phytoplankton by zooplankton under high food availability, should be done with care. In an attempt to assess the relevance of the S.E.H. to biomanipulation theory in hypertrophic systems, a fish exclosure experiment was carried out in which the efficiency of two differently structured zooplankton communities in reducing phytoplankton biomass was examined. By inoculating part of the enclosures with laboratory grown *Daphnia magna*, a community dominated by this large cladoceran species could be compared with a community mainly consisting of *Bosmina* and smaller *Daphnia* species. After the exclusion of fish, there was an exponential increase of total zooplankton biomass. Phytoplankton growth was efficiently suppressed to equal levels in both treatments, though there was a difference in timing: chlorophyll-*a* levels in the enclosures inoculated with *D. magna* dropped one week earlier than in non-inoculated enclosures. The time-lag was even more pronounced when large phytoplankton was considered. In accordance with the S.E.H., the time lags could be explained by differences in population growth potential as well as by differences in zooplankton grazing rates (indirectly measured as the minimal zooplankton biomass needed to suppress phytoplankton growth) and food particle size range.

Introduction

Biomanipulation is one of the most popular methods in attempts to restore eutroficated lakes (Gulati et al., 1990; Reynolds, 1994). The major aim of biomanipulation is a reduction of water turbidity, in favour of submerged macrophytes. The method involves a drastic modification of the foodweb and is supposed to act via several mechanisms (Scheffer et al., 1993), such as a reduction of sediment resuspension and associated internal eutrophication via removal of benthivorous fish and the suppression of phytoplankton growth

via a removal of planktivorous fish (cf. the hypothesis of cascading trophic interactions; Carpenter et al., 1985, Carpenter & Kitchell, 1992). The trophic cascade hypothesis assumes a strong top-down control in lake ecosystems. Removal of size selective planktivorous fish is expected to result in dense populations of large zooplankton, which is supposed to reduce phytoplankton biomass because of its high grazing abilities.

Strong top-down effects of fish on zooplankton have frequently been demonstrated (DeMelo et al., 1992). High fish predation pressure generally results in low zooplankton standing stocks. In addition, the size selective feeding of fish can heavily affect the size structure of zooplankton communities by putting highest mortality rates on large zooplankton individuals and species, favouring small ones (Zaret, 1980). A reduction of predation by the removal of fish is therefore not only expected to result in an increase in total zooplankton biomass but should also allow populations of large zooplankton to develop. Under conditions of low fish predation pressure, the Size Efficiency Hypothesis (S.E.H.) predicts that large zooplankters are superior competitors to small ones because of (1) a broader food particle size range, (2) a higher filtration rate and (3) a higher metabolic efficiency due to a lower respiration rate per unit body weight (Brooks & Dodson, 1965; Hall et al., 1976). This has important implications with respect to the control of phytoplankton biomass in lakes. A high metabolic efficiency should allow populations of large zooplankton to grow fast, while a high filtration efficiency and a broad food particle size range should increase the grazing impact on algae at a given zooplankton biomass. Although an increase of total zooplankton biomass after a reduction of fish may result in a higher grazing pressure exerted on phytoplankton, the shift towards a dominance of large zooplankton might potentially be an even more important factor in a successful biomanipulation.

The superiority of large zooplankton as suppressor of phytoplankton growth is generally taken for granted in many biomanipulation studies (Lammens et al., 1990). However, it should be kept in mind that the S.E.H. originally referred to the relative competitive ability of different zooplankton size classes under food limiting conditions rather than to the ability of zooplankton to graze down algae under high food conditions. Therefore, extrapolations of the S.E.H. to biomanipulation theory should be done with care. Moreover, the general validity of the S.E.H. is still controversial. The competitive superiority of large to small zooplankton species has convincingly been demonstrated under specific laboratory conditions (food limiting conditions, one single food source, steady state food levels; Gliwicz, 1990a), but conflicting results have been obtained (Tessier & Goulden, 1987), especially when different food sources (Gliwicz & Lampert, 1990) or when species with differing feeding modes (DeMott, 1982) were compared. Therefore, due to the complexity of field conditions, the predictive value of the S.E.H. may be low (Gliwicz, 1990a).

So far, very few experiments have specifically been designed to test the potential importance of the S.E.H. with respect to phytoplankton control under field con-

ditions. In enclosure studies designed to investigate foodweb interactions in limnetic environments, differently structured zooplankton communities were most often obtained by the introduction of fish predation as a treatment (De Melo et al., 1992; Christoffersen et al., 1993). Beside their role of size selective predator, however, fish also influence nutrient cycles and can heavily affect the transparency of the water by resuspending sedimented particles (Threlkeld, 1988; Vanni & Findlay, 1990; Vanni & Layne, 1997; Vanni et al., 1997). While such experiments are important to gain insights in general mechanisms of top-down or bottom-up control (Brett & Goldman, 1997), they do not allow a direct comparison between the grazing ability of different types of zooplankton communities. A limited number of studies have established different size structures of zooplankton communities by simple removal (via filtering) or addition of zooplankton, thus avoiding confounding effects by fish. Havens (1993) and Fussmann (1996) compared the grazing effect of microzooplankton (protozoa, rotifera) with macrozooplankton (*Daphnia*, calanoid and cyclopoid copepods), while differences in grazing impact between medium-sized zooplankton (*Bosmina*, small and medium-sized *Daphnia* species) and large zooplankton (*D. pulex*, *D. magna*) have been considered by Bergquist et al. (1985), Dawidowicz (1990) and Sarnelle (1993). These studies, performed in systems of low to moderate productivity, mainly focused on net effects of zooplankton grazing. The outcome of these studies differs substantially and in none of them (except Dawidowicz, 1990), a clear distinction can be made between qualitative and quantitative effects of grazing, because different levels of total zooplankton biomass were applied between treatments.

In the present study, we set out to test the relevance of the S.E.H. to biomanipulation theory under hypertrophic conditions. For this, we carried out an enclosure experiment in the hypertrophic Lake Blankaart, in which we monitored the suppression of phytoplankton by a zooplankton community dominated by *Daphnia magna* and a zooplankton community that mainly consisted of *Bosmina longirostris* and smaller *Daphnia* species. In contrast to similar experiments performed by Bergquist et al. (1985), Dawidowicz (1990) and Sarnelle (1993), our experiments were performed under hypertrophic conditions, and we took treatment specific differences in biomass levels into account. As a partial test of the S.E.H., we made an attempt to explain the differences in the observed net-effects of grazing by characteristics of the zooplankton itself, such as differences in the rate of population biomass increase and in 'critical biomass', being the minimal zooplankton biomass necessary to suppress phytoplankton growth.

Study site

Lake Blankaart is a small (30 ha) and shallow (mean depth: 1.5 m) lake, located in Western Flanders, Belgium. It was originally created by peat digging. In the beginning of the century, it was characterized by transparent water and an extensive vegetation of submerged macrophytes (Massart, 1907). However, along with eutrophication, the lake turned turbid during recent decades. Nowadays, according to the mean annual concentration of chlorophyll-a (240 μg/l) and total phosphorus (860 μg/l), as well as the mean values of Secchi disk transparency (0.23 m), the lake can be classified as hypertrophic (O.E.C.D., 1982). The phytoplankton community is dominated by green algae (*Chlorella, Scenedesmus, Pediastrum, Closterium, Oöcystis*), diatoms (*Cyclotella, Aulacoseira, Stephanodiscus*) and cyanobacteria (*Pseudanabaena, Anabaena, Microcystis, Oscillatoria*). While large cladocerans (*D. magna, D. pulex*) can be occasionally found at very low densities in the lake, the zooplankton community is presently almost exclusively composed of small to medium-sized zooplankton: *Bosmina longirostris, D. galeata, D. parvula* and *Acanthocyclops robustus* (Declerck et al., in prep.). The fish community is dominated by planktivores such as *Rutilus rutilus* and *Scardinius erythrophtalmus* and benthic fish such as *Carassius auratus gibelio, Blicca bjoerkna* and *Abramis brama* (Peeters et al., 1996; De Smedt et al., 1997).

Material and methods

The enclosure experiment was started during early spring (7 May) 1994. Six cylindric enclosures with closed-bottom plastic bags (2 m diameter, 1 m depth; see Van der Werf et al., 1987 for detailed description), were filled with approximately 3000 liters of unfiltered lake water. In three of them, the large cladoceran *D. magna* was introduced at a density of approximately 0.5 ind./l^{-1}. The inoculated *D. magna* consisted of an equal mixture of two clones that were hatched from ephippia collected in the littoral zone of Lake Blankaart (1992). The enclosures were kept fishless. All enclosures differed from the lake situation in the absence of a water-sediment interaction zone, a strong reduction of

wind-induced turbulence, and the absence of fish. The enclosures inoculated with *D. magna* differed from the non-inoculated enclosures in three aspects of the zooplankton community: initial biomass (approximately three times higher in the inoculated than in the non-inoculated enclosures), species composition and size structure.

Sampling was started the second day of the experiment. For a period of 34 days, we sampled for nutrients, phyto- and zooplankton every four days. All samples were taken in the central region of the enclosures. Two samples were taken at each of two depths (0.4 m and 1 m) and pooled. Water for chemical analysis and phytoplankton was sampled using a Ruttner bottle (2 l). The zooplankton was collected with a Schindler-Patalas plankton trap (12 l). Water transparency was measured using a Secchi-disk.

Dissolved nutrient concentrations were analysed after filtering water through Whatman GF/C filter paper. The following analytical methods were used (Greenberg et al., 1985): nesslerization for ammonia, Lombard method for nitrites, chromotropic acid method for nitrates, molybdosilicate method for silicates and stannous chloride method for soluble reactive phosphorus (S.R.P.). Total phosphorus was analysed from unfiltered samples after persulfate digestion, followed by the stannous chloride method. Chlorophyll-a content was analysed on methanolic extracts using the spectrophotometric method proposed by Talling and Driver (1963). We measured total chlorophyll-a content as well as the fraction present in large algae that are retained by a mesh of 30 μm. Phytoplankton was identified to species level and counted according to the Utermöhl technique (Unesco, 1974). Individual cells were measured and biovolumes were estimated via formulae of simple geometric volumes. The zooplankton samples were preserved in a 5% sucrose formalin solution. Cladocera were identified to species level and counted and measured. For each enclosure on each date, the total sample or a subsample was treated, and the body length of at least 30 individuals was measured of each species having a share of more than 10% in the total zooplankton biomass. Biomass of zooplankton populations was estimated from their density, their size distribution and published length *vs.* dry weight regression relationships (Bottrell et al., 1976).

From biomass data, an instantanous rate of population biomass increase (r_b) was calculated for the most important species in the enclosures (*B. longirostris* and *D. magna*).

$$r_b = \frac{\ln B_t - \ln B_0}{t}$$

where

- B_0 = population biomass at the beginning of time interval $[t_0, t_t]$,
- B_t = population biomass at the end of time interval $[t_0, t_t]$,
- t = duration of time interval in days.

For each enclosure, we estimated r_b for one interval, with t_0 = 13 May and t_t = the day on which the greatest reduction in phytoplankton biomass in the given enclosure was observed. The second sampling day (13 May) was chosen as the start of the interval, because zooplankton densities on the first sampling day were too low to allow reliable biomass estimation. The end of the interval was taken as the day on which food conditions change dramatically for the zooplankton populations. Differences in r_b between species were tested by the Mann Whitney U-test.

The minimal zooplankton biomass necessary to suppress chlorophyll-a levels was determined for each enclosure. A 'critical' zooplankton biomass was estimated for each enclosure, as the total zooplankton biomass at the end of the interval during which a phytoplankton crash had occurred. Critical biomasses were determined with respect to total chlorophyll-a levels (CB_{TOT}) as well as with respect to chlorophyll-a present in the fraction of large algae ($CB_{>30}$). Differences between treatments were tested by the Mann Whitney U-test.

Results

Zooplankton

At the start of the experiment, total cladoceran biomass was substantially higher in the inoculated enclosures due to the introduction of *D. magna* (Figure 1A). Apart from *D. magna*, the remaining biomass was similar for both treatments and was comprised of *B. longirostris* and Daphniidae (*D. galeata*, *D. parvula*, *D. pulex*, *D. obtusa* and *Ceriodaphnia dubia*). The latter five species will further be referred to as 'other *Daphnia*'.

During the course of the experiment, total zooplankton biomass increased exponentially in both sets of enclosures (Figure 1A). During the early time intervals, the non-inoculated enclosures lagged one time interval (4 days) behind the inoculated enclosures, due

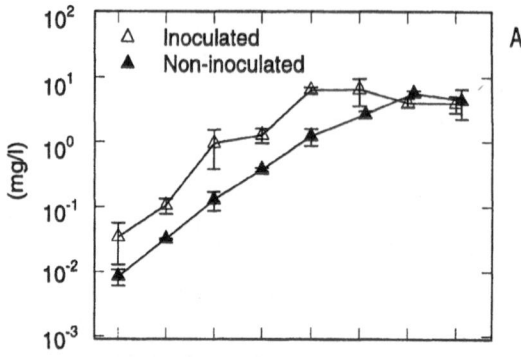

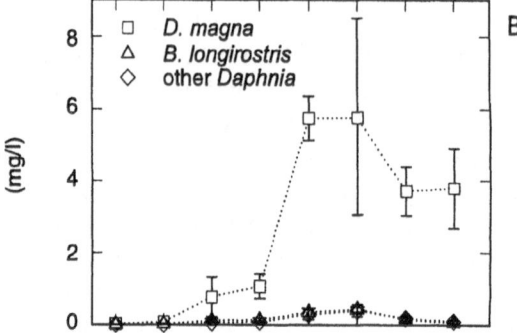

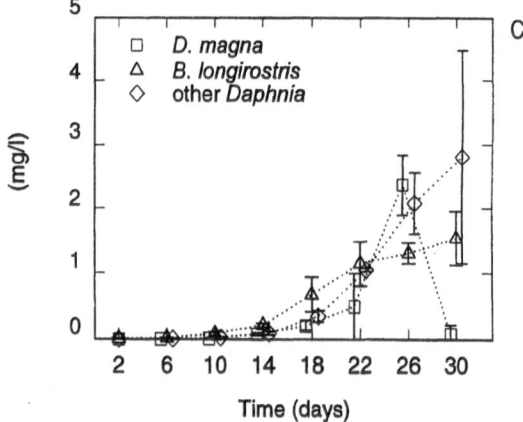

Figure 1. Changes in zooplankton biomass (average ± S.D.) in the enclosures during the course of the experiment. A = Total zooplankton biomass in the inoculated and non-inoculated enclosures; B = Biomass of *Daphnia magna*, *Bosmina longirostris* and other *Daphnia* in the inoculated enclosures; C = Biomass of *Daphnia magna*, *Bosmina longirostris* and other *Daphnia* in the non-inoculated enclosures.

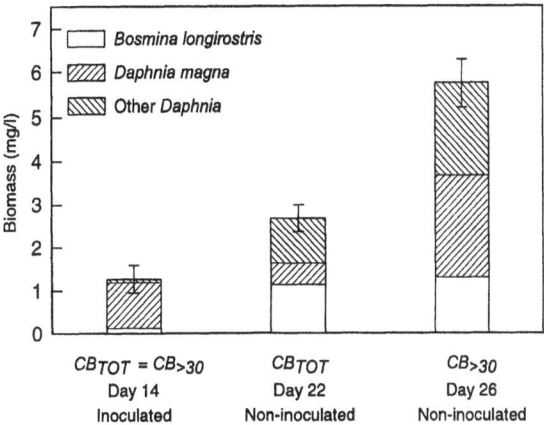

Figure 2. Critical zooplankton biomass (average ± S.D.) needed to suppress total phytoplankton biomass (CB$_{TOT}$) or the biomass of large algae (CB$_{>30}$).

to the differences in initial biomass. The inoculated enclosures were always dominated by *D. magna* (Figure 1B). The non-inoculated enclosures were dominated by *B. longirostris* and Daphniidae other than *D. magna*, but a substantial increase of *D. magna* was noticed towards the end of the experiment (Figure 1C). The appearance of *D. magna* in the non-inoculated enclosures indicates that this species is present in the lake even though it can not be detected by routine sampling (Declerck, pers. obs.).

D. magna was faster in building up population biomass than *B. longirostris*. The mean r_b-value of *D. magna* (0.300 day^{-1}) was 33% higher than the mean value obtained for *Bosmina* (0.226 day^{-1}; Mann Whitney U-test, $P = 0.016$). The critical biomass CB$_{TOT}$, capable of suppressing total phytoplankton biomass, was reached after 14 and 22 days in the inoculated and non-inoculated enclosures, respectively (Figure 2). The mean CB$_{TOT}$ in the inoculated enclosures (1270 μg/l^{-1}) was only 48% of that in the non-inoculated enclosures (2676 μg/l^{-1}; Mann Whitney U-test, $P = 0.05$). The observed differences were even bigger when critical biomasses with respect to the decrease of large algae were considered. In the inoculated enclosures, CB$_{>30}$ equaled CB$_{TOT}$ as the population crash of large phytoplankton species occurred on the same day as that of smaller species. In the non-inoculated enclosures, however, the CB$_{>30}$ was reached only after 26 days. The mean CB$_{>30}$ in the inoculated enclosures was found to be only 22% of that in the non-inoculated enclosures.

Nutrients

Levels of total phosphorus fluctuated around 0.3 mg/l^{-1}, about half of the concentration in the lake (Figure 3B). Initially, 30% of the P was present as S.R.P., but the S.R.P.-levels dropped to about 0.06 mg/l^{-1} after the first sampling interval (Figure 3A). After the initial drop in S.R.P., levels started to increase again, with the onset of this increase starting earlier in the inoculated (day 18) than in the non-inoculated (day 26) enclosures.

Levels of nitrate showed a slight but steady decrease during the course of the experiment, in both inoculated and non-inoculated enclosures (Figure 3C). The decrease was somewhat faster in the non-inoculated than in the inoculated enclosures. After 30 days, nitrate levels reached 77% and 63% of the initial concentration in the inoculated and non-inoculated enclosures, respectively. Only a small fraction of the N was present as ammonia (Figure 3D) and nitrite (data not shown, levels lower than 0.38 mg/l^{-1}). The concentration of ammonia (Figure 3D) showed a pattern similar to the one of S.R.P., with a steep raise after 18 days in the inoculated and after 26 days in the non-inoculated enclosures.

Phytoplankton

Two days after the start of the experiment, total chlorophyll-a averaged 175 μg/l^{-1} in both treatments (Figure 4A), 15% of which was present as large algae (> 30 μm; Figure 4B). The lake water contained total chlorophyll-a levels about two times higher than in the enclosures. During the course of the experiment, a drastic reduction in total chlorophyll-a was observed in all enclosures, with levels dropping to 3 μg/l^{-1} or less (Figure 4A). This decrease in chlorophyll-a levels was clearly associated with an increase of water transparency: whereas Secchi-depth at the start of the experiment was as low as 0.24, the bottom of all enclosure bags was visible by the end of the experiment (Secchi-depth > 1m; Figure 4C). In both sets of enclosures, the reduction of total chlorophyll-a occurred quite suddenly. There was, however, an important time lag with respect to the onset of the drop in chlorophyll-a levels in the two treatments. In the inoculated enclosures, the reduction of chlorophyll-a levels started 10 to 14 days after the start of the experiment. In the non-inoculated enclosures, the reduction in chlorophyll-a levels started after approximately 18 days. The changes in chlorophyll-a levels attributable to large phytoplank-

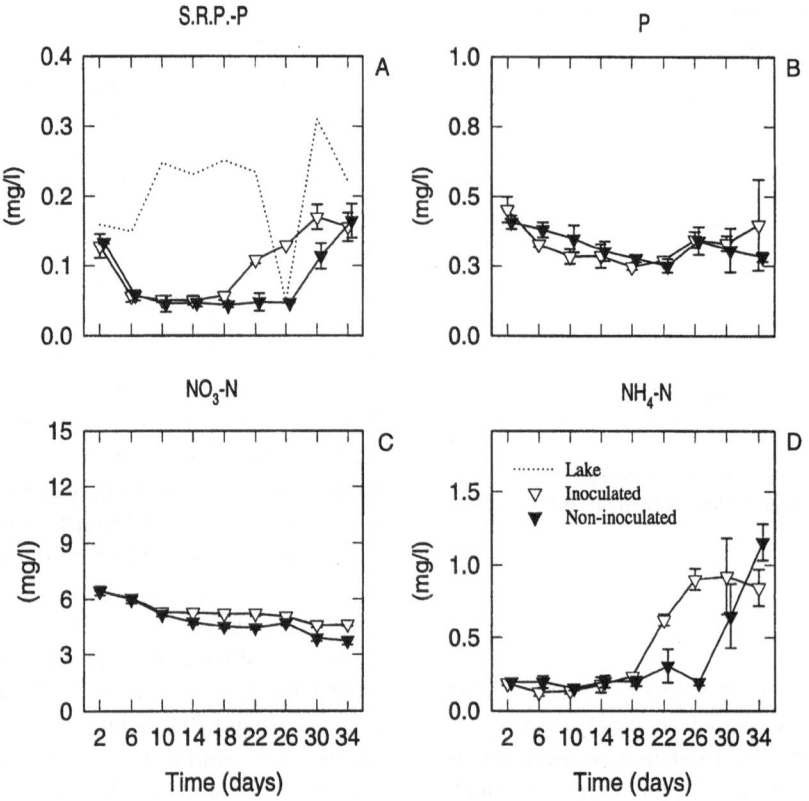

Figure 3. Changes in soluble reactive phosphorus (A), total phosphorus (B), nitrates (C) and ammonia (D) (average ± S.D.) in the inoculated enclosures, in the non-inoculated enclosures and in the lake during the course of the experiment.

ton (> 30 μm) are similar to those of total chlorophyll-*a*, but the time lag observed between inoculated and non-inoculated enclosures was more pronounced, with a dramatic decrease in chlorophyll *a* > 30 μm biomass occurring after about 10 and 22 days respectively. Small and large algae were simultaneously suppressed in the inoculated enclosures. In the non-inoculated enclosures, however, the reduction in biomass of large algae occurred four days after the reduction in total chlorophyll-*a* levels, indicating that the small algae disappeared first.

Phytoplankton communities were initially dominated by Chrysophyta (mainly *Cyclotella chaetoceros*) (Figure 5A and 5B). Cyanobacteria (mainly *Oscillatoria tenuis* and *Pseudanabaena catenata*) were present but their biovolumes were relatively low. During the first week of the experiment, a strong decrease of Chrysophyta and a relative and absolute increase of Chlorophyta (mainly *Pediastrum boryanum* and *Chlorella minutissima*) and Cyanobacteria (*Pseudanabaena catenata*) was observed. This was followed by a major reduction in the biovolume of all phytoplankton taxa. A similar time lag between treatments was

found as for the reductions of chlorophyll-*a*. At the end of the experiment, phytoplankton communities of both treatments had similar phytoplankton compositions and were dominated by the colonial green alga *P. boryanum*.

Discussion

The results of our experiment provide a clear example of top-down control and support the idea of cascading trophic interactions (Carpenter & Kitchell, 1992). Despite high nutrient levels, a strong and negative relationship was found between the biomasses of phyto- and zooplankton. After the exclusion of fish, zooplankton biomass increased exponentially and its grazing pressure resulted in a drastic decrease of the phytoplankton, leading to a high water transparency. The time lag in zooplankton biomass increase between the inoculated and non-inoculated enclosures was reflected in the dynamics of phytoplankton as well as of some nutrients.

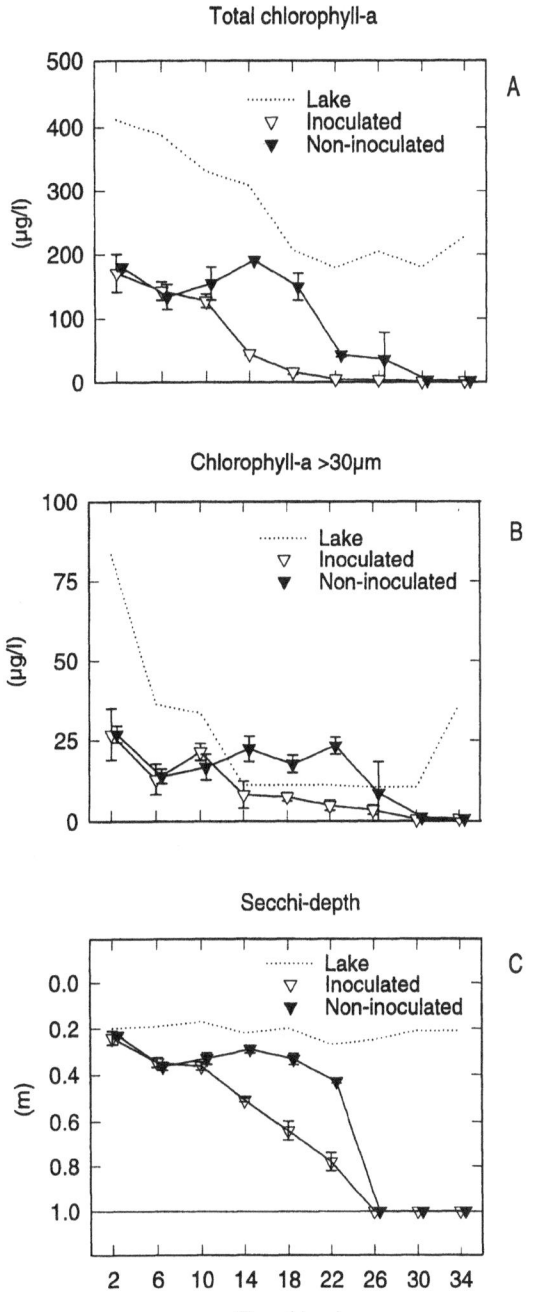

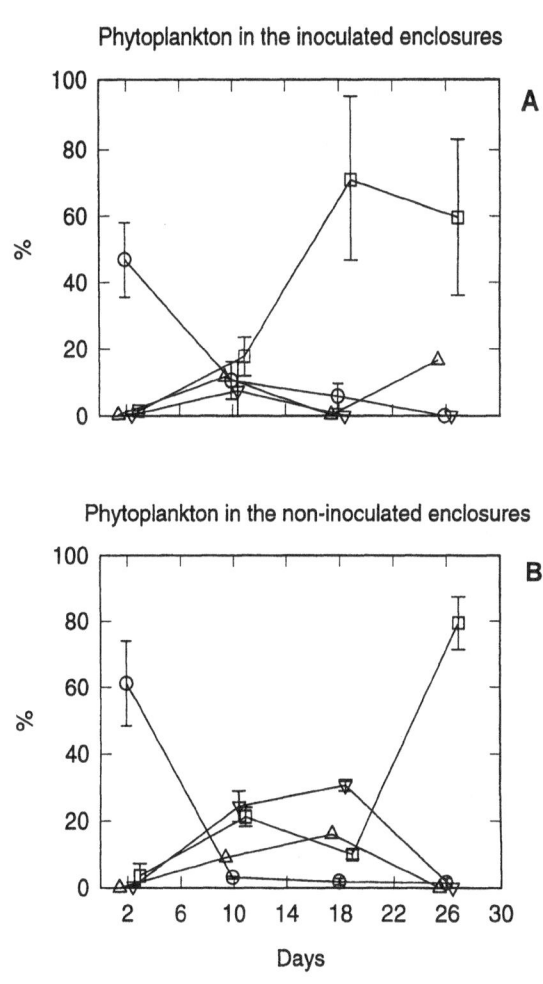

Figure 4. Changes in chlorophyll-*a* (average ± S.D.) and Secchi-depth in the inoculated enclosures, in the non-inoculated enclosures and in the lake during the course of the experiment. A = Total chlorophyll-*a*; B = chlorophyll-*a* in large algae that are retained by a mesh of 30 μm; C = Secchi-depth.

Figure 5. Phytoplankton composition (% biovolume of the most abundant phytoplankton taxa) in the inoculated (A) and non-inoculated enclosures (B).

In all of the enclosures, there was a pronounced reduction of nutrients and chlorophyll-*a* levels at the start of the experiment. Most probably, this was due to a reduction of wind-induced suspension in the enclosed water masses (Uehlinger et al., 1984; Van Donk et al., 1994), resulting in both a reduction of nutrient release associated with sediment resuspension and an enhanced settling of algae. The time lag observed between the phytoplankton dynamics of the different treatments can, however, not be explained by sedimentation effects, because there is no reason for sedimentation rates to differ systematically between enclosures. Furthermore, the reduction of algae during the course of the experiment can not be attributed to a decrease in nutrient availability. Though nutrient levels in the enclosures were lower than in the lake, the algae biomass, after the initial reduction due to sedimentation, was able to maintain a constant level for 10 to 18 days, depending on the treatment. Moreover, levels of phosphate and ammonia increased towards the end of the experiment, probably as a result of zooplankton excretion. Though nitrate levels were reduced, the remaining levels were too high to be limiting. As the difference in time of suppression of the algae between inoculated and non-inoculated enclosures is associated with a difference in the biomass and species composition of the zooplankton community rather than with differences in nutrient levels, our data strongly indicate that the reduction in algal biomass is caused by zooplankton grazing.

In the process of phytoplankton suppression, a distinction can be made between a phase of reduction of phytoplankton community biomass and a phase of keeping the phytoplankton community biomass at a low level. The first phase is characterised by an exponentially growing zooplankton biomass and a high but decreasing phytoplankton biomass. During the second phase, the zooplankton has reached a critical biomass, and the phytoplankton biomass is reduced to a very low level. In this latter phase, the zooplankton experiences very low food concentrations. The factors determining the success of a zooplankton population in suppressing phytoplankton growth differ between both phases. During the first phase factors such as the species specific clearance rate and the rate of population growth are expected to be most important. Species specific clearance rates largely determine the grazing impact of a given zooplankton community on the phytoplankton. The faster a zooplankton population is able to build up biomass under conditions of high phytoplankton densities, the faster it will be able to suppress a phytoplankton bloom. During the second phase, starvation resistance is essential. A grazer will not be able to control phytoplankton growth if its population crashes after the food source has been depressed, giving the algae the opportunity to bloom again. Finally, a broad food particle size range is important during both phases because it lowers the probability of a bloom of large algae (Dawidowicz, 1990; Gliwicz, 1990b). From the Size Efficiency Hypothesis (Brooks & Dodson, 1965; Hall et al., 1976), we expect large zooplankton to be more efficient than small zooplankton during both phases of phytoplankton bloom suppression. Due to a broader food particle size range and a higher filtration rate, large zooplankton should have a higher capacity to graze down algae. The low respi-

ration rate and the high filtration rate lead to a high net biomass increase when food is abundant, a high resistance to starvation, and a low threshold food concentration (Gliwicz, 1990a).

At the end of our experiment, algae were suppressed to equal levels by both medium and large sized zooplankton communities, and large *Pediastrum* colonies progressively dominated the inoculated als well as the non-inoculated enclosures. Nevertheless, our results support several expectations of the S.E.H. In the non-inoculated enclosures, suppression of chlorophyll-*a* lagged approximately eight days behind that of the enclosures inoculated with *D. magna*. This time lag was even larger (12 days) with respect to the suppression of large algae ($> 30 \mu m$). This can only partly be attributed to differences in initial zooplankton biomass because the initial time-lag in total zooplankton biomass amounted to four days only. Our results show that *D. magna* is superior to the smaller zooplankton in both net biomass increase of the population during the exponential growth phase and in clearance rate. The r_b-value estimated for *D. magna* was higher than the value estimated for *Bosmina*. Moreover, the critical biomass (CB_{TOT}) of the community dominated by *D. magna* was lower than that of the community dominated by smaller species, suggesting a higher clearance rate of the larger zooplankton. When suppression of large algae is considered, the difference in critical biomass ($CB_{>30}$) was very pronounced, amounting to a factor of almost five. Our estimate of the difference in the $CB_{>30}$-value of *D. magna* compared to the $CB_{>30}$-value of other, smaller species may even be underestimated because populations of *D. magna* had already considerably developed in the non-inoculated enclosures on day 26 of the experiment.

Only few studies have investigated the differences in grazing impact between medium-sized zooplankton (*Bosmina*, small and medium-sized *Daphnia* species) and large zooplankton (*D. pulex*, *D. magna*), without applying fish predation as a method to obtain differently structured zooplankton treatments (Bergquist et al., 1985; Dawidowicz, 1990; Sarnelle, 1993). The outcomes of these studies differed widely among each other and were not in full accordance with our findings. Whereas large zooplankton was able to suppress an algal bloom, Dawidowicz (1990) found an increase of total phytoplankton biomass as a response to grazing by medium-sized zooplankton, due to the increase of large, less edible algae. In contrast with these findings, Bergquist et al. (1985) observed a relative increase of large algae and a decrease of small algae when grazed by large zooplankton, and the opposite pattern in the case of grazing by small zooplankton. In our experiment, large colonial green algae (*Pediastrum*) tended to dominate after a major reduction of total phytoplankton biovolume, and no substantial qualitative differences in species composition of the phytoplankton community were found between treatments. The lack of coherence in the results of these studies can be explained in several ways. First, it has been shown that differences in total zooplankton biomass can lead to differences in phytoplankton biomass and species composition, irrespective of qualititative differences between zooplankton communities. Bergquist & Carpenter (1986) showed that the response of phytoplankton communities to grazing by zooplankton may differ among treatments having the same zooplankton composition but different zooplankton biomasses. When studying the effect of zooplankton composition or size distribution on the phytoplankton community structure, it is therefore important to work with comparable levels of total zooplankton biomass in all of the treatments involved. In our study, differences in initial levels of total zooplankton biomass were partially taken into account by our estimate of the critical biomass. Secondly, the trophic state of the system under study may be an important factor determining the outcome of the enclosure experiments. The enclosure experiment by Dawidowicz (1990) was performed under mesotrophic conditions. The increase of large algae in the presence of medium-sized zooplankton was explained by a lower elimination rate, combined with a transfer of nutrients from small, edible algae to a large, less edible phytoplankton fraction. Such a nutrient transfer mechanism is probably much less important under hypertrophic conditions as nutrients are generally not limiting. In our experiment we indeed did not observe growth stimulation of large algae in the non-inoculated enclosures. It is well conceivable, however, that the fact that phytoplankton biovolumes dropped to equal low levels in both the inoculated and non-inoculated treatment and that large algae were not more abundant in the non-inoculated enclosures than in the inoculated enclosures, may be due to an increase of *D. magna* in the non-inoculated enclosures at the end of the experiment (day 26).

Conclusions

Although phytoplankton in our enclosure experiment was suppressed to equal levels by both intermediate-and large-sized zooplankton communities, several expectations of the Size Efficiency Hyptohesis were supported by our observations. First, the population biomass growth rate of large zooplankton (*D. magna*) was higher than that of small zooplankton (*Bosmina longirostris*). Secondly, critical biomass needed to suppress phytoplankton populations was lower in a zooplankton community dominated by the large-bodied *D. magna* than in a zooplankton community in which *D. magna* was less abundant. As a result, the time frame within which the zooplankton can control phytoplankton growth efficiently is much reduced when large-bodied species such as *D. magna* are abundant. The difference in time frame between the treatments was more pronounced when large phytoplankton was considered, an observation which is consistent with the broad food particle size range of *D. magna*. It has been suggested that filamentous or toxic Cyanobacteria are harmful to zooplankton and may weaken top-down effects. However, at low densities, their negative effects are minor. As long as Cyanobacteria densities are not too high, their populations can be kept under control by zooplankton grazing (Dawidowicz, 1990). Therefore, the importance of high zooplankton densities early in the growing season has been stressed (Gliwicz, 1990b). At the beginning of spring, zooplankton biomasses are often very low, however. The potential of a high population biomass increase combined with a high grazing impact, especially on large algae, might therefore be critical in determining the success of a zooplankton community in preventing a bloom of nuisance phytoplankton. In lakes or ponds in which larger species such as *D. magna* do not occur, it may be useful to inoculate them early in the growing season as an additional measure accompagnying mass removal of planktivorous and benthivorous fish. In doing that, however, one preferentially should use clones isolated from similar or nearby habitats, because it is likely that these clones will be better adapted to cope with habitat-specific conditions (De Meester, 1996a; De Meester, 1996b; Declerck et al., in prep.). Densities must not be unrealistically high to be effective: our results suggest that the presence of *D. magna* in Lake Blankaart may have been important in phytoplankton control in the non-inoculated enclosures, yet densities of *D. magna* in Lake Blankaart at the start of the experiment were certainly less than 1 individual per 100 liters. For a quick restoration of a small lake of 1 ha surface and an average depth of 1 m, the inoculation of 10000 to 100000 *Daphnia magna* (1 ind./100 l) might be sufficient.

Acknowledgements

This research was supported by a scholarship provided to S. Declerck by the Flemish Institute for the stimulation of Scientific and Technological Research in the Industry (I.W.T.). L. De Meester is a Postdoctoral Researcher with the Fund for Scientific Research-Flanders. Financial support was also provided from a PP grant (University of Granada, Spain) to J. M. Conde-Porcuna. We thank the Centre for Limnology (The Netherlands) and in particular Dr J. Vijverberg and Dr R. Gulati for allowing us to use their enclosure frames, Natuurreservaten v.z.w. for permitting us access to the nature reserve 'De Blankaart', the Institute of Nature Conservation (I.N.) for logistic support, and the Flemish Water Supply Company (V.M.W.) for chemical analyses, and three anonymous reviewers for comments on an earlier version of the manuscript.

References

Bergquist, A. M. & S. R. Carpenter, 1986. Limnetic herbivory: effects on phytoplankton populations and primary production. Ecology 67: 1351–1360.

Bergquist, A. M., S. R. Carpenter & J. C. Latino, 1985. Shifts in phytoplankton size structure and community composition during grazing by contrasting zooplankton assemblages. Limnol. Oceanogr. 30: 1037–1045.

Bottrell, H. H., A. Duncan, Z. M. Gliwicz, E. Grygierek, A. Herzig, A. Hillbricht-Ilkowska, H. Kurasawa, P. Larsson & T. Weglenska, 1976. A review of some problems in zooplankton production studies. Norw. J. Zool. 24: 419–456.

Brett, M. T. & C. R. Goldman, 1997. Consumer versus resource control in freshwater pelagic food webs. Science 275: 384–386.

Brooks, J. H. & S. I. Dodson, 1965. Predation, body size, and composition of plankton. Science 150: 28–35.

Carpenter, S. R. & J. F. Kitchell, 1992. Trophic cascade and biomanipulation: Interface of research and management-A reply to the comment by DeMelo et al. Limnol. Oceanogr. 371: 208–213.

Carpenter, S. R., J. F. Kitchell & J. R. Hodgson, 1985. Cascading trophic interactions and lake productivity. BioScience 35: 634–638.

Christoffersen, K., B. Riemann, A. Klysner & M. Sondergaard, 1993. Potential role of fish predation and natural poopulations of zooplankton in structuring a plankton community in eutrophic lake water. Limnol. Ocanogr. 38: 561–573.

Dawidowicz, P., 1990. Effectiveness of phytoplankton control by large-bodied and small-bodied zooplankton. Hydrobiologia 200/201: 43–47.

De Meester, L., 1996a. Evolutionary potential and local genetic differentiation in a phenotypically plastic trait of a cyclical parthenogen, Daphnia magna. Evolution 50: 1293–1298.

De Meester, L., 1996b. Local genetic differentiation and adaptation in freshwater zooplankton populations: Patterns and processes. Ecoscience 3: 385–399.

DeMelo, R., R. France & D. J. McQueen, 1992. Biomanipulation: Hit or myth? Limnol. Oceanogr. 37: 192–207.

DeMott, W. R. & W. C. Kerfoot, 1982. Competition among cladocerans: nature of the interaction between Bosmina and Daphnia. Ecology 63: 1949–1966.

De Smedt, P., W. Rommens, S. Declerck, W. Vyverman, C. Belpaire, B. Denayer, L. De Meester, F. Ollevier & J. Van Assche, 1997. Ecologisch onderzoek in en rond het erkend natuurreservaat 'De Blankaart', met inbegrip van Actief Biologisch Beheer van Kasteel- en Visvijver. Studieopdracht van AMINAL (afdeling Natuur) & Ecologisch Impulsgebied Ijzervallei.

Fussmann, G., 1996. The importance of crustacean zooplankton in structuring rotifer and phytoplankton communities: an enclosure study. J. Plankt. Res. 18: 1897–1915.

Gliwicz, Z. M., 1990a. Food thresholds and body size in cladocerans. Nature 343: 638–640.

Gliwicz, Z. M., 1990b. Why do cladocerans fail to control algal blooms? Hydrobiologia 200/201: 83–97.

Gliwicz, Z. M. & W. Lampert, 1990. Food thresholds in Daphnia species in the absence and presence of blue-green filaments. Ecology 71: 691–702.

Greenberg, A. E., R. R. Trussell & L. S. Clesceri (eds), 1985. Standard Methods for the Examination of Water and Wastewater. APHA, AWWA, WPCF, 16th edn., Washington, U.S.A.

Gulati, R. D., E. H. R. R. Lammens, M.-L. Meijer & E. Van Donk (eds), 1990. Biomanipulation – Tool for Water Management. Dev. Hydrobiol. 61. Kluwer Academic Publishers, Dordrecht, 628 pp. Reprinted from Hydrobiologia 200/201: 619–627.

Hall, D. J., S. T. Threlkeld, C. W. Burns & P. H. Crowley, 1976. The size-efficiency hypothesis and the size structure of zooplankton communities. Ann. Rev. Ecol. Syst. 7: 177–208.

Havens, K. E., 1993. An experimental analysis of macrozooplankton, microzooplankton and phytoplankton interactions in a temperate eutrophic lake. Arch. Hydrobiol. 127: 9–20.

Lammens, E. H. R. R., R. D. Gulati, M.-L. Meijer & E. Van Donk, 1990. The first biomanipulation conference: a synthesis. Hydrobiologia 200/201: 619–627.

Massart, J., 1907. Essai de géographie botanique des districts littoraux et alluviaux de la Belgique. Henri Lamertin, Brussel.

O.E.C.D. Eutrophisation des eaux. Méthodes de surveillance, d'évaluation et de lutte. O.E.C.D., Paris.

Peeters, B., L. De Meester, B. Denayer, P. De Smedt, K. Nuydens & F. Ollevier, 1996. Inventarisatie van het visbestand van de Blankaartvijver en omliggende waterlopen met afvissing van de Kasteel- en Visvijver. Beschrijving van de visstand en voorstellen inzake actief biologisch beheer. Studieopdracht van AMINAL (afdeling Natuur) & Ecologisch Impulsgebied Ijzervallei.

Reynolds, C. S., 1994. The ecological basis for successful biomanipulation of aquatic communities. Arch. Hydrobiol. 130: 1–33.

Sarnelle, O., 1993. Herbivore effects on phytoplankton succession in a eutrophic lake. Ecol. Monogr. 63: 129–149.

Scheffer, M., S. H. Hosper, M.-L. Meijer, B. Moss & E. Jeppesen, 1993. Alternative equilibria in shallow lakes. Trends Ecol. Evol. 8: 275–279.

Talling, J. F. & D. Driver, 1963. Some problems in the estimation of chlorophyll-a in phytoplankton. Proc. Conference on Primary Productivity Measurements, Marine and Freshwater. US Atomic Energy. Comm. TID-7633: 142–146.

Tessier A. J. & C. E. Goulden, 1987. Cladoceran juvenile growth. Limnol. Oceanogr. 32: 680–686.

Threlkeld, S. T., 1988. Planktivory and planktivore biomass effects on zooplankton, phytoplankton, and the trophic cascade. Limnol. Oceanogr. 33: 1362–1375.

Uehlinger, U., P. Bossard, J. Bloesch, H. R. Bürgi & H. Bührer, 1984. Ecological experiments in limnocorrals: Methodological problems and quantification of the epilimnetic phosphorus and carbon cycles. Verh. int. Ver. Limnol. 22: 163–171.

UNESCO, 1974. A review of methods used for qualitative phytoplankton studies. Unesco tec. Pap. Mar. Sci. 18.

Van der Werf, B., J. Schrotenboer, A. F. Richter, J. R. Moed, H. L. Hoogveld & H. De Haan, 1987. A durable and transportable limnetic enclosure system suitable for wind-exposed lakes. Can. J. Fish. aquat. Sci. 44: 1649–1652.

Van Donk, E., M. P. Grimm, P. G. M. Heuts, G. Blom, K. Everards & O. F. R. van Tongeren, 1994. Use of mesocosms in a shallow eutrophic lake to study the effects of different restoration measures. Arch. Hydrobiol. Beih. Ergebn. Limnol. 40: 283–294.

Vanni M. J. & C. D. Layne, 1997. Nutrient recycling and herbivory as mechanisms in the 'top-down' effect of fish on algae in lakes. Ecology 78: 21–40.

Vanni M. J. & D. L. Findlay, 1990. Trophic cascades and phytoplankton community structure. Ecology 71: 921–937.

Vanni, M. J., C. D. Layne & S. E. Arnott, 1997. 'Top-down' trophic interactions in lakes: effects of fish on nutrient dynamics. Ecology 78: 1–20.

Zaret, T. M., 1980. Predation and Freshwater Communities. Yale University Press.

Hydrobiologia 360: 277–285, 1997.
A. Brancelj, L. De Meester & P. Spaak (eds), Cladocera: The Biology of Model Organisms.
©1997 Kluwer Academic Publishers.

Chydorid assemblages in the sedimentary sequence of Lake La Cruz (Spain) subject to water level changes

F. Mezquita & M. R. Miracle
Dep. de Microbiologia i Ecologia, Fac. C. Biològiques, Univ. de València. E-46100, Burjassot (València) Spain
(e-mail: francesc.mezquita@uv.es, rosa.miracle@uv.es)

Key words: Chydoridae, paleolimnology, diversity, species associations, cluster analysis, principal components analysis

Abstract

Changes among chydorid assemblages in Lake La Cruz sediments, from a core of 178 cm length, have been studied using community structure indices and multivariate methods (cluster analysis, principal components analysis).

The results revealed the existence of two main sources of variation in these assemblages. One (axis I of PCA1) is associated with the trophic level of the lake, which is hypothetised to be greater in the bottom and upper part of the core. Both zones are characterised by an increase in the relative frequency of *Chydorus sphaericus*, accompanied by a marked reduction in the relative frequency of the most abundant species throughout the history of the lake, *Acroperus neglectus*.

The other source of variation (axis II of PCA1) may be interpreted as the alternation of periods of dryer and wetter weather. Our analyses show the separation of two assemblages which alternatively prevail in the sedimentary sequence, one constituted by *Pleuroxus laevis, Alona guttata, Graptoleberis testudinaria* and, sometimes, *Alona rectangula*, and another dominated by *Chydorus sphaericus, Alona quadrangularis, Alona affinis* and sometimes *Leydigia* species. According to the ecological preferences of these species, the dominance of the first chydorid assemblage can be related to episodes of higher temperature and with very low and fluctuating water level, or with a high water level but a very reduced extent of shallow water in the lake. Prevalence of the second group corresponds to episodes with a colder environment and permanent waters, the level of which could be variable but in such a way that a benthic sublittoral zone may develop. The structure of the community is strongly affected by changes in lake level. High indices of fluctuation in community structure and small mean body size of the chydorid assemblage marked the period with more extreme dry conditions, characterised by *Alona rectangula* maxima. The peaks of *C. sphaericus* coincided with peaks of the index of fluctuation D_0 and prevalence of smaller species of the chydorid assemblage.

Introduction

The analysis of chydorid assemblages in sediments is not only a paleolimnological tool to indicate past conditions but also a way to study the ecology of the assemblages of these taxa, integrating both seasonal and spatial distributions. A lake which can provide a stable and peaceful environment of deposition is specially suitable for this study. This is the case of Lake La Cruz, a small and deep sink hole with a high relative depth, sheltered from the wind and without any other superficial flux of water than that derived from rain and evaporation. It is mainly fed by subterranean springs which are most likely located quite deep in the lateral walls. At present, the lake is meromictic (Vicente & Miracle, 1988) and the permanent absence of oxygen prevents bioturbation and helps in the preservation of remains, although this condition is quite recent. A whiting phenomenon due to tumultuous calcite precipitation occurs every summer and results in a varved sediment (Rodrigo et al., 1993).

In June 1994 a sediment core of 178 cm length was taken in the centre of the lake and it was observed that the varves occur only in the top 24 cm of the sediment and are related to the nowadays meromictic condition of the lake. An integrated study of this core is being published elsewhere (Julià et al., in press). We here concentrate on the analysis of the sedimentary sequence of chydorid remains. The aim of the paper is to typify the different chydorid communities that occurred during the history of this lake. Shifts in the predominant groups of species, explored by means of multivariate statistics, along with calculations of diversity and stability indices will be used to discover general trends in the evolution of the chydorid assemblages.

Material and methods

Lake La Cruz is located in the Iberian Ranges near Cuenca (Central Spain) at 1000 m a.s.l. inside a doline surrounded by steep walls of 20–25 m high. It is a solution sink with abrupt slopes, which at the time of the coring (1994) had a mean diameter of 132 m and a maximum depth of 23.5 m. The main limnological features of this lake, as well as data on the community, were published elsewhere (Vicente & Miracle, 1988; Dasi & Miracle, 1991; Miracle et al., 1992; Rodrigo et al., 1993; Armengol-Díaz et al., 1993).

The lake is now characterised by a biogenic iron meromixis with a permanent anoxic monimolimnion. In the last years the oxicline varied seasonally from 12 m to 18 m depth. The lake water is rich in bicarbonates, the mean annual value for alkalinity being about 5 meq l^{-1} in the mixolimnion. The mixolimnetic annual mean conductivity is approximately 500 μS cm^{-1} and the annual mean pH is approximately 8.

The studied core was taken at the location of maximum depth of the lake, in June 1994, with a modified Livingstone corer (Montserrat, 1992). Information on core extraction, sectioning and dating are given in Julià et al. (in press). Here we will briefly indicate that by counting the varves observed in the uppermost part of the sediment, it was deduced that the first varve was deposited about 300 years BP. In addition, the core was dated at a depth of 77–78 cm by ^{14}C analysis of a tree efflorescence (probably from *Alnus* sp.), which yielded an age of 640 ± 60 years BP.

Cladoceran microfossils were recovered from sediments according to the methods developed by Frey (1979). One ml of fresh sediment was taken from 1 cm thick slices which were previously cut from the core at 5 or 10 cm intervals. After weighing, it was heated in 10% KOH. This was carried out under constant agitation and for the time necessary to deflocculate without damaging microfossils. The remaining material was then sieved through a 30 μm mesh Nytal filter and washed several times with water. The filtrate was subsequently submerged in 5% HCl in order to dissolve the abundant calcium carbonate particles. After washing, the filtrate was diluted in distilled water and a few drops of cotton-blue in lactophenol were added. Several 50 μl subsamples were examined under an inverted microscope until at least 200 fragments of the most abundant chydorid species were counted. The total number of individuals was estimated as the maximum count of head shields, postabdomens or valves (calculated using algorithms for valve fragments), following Frey (1979).

Identifications and ecological characteristics of the species found were obtained from Frey (1959, 1962), Amoros (1984), Alonso (1985, 1996) and Margaritora (1985).

The Shannon index

$$H = - \sum_{i=1}^{s} p_i \log_2 p_i$$

was used to calculate the diversity of chydorid assemblages. In addition, we used the index of fluctuation D_0, formulated by Dubois (1973) as the Taylor's expansion of the diversity index H around a reference state

$$D_0 = \sum_{i=1}^{s} p_i \log_2 p_i / \bar{p}_i,$$

where p_i = proportions of species i, $\bar{p}_i$ = mean p_i. In this sense, D_0 is a function of time representing the deviation of the species proportions through time from an average state. The other indices used were evenness ($E = H'/H'_{max}$) and the average body size of species of the chydorid community using maximum lengths given by Alonso (1996).

Graphic representation of relative frequencies of chydorids and the corresponding dendrogram were obtained with the software TILIA v. 1.12 (Grimm, 1993), which uses CONISS (Grimm, 1987) for the constrained cluster analysis with the Edwards and Cavalli-Sforza distance. Other statistical analyses were performed using SPSSwin 5.02 and CANOCO 3.1 (ter Braak, 1988). Arcosinus of the square root of percentages transformation was used for the principal components analysis (PCA1) and for the calculation of Pearson correlation coefficients.

Results

Fifteen species and one subspecies of chydorids were found in the sedimentary record of Lake La Cruz, with densities (Figure 2) ranging from 260 to 3520 ind ml^{-1} of fresh sediment (180 to 5760 ind g^{-1} of dry sediment). The most abundant species in the whole sequence was *A. neglectus*, attaining densities from 100 to 1290 ind ml^{-1} and a relative frequency varying from 20 to 51.6% (Figure 1).

Figure 1 shows the changes in the relative abundance (%) for each species, as well as the zonation established by a constrained cluster analysis with these data, which led to the partitioning of the whole record in 8 different zones and subzones. Looking at the changes in species composition that define the zones, two trends can be observed: (i) main reduction in the relative abundance of the most common species, *A. neglectus*, are often coupled to a sudden rise in the relative abundance of *C. sphaericus*, specially in zones 4 and 6, and (ii) two different assemblages alternatively dominate the chydorid community: group (A) formed by *G. testudinaria, A. guttata, P. laevis* and *A. rectangula* and group (B) with *A. quadrangularis, A. affinis,* and *C. sphaericus*. These two points are apparent from Pearson correlation coefficients between the most abundant species (Table 1) as well as by a cluster analysis without the stratigrafic sequence constriction (Figure 3). As can be observed, bindings are stronger inside group (A) than inside group (B).

The total amount of chydorid remains varies in the different zones. Zones 2 and 3 are characterised by a marked reduction in the number of remains (Figure 2); this is still more apparent when numbers per dry weight are considered. Moreover, there is a relationship between the abundance of remains and the conversion factor from volume (ml) to dry weight (g) of sediment subsamples: a small volume/weight ratio (around 1 or below it) is due to low water content and corresponds to low numbers of chydorid remains. Taking into account the lithological description of the core in Julià et al. (in press), these are strata (especially samples at 65 cm and those of zone 3) with an important sandy detrital fraction. We conclude that chydorid remains are diluted by the allochthonous mineral particles in what we think are periods of lower level temporary waters.

With respect to species diversity (Figure 2), the minimum H' values are found in the samples corresponding to changing conditions, i.e. samples in the limits of main zones, such as 25 and 65 cm core depth. They coincide with the samples in which the relative percentage of *A. neglectus* is higher. In region 3, at the point of minimum number of remains, diversity shows also a relative minimum. On the other hand, maximum H' values (>3bits ind^{-1}) appear in the top zone along the varved sequence (6a and 6b), during the period in which a meromictic condition, with summer precipitation of calcite, was established. Trends of H' and evenness are similar, and mark the periods of change between zones. However, in periods of extreme conditions where diversity is low (e.g. zone 3), evenness may show high values. This is due to the fact that in these periods there is a reduced number of well-adapted species, and the community must not be regarded as an immature stage but as a successional stage induced by the extreme environmental conditions.

D_0 can be used as a stability index, as it relates each community stage to a reference state, defined as the averaged species proportions in the whole sequence. High values indicate major fluctuations of the community and maximum deviations from the state to which the structure of the community is approaching (Miracle, 1978). The index is zero when the species proportions are equal to their respective means through time, and the index is usually inversely correlated to indices of diversity. In the sedimentary sequence of Lake La Cruz, zones 1, 3 and 6a are the periods with the strongest changes occurring in the lake with respect to the chydorid assemblage (Figure 2). The peaks in zones 1 and 6a (and to a lesser extent in 5b) coincide with relative maxima of *C. sphaericus* and are associated with a higher trophic condition of the lake. In the case of zone 3, the high values of the fluctuations index indicate another extreme situation, in this case related to dryness. This period is characterised by the minimum values of chydorid density and maximum relative frequencies of *P. laevis* (lower part of zone 3) and *A. rectangula* (higher part of zone 3).

The mean size of the species of the chydorid assemblage shows great changes with time (see Figure 2). Assemblages consisting of relatively small species occur in the driest period (zone 3). In addition, a relatively smaller body size also occurs in zone 2, which is also suspected to be quite dry, as well as in recent times, when the lake is meromictic. In all these periods small species of *Alona* or *C. sphaericus* have high relative abundances.

280

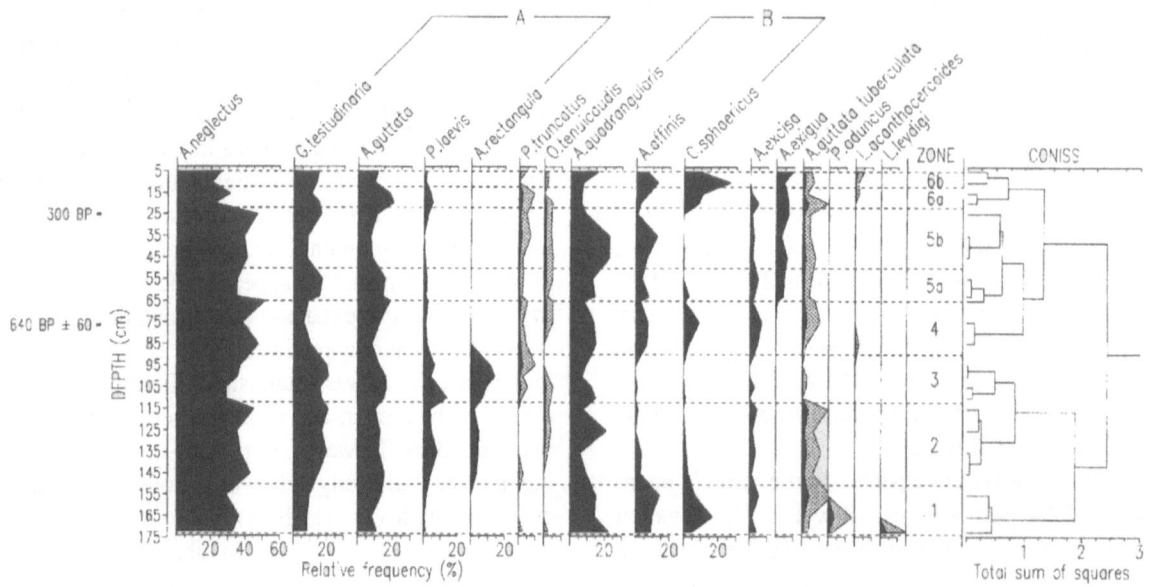

Figure 1. Relative abundance diagram for the different chydorid species found in the core of Lake La Cruz, and zones established from a constrained cluster analysis using CONISS (adjacent dendrogram). Stippled areas are scale factor 4 × expansions of the solid area.

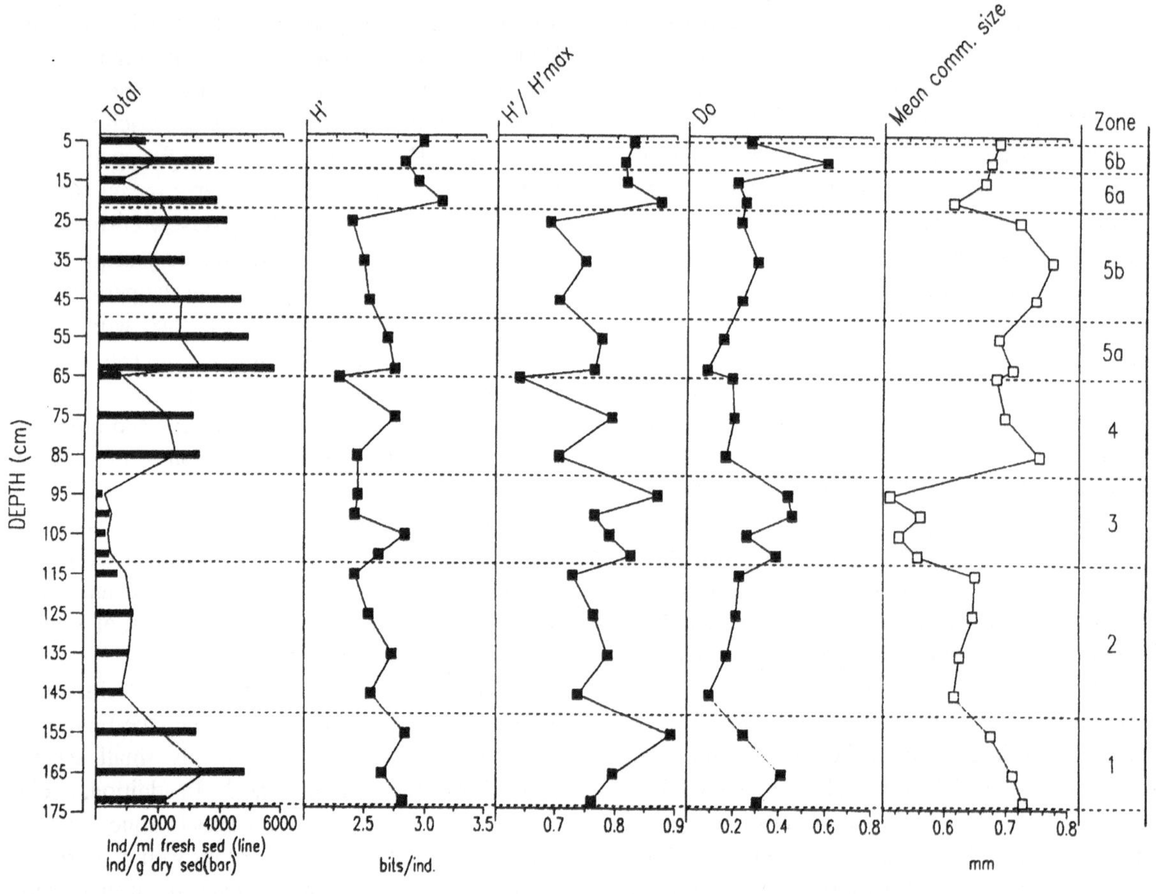

Figure 2. Diagram of total abundance of chydorids for Lake La Cruz per fresh and dry sediment, together with some community indices: Shannon diversity index H', evenness index H'/H_{max}, Dubois fluctuation index D_0 and average body size of species of the community.

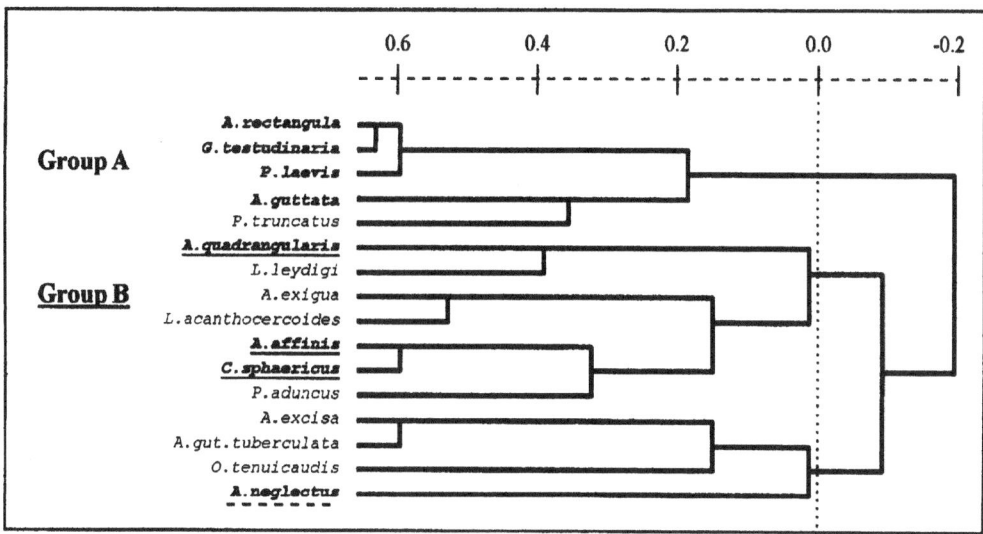

Figure 3. Unconstrained cluster analysis of the relative abundance for all species, using UPGMA method and Pearson correlation distance (scale). The most frequent species are typed in bold italics; underlined and not underlined bold types indicate belonging to a particular group, as shown in Table 1.

Table 1. Pearson correlation coefficients and significance (*p*-values) for the relative abundances of the most important species of the chydorid community. Non significant values are not given.

	Group A				Group B
	A. rectangula	G. testudinaria	P. laevis	A. guttata	C. sphaericus
Group A					
G. testudinaria	0.65***				
P. laevis	0.64***	0.57***			
A. guttata			0.43*		
Group B					
A. affinis	−0.64***	−0.69***	−0.66***	−0.43*	0.60***
A. quadrangularis		−0.52*		−0.53*	
A. neglectus					−0.56**

***p<0.005; **p<0.01; *p<0.05

Principal components analysis was performed using both species percentages and absolute abundances. This linear multivariate statistics method is preferred to gaussian methods (such as correspondence analysis) when the gradients are not too wide (ter Braak & Prentice, 1988). This condition fits well with our data, because most abundant species are present along the whole sequence. The results from the PCA1 using species percentages are shown in Figure 4. The first factor (FI) accounted for 40.3% of the total variance, and the second (FII) for 28.2%. The species with a major weight on FI are *A. neglectus* in the negative part and *C. sphaericus* and *A. affinis* in the positive. *G. testudinaria, P. laevis, A. rectangula* and *A. guttata*

are the most important species for FII in the positive part and *A. affinis* and *A. quadrangularis* in the negative part (Figure 4). Most of the zones that were established by the clustering method, are arranged separately in the PCA analysis, with some exceptions like zones 4 and 5.

In the case of the PCA2 using absolute abundances, the first factor (FI) accounts for 41.6% of the variance and is related to the abundance of the species, which, with the exception of *A. rectangula* and *P. laevis*, show a steep decline in zones 2 and 3. FI strongly separates zones 2 and 3 in the positive part of the axis, from all the other zones, which lie in the negative part of the first axis. *A. rectangula* is the only species with a

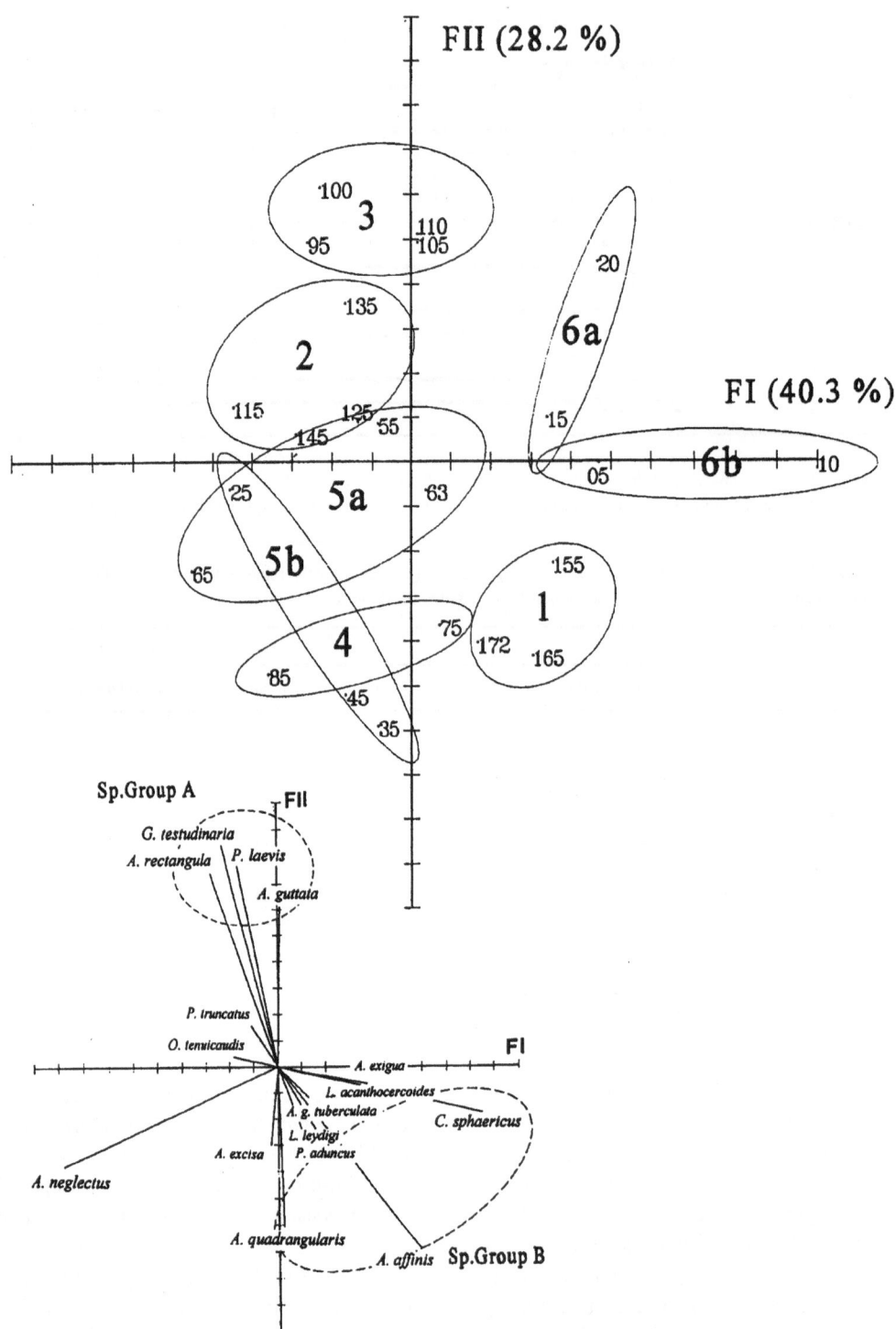

Figure 4. Principal Components Analysis (PCA1) using all chydorid species found in the sedimentary sequence of Lake La Cruz. A: Samples represented by their depth in the sediment core are ordinated in the first two factors space. The zones as established in Figures 1 and 2 are marked with ellipses. B: Chydorid species are positioned in the first two factors space and groups of species A and B are delimited by a dashed ellipse.

positive loading on the first axis, and is clearly separated from all the other species. The second factor (FII, 13,7%) has higher loading with 'rare' species that are not present all along the core. Further PCA with absolute abundances of the ten most frequent chydorid species was performed in order to avoid the effect of those rare species. In the resulting PCA, the first factor was very similar to FI of PCA2 and the second one separated species and samples in a way similar to FII of PCA1, but with a less high percentage of variance explained.

Discussion

The existence of some parallel patterns of the PCA1 and D_0 index is remarkable. Samples with a low D_0 index are located near the centre of the two first axes of PCA, whereas those with high D_0 values, are located far from the centre of the axes. This indicates that PCA extracts the factors which account for the higher deviations or shifts in the community structure. PCA is able to determine the discontinuities or main perturbations in the system which subsequently change gradually to the average state. Species having both higher loadings in PCA factors and peaks coinciding with D_0 index maxima were *C. sphaericus* and *A. rectangula*, which are described as ubiquitous and tolerant to stress conditions (Whiteside, 1970; Amoros & Jacquet, 1987).

The main source of variation of the chydorid community throughout the core sequence may be related to changes in: (1) the trophic state of the lake (Axis FI of PCA1) and (2) the water mean temperature and/or level which determined shifts on the type of littoral zone (axis FII of PCA1).

The highest trophic level of the lake occurred in the period corresponding to the upper part of the core (zone 6) followed by the zone at its bottom (zone 1). The highest trophic level in zone 6 is supported by the varved structure of sediments present in that section, which indicate a meromictic phase of the lake in the most recent times (Julià et al., in press). Zone 1 is supposed to correspond to a shallow pond which probably had a higher trophic level due to its reduced mean depth.

Our results show that these zones are characterised by a high relative abundance of *C. sphaericus*. *C. sphaericus* is a cosmopolite species 'which often becomes superabundant in extreme eutrophication' (Frey, 1979) and its accumulation rate showed a highly significant correlation with chlorophyll a in fifteen lakes of Florida (Binford, 1986). In addition,

its biomass increased during periods of rising eutrophy in Tjeukemeer (Vijverberg & Boersma, 1997). The species becomes planktonic under eutrophic conditions or in the presence of algal blooms (Gannon, 1972; Meyers, 1984), thus leading to an additional increase of its relative abundance when the whole lake is considered. Moreover, *C. sphaericus* attains 'maximum concentrations in the eutrophic and mesotrophic regions of the Great Lakes', together with *A. affinis* (Evans & Stewart, 1977). These last two species show a strong positive correlation in our study (Table 1, Figure 3) and *C. sphaericus* and *A. affinis* represent the maximum and second maximum weight, respectively, in the positive part of the first factor extracted by PCA1. This corresponds most probably to a more eutrophic character of the lake for zones 1 and 6 (FI, Figure 4). The index of fluctuation D_0 indicates that in those episodes the lake was far from a hypothetical average state. Moreover, the peaks of this index in zones 1 and 6 may indicate pulses of high productivity as has been observed for zooplankton annual cycles (Miracle, 1978). In our sediment core the two most pronounced maxima of *C. sphaericus* clearly coincide with D_0 peaks, which show relative minima of diversity. It can also be observed that the mean body size of the species of the chydorid assemblage in zones 1 and 6 is moderate, as can be expected from the overall tendency in that eutrophic conditions favour smaller species with high rates of increase, which are able to quickly colonise disturbed habitats (Margalef, 1983).

The striking fact that H' diversity index shows its greatest values in periods of high trophic level, especially in zone 6 (Figure 2), has also been found by Goulden (1964) in Esthwaite Water sedimentary record for cladoceran remains. However, this increase in diversity is only observed in conditions of intermediate trophy. It was found in Indiana lakes that diversity is related to productivity except in highly eutrophic lakes, in which diversities fall down to very low values (Whiteside & Harmsworth, 1967). Lake La Cruz, due to its situation, morphometry and water hardness, is at present oligo-mesotrophic and probably has never gone very far beyond the mesotrophic state. The increase of diversity is also related to the extent and slope of the lake shore, which has changed along the lake's history as will be commented below.

The second source of variation (FII of PCA1, Figure 4) is thought to come from climatic changes, more specifically with respect to precipitation and temperature, which have influenced the water level of the lake. The ecological preferences, based on the literature (e.g.

Fryer, 1968, 1993; Williams, 1982), of the two groups of species that were delimited in the results (Table 1, Figures 1, 3 and 4) can be characterised as follows: species of group (A) are more or less depending on macrophytes or very shallow zones, whereas species of group (B) are essentially bottom dwellers. The changes in water level can explain the alternating prevalence of these two groups of species. These changes involve an important shift in littoral development, due to the peculiar morphometry of Lake La Cruz, which is a circular sink with quite vertical slopes. A low water level results in a wider extent of the shallow water zone whereas a further lowering will lead to a very shallow water or even a temporary nature of the water body. In the first case the species of group (B) could increase, while in the second case species of group (A) are more suited. A high water level could also result in two developments depending on the length of the sublittoral environment. If the water covers long gradual slopes, species of group (B) are favoured, whereas when gradual slopes are shorter, species of group (A) predominate. More specifically, prevalence of species of group (B) was found in zones 1, 4 and 6. On the basis of the abundance of planktonic diatoms (Julià et al., in press), zone 1 corresponds to a shallow pond while a high volume seems to be characteristic for the upper part of the sediment core (zone 4 and 6). In both cases a littoral plus a sublittoral zone could be developed. On the other hand, the relative abundance of group (A) has been found under the drier conditions of zones 2 and 3, with very shallow and fluctuant waters, but also in zone 5 when the lake probably had a high water level but a very narrow edge delimiting the sink. Presently, the slopes are very steep after one or two meters, and studies on the living littoral communitiy show the almost exclusive presence of the species of group (A) (Boronat & Miracle, upublished data).

In Banyoles lake, a karstic lake with different basins, Rieradevall (1991) also divided the species in two groups: a group of littoral species (*A. rectangula* and *G. testudinaria*) mainly found at depths above 5 m and equivalent to our group (A), and a group of sublittoral species (*A. affinis, A. quadrangularis, L. leydigi* and *L. acanthocercoides*) found at depths below 5 m and equivalent to our group (B). Moreover, in the basins where there is a smooth slope providing both littoral and sublittoral environments, the species of the second group increase.

Nevertheless, not only water level or permanence, but also water temperature may be involved in those trends. Some species belonging to our group (A),

such as *G. testudinaria, A. rectangula* and *A. guttata* are considered by Goulden (1964), who mainly based his analysis on latitudinal distribution and abundance, as species preferring higher temperatures. Species belonging to group (B), like *C. sphaericus* or *A. affinis,* are considered as species inhabiting mainly regions with colder climate. The same author also found *A. quadrangularis* to be a species which can dwell in cold regions. In the same way, De Costa (1964) considered *A. quadrangularis* and *A. excisa* as 'northern' species. In addition, *C. sphaericus* and *A. affinis* are the most frequent chydorid species found in high mountain lakes of Spain (Miracle, 1982), and Keen (1973) found *C. sphaericus* being abundant in winter months, even under ice cover. All these findings support a hypothetical temperature gradient in the positive direction of FII in PCA1 (Figure 4).

For zone 3 and to a lesser extent zone 2, the appearance of detritic sandy dolostone bands (Julià et al., in press) may indicate periods of severe dryness, which favoured land-slides. In these zones, especially in zone 3, we have observed an increase in the frequency of *A. rectangula*, together with a very low density of total chydorid remains (FI of PCA2). The extreme conditions which the lake faced during zone 3 are also strikingly marked by the high values of D_0 and the very small diversity and mean body size of the chydorid assemblage (Figure 2). All this suggests that we are dealing with episodes in which water level was very low and fluctuant, the lake being probably temporary in nature, since *A. rectangula* is the most frequent chydorid in temporary mineralised waters of Spain (Alonso & Comellas, 1984) as well as in rice fields (Alfonso, 1996).

Acknowledgments

This survey was supported by the CICYT I+D project CLI 95-1905-C03-02. We are most grateful to all those who have been involved in this project: R. Julià, E. Vicente, J. R. Roca, G. Seret, F. Burjachs, M. J. Dasí and S. Giralt, particularly the first two who contributed to core extraction and sample preparation. We also thank A. Sanz-Brau and S. Sanz for advice on computational methods and suggestions to an earlier version of the manuscript. Luc De Meester and two anonymous referees greatly improved the English text and other aspects of this paper. F. Mezquita is recipient of an FPI fellowship of the Conselleria d'Educació i Ciència de la Generalitat Valenciana.

References

Alfonso, M. T., 1996. Estudio de las comunidades zooplanctónicas de los ecosistemas acuáticos del Parque Natural de la Albufera de Valencia. Ph D. Thesis. Univ. València, 310 pp.

Alonso, M., 1985. Las lagunas de la España peninsular: taxonomía, ecología y distribución de los Cladóceros. Ph. D. Thesis. Univ. Barcelona, 795 pp.

Alonso, M., 1996. Fauna Ibérica: Branchiopoda. Museo de Ciencias Naturales. C.S.I.C. Madrid, 486 pp.

Alonso, M. & M. Comellas, 1984. A preliminary grouping of the small epicontinental water bodies in Spain and distribution of crustaceans and Charophytes. Verh. int. Ver. Limnol. 22: 1699–1703.

Amoros, C., 1984. Crustacés Cladocères. Introduction pratique à la systematique des organismes des eaux continentales françaises. Bull. Soc. Linn. Lyon 53, 63 pp.

Amoros, C. & C. Jacquet, 1987. The dead-arm evolution of river systems: A comparison between the information provided by living Copepoda and Cladocera populations and by Bosminidae and Cydoridae remains. Hydrobiologia 145: 333–341.

Armengol-Díaz, J., A. Esparcia, E. Vicente & M. R. Miracle, 1993. Vertical distribution of planktonic rotifers in a karstic meromictic lake. Hydrobiologia 255/256: 381–388.

Binford, M. W., 1986. Ecological correlates of net accumulation rates of Cladocera remains in lake sediments. Hydrobiologia 143: 123–128.

Dasí, M. J. & M. R. Miracle, 1991. Distribución vertical y variación estacional del fitoplancton de una laguna cárstica meromíctica, la Laguna de La Cruz (Cuenca, España). Limnética 7: 37–59.

De Costa, J., 1964. Latitudinal distribution of chydorid Cladocera in the Mississippi Valley, based on their remains in surficial lake sediments. Invest. Indiana Lakes and Streams 6: 65–101.

Dubois, D. M., 1973. An index of fluctuations, Do, connected with diversity and stability of ecosystems: applications in the Lotka – Volterra model and in an experimental distribution of species. Rapport de sythèse III, Programme National sur l'environment Physique et Biologique, Project Mer. Commision Interministérielle de la Politique Scientifique. Liège.

Evans, M. S. & J. A. Stewart, 1977. Epibenthic and benthic microcrustaceans (copepods, cladocerans, ostracods) from a nearshore area in southeastern Lake Michigan. Limnol. Oceanogr. 22: 1059–1066.

Frey, D. G., 1959. The taxonomic and phylogenetic significance of the head pores of the chydoridae (Cladocera). Int. Revue ges. Hydrobiol. 44: 27–50.

Frey, D. G., 1962. Supplement to: The taxonomic and phylogenetic significance of the head pores of the Chydoridae (Cladocera). Int. Revue ges. Hydrobiol. 47: 603–609.

Frey, D. G., 1979. Cladocera analysis. In Berglund, B. E. (ed.), Paleohydrological Changes in the Temperate Zone in the Last 15000 Years. Dept. Quaternary Geol., Lund. Sweden: 227–257.

Fryer, G., 1968. Evolution and adaptative radiation in the Chydoridae: a study in comparative functional morphology and ecology. Phil. Trans. r. Soc. London. B254: 221–385.

Fryer, G., 1993. The Freshwater Crustacea of Yorkshire. Yorkshire Naturalists Union & Leeds Phylosophical and Literary Society, 312 pp.

Gannon, J. E., 1972. Effects of eutrophication and fish predation on recent changes in zooplankton Crustacea species composition in Lake Michigan. Trans. Am. Micros. Sci. 91: 82–85.

Goulden, C. E., 1964. The history of the cladoceran fauna of Esthwaite Water (England) and its limnological significance. Arch. Hydrobiol. 60: 1–52.

Grimm, E. C., 1987. CONISS: A FORTRAN 77 program for stratigraphycally constrained cluster analysis by the method of incremental sum of squares. Comp. Geosci. 13: 13–35.

Grimm, E. C., 1993. TILIA Diagramming Program. Illinois State Museum. Springfield.

Julià, R., F. Burjachs, M. J. Dasí, F. Mezquita, M. R. Miracle, J. R. Roca, G. Seret & E. Vicente (in press). Recent evolution of a meromictic lake in the Cuenca Mountains (Spain) based on palaeoecological data. Aquat. Sci.

Keen, R., 1973. A probabilistic approach to the dynamics of natural populations of the chydoridae (Cladocera, Crustacea). Ecology 54: 524–534.

Margalef, R., 1983. Limnología. Ed. Omega, Barcelona, Spain, 1010 pp.

Margaritora, F. G., 1985. Fauna d'Italia v. XXIII: Cladocera. Ed. Calderini, Bologna, Italy, 399 pp.

Meyers, D. G., 1984. Habitat shifting, feeding mode versatility, and alternate resource exploitation by herbivorous cladoceran zooplankton in a montane lake. In Meyers, D. G. & J. R. Strickler (eds), Trophic Interaction within Aquatic Ecosystems. AAAS Selected Symposium 85: 309–345.

Miracle, M. R., 1978. Organització del zooplàncton d'aigua dolça durant un cicle anual: aplicació d'un índex de fluctuacions. Col·loquis de la Societat Catalana de Biologia X: 183–193.

Miracle, M. R., 1982. Biogeography of the freshwater zooplanktonic communities of Spain. J. Biogeogr. 9: 455–467.

Miracle, M. R., E. Vicente & C. Pedrós Alió, 1992. Biological studies of Spanish meromictic and stratified lakes. Limnética 8: 59–77.

Montserrat, J., 1992. Evolución glaciar y postglaciar del clima y la vegetación en la vertiente sur del Pirineo: Estudio palinológico. Monografías del Instituto Pirenaico de Ecología, 6. C.S.I.C. Jaca.

Rieradevall, M., 1991. Ecology and production of benthos of Lake Banyoles. Ph. D. Thesis. Univ. Barcelona, 223 pp.

Rodrigo, M. A., E. Vicente & M. R. Miracle, 1993. Short term calcite precipitation in the karstic meromictic Lake La Cruz (Cuenca, Spain). Verh. int. Ver. Limnol. 25: 711–719.

ter Braak, C. J. F., 1988. CANOCO – a FORTRAN Program for Canonical Community Ordination. Microcomputer Power, Ithaca, New York, USA.

ter Braak & I. C. Prentice, 1988. A theory of gradient analysis. Adv. ecol. Res. 18: 271–317.

Vicente, E. & M. R. Miracle, 1988. Physicochemical and microbial stratification in a meromictic karstic lake of Spain. Verh. int. Ver. Limnol. 23: 522–529.

Vijverberg, J. & M. Boersma, 1997. Long-term dynamics of small-bodied and large-bodied cladocerans during the eutrophication of a shallow reservoir, with special attention for Chydorus sphaericus. Hydrobiologia 360: 233–242.

Whiteside, M. C., 1970. Danish chydorid Cladocera: modern ecology and core studies. Ecol. Monogr. 40: 79–118.

Whiteside, M. C. & R. V. Harmsworth, 1967. Species diversity in chydorid (Cladocera) communities. Ecology 48: 664–667.

Williams, J. B., 1982. Temporal and spatial patterns of abundance of the chydoridae (Cladocera) in Lake Itasca, Minnesota. Ecology 63: 345–353.

Hydrobiologia **360**: 287–289, 1997.
A. Brancelj, L. De Meester & P. Spaak (eds), Cladocera: The Biology of Model Organisms.
©1997 *Kluwer Academic Publishers.*

The future of cladoceran research

Petter Larsson[1] & Maria Rosa Miracle[2]
[1]*Department of Zoology, University of Bergen, Allégt. 41, N-5007 Bergen, Norway*
[2]*Department of Ecology, University of Valencia, E-46100 Burjassot (Valencia), Spain*

The cladocerans are a group of Crustacea used as model organisms in many fields of biology. In systematics, ecology, physiology and genetics, the cladocerans appear as frequently used examples in case studies. Their reproductive system, switching between parthenogenesis and sexual reproduction, makes them particularly suited for studies dealing with genetic and environmentally induced variability. It is on the whole a particularly handy group of animals that can be used for studying almost all kinds of evolutionary processes.

It is, however, from time to time necessary to look upon the biology of the cladocerans as such, and to set the biology of the group in focus. Even model organisms have their own biology and specialities. One has to be aware of these characteristics to avoid erroneous generalising. The 'Symposia on Cladocera' offer an ideal opportunity for people with different scientific training to discuss on their common interest, namely the group of animals they use in their research. The Fourth International Symposium on Cladocera held in Postojna hosted 26 papers dealing with the genus *Daphnia*, 5 papers dealing with *Moina* and 23 of these papers also included information on other genera. This indicates that as it happen in other groups of organisms, there is one genus that attrack more attention because its wide dominance in some environments and its easy cultivation under controlled condition.

The Fourth International Symposium on Cladocera deepened our insight into many aspects of cladoceran life, but it also clarified gaps in our understanding. Although cladocerans are a relatively well-studied group of invertebrates, there is still so much to study. There are several areas of research for which cladocerans are excellent model organisms. Moreover, as important elements of aquatic ecosystems they illustrate ecological processes and they are guides in the sediment cores when the history of lakes and climate is studied. In our opinion, the most important processes to be studied in the future should take advantage of the special features of the cladocerans, i.e. their special reproductive system combined with their flexible life history, morphology and behaviour, their importance as grazers on the algal communities and their significance as prey for vertebrate and invertebrate predators.

The power of molecular genetic techniques in ecology and taxonomy

Modern molecular genetic techniques are able to discriminate between clones within and between localities and they can give information about genetic distances between populations and species. Due to the asexual reproduction resulting in the formation of clones, phenotypic variability can be studied in relation to specific genotypes (Lynch, 1984). In cladocerans, one can thus study the relationship between phenotype and genotype in a much more straightforward way than in organisms with obligatory sexual reproduction.(Hebert & Crease, 1980; Lynch, 1984; Weider & Hebert, 1987; Mort, 1991; De Meester et al., 1995; Spaak & Hoekstra, 1995; Spitze, 1992 and several others). If molecular genetic techniques are properly incorporated in taxonomical, ecological, behavioural and physiological studies, evolutionary processes can be studied in much detail using cladocerans as model organisms. There remain problems to be solved, however, such as the difficulty to carry out extensive crossing experiments with cladocerans. The difficulty to control the formation of sexual eggs hampers the establishment of controlled crosses among genotypes. In addition, the controlled hatching of sexual diapause eggs is not without problems (Schwartz & Hebert, 1987; Larsson, 1991). Further research is needed to overcome these problems, as this will open unknown possibilities for testing evolutionary processes. Also the comparison of the molecular study of populations that have lost sexuality, with that of cyclic parthenogens may give insight into the mechanisms involved in suppression of meiosis.

Although there are pelagic as well as littoral cladocerans, the number of littoral species is much higher than the number of pelagic species. When Frey (1973, 1986) provided data demonstrating that the earlier belief in the cosmopolitanism of Chydoridae was unjustified, it inspired new studies focusing on the systematics and the distribution of species (Korovchinsky, 1996). The systematics of cladocerans has a long history, but it has always been difficult to classify species based on morphological characters alone, especially for pelagic taxa. Molecular techniques and detailed morphological knowledge have to be combined in future taxonomic and phylogenetic research. The application of molecular genetic techniques has already greatly improved our understanding of the taxonomy and systematics of problematic taxa such as *Daphnia* (Hebert et al.,1989; Weider & Hobæk, 1994; Hebert, 1995; Colbourne & Hebert 1996; Schwenk, 1997).

Phenotypic plasticity and the evolution of inducible defences

Phenotypic variability has since long been a central theme in the studies on cladocerans, as is evidenced by the many publications on cyclomorphosis (e.g. reviewed by Jacobs, 1987). Since the first papers demonstrating the existence predator-induced morphological changes in *Daphnia* back in 1981 (Grant & Bayly, 1981; Krueger & Dodson, 1981), there has been a renewed interest in the ecology and evolution of phenotypic plasticity in cladocerans, and *Daphnia* in particular (review by Larsson & Dodson, 1993). How environmental factors affect the behaviour, life history and morphology of cladocerans may vary between clones and populations. Here is another large field for reserach. There are few other groups of organisms where the relationship between the genotype and the effects of the environment can so easily be studied. The phenotypic plasticity of daphnids and their capability to adjust to constraints imposed by the environment is intriguing. It is surprising that animals with such a small nerve ganglion are able to discriminate between the smell and other signals from food, conspecifics and predators, and are able to show adaptive shifts in behaviour, morphology and life history in response to this multitude of stimuli. Although many aspects of these responses have already been studied, many more surprising insights into the evolution of inducible defences and the subtleties of predator-prey arms races are expected in the future. There are, however, some methodological difficulties slowing down progress in this field too. The chemical identity of predator kairomones and other signals to which the cladocerans react is still virtually unknown (Tollrian & Elert, 1994; Elert & Loose, 1996). Although many experiments can be done with water that has been conditioned by a known number of specific organisms or with extracts from such conditioned water, more elegant and controlled experiments await the characterisation of the chemical nature of the cues involved.

Sensory capacities

To be able to react on environmental stimuli, animals need a sensory apparatus to handle the information. There is a need for detailed anatomical studies and studies on the sensory physiology of cladocerans. We know something about the vision of cladocerans (e.g. Ringelberg, 1987), but the capacity the animals have to recognise chemical signals, gut filling and hydrodynamic waves are almost blank fields that need careful study. Detailed micro-anatomical studies of cladocerans, such as presented by Dumont & Silva-Briano (1997) should be complemented by ecophysiological research on the function of specific sense organs. There is a whole scope of behavioural and ecological experiments that need to be done to fully appreciate the sensory capacities of cladocerans and the consequences for their ecology (e.g. Van Gool, 1997; Larsson, 1997; De Meester & Cousyn, 1997; Weber & Declerck, 1997).

Trophic interactions and community structure

Large cladocerans are known to be very efficient grazers in the pelagic zone of lakes, but they are particularly vulnerable to predation. Since the classic paper by Brooks & Dodson (1965), many field and laboratory studies have been carried out to test the Size Efficiency Hypothesis (e.g. Hall et al., 1970; Gliwicz, 1990). Given their key role as prey for fish and invertebrate predators and grazers of phytoplankton, cladocerans are considered key organisms in biomanipulation experiments aimed at reducing algal blooms in eutrophic lakes. Although biomanipulation measures have certainly not always been successful (Gulati et al., 1990), the scientific data generated by whole-lake and encosure experiments have been very valuable and have provided us with insight into the reactions of the whole ecosystem upon manipulations of the food web. Cladoceran researchers may thus take advantage of the results from biomanipulation experiments to gain a better understanding of ecosystem functioning. Whereas most of the modern ecology is focused on population ecology, the popularity of biomanipulation attempts can be used to rigorously test new hypotheses on the ecology of communities (e.g. Carpenter & Kitchell, 1993).

Epilogue

In spite of the role cladocerans play as model organisms, very little general ecological theory has resulted from cladoceran research. Up till now much of the research on cladocerans indeed has been phenomenological and anecdotal. Many interesting observations have been done, which have, however, been insufficiently incorporated into new general theories. Researchers dealing with cladocerans face a big challenge in trying to use their unique findings to a better understanding of the evolutionary processes in general.

References

Brooks, J. L. & S. I. Dodson, 1965. Predation, body size, and compostion of plankton, Science 150: 28–35.

Carpenter, S. R. & J. F. Kitchell (eds), The Trophic Cascade in Lakes. University Press, Cambrigde, 385 pp.

Colbourne, J. K. & P. D. N. Hebert, 1996. The systematics of North-American *Daphnia* (crustacea, anomopoda) – a molecular phylogenetic approach. Phil. Trans. r. Soc. Lond. B 351: 349–360.

De Meester, L. & C. Cousyn, 1997. The change in phototactic behaviour of a *Daphnia magna* clone in the presence of fish kairomones: the effect of exposure time. Hydrobiologia 360: 169–175.

De Meester, L., L. J. Weider, R. Tollrian, 1995. Alternative antipredator defences and genetic polymorphism in a pelagic predator-prey system. Nature 378: 483–485.

Dumont, H. J. & M. Silva-Briano, 1997. Sensory and glandular equipment of the trunk limbs of the Chydoridae and Macrothricidae. Hydrobiologia 360: 33–46.

Frey, D. G., 1973. Comparative morphology and biology of three species of Eurycercus (Chydoridae, Cladocera) with a description of *Eurycercus macracanthus* sp. nov. Int. Revue ges. Hydrobiol. 58: 221–267.

Frey, D. G., 1986. The non-cosmopolitanism of chydorid Cladocera: Implications for biogeography and Evoultion. In Gore, R. H. & K. L. Heck (eds), Crustacean Biogeography, Balkema, Rotterdam: 237–256.

Gliwicz, M. Z., 1990. Food tresholds and body size in cladocerans. Nature 343: 638–640.

Grant, J. W. G. & I. A. E. Bayly, 1981. Predator induction of crests in morphs of the *Daphnia carinata* King complex. Limnol. Oceanogr. 26: 201–218.

Gulati, R. D., E. H. R. R. Lammens, M.-L. Meijer & E. van Donk (eds), 1990. Biomanipulation – Tool for Water Management. Dev. Hydrobiol. 61, Kluwer Academic Publishers, Dordrecht, 628 pp. Reprinted from Hydrobiologia 200/201.

Hall, D. J., W. E. Cooper & E. E. Werner, 1970. An experimental approach to the production dynamics and structure of freshwater animal communities. Limnol. Oceanogr. 15: 839–928.

Hebert, P. D. N., 1995. The *Daphnia* of Northern America: An Illustrated Fauna. CD-ROM, distributed by the author. Department of Zoology, University of Guelph, Guelph, Ontario.

Hebert, P. D. N. & T. J. Crease, 1980. Clonal coexistence in *Daphnia pulex*, another planktonic paradox. Science 207: 1363–1365.

Hebert, P. D. N., S. S. Schwartz & J. Hrbácek, 1989. Patterns of genetic diversity in Czecheslovakian *Daphnia*. Heredity 62: 207–216.

Jacobs, J., 1987. Cyclomorphosis in *Daphnia*. In Peters, R. H. & R. De Bernardo (eds), *Daphnia*. Mem. Ist. ital. Idrobiol. 45: 325–352.

Korovchinsky, N. M., 1996. How many species of Cladocera are there? Hydrobiologia 321: 191–204.

Krueger, D. A. & S. I. Dodson, 1981. Embryonical induction and predation ecology in *Daphnia pulex*. Limnol. Oceanogr. 26: 219–223.

Larsson, P., 1991. Intraspecific variability in response to stimuli for male and ephippia formation in *Daphnia pulex*. Hydrobiologia 225: 281–290.

Larsson, P., 1997. Ideal free distribution in *Daphnia*? Are daphnids able to consider both the food patch quality and the position of competitors? Hydrobiologia 360: 143–152.

Larsson, P. & S. I. Dodson, 1993. Chemical communication in planktonic animals. Arch. Hydrobiol. 129: 129–155.

Lynch, M., 1984. The limits of life history evolution in *Daphnia*. Evolution 38: 465–482.

Mort, M. A., 1991. Bridging the gap between ecology and genetics: the case of freshwater zooplankton. Trends Ecol. Evol. 6: 41–45.

Ringelberg, J., 1987. Light induced behhaviour in *Daphnia*. In Peters, R. H. & R. De Bernardo (eds), *Daphnia*. Mem. Ist. ital. Idrobiol. 45: 285–323.

Spaak, P. & J. R. Hoekstra, 1995. Life history variation and coexistence of a *Daphnia* hybrid with its parental species. Ecology 76: 553–564.

Spitze, K., 1992. Predator mediated plasticity of prey life history and morphology: *Chaoborus americanus* predation on *Daphnia pulex*. Am. Nat. 139: 229–247.

Schwartz, S. S. & P. D. N. Hebert, 1987. Methods for activation of resting eggs of *Daphnia*. Freshw. Biol. 17: 373–379.

Schwenk, K., 1997. Evolutionary genetics of *Daphnia* species complexes – hybridism in syntopy. Ph.D. thesis, University of Utrecht, The Netherlands, 141 pp.

Tollrian, R. & E. von Elert, 1994. Enrichment and purification of *Chaoborus* kairomone from water: Further steps toward its chemical characterization. Limnol. Oceanogr. 39: 788–796.

Van Gool, E., 1997. Light-induced swimming of *Daphnia*: can laboratory experiments predict diel vertical migration. Hydrobiologia 360: 161–167.

Von Elert, E. & C. Loose, 1996. Predator-induced diel vertical migration in *Daphnia*: Enrichment and preliminary chemical characterization of a kairomone exuded by fish. J. chem.. Ecol. 22: 885–895.

Weber, A. & S. Declerck, 1997. Phenotypic plasticity *Daphnia* life history traits in response to predator kairomones: genetic variability and evolutionary potential. Hydrobiologia 360: 89–99.

Weider, L. J. & P. D. N. Hebert, 1987. Ecological and physiological differentiation among low-arctic clones of *Daphnia pulex*. Ecology 68: 188–198.

Weider, L. J. & A. Hobaek, 1994. Molecular biogeography of clonal lineages in a high-arctic apomictic *Daphnia* complex. Mol. Ecol. 3: 497–506.

Weider, L. J., A. Hobaek, T. J. Crease & H. Stibor, 1996. Molecular characterization of clonal population structure and biogeography of arctic apomictic *Daphnia* from Greenland and Iceland. Mol. Ecol. 5: 107–118.

Hydrobiologia **360**: 291–294, 1997.
© 1997 *Kluwer Academic Publishers.*

Some suggestions for future cladoceran research

Joop Ringelberg
*Netherlands Institute of Ecology, Centre for Limnology, Rijksstraatweg 6, 3631 AC Nieuwersluis, The Netherlands
and Department of Aquatic Ecology, University of Amsterdam (e-mail address: ringelberg@cl.nioo.knaw.nl)*

Abstract

The topics dealt with at the Cladoceran Symposium showed a large diversity. A high diversity is what characterises the world we study, consisting of many different habitats and a rich species set. This easily leads to anecdotal research which then prevents the building of a consistent and sophisticated body of knowledge. The contribution of cladoceran research to general ecological theory is limited as a look in textbooks reveals. Therefore, choices have to be made for future research and a few suggestions to this extent are shortly mentioned in this essay in consequence of the Symposium on Cladocera held in Postojna (Slovenia), August 1996.

Taxonomy and a modern synthesis

The research discussed at the Cladocera symposia has gradually shifted from predominantly object-oriented to problem-oriented. The number of papers on species descriptions and on the distribution of cladocerans in particular lakes has decreased and more attention has been given to aspects of physiology, behaviour and genetics. This development must be pursued, but at the same time the important role of taxonomy in the understanding of cladoceran ecology must not be forgotten. The Cladocera are rich in species and of several taxa the precise taxonomic status is not yet clear. This even holds for intensively studied species such as those belonging to the *Daphnia longispina* group. Species names have often been changed in the past and to the non-taxonomist this has been confusing. It is inevitable, however, because concepts in taxonomy change as they do in other branches of biology. If contributions of ecology but especially genetics are incorporated, the status of the traditional species will probably change again. It must be a challenge to the taxonomist, the geneticist and the ecologist to work together on a new systematics of, for instance, that problematic *D. longispina* group. The hybrids of *D. hyalina, D. cucullata* and *D. galeata* are very common in lakes and often more dominant than the parental species (Schwenk & Spaak, 1995). Evidently, the relative fitness of these hybrids is then higher compared to that of the parents for commonly occurring environmental circumstances. Spaak & Hoekstra (1995) called the phenomenon a temporary hybrid superiority. How temporary is it? Is it conceivable that we witness a rapid evolution within this group now that our lakes are changing fast in physical, chemical and biological composition due to man's actions? Ecophysiology, behaviour and population dynamics of these various taxa are indeed different. Although characterisation by molecular techniques is not so difficult, distinction is a problem to the ecologist. As yet morphological distinction alone might fail, but it would be important for the ecologist if an expert system with easy morphological and morphometric data as input could be used to distinguish parents and hybrids on a routine base (Schwenk, pers. comm.).

It is evident to me that much problem-oriented research can be done within a taxonomical context that is very important to all students of Cladocera. Perhaps the next meeting can have a session titled 'The taxonomy of the *D. longispina* complex, a modern synthesis' to which taxonomists, geneticists, physiologists and ecologists can contribute in an integrated way. This group has received considerable attention and could be an example for the study of other problematic species complexes, as, for example, the *D. pulex* complex in Europe, and the *Bosmina* species complex.

Life histories, population dynamics and competition

Recently, many life table studies on *Daphnia* species were performed to quantify the influence of several environmental factors on life history characteristics. At the Postojna symposium examples were given by Boersma (1997) and Weber & Declerck (1997). Life history studies on *Daphnia* have contributed to life history theory and this will become apparent in books on this topic (Roff, 1992; Stearns, 1992) in the near future. The studies were performed, amongst others, to interpret population dynamics and seasonal succession in nature (Boersma, 1994; Spaak, 1994). However, culturing individual animals in small tubes might have limitations in this respect. One level of study in between the solitary existence in a 'life history tube' and population dynamics or intra- and interspecific interaction in a lake is realised with the population and competition experiment (e.g. Reede, 1997). These experiments are challenging but not without problems. Competition for food is most important, thus the carrying capacity of the experimental set-up for both competitors must be first determined. In the experiments, factors can be kept constant but experiments with algae (Sommer, 1985; Spijkerman & Coesel, 1996) have demonstrated that the outcome in terms of competitive exclusion or coexistence changes when resource concentration is pulsed. At the Symposium, Boersma started his lecture with a short theoretical introduction and I think it is a good idea for a future symposium to devote a workshop-like session to the theory of population growth and competition.

Chemoperception and the role of infochemicals

Another challenging, modern field is that of intra- and interspecific communications. As in the macroscopic world of insects, birds and mammals and in the microscopic world of cells, cladocerans use complex signals to maximise fitness. Community composition and ecosystem flow of matter are also steered by the information these signals contain. In contrast to the groups mentioned, cladocerans and copepods are less well studied. Communication between zooplankton species in general and, probably, between benthic species as well, seems largely to involve chemical signals. Intraspecific signals (called pheromones: see for terminology Dicke & Sabelis, 1992) are probably used in swarming behaviour and in mate-finding although definite proof of their existence and role in communi-

cation must still be given. Of the interspecific signals, kairomones produced by predators have received most attention. These predator exudates induce changes in morphology, life history characters as well as behaviour of prey animals and examples were presented at the symposium. Until now, morphological changes were described predominantly for *Daphnia* species inhabiting small water bodies where *Chaoborus* and *Notonecta* cause the induction of neckteeth. However, neckteeth can also be induced in *D. hyalina* and *D. longispina* as was illustrated in posters by Čelhar & Brancelj and by Miracle & Boronat, respectively. One of the questions, however, is to what extent these tiny neckteeth in the pelagic daphnids are effective as predator defences. Experimental studies must reveal their quantitative role.

The examples given above mark the beginning of a promising field of research and many interesting problems in chemical communication in the cladocerans need our attention. To start with, what chemical sense organs are present and where are they situated? Sensory structures need to be described in the first place and, as Dumont & Silva-Briano (1997) demonstrate, with scanning electron microscopy informative pictures can be obtained. However, to elucidate the function of these minute structures, studies of behaviour have to be initiated. These will not be easy to perform because the animals are small and special methods have to be developed. Nevertheless, simple observations on individuals in the laboratory should be a good start. Examples are the amusing videos of mating behaviour in Moina (Forro, 1997) or of different swimming modes in *Daphnia* (Larsson, 1997).

To start with, a comparison of the observed structures found in cladocerans with those described for other crustacean will certainly provide ideas of their function. It would be strange if cladocerans had evolved sense organs completely different from other crustaceans. For instance, some of the pore structures shown by Dumont & Silva-Briano (1997) look like the force-receptors that regulate walking in crabs (Zil & Seyfarth, 1996). Perhaps some of them play a role in the synchronisation of thoracal limb movements. Interesting as they are, these studies easily become pure physiology and too detached from the field of ecology. Moreover, other problems await the ecophysiologist and the ecologist that explore the field of communication, chemical and otherwise, that are of direct importance for the understanding of community structure and dynamics of material flow. Experiments on choice behaviour by Van Gool & Ringelberg

(1995) demonstrated that *Daphnia* is able to differentiate between different species of food algae, which is of paramount ecological importance. Nevertheless, it remains to be demonstrated in what way this chemoperception is used in nature. If near-field chemoperception is involved, a short range differentiation between edible and non-edible particles might lead to selection. Although *Daphnia* is generally considered an a-selective filter feeder, near-field chemoperception could be used to change filtering rate rapidly and thereby avoiding filtering noxious particles. Suggestions of algal patch finding have been made but the then necessary far-field chemoperception is not without problems. The question is to what extent chemical plumes are present in the open water zone and whether *Daphnia* and other planktonic animals are capable of oriented swimming with regard to a chemical source or not. Swimming by trial and error is inefficient in a three-dimensional space where distances to be covered are relatively large in relation to swimming speed. An environmental factor with an inherent directional aspect might be used if the direction is highly correlated with the non-directional factor. Light is such a factor. For instance, the angular light distribution is used in positive phototaxis to reach near-surface water layers if oxygen concentration at the place of the animal is low or carbon dioxide concentration is high (Loeb, 1904). Likewise, *Daphnia* cannot react to predator kairomones with oriented swimming and, again, phototaxis is used to execute movements away from the zone of predation (Ringelberg, 1991). It remains to be discovered by what means chemoperception is used to find patches of algae.

Infochemicals, community structure and flow of matter in ecosystems

Intra- and interspecific communication, especially by infochemicals, is not just another interesting topic of research in aquatic ecology. I already indicated that it is important in the understanding of community structure and the flow of matter in ecosystems. I will in short indicate what is to be understood by this. At the individual level, it amounts to increasing relative fitness as realised by producing the highest possible number of reproducing offspring. At the population level, it means that the relative rate of population growth is affected by infochemicals. A well-studied example is the enhancement of diel vertical migration by fish kairomones, realising a reduced death rate by

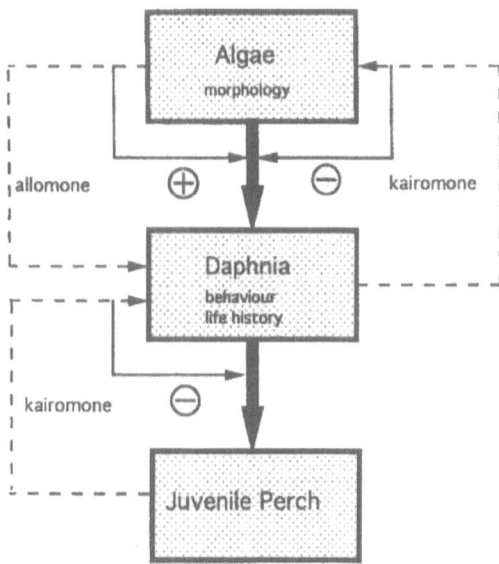

Figure 1. Scheme of a simple tritrophic foodchain with flows of matter between compartments represented by heavy arrows and flows of information by dashed lines. The *Daphnia* kairomone has probably a negative effect on the flow from algae to *Daphnia* (Hessen & Van Donk, 1993), the allomone a positive effect (Van Gool & Ringelberg, 1995) and the perch kairomone negatively influences the amount of *Daphnia* eaten by the perch (Ringelberg et al., 1991).

predation but, at the same time, most often causing a lower birth rate. The net result is the best possible relative rate of increase at the ambient environmental circumstances. If the death rate, instead of expressing it as numbers eaten by the predator, is calculated as the biomass transferred from the prey population to the predator population, we use the parlance of systems ecology. Schemes of flow of matter or of energy through trophic compartments are in fact schemes of condensed foodwebs. It can be envisaged that nature is composed of two networks superimposed on one another: the material flow network that pictures the quantitative relations between species and the communication network which consists of streams of information from one species to another. The latter regulates the material flow pattern. Information is immaterial, but it must have a material vehicle for its transport. In this respect infochemicals are important information carriers. Although it is not always clear whether we are dealing with a flow of matter or a flow of information, an important criterion of distinction is a high or a (very) low energy content, respectively. Traits of emitting and receiving information are especially prone to selection and evolution proceeds for a large part along these lines.

To concretise the two superimposed networks, a simple tritrophic relation between algae, *Daphnia* and fish is given in Figure 1. It was demonstrated by Hessen & Van Donk (1993) and Lampert et al. (1994) that a chemical, mediated by *Daphnia*, changes cell walls and induces colony formation in *Scenedesmus* and probably also in other algae such that edibility by *Daphnia* is affected. Thus, this infochemical influences grazing by *Daphnia* and probably lowers transfer of matter from the algae to the cladoceran. Likewise, by inducing diel vertical migration, the fish kairomone diminishes the material flow between *Daphnia* and fish predator (Ringelberg et al., 1991). The quantification of the two networks offers a means for a more complete understanding of the functioning of nature which is a real challenge to the ecologist.

Epilogue

The past Symposia on Cladocera, although very interesting, have been broad, diverse and, therefore, lacking in coherence. We can do better in future. If problem-oriented research is pursued and part of individual research is directed to a limited number of research topics, mutual interest will increase and lead to better discussions and better science. A mild co-ordination might be necessary. Utopia? Then we failed.

Acknowledgments

I thank my colleagues Dr Ramesh Gulati and Dr Koos Vijverberg cordially for suggested improvements to the manuscript.

References

Boersma, M., 1994. On the seasonal dynamics of *Daphnia* species in a shallow eutrophic lake. Doctoral Thesis, University of Amsterdam.

Boersma, M., 1998. Offspring size in *Daphnia*: does it pay to be overweight? Hydrobiologia 360: 79–88.

Dumont, H. J. & M. Silva-Briano, 1997. Sensory and glandular equipment of the trunk limbs of the Chydoridae and Macrothricidae (Crustacea: Anomopoda). Hydrobiologia 360: 33–46.

Dicke, M. & M. W. Sabelis, 1992. Costs and benefits of chemical information conveyance: proximate and ultimate factors. In B. D. Roitberg & M. B. Isman (eds), Insect Chemical Ecology, an Evolutionary Approach. Chapman & Hall, New York: 123–155.

Forró, L., 1997. Mating behaviour in *Moina brachiata* (Jurine, 1820) (Crustacea, Anomopoda). Hydrobiologia 360: 153–159.

Hessen, D. & E. Van Donk, 1993. Morphological changes in *Scenedesmus* induced by substances released from *Daphnia*. Arch. Hydrobiol. 127: 129–140.

Lampert, W., K. O. Rothhaupt & E. von Elert, 1994. Chemical induction of colony formation in a green alga (*Scenedesmus acutus*) by grazers (*Daphnia*). Limnol. Oceanogr. 39: 1543–1550.

Larsson, P., 1997. Ideal free distribution in *Daphnia*? Are daphnids able to consider both the food patch quality and the position of competitors? Hydrobiologia 360: 143–152.

Loeb, J., 1904. The control of heliotropic reactions in fresh water crustaceans by chemicals, especially C02. Univ. Calif. Publ. Physiol. 2: 1–3.

Reede, T., 1997. Preliminary experiments on resource competition between a migrating and a non-migrating clone of the hybrid *D. galeata* × *hyalina*. Hydrobiologia 360: 109–115.

Ringelberg, J., 1991. A mechanism of predator-mediated induction of diel vertical migration in *Daphnia hyalina*. J. Plankton Res. 13: 83–89.

Ringelberg, J., B. J. G. Flik, D. Lindenaar & K. Royackers, 1991. Diel vertical migration of *Daphnia hyalina* (sensu latiori) in Lake Maarsseveen: Part 2. Aspects of population dynamics. Arch. Hydrobiol. 122: 385–401.

Roff, D. A., 1992. The Evolution of Life Histories, Theory and Analysis. Chapman & Hall, New York, 535 pp.

Schwenk, K. & P. Spaak, 1995. Evolutionary and ecological consequences of interspecific hybridization in cladocerans. Experientia 51: 465–481.

Spaak, P., 1994. Genetical ecology of coexisting *Daphnia* hybrid species complex. Doctoral Thesis, University of Utrecht.

Spaak, P. & J. R. Hoekstra, 1995. Life history variation and the coexistence of a *Daphnia* hybrid with its parental species. Ecology 76: 553–564.

Spijkerman, E. & P. F. M. Coesel. Competition for phosphorus among planktonic desmid species in continuous-flow culture. J. Phycol. 32: 939–948.

Stearns, S., 1992. The Evolution of Life Histories. Oxford University Press, New York, USA, 249 pp.

Sommer, U., 1985. Comparison between steady state and non-steady state competition: Experiments with natural phytoplankton. Limnol. Oceanogr. 30: 335–346.

Van Gool, E. & J. Ringelberg, 1995. Daphnids respond to algae-associated odours. J. Plankton Res. 189: 197–202.

Weber, A. & S. Declerck, 1997. Phenotypic plasticity of *Daphnia* life history traits in response to predator kaironomes: genetic variability and evolutionary potential. Hydrobiologia 360: 89–99.

Zill, S. N. & E. A. Seyfarth, 1996. Exoskeletal sensors for walking. Sci. Am. 275: 70–75.

Hydrobiologia **360**: 295–299, 1997.
A. Brancelj, L. De Meester & P. Spaak (eds), Cladocera: The Biology of Model Organisms.
©1997 *Kluwer Academic Publishers.*

The future of cladoceran genetics: methodologies and targets

Paul D. N. Hebert & Derek J. Taylor
Department of Zoology, University of Guelph, Guelph, Ontario, NlG 2W1, Canada

Abstract

Molecular studies are poised to resolve areas of uncertainty which have existed for nearly 200 years concerning the taxonomy, evolution and phylogenetic relationships of the cladoceran crustaceans. This paper aims to both summarise some of the key methodological advances which have set the stage for rapid progress and to discuss four areas of work which will profit from their application.

Methodological considerations

From 1970 to 1990, genetic studies on cladocerans were almost entirely reliant upon allozyme characterisations. Over this interval, the primary methodological advance involved the adoption of higher sensitivity electrophoretic approaches (Hebert & Beaton, 1989), aiding both the acquisition of multilocus genotypic data and enabling studies on even the smallest cladocerans.

Since 1990 there has been an extraordinary diversification of techniques based on the polymerase chain reaction (PCR) which allow the detection of DNA sequence diversity. PCR-based analyses on specific genes do suffer from the disadvantage that primer sequences must be identified. Fortunately this quest has been simplified by the identification of 'universal' primers (designed from conserved regions and often including 'wildcard' sites) (Kocher et al., 1989) that allow amplification of nuclear and mitochondrial genes for most, if not all, cladocerans. Long PCR (Cheng et al., 1994) now permits the amplification of fragments of more than 16.0 kb in length, making analysis of the entire mtDNA genome straightforward. Aside from the analysis of coding sequences, efforts have been made to identify both mini- and microsatellite markers (Tautz, 1989; Queller et al., 1993) because of the hypervariability typical of these segments of the genome.

The analytical approaches employed to recognize sequence diversity in PCR amplified fragments of the genome have themselves shown rapid transition. Restriction fragment analysis was initially used to generate partial sequence information (Dowling et al., 1996), but recent studies typically employ either automated sequencing to provide complete sequence characterisation (Palumbi, 1996) or techniques such as single stranded conformational polymorphism analysis for quick screening (Hayashi, 1991).

Given the rapid progress in the study of DNA diversity, it might be expected that allozyme analyses would have little further role. In fact, this approach is often the most effective means for surveying genetic variation in single copy nuclear loci, and hence remains useful for examining processes impacted by sexual reproduction. For example, allozyme studies remain the best approach for the determination of species boundaries, unless the taxa under study have a very recent origin. Allozyme analyses are also ideal for the recognition of interspecific hybridization, and for the investigation of key biological attributes of single species

Allozyme data are not a panacea. They are, for instance, of limited value in phylogenetic analyses, because the number of characters (typically <20 loci) is small. By contrast, studies of sequence variation in the genome can potentially examine several billion nucleotide pairs, although, for practical reasons, sequences are rarely determined at more than 1000 nucleotide sites. The clarity of the phylogenetic signal is, of course, not simply dependent upon the number of characters examined, but also upon the nature of character state variation. Genes with no variation are uninformative, but so too are genes which show such rapid evolution that shared character states reflect secondary convergence rather than shared ancestry (Swofford et al., 1996). Because genes vary in their rate of evolution and taxon assemblages vary in their age, few genes are useless in all phylogenetic contexts and no gene is invariably valuable. However, as their rates of sequence divergence are known, specific genes can be targeted for analysis depending upon the age of the taxonomic assemblage under investigation. For example, nuclear ribosomal genes (18S, 28S) are useful in the study of phylogenetic relationships among taxa which last shared common ancestry 100 million or more years ago (Spears et al., 1992), while their mitochondrial counterparts (12S, 16S), which evolve more rapidly (0.4–0.9% per my versus 0.1% per my), provide insights concerning phylogenetic divergence over intervals of 10–100 my (Cunningham et al., 1992; Sturmbauer et al., 1996). Finally, to resolve relationships among more closely allied lineages, it is necessary to survey sequence diversity in more rapidly evolving genes, a task which is often complicated by the need to identify primer sequences. Admittedly, because the cladocerans lack a detailed fossil record, there is often uncertainty in the a priori selection of a gene.

Questions

Resolving multispecies relationships – phylogenetic analysis

Until recently the acquisition of data suitable for phylogenetic analysis was problematic. However, it is now straightforward to obtain sequence information for slowly evolving genes, although the alignment of sequences and subsequent data analysis present a greater challenge. Decisions on character homology are complicated by insertion/deletion events which are frequent in ribosomal DNA, the commonest target of this work. Moreover, the processes of lineage sorting, interspecific gene flow, and gene duplications are known to create discrepancies between species trees and gene trees (Maddison, 1996). Future efforts will reduce these difficulties by obtaining phylogenetic information from multiple genes and by developing primers for single copy protein-coding genes.

An important phylogenetic controversy to readers of this volume that may be resolved in the near future involves the question 'What is a water flea?' Cladocerans do share a distinctive breeding system (cyclic parthenogenesis with environmental sex determination), but their striking morphological diversity

has led several researchers (Starobogatov, 1986; Fryer, 1987) to contest their monophyly. Although this conclusion has gained general acceptance (e.g. Dodson & Frey, 1991), sequence analysis of 28S rDNA (Taylor, in prep.) and secondary structure of 18S rDNA (Crease & Taylor, in prep.) both convincingly support cladoceran monophyly. Other studies have used sequence diversity in 12S rDNA to clarify the boundaries of and taxonomic affinities within the genus *Daphnia* (Taylor et al., 1996; Colbourne et al., 1996). The success of these investigations indicates that it will be possible to make rapid progress in defining phylogenetic relationships among the 79 genera which comprise the Cladocera. Perhaps more significantly, these studies will set the stage for investigations of both the patterns of character state evolution and the time frames over which diversification of this group has occurred (Colbourne et al., 1997). Admittedly, efforts to gain a highly resolved understanding of the time scale of cladoceran evolution will have to solve the Gordian knot of calibration with a poor fossil record and few well-characterized vicariance events.

Delineating species boundaries – the new taxonomy

The fluid taxonomy of cladocerans provides a testament to the difficulties involved in determining species boundaries solely through morphological criteria (Korovchinsky, 1996). It is now clear that genetic studies can resolve much of this uncertainty. The historical trend in cladoceran taxonomy is one of examining characters at an ever decreasing scale – first body shapes and limb morphology, then mandible anatomy, ephippial microsculpturing, and spermatozoal ultrastructure. The new taxonomy extends this trend by examining the 'morphology of molecules', but with the following advantages: a known genetic basis, numerical abundance, character independence, and relatively weak selective pressures (compared to traditional morphological characters).

If the results of past work are indicative, current taxonomic systems recognise less than half of cladoceran species, even in the best-studied groups. For example, genetic studies have indicated that *Daphnia* fauna of North America includes at least 34 species rather than the 15 species discriminated by earlier morphological studies (Hebert, 1995), and the global total for the genus is likely closer to 200 species than to the 11 taxa recognised by Wagler (1936).

The new taxonomy means that waterfleas will become one of the few ecologically important groups of arthropods for which a comprehensive global taxonomy and phylogeny is tractable. All aspects of cladoceran comparative biology will be enriched by this new information. For example, the discipline of conservation biology has largely ignored cladocerans. Yet, recent genetic studies have revealed the existence of endemic cladoceran species occupying rare or threatened habitats. In addition, the extent and impact of biological invasions can now be assessed. Many cladoceran groups consist of taxa that are 'cryptogenic' – species for which the native or introduced status is unknown (Carlton, 1996). Genetic markers have shown that cryptic invaders exist in the Cladocera and that they can impact native populations by hybridization and introgression (Taylor & Hebert, 1993a).

The comprehensive revision of cladoceran taxonomy through genetic analysis is a substantial task, and efforts will need to be prioritized. Taxonomic diversity should be investigated at a global level for at least a few genera and *Daphnia* is one obvious candidate. Other work might target areas of cladoceran endemism, groups whose taxonomic diversity seems low, or habitats which are threatened. Finally it seems important to direct work towards benthic as well as the planktonic genera which have been the focus of past work. From a methodological perspective there will be an increasing move towards tests of the robustness of taxonomic systems, based upon sequence analysis in a range of nuclear and mitochondrial genes, as well as allozyme data.

Hybridization and its role in evolutionary diversification

Although morphologists long suspected its occurrence, allozyme studies first documented hybridization in cladoceran populations, revealing, as well, the extraordinary prevalence of hybrids in some habitats (Hebert, 1985; Mort, 1991; Schwenk & Spaak, 1997). Phylogenetic studies on North American *Daphnia* have shown that hybridization occurs only between closely allied species, suggesting that hybridization may be absent in genera with only a few divergent taxa (Colbourne & Hebert, 1996). Nonetheless hybrids have now been detected in a number of other cladoceran genera including *Simocephalus* (Hann & Hebert, 1982), *Bosmina* (Little et al., 1997) and *Daphniopsis* (Hebert & Wilson, in prep.). In the latter two cases, hybrids reproduce by obligate parthenogenesis and may have been derived from a single hybridization event. However, other hybrids lack the ability to produce resting eggs

and have evidently been synthesized independently in each habitat where they occur.

There is a need for a broader examination of the incidence of hybridization, especially in benthic cladocerans. It is possible that reproductive isolation evolves more rapidly in benthic than pelagic forms, perhaps as a result of microhabitat preferences, explaining the apparent rarity of hybrids in groups such as the chydorids (Frey, 1982). Another key area for investigation involves the nature of interactions between parental taxa and hybrids. The dominance of hybrids indicates that they are superior to their parents in some settings. However, it is unclear if hybrid superiority is a general property or an attribute possessed by a small minority of hybrid clones, an issue which could be resolved through experimental studies on parental taxa and a random array of their hybrid derivatives. There is, as well, a need for more detailed habitat characterisation, to learn if the superiority of hybrids reflects their occupancy of an 'intermediate' microhabitat. The fate of hybrid clones also requires further investigation. Studies have shown that when hybrids of the *D. longispina* complex occur in North American lakes they are typically dominated by a single clone whereas hybrid diversity is much higher in European lakes (Taylor & Hebert, 1993b; Schwenk & Spaak, 1995). This difference might reflect the differential persistence of individual clones. Aging hybrid clones is not likely to be simple, but the analysis of multigenic families such as 18S rDNA may offer a valuable approach. If hybrids are of recent origin, they should show an equal number of the gene copies derived from each parent, while an unbalanced number of gene copies should characterise older hybrids because of the process of molecular drive. Obligately asexual clones derived from the hybridization of *D. pulex* and *D. pulicaria* show evidence of variation in the copy number of 18S genes derived from each parental species, suggesting they are an admixture of recent and ancient clones (Crease & Lynch, 1991). There is, as well, a need for more detailed genetic characterization of hybrid systems, exploiting both mitochondrial DNA to determine the directionality of hybridization and a range of nuclear markers to determine the evolutionary role of hybridization. There is now evidence that hybrid clones often backcross with their parents (Spaak, 1996), suggesting the possibility of introgression. In fact at least one species, *D. mendotae*, appears to have evolved as a direct and rapid consequence of introgression (Taylor & Hebert, 1993b). This mode of reticulate speciation has long been thought to be important in plants, but may also be significant in cladocerans, because the prevalence of hybrids enhances gene exchange between species.

Probing intraspecific diversity

Genetic analyses are also positioned to greatly extend our understanding of reproductive systems and patterns of genetic diversity in cladoceran populations. Allozyme studies have confirmed that most cladocerans produce their diapausing eggs through standard sexual reproduction, but two other breeding systems have also been detected. Polar populations of *Holopedium gibberum* produce their resting eggs through self-fertilization, a breeding system previously unknown in cladocerans (Hebert et al., in prep.). Allozyme studies have also confirmed earlier conclusions that some cladocerans produce their diapausing eggs asexually, but have shown that these lineages are not restricted to the arctic as initially thought (Hebert, 1987). These asexuals have a variable genesis; some are polyploids of hybrid origin, while others are diploids derived from a single parental taxon. Past work has indicated that these asexuals ordinarily show high levels of genetic diversity, but further studies are needed to more fully explain its origin (Weider et al., 1997).

Aside from deepening knowledge of breeding system evolution, genetic analysis is positioned to extend our understanding of the patterning of genetic divergence among populations (De Meester, 1996). Initial allozyme studies challenged to the view that gene flow was sufficient to minimise genetic divergence among cladoceran populations on a continental scale. In fact, evidence soon accumulated that local populations showed marked divergence (Hebert, 1987; Mort, 1991). Although there is little information on the broader spatial patterning of diversity, it appears that there is no simple increase in genetic divergence with distance. Instead populations of many widely distributed species consist of a small number of genetically divergent groups, showing homogeneity over large areas and abrupt divergence at their zones of contact (Hebert & Finston, 1996). The factors responsible for this patterning of diversity remains unclear; perhaps modern populations derive from a small number of refugial lineages which persisted in isolation throughout the Pleistocene. If so, a detailed analysis of mtDNA diversity would likely clarify this fact as it has in studies of freshwater fishes (Bernatchez & Wilson, 1997).

Aside from revealing the spatial patterning of gene frequencies, allozyme studies provided the first indication of temporal shifts in the genotypic composition of populations (Hebert, 1987; Carvalho & Crisp, 1987). This work established the occurrence of surprisingly rapid shifts in genetic composition, signalling the presence of large fitness differences among genotypes. There remains, however, little understanding of the selective forces which underpin these fitness differences. Do they, for example, reflect the varying competitive abilities of clones (Spaak, 1996) or their varying resistance to parasites and predators (Little & Ebert, 1997)? There is, as well, a need for studies which examine the basis of variation in life history traits and other fitness components (Lynch & Spitze, 1994). Both of these areas of uncertainty should profit from the increased resolution of genotypic diversity that can be gained through the analysis of microsatellite markers. PCR-based techniques have also set the stage for studies which examine shifts in genotypic composition over long periods, through the amplification of DNA from ephippial eggs recovered from lake sediments. Depending upon the success of these efforts, it may be possible to reconstruct genetic change in cladoceran populations over several thousand years.

Conclusions

It is apparent that the future contributions of genetics to cladoceran biology will be substantial, as existing techniques are applied on broader taxonomic and geographic scales. The synthesis of information on phylogeny, taxonomy and the patterning of genetic diversity will not only make possible the first detailed understanding of evolutionary diversification and speciation in the Cladocera, but also reveal new opportunities for work on the developmental biology, ecology and physiology of these organisms.

References

Bernatchez, L. & C. C. Wilson, 1997. Comparative phylogeography of Nearctic and Palearctic fishes. Mol. Ecol. in press.
Carlton, J. T., 1996. Biological invasions and cryptogenic species. Ecology 77: 1653–1655.
Carvalho, G. R. & D. J. Crisp, 1987. The clonal ecology of *Daphnia magna*. I. Temporal changes in the clonal structure of a natural population. J. anim. Ecol. 56: 453–468.
Cheng, S., R. Higuchi & M. Stoneking, 1994. Complete mitochondrial genome amplification. Nat. Gen 7: 350–351.
Colbourne, J. K., P. D. N. Hebert, 1996. The systematics of North American *Daphnia* (Crustacea: Anomopoda): a molecular phylogenetic approach. Phil. Trans. r. Soc., Lond. B 351: 349–360.

Colbourne, J. K., P. D. N. Hebert & D. J. Taylor, 1997. Evolutionary origins of phenotypic diversity in *Daphnia*. In T. J. Givnish & K. Systema (eds), Molecular Evolution and Adaptive Radiation, Cambridge University Press: 163–189.

Crease, T. J. & M. Lynch, 1991. Ribosomal variation in *Daphnia pulex*. Mol. Biol. Evol. 8: 620–640.

Cunningham, C. W., N. W. Blackstone & L. W. Buss, 1992. Evolution of king crabs from hermit crab ancestors. Nature 355: 539–542.

De Meester, L., 1996. Local genetic differentiation and adaptation in freshwater zooplankton populations: patterns and processes. Ecoscience 3: 385–399.

Dodson, S. I. & D. G. Frey, 1991. Cladocera and other Branchiopoda. In J. H. Thorp & A. P. Covich (eds), Classification of North American Freshwater Invertebrates, Academic Press: 723–786.

Dowling T. E., C. Moritz, J. D. Palmer & L. H. Rieseberg, 1996. Nucleic acids III: analysis of fragments and restriction sites. In D. M. Hillis, C. Moritz & B. K. Mable (eds), Molecular Systematics, Sinauer Press, Sunderland, Massachusetts: 249–320.

Frey, D. G., 1982. Contrasting strategies of gamogenesis in northern and southern populations of Cladocera. Ecology 63: 223–241.

Fryer, G., 1987. A new classification of the branchiopod Crustacea. Zool. J. linn. Soc. 91: 357–383.

Hann, B. J. & P. D. N. Hebert, 1982. Re-interpretation of genetic variation in *Simocephalus* (Cladocera, Daphniidae). Genetics 102: 101–107.

Hayashi, K., 1991. PCR-SSCP: A simple and sensitive method for detection of mutations in genomic DNA. PCR Methods Application 1: 34–38.

Hebert, P. D. N., 1985. Interspecific hybridization and cyclic parthenogenesis. Evolution 39: 216–220.

Hebert, P. D. N., 1987. Genotypic characteristics of the Cladocera. Hydrobiologia 145: 183–193.

Hebert, P. D. N., 1995. The *Daphnia* of North America: An Illustrated Fauna. CD-ROM, University of Guelph, Guelph, Ontario.

Hebert, P. D. N. & M. J. Beaton, 1989. Methodologies for allozyme analysis using cellulose acetate electrophoresis. Helena Laboratories, Beaumont, Texas.

Hebert, P. D. N. & T. J. Finston, 1996. Genetic differentiation in *Daphnia obtusa*: a continental perspective. Freshwat. Biol. 35: 311–321.

Kocher, T. D., W. K. Thomas, A. Meyer, S. V. Edwards, S. Paabo, F. X. Villablanca & A. C. Wilson, 1989. Dynamics of mitochondrial DNA evolution in animals: amplification and sequencing with conserved primers. Proc. natn. Acad. Sci. U.S.A. 86: 6196–6200.

Korovchinsky, N. M., 1996. How many species of Cladocera are there? Hydrobiologia 321: 191–204.

Little, T. J. & D. Ebert, 1997. Genetic variability for parasite susceptibility in natural populations of *Daphnia* (Cladocera: Crustacea). in review.

Little, T. J., R. De Melo, D. J. Taylor & P. D. N. Hebert, 1997. Genetic characterization of an arctic zooplankter: insights into geographic polyploidy. Proc. r. Soc., Lond. B. 264: 1363–1370.

Lynch, M. & K. Spitze, 1994. Evolutionary genetics of *Daphnia*. In L. A. Real (ed.), Ecological Genetics, Princeton University Press: 109–128.

Maddison, W. P., 1996. Molecular approaches and the growth of phylogenetic biology. In J. D. Ferraris & S. R. Palumbi (eds), Molecular Zoology: Advances, Strategies and Protocols, Wiley-Liss: 47–63.

Mort, M. A., 1991. Bridging the gap between ecology and genetics: the case of freshwater zooplankton. Trend. Ecol. Evol. 6: 41–44.

Palumbi, S. R., 1996. Nucleic acids II: the polymerase chain reaction. In D. M. Hillis, C. Moritz & B. K. Mable (eds), Molecular Systematics, Sinauer Press, Sunderland, Massachusetts: 205–247.

Queller, D. C., J. E. Strassman & C. R. Hughes, 1993. Microsatellites and kinship. Trend. Ecol. Evol. 8: 285–288.

Schwenk, K. & P. Spaak, 1995. Evolutionary and ecological consequences of interspecific hybridization in cladocerans. Experientia 51: 456–481.

Schwenk, K. & P. Spaak, 1997. Ecological genetics of interspecific hybridization in *Daphnia*. In B. Streit, T. Städler & C. M. Lively (eds), Evolutionary Ecology of Freshwater Animals, Birkhäuser Verlag, Basel: 199–229.

Spaak, P., 1996. Temporal changes in the genetic structure of the *Daphnia* species complex in Tjeukemeer, with evidence for backcrossing. Heredity 76: 539–548.

Spears, T., L. G. Abele & W. Kim, 1992. The monophyly of Brachyuran crabs: a phylogenetic study based on 18S rRNA. Syst. Biol. 41: 446–461.

Starobogatov, Y. I., 1986. Systema racobraznyh. Zoologitchesky J. 65: 1769–1781.

Sturmbauer, C., J. S. Levinton & J. Christy, 1996. Molecular phylogeny analysis of fiddler crabs: test of the hypothesis of increasing behavioral complexity in evolution. Proc. natn. Acad. Sci. U.S.A. 93: 10855–10857.

Swofford, D. L., G. J. Olsen, P. J. Waddell & D. M. Hillis, 1996. Phylogenetic inference. D. M. Hillis, C. Moritz & B. K. Mable (eds), Molecular Systematics, Sinauer Press, Sunderland, Massachusetts: 407–514.

Tautz, D., 1989. Hypervariability of simple sequences as a general source for polymorphic DNA markers. Nucl. Acids Res. 17: 6463–6471.

Taylor, D. J. & P. D. N. Hebert, 1993a. Cryptic intercontinental hybridization in *Daphnia* (Crustacea): the ghost of introductions past. Proc. r. Soc., Lond. B 254: 163–168.

Taylor, D. J. & P. D. N. Hebert, 1993b. Habitat dependent hybrid parentage and differential introgression between neighbouringly sympatric *Daphnia* species. Proc. natn. Acad. Sci. U.S.A. 90: 7079–7083.

Taylor, D. J., P. D. N. Hebert & J. K. Colbourne, 1996. Phylogenetics and evolution of the *Daphnia longispina* group (Crustacea) based on 12S rDNA sequence and allozyme variation. Mol. Phyl. Evol. 5: 495–510.

Wagler, E., 1936. Die Systematik und geographische Verbreitung des Genus *Daphnia*. O. F. Müller mit besonderer Berucksichtigung der sudafrikanischen Arten. Arch. Hydrobiol. 30: 505–556.

Weider, L. J., A. Hobaek, F. Dufresne, J. K. Colbourne, T. J. Crease & P. D. N. Hebert, 1997. Circumarctic phylogeography of an asexual species complex: mtDNA variation in arctic *Daphnia*. in review.

Hydrobiologia **360**: 301–303, 1997.
A. Brancelj, L. De Meester & P. Spaak (eds), Cladocera: The Biology of Model Organisms.
©1997 *Kluwer Academic Publishers.*

Cladoceran studies: where do we go from here?

Henri J. Dumont
Institute of Animal Ecology, University of Ghent, Ledegansckstraat 35, B-9000 Gent, Belgium

Abstract

For ecology and genetics of the Cladocera to progress more rapidly, it would be beneficial if efforts were concentrated on a single taxon. *Daphnia magna* is probably the best candidate. But what are the Cladocera? Upon inspection, confusion is found to rage at any level. We do not know whether the group as a whole is monophyletic, the number of families and genera is uncertain, and we have only a faint idea of the total number of species. Neither do we know much about the geographical distribution of most species, which may have been blurred by human dispersal since the appearance of large-scale navigation. Currently, introductions and invasions continue to occur. These are well monitored, but events in the past could also be reconstructed, using the (sub)fossil record. There is thus a good future for the use of cladoceran remains in paleolimnology.

Model organism

Others in this section discuss aspects of ecology and genetics, so I will be brief on these topics. Suffice it to say that I strongly advocate a conduct similar to that of our colleagues in classical genetics and in molecular biology, which is that we should select and concentrate on one or few taxa about which we could try to learn everything there is to know, instead of going in all directions at once. A candidate for this role of model organism is to be looked for in the genus *Daphnia*. I would recommend *Daphnia magna* as a candidate to become the *Drosophila melanogaster* of aquatic ecology but realise that, like *D. melanogaster, magna* might not be everybody's choice. It is not known to hybridise with other *Daphnia*, which is a disadvantage, but it also has a number of distinct advantages: it occurs over a broad geographic range, in a wide variety of water types, and it is easy to culture.

In particular, as nucleotide sequencing facilities become widely available, more rapid, and less expensive, it might soon be realistic to think of compiling a full record of *Daphnia magna*'s genome. Apart from revealing how a crustacean is to be assembled, organisms like *Daphnia* (we could use a few others, from other major branches of the zoological tree) could be used to tell us how robust (or weak) single-locus (or even multi-locus) phylogenies really are. With yeast and the human genome already partly available, *Daphnia* should rank high among the candidates next in line. It strikes me as odd that, as far as physical gene mapping is concerned, *Daphnia* should lag so far behind that other branchiopod, *Artemia*. I can see no reason for this other than that *Artemia* workers have benefited from about two decades of concerted action, which translates into a number of sizeable monographs. In contrast, it took until 1987 for a book, boldly entitled '*Daphnia*', to see the light (Peters & De Bernardi, 1987). I'd like to believe that nobody will regard this as the definitive work; rather, it is a useful start. About every chapter merits to be worked out to a full-sized volume in its own right.

Morphology and systematics

Which brings me to morphology and systematics (or classification). Is there reason to continue working on these traditional disciplines, and if so, what subjects should be prioritised? In working with Stefan Negrea from Bucharest on the introductory volume 'Cladocera' for my series of zooplankton identification guides, it struck me that we are currently in no position to even define what cladocerans really are. Geoffrey Fryer (1987, 1995) started this off, and recently new voices were raised (e.g. Martin & Cash-Clark, 1995) in what

I expect to become a debate. The problem is not so much to work out whether the aggregate known as the Cladocera is a monophyletic group, than at what level the cut-off points between branches are situated, hence what are the ranks of the four constituent groups. Clearly, they are not at the same level. A *Diaphanosoma* is closer to a true anomopod like *Daphnia* than to a *Bythotrephes* or to *Leptodora*. But that is a crude evaluation, in need of much refinement. Progress can be expected from explicit comparisons between one or more homologous genes. Current favourites are 18s and 28s RNA genes, tubulin genes, cytochrome b genes, and the list is growing continually. The techniques for this are available and straightforward, yet work on the cladocerans lags behind that in other groups. Even if it might require only few years of work to close this gap, I am not aware of anyone having embarked on the subject. Of course, such phylogenies can also be established on morphological grounds, given that a sufficient number of character states can be evaluated. Here, too, cladocerans sadly lag behind. Trunk limb structure, for example, rich in interesting traits, is not much better known today than in the pioneer days of Behning, Eriksson and others. A notable exception is the recent book by Alonso (1996), where the study of trunk limb morphology was finally taken seriously, as well as some recent papers by Alexei Kotov on the Bosminidae (e.g. in this volume).

Much to my dismay, I recently discovered that within the anomopods, family boundaries are unclear too. Thus, Smirnov (1992) correctly took the Ilyocryptidae out of the Macrothricidae. Ilyocryptids are in fact so different from all other anomopods in trunk limb structure that they assume an isolated position within the order. But numerous other problems at the family level await a solution: neither the chydorids nor the macrothricids are in fact homogeneous (or, in the current jargon, monophyletic).

Down one more step then, to genus level. Let me first state that the recent avalanche of papers on natural hybridisation within *Daphnia* lends support to the controversial view of the genus as an objective taxon by A. Dubois. Dubois (1988) claims that a genus is the collection of all species capable of interbreeding. This definition (what can't interbreed belongs to different genera), however intellectually pleasing, unfortunately is of limited practical use in cyclical parthenogens, most of which have only been seen in pickled samples. Yet, it should free us of the fear that 'new species', described on morphological grounds only (a practice that is likely to be with us for a while), need to be 'well demarcated' from all their congeners. I rather consider that minute morphological characters may be quite significant, if it can be shown that they are consistent. With this in mind, I would not be surprised if future work would show that some of the more specious anomopod genera need to be split. Imagine indeed some six co-existing species of *Alona* or *Pleuroxus* to go sexual and all interbreed. Could so much reproductive waste be absorbed without introgression?

And, finally, at the bottom line there is the classical question: how many species of Cladocera are there? Korovchinsky (1996) recently attempted to answer it, but was frustrated by the number of ill-described species. With that, I agree. My own estimate, based more on a gut-feeling than on anything else, is 'around 500', but that refers to known species, not to the world's total. Part of Korovchinsky's and my own uncertainty stems from the flood of synonyms – both old and new – that bedevils all such estimates. This is not the place to judge the taxonomic work of others, but it seems that the cripple descriptions of our forefathers are currently replaced by cripple descriptions by poorly trained workers from underequipped and underfunded laboratories in the developing world. In my own limited way, I have attempted to do something about this situation, by organising an in-depth training course on zooplankton taxonomy (which I had to sell to my sponsor, the Belgian minister of cooperation, as a course in lake management). For the past eight years, we yearly trained about ten people from Africa, Asia, and Latin America per session. We send them home with a wagon load of literature, and often even with a plankton net. But will it help? Sometimes, when I hear that some alumni of the course have started a course of their own, I feel encouraged. Sometimes, when I receive manuscripts for Hydrobiologia from those same countries, written in incredibly poor language, and with unspeakable figures, I feel discouraged.

In our days of a renewed interest in the world's biodiversity, taxonomists around the world hope that more money will start flowing their way. However, for this hope to be fulfilled, much will depend on the quality of their work. In the case of cladoceran taxonomy, uplifting its caliber is a major priority for the future. To get there, we may, *inter alia*, need to specify what the minimum requirements for a species description are. Some standard description format should be developed in the near future.

But what of the ultimate number of cladoceran species? Possibly no quest is more academic in nature than this, and surely no ecosystem will collapse for the loss of a cladoceran species. Yet, it is undeniable that there is something in human curiosity that simply demands to get an answer to this sort of question. Let us therefore continue to spend modest amounts of money and energy on it.

I would dare to speculate that we currently know about half of the species, but where are the others? Exploring some type of environment that was left untouched by our predecessors is one way of extending the list. Thus, various types of underground waters have been successfully explored of late (Brancelj, 1990; Dumont & Brancelj, 1994), but temporary waters in the dry tropics and small waterbodies in equatorial forests are high on my list of priorities too.

If we don't know how many species there are, we often don't know where they are either. It is probably no longer true that the distribution of cladocerans reflects that of cladocerologists (there are in fact too few cladocerologists for that), but clearly, now that we no longer believe in cladoceran cosmopolitanism (Frey, 1987), we need to substantiate non-cosmopolitanism. What are the ranges of all these species, and can we ever understand why they live where they live? In the light of our bad habit of dispersing living things all over this planet, we have doubtlessly created numerous artefacts of distribution already. We are well aware of some of the more recent ones, and they get adequately documented, but what of the earlier ones? Possibly we will never know, and possibly some of the residual cases of cosmopolitanism reflect involuntary human phoresis (like by the thousands of ships that circled the earth between the fifteenth and the nineteenth century, and were dependent for their drinking water on wooden casks, that were rinsed and refilled at each stop, a perfect way of dispersing cladoceran resting stages). But there is one way of approaching at least some aspects of this question: many cladocerans, contrary to copepods, fossilize well, and their faunas leave a rich signature in the sediments of the waters they inhabit. Cladoceran microfossils thus have a story to tell, which may enlighten us about past environmental change, but also, in a variety of cases, about the date of their arrival at a given site. Browsing through the pages of e.g. the Journal of Paleolimnology, I find little evidence of this approach being used to potential. It definitely merits more attention.

References

Alonso, M., 1996. Crustacea, Branchiopoda. Fauna Iberica 7: 486 pp.

Brancelj, A., 1990. *Alona hercegovinae* n.sp,. (Cladocera: Chydoridae), a blind, cave-inhabiting cladoceran from Hercegovina. Hydrobiologia 199: 7–16.

Dubois, A., 1988. The genus in zoology: a contribution to the theory of evolutionary systematics. Mem. Mus. natn. Hist. nat. Paris (A) 140, 124 pp.

Dumont, H. J. & A. Brancelj, 1994. *Alona alsafadii* n.sp. from Yemen, a primitive groundwater-dwelling member of the *A. karua*-group. Hydrobiologia 145: 5–7.

Frey, D. G., 1987. The taxonomy and biogeography of the Cladocera. Hydrobiologia 145: 5–17.

Fryer, G., 1987. A new classification of the branchiopod Crustacea. Zool. J. linn. Soc. 91: 357–383.

Fryer, G., 1995. Phylogeny and adaptive radiation within the Anomopoda: a preliminary exploration. Hydrobiologia 307: 57–68.

Korovchinsky, N. M., 1996. How many species of Cladocera are there? Hydrobiologia 321: 191–204.

Kotov, A. A., 1997. Structure of thoracic limbs in *Bosminopsis deitersi* Richard, 1895. (Anomopoda, Branchiopoda). Hydrobiologia 360: 25–32.

Martin, J. W. & C. E. Cash-Clark, 1995. External morphology of the onychopod 'Cladoceran' genus Bythotrephes. Zool. Scr. 24: 61–90.

Peters, R. H. & R. de Bernardi (eds), 1987. *Daphnia*. Mem. ist. Ital. Idrobiol. 45, 502 pp.

Smirnov, N. N., 1992. The Macrothricidae of the world. Guides to the Identification of the Microinvertebrates of the Continental Waters of the World 1. SPB Academic Publishing, The Hague, 143 pp.